GENERATION, TRANSMISSION AND UTILIZATION OF ELECTRICAL POWER

GENERATION, TRANSMISSION AND UTILIZATION OF ELECTRICAL POWER

BY

A. T. STARR

M.A., PH.D., M.I.E.E.

FOURTH EDITION

LONDON

SIR ISAAC PITMAN & SONS, LTD.

First published 1937
Second edition 1941
Reprinted 1945
Reprinted 1946
Reprinted 1949
Third edition 1953
Fourth edition 1957

SIR ISAAC PITMAN & SONS, LTD.
PITMAN HOUSE, PARKER STREET, KINGSWAY, LONDON, W.C.2
THE PITMAN PRESS, BATH
PITMAN HOUSE, BOUVERIE STREET, CARLTON, MELBOURNE
27 BECKETTS BUILDINGS, PRESIDENT STREET, JOHANNESBURG

ASSOCIATED COMPANIES
PITMAN MEDICAL PUBLISHING COMPANY, LTD.
45 NEW OXFORD STREET, LONDON, W.C.1
PITMAN PUBLISHING CORPORATION
2 WEST 45TH STREET, NEW YORK
SIR ISAAC PITMAN & SONS (CANADA), LTD.
(INCORPORATING THE COMMERCIAL TEXT BOOK COMPANY)
PITMAN HOUSE, 381–383 CHURCH STREET, TORONTO

MADE IN GREAT BRITAIN AT THE PITMAN PRESS, BATH
E7—(T.54)

PREFACE TO THE FOURTH EDITION

THE use of electrical power is increasing rapidly and resources of coal and oil are not likely to meet requirements for more than fifty to a hundred years. For these and other reasons there is an overwhelming drive to use nuclear energy, and massive development is under way. An additional Appendix gives a qualitative explanation of the present method of utilizing nuclear (or atomic) energy, and there is an outline of the Calder Hall station in Chapter I.

The other main alterations are in Chapter I and are concerned with generation.

A. T. STARR

NEW BARNET, 1956

PREFACE TO THE FIRST EDITION

THE main purpose with which this book was written was to present the student, who is studying for the Engineering Degrees, the Faraday House Diploma, the National Diploma, or the National Certificate, with a single book containing a description of the syllabus known as "Electric Power." The syllabus is so wide that a choice must be made, not only as regards the matter to be included, but also as to the depth of treatment of those parts that have been dealt with. The author hopes that his choice is reasonable, and that the inclusion of many illustrations and much descriptive matter will make the book interesting and acceptable.

The author wishes to thank the various manufacturers, editors, and writers for permission to include published material; and his colleagues Messrs. A. Regnauld, S. O. Pearson, A. N. Arman, and E. J. Keefe for much help by reading, criticizing, and correcting the manuscript and proofs.

Finally, the author desires to thank the Technical Editor to Sir Isaac Pitman & Sons and Mr. W. F. Floyd for eliminating obscurities and errors from the text.

A. T. STARR

FARADAY HOUSE
July, 1937

CONTENTS

PAGE

PREFACE TO THE FOURTH EDITION v

PREFACE TO THE FIRST EDITION vi

CHAPTER I

GENERATION OF ELECTRICAL ENERGY 1

Sources of energy—Heat and mechanical energy—Heat value of coal, oil, gas, and uranium—Energy resources and production—Electrical transmission of energy—Steam power stations—Choice of site—Prime movers—Arrangement of plant—Operating costs of a steam power station—Hydro-electric stations—Choice of site—Prime movers—Arrangement of hydro-electric installation—Operating costs of a hydro-electric station—Nuclear power stations—The Central Electricity Authority and the Grid—Examples

CHAPTER II

TRANSMISSION OF ELECTRICAL ENERGY 31

Transmission by low-voltage direct current—The radial system—The ring system—Three-wire system—Balancers—Boosters—Advantage of high voltage—Transmission by alternating current—Alternating current systems—Copper efficiencies—High voltage, direct current—Examples

CHAPTER III

OVERHEAD LINES: MECHANICAL DESIGN 46

Introduction — Regulations — Stranded conductors — Conductor materials—Calculation of sag—Effects of wind and ice—Tension and sag at erection—Spans in hilly country—Trolley wires—Poles and towers—Insulators—Joints and clamps—Guards—Earthing—Vibration and dampers—Examples

CHAPTER IV

OVERHEAD LINES: ELECTRICAL DESIGN 81

Resistance of overhead lines—Inductance of an overhead line—Inductance of three-phase lines—Capacitance of overhead lines—Effect of earth—Capacitance of three-phase line—Electric stress—Corona discharge—Disruptive critical voltage—Visual critical voltage—Corona power loss—Avoidance of corona—Current effects of corona—Insulators—Suspension-type insulators—Examples

CHAPTER V

PAGE

UNDERGROUND CABLES 98

Introduction—Types of cable—Manufacture of paper-insulated cables — Laying — Jointing — Insulation resistance — Stress and capacitance of single-core cable—Capacitance grading—Power factor of single-core cable—Capacitance of three-core, belted type cable—Stress in a three-core cable—Inductance of cables—Sheath effects—Currents in bonded sheaths—Breakdown voltage and mechanism of breakdown—Thermal characteristics of cables—Examples

CHAPTER VI

TRANSMISSION CALCULATIONS 137

Short lines—Regulation—Line and Machine Charts—Maximum power with a given regulation—Mixed conditions—Medium lines—Constants of the general network—Two networks in tandem—Rigorous solution of long transmission line—Equivalent network of a long line—Very long line—Ferranti effect—Losses in open-circuited line—Tuned power lines—Interconnected systems—Voltage drop in distribution—Calculation of interconnected systems—Direct application of Kirchhoff's laws—Superposition of currents—Network simplification—Circulating-currents method—Use of Thévénin's theorem—A.C. interconnected systems—Examples

CHAPTER VII

VOLTAGE REGULATION STABILITY 177

Voltage control—Voltage control by generator excitation—Tap-changing transformers—Calculation of transformer ratios—Booster transformers—Phase angle control—Phase-shifting transformer—Induction regulator—Parallel operation of generators—Voltage control by power factor—Static phase advancers—Synchronous condenser—Power-angle diagram—Stability—Examples

CHAPTER VIII

SHORT CIRCUITS: SYMMETRICAL COMPONENTS 209

Introduction—Percentage reactance and short-circuit currents—Symmetrical short-circuit currents—Short-circuit current of alternators—Current-limiting reactors: sectionalization of networks—Method of locating reactors—Choice of interconnection to limit currents—Protection of switchgear by reactors—Unsymmetrical short circuits: symmetrical components—Application of symmetrical components—Impedances to the various phase-sequence currents—General method using symmetrical components—Interference with communication circuits—Examples

CHAPTER IX

SWITCHGEAR AND PROTECTION 237

Introduction—Principles of arc extinguishing—Rating of circuit-breakers — Air circuit-breakers — Fuses — Relays — Protection of alternators and transformers—Protection of lines and cables—Overload protection—Balanced protective systems—Distance or impedance protection—Bus-bar arrangements—Testing of transmission systems—Fault location—Examples

CHAPTER X

PAGE

VOLTAGE TRANSIENTS AND LINE SURGES 275

Introduction—Resonance—Switching—Transients in circuits with lumped constants—Switching surges—Energy considerations—Sudden interruption of a circuit—Insulation failure or earthing of a line—Reflection and transmission of travelling waves—Junction of two lines—Effect of a capacitance—Lightning—Arcing earths—Petersen coil—Lightning and over-voltage protection—Condensers—Surge absorber—Recording of transmission line surges—High-voltage cathode ray oscillograph—Klydonograph—Surge crest ammeter—Examples

CHAPTER XI

ILLUMINATION 312

Introduction—Light—The spectrum of light—Standards of light—Values of illumination—Polar curves—Rousseau diagram and Russell angles—Light fittings and the polar curve—Reflection and absorption—Lighting calculations—Calculation of light required—Sources of light—The incandescent filament lamp—Arc lamps—Gas discharge electric lamps—Summary of sources of light—Examples

CHAPTER XII

ELECTRIC TRACTION 343

Systems of electric traction—Advantages and disadvantages of electric traction—Electrification systems—Mechanical considerations—Transmission of drive—Mechanics of train movement—Speed-time curves—Effect on schedule speed of acceleration, braking and distance—Traction motors—Starting and speed control of d.c. motors—Notching—Series-parallel control—The metadyne—Field weakening or tapped field control—Starting and speed control of a.c. motors—Three-phase motors—Single-phase motors—Electric braking—Trolley bus control equipment—Master controllers—Multiple-unit control—Overhead construction for tramways and trolley buses—Feeding and distributing system for tramways—Overhead collection for locomotives—Examples

CHAPTER XIII

INDUSTRIAL UTILIZATION 396

Introduction—Group and individual drive—Choice of drive and motor—Properties of motors and applications—D.C. shunt motor—D.C. series motor—Compound-wound d.c. motor—Single-phase series motor—Repulsion motor—Synchronous motor—Squirrel-cage induction motor—Slip-ring induction motor—Pole-changing induction motor—Cascaded induction motor—Induction motor cascaded with a.c. commutator motor—Synchronous-induction motor—Schrage motor—The N-S motor—Motors for given services—Domestic uses—Machine tools—Cranes—Lifts—Textile machinery—Printing machinery—Paper-making machinery—Iron and steel works—Mining—Examples

CHAPTER XIV

PAGE

ECONOMIC CONSIDERATIONS 416
Introduction—Interest and depreciation—Generation—Maximum demand—Diversity factor—Transmission—Size of conductors: Kelvin's law—Current density—Transmission voltage—Power factor—Power factor improvement—Distributors—Transformer losses—Examples

APPENDIXES

I. HYPERBOLIC SINE AND COSINE 436
Linear differential equation—Table of hyperbolic functions

II. ALTERNATING CURRENTS 440
Mean value—Root mean square value—Wave form and form factor—Vector representation of alternating currents—Power—Vector method applied to simple circuits—Complex numbers—Algebraic method for alternating currents—Power in alternating-current circuits—Vector power—Polyphase currents—Three-phase loads—Examples

III. TRANSFORMERS 471
Percentage impedance—Transformers in parallel—Three-phase transformers

IV. THÉVÉNIN'S THEOREM 478

V. METERING, DISTANCE INDICATION (TELEMETERING) AND REMOTE CONTROL 480
Measurement of kWh. and reactive kVAh—Summation metering—Midworth distant repeater—Direct distant indication—Indirect distant indication—Data-transmission systems

VI. CONVERTING PLANT 486
Gas-filled valves—Gas-filled thermionic diode—Gas-filled thermionic triode—Mercury-pool rectifiers—Polyphase rectifiers—Grid control of thyratrons—Inverters—Examples

VII. HEATING, WELDING, AND ELECTROCHEMICAL PROCESSES . 503
High-frequency heating—Induction heating—Dielectric heating—Welding—Electrochemical processes

VIII. PRINCIPLES OF NUCLEAR ENERGY 511
Structure of atoms—Periodic table—Radioactivity—Transmutation—Nuclear fusion—Nuclear fission—Reactor with natural uranium—Enriched reactors

ANSWERS TO EXAMPLES 522

INDEX 525

FIG.
5. Layout Plan of Battersea Power Station and Site . *facing* 12
6. West Elevation of Battersea Power Station 12
7. Pulverized-fuel-fired Boiler at Hams Hall, Birmingham . . 12
8. Cross-section of Boiler House at Battersea 12
10. Geographical Arrangement of the Grampian Scheme . . 20
11. Civil Engineering Works of the Rannoch Plant . . . 20
12. Cross-section of the Rannoch Power Station . . . 20

CHAPTER I

GENERATION OF ELECTRICAL ENERGY

Sources of Energy. Modern civilization differs from earlier civilizations in the enormous use of energy produced by machines. The results are a diminution of mechanical drudgery, a shorter working day, a higher standard of living, a healthier and more balanced diet (since food can be transported easily from one end of the world to another), and freedom to a large extent from local famines. There is a close relation between the energy used per person and the standard of living.

Within the last thirty years there has been a rapid increase in the generation of electrical energy, and the following table gives values for Great Britain.

TABLE I

ELECTRICITY CONSUMPTION IN GREAT BRITAIN

(In Milliard units, i.e. 10^9 kWh.)

	1925	1950	1954 (estimated)	1965 (forecast)	1975 (forecast)
Domestic and agricultural . .	0·6	14·9	19·6	37	63
Industrial and traction . .	5·1	31·0	42·9	79	138
Total sales	5·7	45·9	62·5	116	201
Total generated . . .	6·4	51·9	69	130	223

In 1953–4 the consumption per head in Great Britain was about 1 300 kWh./annum, whilst in Norway it was 4 300, and in the U.S.A. 2 900. It is fairly certain, therefore, that the rate of increase shown in the table is conservative.

The primary source of energy is the Sun. Direct utilization of solar radiation, which has been estimated to be equivalent to 5 000 h.p. per acre at noon in summer, has been made in Egypt, where in a certain installation the mirror to boiler surface ratio is $4\frac{1}{2} : 1$, the boiler efficiency is 40 per cent, and the plant develops 63 b.h.p. per acre. The method is inconvenient as it requires a large area and the absence of clouds, but has probably a sphere of use in irrigation and other pumping services.

The energy of winds, produced by the Sun, has been used for many hundreds of years in windmills, and can be used to drive small generators, which charge up storage batteries for continuous use. Efficient generators are being designed in this country.

The main sources of energy are fuels, viz. coal and oil, and water power, viz. water at a high level and tides. The utilization of the energy of fuel is possible by means of steam and internal combustion engines and the turbine; but the harnessing of water power has been made possible by the electrical transmission of energy which alone can transmit the energy from the places where the water power occurs to where the energy is needed.

The internal combustion engine using petrol is suitable for mobile units, such as road traffic and small craft for river and seashore work, and recently the Diesel engine using heavy oil has been adapted to the same purpose. The latter is also replacing coal- and oil-fired steam plants for marine propulsion, but even so it has been found advantageous to use an electrical conversion in some cases.

A new form of liberating energy has been devised in World War II by the fission (or nuclear disintegration) of uranium. The energy released in this way is converted to heat, which may then be used to drive steam turbines. Development is proceeding in the U.S.A. and Great Britain, but it is not expected that bulk supplies of energy will be available before 1960 or later. The method is known as that of *atomic energy*.

The conversion of the heat of combustion of coal and oil and of nuclear fission into electrical energy, and the distribution of the latter, afford advantages which will be discussed later.

Heat and Mechanical Energy. There are several units of heat and of mechanical energy, and it is useful to list their values and relationships. The *calorie* (cal.) is the amount of heat required to raise the temperature of 1 g. of water by 1° C.: the *Calorie* or *kilo-calorie* (k.cal. or Cal.) is 1 000 times as great, and is the unit used when nutrition is discussed, e.g. an intake of 2 000 to 3 000 Calories per day is desirable for an adult. The British unit of heat is the *British Thermal Unit* (B.Th.U.) which raises the temperature of 1 lb. of water by 1° F. Since 1 lb. = 453·6 g. and 1° F. = (5/9)° C., it follows that

$$1 \text{ B.Th.U.} = 252 \text{ cal.} = 0{\cdot}252 \text{ k.cal.} \qquad (1.1)$$

In dealing with large resources of energy it has been found convenient to introduce a large unit of heat denoted by Q, where

$$1 \text{ Q} = 10^{18} \text{ B.Th.U.} \qquad (1.2)$$

The experiments of Joule and Robert Mayer established the fact that heat and mechanical energy are interchangeable, one unit of heat producing a fixed amount of mechanical energy, and vice versa. The relation between them is called the *mechanical equivalent of heat*, and is given as

$$\left.\begin{array}{ll} & 1 \text{ B.Th.U.} = 778 \text{ ft. lb.} \\ \text{or} & 1 \text{ cal.} = 4{\cdot}18 \,.\, 10^7 \text{ erg.} = 4{\cdot}18 \text{ J (joule)} \end{array}\right\} \qquad (1.3)$$

The units of power are the *watt* (W.) and the *horse power* (h.p.)

$$\left.\begin{array}{ll}\text{where} & 1\ \text{W.} = 10^7\ \text{erg./sec.} = 1\ \text{J/sec.} \\ \text{and} & 1\ \text{h.p.} = 550\ \text{ft. lb./sec.} = 746\ \text{W.}\end{array}\right\} \quad . \quad . \quad (1.4)$$

The relation between the h.p. and the watt follows from the facts that 1 lb. = 453·6 g. and 1 ft. = 12 × 2·54 cm. The electrical unit of energy is the *kilowatt-hour* (kWh.) and it is seen that

$$\left.\begin{array}{l}1\ \text{kWh.} = (3\,600)\ (1\,000)\ \text{W. sec.} = 3{\cdot}6\ .\ 10^6\ \text{joule} \\ \qquad = 8{\cdot}6\ .\ 10^5\ \text{cal.} = 3\,412\ \text{B.Th.U.}\end{array}\right\} (1.5)$$

$$1\ \text{milliard} = 10^9\ \text{kWh.} = 3{\cdot}4\ .\ 10^{12}\ \text{B.Th.U.}$$

Heat Value of Coal, Oil, Gas, and Uranium. The heat value of coal varies with the grade: in this country an average value is 12 000 B.Th.U. per lb. If all of the heat could be converted into electrical energy, 1 kWh. would require (3 412/12 000) = 0·28 lb. of coal: the thermal efficiency in a power station had an average value of 23·7 per cent in 1954, so that 1·2 lb. coal is used per kWh. generated. The highly efficient Portobello H.P. had an efficiency of 31·27 per cent in 1954, so that it requires 0·89 lb. coal per kWh. Lignite has a heating value of 4 800 B.Th.U. per lb. and peat has 6 300 B.Th.U. per lb. It is common to express large heat units in terms of tons of coal of heat value 12 000 B.Th.U. per lb.; thus

$$\left.1\ \text{Q} = \frac{10^{18}}{12\,000 \times 2\,240}\ \text{tons} = 37\,000\ \text{megatons}\right\} \quad . \quad (1.6)$$

$$1\ \text{milliard} = 0{\cdot}55\ \text{megatons or about}\ 0{\cdot}6\ \text{megatons}$$

Heavy oils have a heat value of 20 000 B.Th.U. per lb., so that only three-fifths by weight is required as compared with coal. Quite a number of oil-burning stations produce 1 kWh. per 0·65 lb. of oil, the thermal efficiency being between 25 and 30 per cent, which is greater than for all but the best coal-burning stations. Light oils have a heat value of about 15 000 B.Th.U. per lb.

Coal gas has a heat value of 550 B.Th.U. per ft.^{3} and Mond producer gas has 160 B.Th.U. per ft.^{3}

The heat value of uranium depends upon the method used. Natural uranium contains 99·3 per cent of ^{238}U and 0·7 per cent ^{235}U, the latter being spontaneously fissile while the former is split by the slow released neutrons, in the way described in Appendix VIII. In the simplest method one ton of uranium will produce 3 000 MW. days of energy, which is equivalent to the heat value of about 10 000 tons of coal. By re-cycling the uranium and the produced plutonium, one ton of uranium should produce heat equivalent to 100 000 tons of coal. Eventually the use of fast breeder reactors should make one ton of uranium equivalent to 1 000 000 tons of coal.

Energy Resources and Production. The world supply of coal and oil is not known with any degree of certainty, and there is a tendency to be conservative in estimates. An estimate in 1920 was of the order of $7{\cdot}3 \,.\, 10^{12}$ tons or 200 Q, but it is becoming increasingly difficult to mine much of these resources. It is estimated that the world reserves of uranium and thorium have a heat value of 1 700 Q, which is about ten times that of coal, oil and gas combined. Table II shows the demand for primary fuel in Great Britain, and Table III shows the coal equivalent of the electricity demand based on Table I.

TABLE II

Demand for Primary Fuel
(Megatons Coal Equivalent)

	1954	1965	1975	1985
Industry	118	156	196	237
Railways and transport	32	35	36	38
Domestic	65	73	83	94
Total	254	311	373	440

TABLE III

Electricity Demand

	1954	1965	1975	1985
Equivalent megatons	42	71	104	145
Plant MW.	20 000	38 000	58 000	82 000

In 1985 about 450 megatons will be required. It is expected that coal may provide 225 megatons, oil 150 megatons, hydro-electric power 5 to 10 megatons, and so nuclear power will have to supply about 70 megatons, which corresponds to about 38 000 MW. of nuclear power station capacity. Table IV shows the power and energy required from nuclear fission (estimated September, 1955).

TABLE IV

Nuclear Fission Power and Energy Required

	1965	1970	1975	1980
Energy (milliard)	25	40	86	146
Power (MW., 75% load factor)	3 900	6 200	13 200	23 000

Electrical Transmission of Energy. The conversion of heat into mechanical energy and the electrical transmission of the latter afford many advantages. Electrical transmission is convenient, clean, cheap, and extremely flexible. The extreme cleanliness accompanying the use of electrical energy is very important, when one bears in mind the damage due to smoke and soot. The considerably decreased severity of fogs in London during the last decade or two has been ascribed in part to the decreased burning of coal consequent upon the increase in the use of electricity for heating. The all-electric house keeps much cleaner and requires less frequent re-decoration than a house which uses gas and coal, and these facts must be remembered when one compares the costs of using coal or gas or electricity.

In the realm of lighting, for streets and home, electricity is unrivalled in convenience and cheapness.

For cleanliness, ease of manipulation, and flexibility the electric motor is supreme, so that the use of electrical energy in industry is rapidly increasing.

Electric traction is particularly suitable for dense suburban traffic where quick starting and braking are imperative, whilst it is essential for underground traffic.

Water power is fairly easily converted into electrical power which can then be transmitted to places distant from the source; by proper choice of the site of the hydro-electric installation the power is continuously available. A new and inexhaustible source of power is thus opened up, of which a rapidly increasing use is being made. Oil is needed for road transport and shipping, and coal is likely to be required for the production of oil by hydrogenation, so that the economic advantages of hydro-electric generation will probably increase in the future.

STEAM POWER STATIONS

In a steam power station the fuel, which is coal or lignite or peat, gives up its heat of combustion to a boiler which delivers steam at a high temperature and a high pressure to the steam turbines. The steam loses heat energy in driving the turbine, which is coupled directly or through gearing to an electrical generator. The *thermal efficiency* is the ratio of the heat equivalent of the mechanical energy transmitted to the turbine shaft, to the heat of combustion; it may reach a value of 30 per cent in a very efficient plant. Then there are the losses in the alternator, so that the *overall efficiency*, which is the ratio of the heat equivalent of the electrical output to the heat of combustion, is slightly lower.

As an example we may take the case of the Battersea Power Station. 1 lb. of coal has a heat value of 11 500 B.Th.U. and delivers 3 498 B.Th.U. to the coupling, so that the thermal efficiency is

3 498/11 500 = 30·4 per cent. The electrical output is 3 394 B.Th.U., so that the overall efficiency is 3 394/11 500 = 29·5 per cent. This figure is for the economical load and no operating losses. For the year ending 31st December, 1934, the thermal efficiency was 27·24 per cent, owing to the facts that the *load factor*, which is the ratio of the average output to the maximum output, was 53·7 per cent and that the load was not always the optimum.

Table V gives the twenty most efficient steam power stations in 1954.

TABLE V

Station	Thermal Efficiency (%)	Energy Sent Out (Milliards)	Pressure (lb./in.^{2})
Portobello H.P.	31·27	0·506	1 350
Stourport "B," L.P.	30·78	0·453	1 250
Littlebrook "B"	30·39	0·753	1 235
Dunston "B," II	29·89	0·722	600 R
Bromborough	29·23	1·320	900
Skelton Grange	28·99	1·428	900
North Tees "C"	28·98	1·181	900
Brunswick Wharf	28·71	1·060	900
Uskmouth	28·57	1·161	900
Poole	28·50	1·292	900
Brighton "B"	28·37	0·700	900
Braehead	28·37	1·250	900
Littlebrook "C"	28·26	0·706	900
Carrington	27·98	0·557	900
Keadby	27·87	1·104	900
Barking "C"	27·84	0·617	900
Huncoat	27·75	0·428	600
Battersea "B"	27·71	1·004	1 350
Carmarthen Bay	27·70	0·758	900
Agecroft H.P.	27·06	0·628	600
All stations	23·72	66·308	

The different ways in which the losses occur are shown diagrammatically in Fig. 1* and in tabular form in Table VI; the data are for the optimum working of the Battersea plant.

Hence the overall efficiency at economical load and no losses is $(3\,394/11\,500) \times 100 = 29{\cdot}5\%$. At 95% operating efficiency, the overall efficiency is $0{\cdot}95 \times 29{\cdot}5 = \underline{\underline{28{\cdot}0\%}}$.

* The author wishes to acknowledge gratefully the use of the figure and table from Hutchinson's *Technical and Scientific Encyclopaedia*, pages 797 and 799, First Edition.

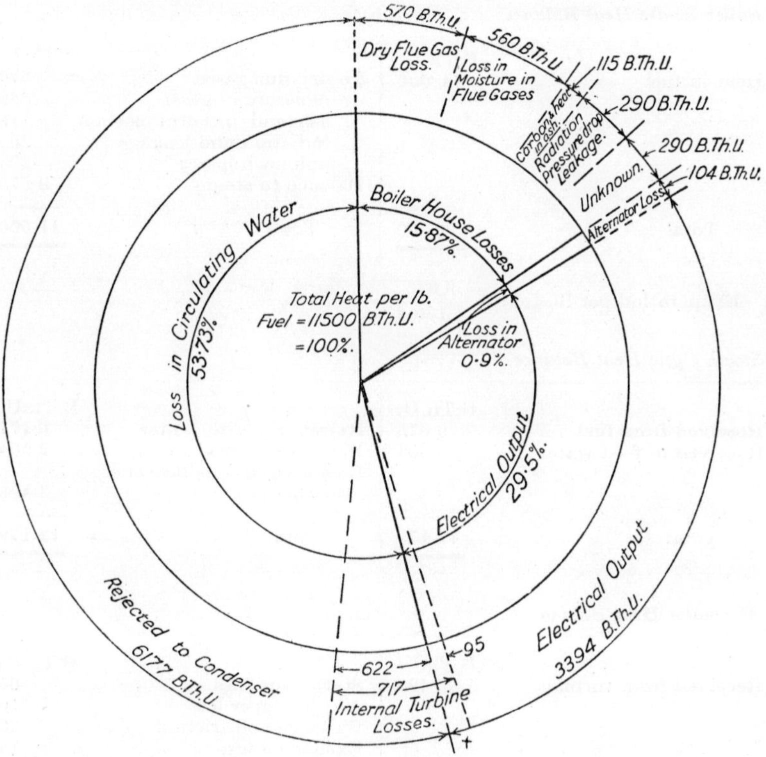

FIG. 1. DIAGRAMMATIC HEAT BALANCE OF MODERN POWER STATION

(Hutchinson's Technical and Scientific Encyclopaedia)

TABLE VI

HEAT BALANCE PER LB. OF COAL AT ECONOMICAL LOAD

Coal	11 500 B.Th.U. as fired, 4·3% H_2, 8% O_2, 10% moisture.
Flue gases . . .	Exit temperature 250° F., 14% CO_2.
Air	60° F.
Circulating water .	60° F.
Heat of evaporation from feed water at 350° F.	1 120·3 B.Th.U. per lb.

Boiler House Heat Balance

	B.Th.U.		B.Th.U.
Heat in fuel . . .	11 500	To dry flue gases . .	570
		,, moisture in gases . .	560
		,, ash and unburnt carbon	115
		,, radiation and leakage .	290
		,, unknown losses . .	290
		Balance to steam . .	9 675
Total	11 500	Total	11 500

$$\text{Evaporation per lb. fuel} = \frac{9\ 675}{1\ 120{\cdot}3} = 8{\cdot}636 \text{ lb.}$$

Steam Cycle Heat Balance

	B.Th.U.		B.Th.U.
Received from fuel . .	9 675	Rejected to circ. water .	6 177
Received in feed water .	2 504	Returned to feed water .	2 504
		Balance to mechanical energy at coupling . . .	3 498
Total	12 179	Total	12 179

Alternator Heat Balance

	B.Th.U.		B.Th.U.
Received from turbine .	3 498	Stator iron loss . . .	60
		Stator copper loss . .	10
		Windage and friction . .	20
		Excitation loss . . .	14
		Balance to electrical output	3 394
Total	3 498	Total	3 498

Coal consumption at economical load and 95% operating efficiency is

$$\frac{3\ 412}{11\ 500 \times 0{\cdot}28} = \underline{1{\cdot}06} \text{ lb. per kWh.}$$

The losses may thus be summarized as follows—

Boiler house losses	15·87%
Rejected to condenser . . .	53·73
Alternator losses	0·9
Output	29·5
TOTAL	100·00

The main loss, it is seen, is that due to the heat rejected to the condenser and is unavoidable for thermodynamic reasons, viz. that heat cannot be converted into mechanical energy without a drop in temperature, and the steam in the condenser is at the lowest temperature.

Fig. 2 shows a block schematic of the flow of power and supply

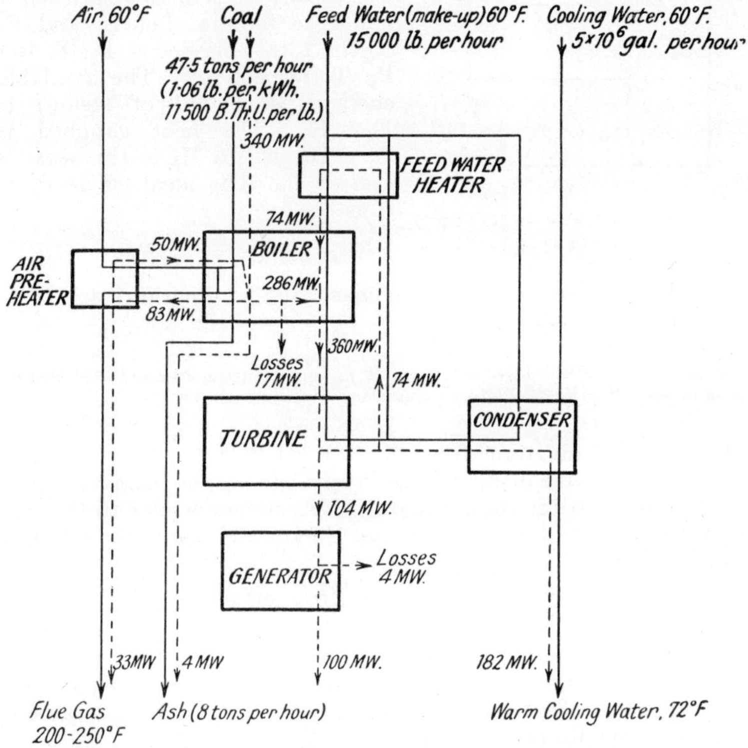

FIG. 2. SCHEMATIC OF HEAT BALANCE OF A 100 MW. STATION (BATTERSEA, 1934)

of materials for the Battersea plant delivering a power of 100 MW.

There are three controllable factors in the steam turbine cycle, the pressure (P_1) and absolute temperature (T_1) of the input steam and the pressure (P_2) in the condenser. The Rankine cycle corresponding to the action is shown in Fig. 3, which is the temperature-entropy or T-ϕ diagram. At *a* the water in the condenser is at temperature T_2, and at pressure P_2. The water is heated in the boilers to the boiling point (T_e) corresponding to the pressure P_2 (path *ab*) and is then evaporated into steam (path *bc*). The steam is then superheated to the temperature T_1 (path *cd*), when it is admitted to the turbine. It expands adiabatically (in the ideal case) along the path *de*, and then condenses in the condensers along path *ea*.

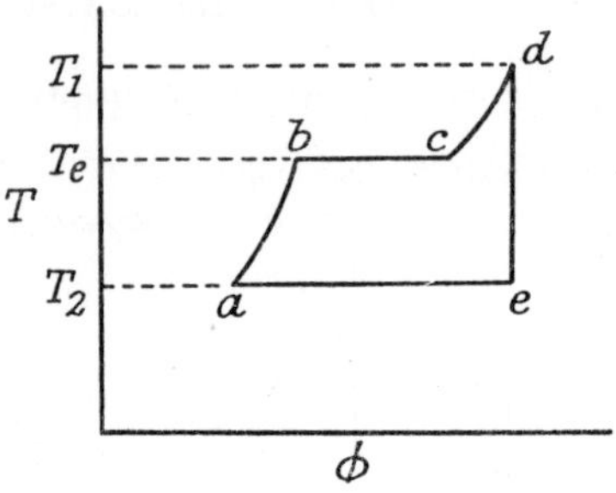

FIG. 3. T-ϕ DIAGRAM OF RANKINE CYCLE

The *ideal* or *standard efficiency* is found as follows. Let E_1 and E_2 be the total energies at P_1, T_1 and P_2, T_2 respectively. The available energy (per pound of steam) is $E_1 - E_2$. The heat supplied is $E_1 - H_2$, where H_2 is the *sensible heat* at T_2. The ideal efficiency is thus

$$\eta_{ideal} = (E_1 - E_2)/(E_1 - H_2).$$

E_1 and E_2 are given on the steam chart (Mollier diagram), and H_2 is given in steam tables.

EXAMPLE. Find the ideal efficiency of an engine using steam at 180 lb./in.^{2} abs. and 480° F., the condenser exhausting at 3 lb./in.^{2} abs.

The Mollier diagram gives $E_1 = 1\,264$ B.Th.U. and $E_2 = 964$ B.Th.U. The value of E_2 is found by drawing a line, from the point corresponding to 180 lb./in.^{2} and 480° F., perpendicular to the axis of ϕ until it meets the 3 lb./in.^{2} line. Steam tables give $H_2 = 109$, so that

$$\eta_{ideal} = \frac{1\,264 - 964}{1\,264 - 109} = 25{\cdot}8\%.$$

Lowering the pressure, P_2, in the condenser decreases E_2 and has a very marked effect in increasing the efficiency. A usual value of P_2 is about 0·6 lb. per in.^{2}

Increasing P_1 and T_1 also increases the efficiency, although an increase of T_1 has much less effect than an increase of P_1; in fact, the ideal efficiency is determined more by the saturation temperature T_e than by T_1. For this reason high pressures are used, 900 lb. per in.^{2} being quite usual and 1 350 lb. per in.^{2} being also in actual use. On the Continent still higher pressures are now used. The increase in efficiency due to superheating is not the reason for the high degree of superheating employed. The advantages are the reduced quantity of steam required for a given output of energy, the highly desirable dryness of the turbine blades with the consequent decrease of mechanical resistance and absence of corrosion, and the decreased loss of heat from the steam pipes.

It may be mentioned that as the critical temperature of steam is 705° F., any steam above this temperature is superheated.

Reheating the steam between turbine stages increases the efficiency, but is somewhat inconvenient. Bleeding the steam, which will be described later, also raises the efficiency slightly.

Successful attempts have been made to use mercury vapour (and other fluids) instead of steam, as much higher ideal (and actual) thermal efficiencies are obtainable with lower pressures because of the thermal properties of mercury vapour. The cost is, however, prohibitive, and the poisonous nature of the vapour is a most undesirable feature.

Choice of Site. In order to keep the costs of distribution as low as possible it is best to have the station near the centre of its district of supply. This is especially the case when the transmission is by direct current and voltage transformations are not possible. If the transmission is by alternating current it is not so necessary for the site to be in the midst of the district served, as the energy can be led easily at high voltage into the district and then transformed down for use at a number of convenient points.

The advent of the Grid has an important effect on the site and size of plant. Attention is focused on the facilities for generation rather than on distribution, although the latter cannot be disregarded. The site is chosen so that an abundant supply of cooling water for the condenser is available, and there is a low cost for the transport of fuel. The basic idea is that energy should be collected at places where it is cheap, fed into the high voltage network or *grid* and then drawn off at various tapping points.

The price of land, the precautions necessary to keep the atmosphere in populated districts unpolluted by fumes and by the residue from pulverized fuel, and the reservation of land for future developments all tend to shift the site away from towns. After having paid consideration to these points, that site which is nearest the centre of the load is chosen. It is clear that the site must be suitable for supporting a large building with heavy machinery; frequently the cost of foundations is a main item and must be carefully considered.

A 25 MW. station occupies about 10 000 square yards and a 100 MW. station about 36 000 square yards; with natural cooling 20 000 and 80 000 gal. per min. of cooling water are required respectively, whereas with towers the figures are 4 500 and 15 000; at 40 per cent load factor the coal per day is 160 tons and 530 tons. These figures allow a preliminary estimate to be made of the size of the site, the quantity of cooling water needed, and the required transport facilities; but a very liberal allowance must be made for future extensions.

Prime Movers. The prime movers in large steam power plants are turbines, although in small plants the reciprocating steam engine is still used. The turbines are of the impulse and reaction types.

If the steam were used up in driving a single-blade ring, the rotor speed would be about 30 000 r.p.m., which is much too high for practical purposes. The speed is lowered to 3 000 or 1 500 r.p.m. by letting the steam act on a series of rotors connected in tandem to the same shaft. Types of turbines differ in the ways that the pressure and velocity of the steam are controlled by the moving and fixed blades (or nozzles). In the impulse turbine the steam is expanded in the nozzles (or fixed blades) only, the pressure over the moving blades remaining constant; examples are the De Laval (single-blade wheel), the Curtis, Rateau and Zoëlly. In the reaction

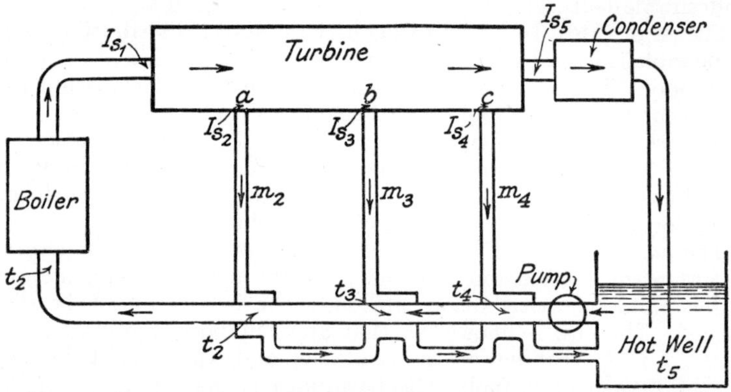

FIG. 4. DIAGRAM OF ARRANGEMENTS FOR BLEEDING A TURBINE
(*Thermodynamics Applied to Heat Engines* (Lewitt))

type, developed by Parsons, the pressure drops continuously over the fixed and moving blades; the blade velocity is relatively low.

An increase in efficiency is obtained by *bleeding*, which is the process of draining a small quantity of steam from the turbine at certain points and circulating this around the feed water pipe leading from the hot-well to the boiler. Fig. 4 shows diagrammatically how this is done.

Arrangement of Plant. Most large power stations are situated by a river so that they can receive water-borne coal, and some can receive both by water and rail, such as Battersea and Brimsdown "B." If the coal is brought by barge or ship, grab cranes take the coal and put it on belt conveyers, which run the coal to the weighing machines. After the coal is weighed and samples taken for analysis, it is carried by conveyer belts straight to the boiler house bunkers or to the store, which is generally a large pit; belt conveyers run also from the store to the bunkers. Rail-borne coal is weighed and then tipped from the wagons and conveyed by belt to the store or bunkers. Fig. 5 gives the general layout plan and Fig. 6 the west elevation of the Battersea Power Station. Most of the coal is brought by sea in the company's colliers, which are unloaded at the jetty on to conveyer belts which deliver the coal to the control tower where it is weighed. Some of the coal is brought by rail, passes through the weighbridge, and is discharged at the wagon tippler house.

Fig. 7 shows the pulverized-fuel-fired boiler of Hams Hall, Birmingham, and Fig. 8 a cross-section of the boiler-house at Battersea.

The disposition of the boilers and turbines is determined by the sizes of the turbines and the available boiler units. Smaller turbines have one boiler each, but the larger have two or more. The disposition in the Battersea station of 1937 is shown in Fig. 5. The three turbines have maximum continuous ratings of 69, 69, and 105 MW. respectively. The former two have an economical load of 52·8 MW.; each has three cylinders, the low pressure cylinder being double flow. There are 19 impulse stages in the high pressure cylinder, 22 in the intermediate, and 6 in each half of the low pressure cylinder, making a total of 47 stages in the expansion. The maximum load is obtained by by-passing the first 9 high-pressure stages. In the largest turbine the high pressure cylinder contains a velocity compounded stage followed by 16 impulse stages. For loads up to 60 per cent of the maximum continuous rating, viz. 63 MW., steam is admitted before the velocity stage; for loads up to 80 per cent it is admitted after the velocity stage; whilst to attain the maximum continuous rating of 105 MW. the steam is admitted after the eighth impulse stage so that it by-passes the velocity stage and the first impulse stages.

There are six boilers, two being spares, each with a maximum continuous evaporation of 312 000 lb. per hr.; the pressure and temperature in the superheater are 650 lb. per in.^{2} and 875–900° F. respectively. The boilers are each fired by a 20-retort Taylor stoker. The grit and soot are removed before the gas is passed into the washing chambers, which remove more than 90 per cent of the oxides of sulphur. The liquid from the washing chambers contains a large percentage of sulphuric acid, which is removed by filtering and aeration before the water is returned to the river. Four forced and four induced draught fans are provided for each boiler, the speeds being regulated by hydraulic couplings.

The steam piping is standardized by the universal use of 9 in. diameter pipes, the main turbine steam supply being fed through four such pipes in parallel.

The two condensers have a total surface of 60 000 ft.^{2} and use nearly 3 million gallons of circulating water per hour. The vacuum at economical rating is 29·1 in. mercury (30 in. barometer); there are four circulating pumps, each with a capacity of 2 million gallons per hour against a head of 30 ft. The condensed water leaves the drain cooler and passes through five feed-water heaters supplied by five bleed points.

Recently another turbo-alternator set of 100 MW. has been installed, working at 950° F. and 1 350 lb. per in.^{2}

The turbines are each coupled to the main generator, house service generator, and exciter group; the first is rated at 80 000 kVA. at 0·8 power factor and 11 kV., three-phase, 50 cycles; the second at 6 250 kVA., 0·8 p.f., 3 300 volts. (It may be mentioned that an alternator at Brimsdown "B" is generating at 33 kV., and Messrs. Parsons are investigating generation at 66 kV.; the machine at Brimsdown has given excellent service for some years.) There is a transformer, 11 kV. : 66 kV., for each generator, with one spare unit; the connections are delta on the generator side and star on the 66 kV. side. There are four transformers for distribution of 20 000 kVA. each, ratio 66 kV. : 22 kV., delta connected on the 66 kV. side and star on the other. The bus-bar reactances are single-phase, oil-immersed type; three sets of these are installed in chambers near the main transformers. The main switchgear is placed on three floors; the tie bus-bars are on the lowest floor and are connected to the transformer on the ground floor; the main and transfer bus-bars are on the middle floor in fire-proof chambers; and the circuit-breakers are on the top floor which is divided into three compartments by steel partitions and rolling shutters. Each generator transformer has duplicate oil circuit-breakers. Bus-bars and connections are of tubular copper insulated by condenser bushings; all ends, joints, and tees are in oil-filled chambers.

In Drakelow "A," near Burton-on-Trent, there are four 60 MW. sets, the turbo-alternators being hydrogen-cooled. The steam is at 1 050° F. and 1 500 lb./in.^{2} and the condenser is at a vacuum of 28·75 in. Hg. The alternators generate at 11·8 kV. The main building is 575 ft. × 340 ft. with two 360 ft. chimneys.

Recently the English Electric Company has made a turbo-generator for 200 MW., working at 2 450 lb./in.^{2} and 1 060° F.

Operating Costs of a Steam Power Station. There are three kinds of costs, a fixed cost which is independent of the maximum output or the energy output, a semi-fixed cost which depends upon the maximum demand but is independent of the total energy output, and a cost which is proportional to the output. If we call these costs £A per annum, £B per kW. maximum demand per annum, and £C per kWh., respectively, the annual cost of the station is

$$£(A + B \times \text{kW.} + C \times \text{kWh.}).$$

A station of 10 000 kW. maximum may possess the formula

$$£(10\,000 + 0{\cdot}8 \times \text{kW.} + 0{\cdot}0015\ \text{kWh.}),$$

whilst a 100 000 kW. station may have the formula

$$£(70\,000 + 0{\cdot}5 \times \text{kW.} + 0{\cdot}00060\ \text{kWh.}).$$

The fixed cost represented by A is due to the cost of the central organization, salaries of the higher officials, capital cost of land (especially if some land is held for future developments and is not being used for present kilowatt demand), etc. The semi-fixed charge represented by B is due to the salaries of the charge engineers and maintenance staff, the cost of the buildings and equipment including spares, and normal running losses and costs. The charge C is mainly on account of fuel, water, etc.

It is difficult to separate the charges A and B, and for a given station they may be lumped together to form a charge $D \times$ kW. Thus for the smaller station given above the formula £(10 000 + 0·8 × kW.) can be replaced by £(1·8 × kW.), and for the larger £(70 000 + 0·5 × kW.) by £(1·2 × kW.).

The pre-war cost of installation of a large steam power station was £12 to £14 per kW., and was divided approximately as follows—

Buildings	35%
Fuel handling, boilers, etc.	20%
Turbines, generators	25%
Switchgear	10%
Transformers, cables, sundries	10%

Interest has to be allowed at about 4 per cent, and a sinking fund provided to replace the installation at the end of its useful life, which is estimated at about twenty years. If the interest rate is r (= 0·04 for 4 per cent), the interest per kW. is £(12 × r) per annum and the sinking fund is

$$\pounds\left\{\frac{12 \times r}{(1 + r)^{20} - 1}\right\} \text{ per annum per kW.}$$

At 4 per cent, the interest is £0·48 and the sinking fund £0·40 per kW. per annum, at 5 per cent the figures are £0·60 and £0·36, the totals being £0·88 and £0·96 respectively. The sinking fund charge has been calculated on the assumption that at the end of twenty years the cost of replacement is the full value; it may be less if the apparatus has some market value, or more if it has none and requires disposal.

The advantage of a high load factor can be easily seen by considering the case of the larger steam power station, whose annual running cost is

$$\pounds(1{\cdot}2 \times \text{kW.} + 0{\cdot}00060\ \text{kWh.}).$$

The fixed annual charge is £120 000 and must be shared by the units generated. By definition the units are proportional to the load factor, so that the fixed charges contribute a cost per unit that is inversely proportional to the load factor. Thus at a load factor of 100 per cent the units generated are 100 000 × 8 760 = 876 million kWh., and the fixed charges represent

$$£\frac{120\ 000}{876\ 000\ 000}\text{ per kWh.} = 0{\cdot}0328\text{d. per kWh.}$$

If the load factor is 50 per cent, there are half as many units to share the cost and each therefore carries 0·0656d. The variable or running cost per unit is £0·00060 = 0·144d. The total cost per unit at 50 per cent load factor is thus 0·2096d.

The cost per unit in Great Britain in 1938, 1948, and 1954 is as follows—

	Fuel	Freightage	Fuel Handling	Repairs	Salaries, etc.	Total Cost
1938	0·108	0·037	0·009	0·024	0·022	0·200
1948	0·322	0·074	0·022	0·045	0·033	0·496
1954	0·462		0·024		0·036	0·565

HYDRO-ELECTRIC STATIONS

It has already been stated that water power has been harnessed reliably by the use of the electrical transmission of energy. Countries such as Sweden, Norway, and Switzerland have been developing large hydro-electric stations for the past forty years because they have not a sufficient supply of coal, especially in times of emergency such as war. The tendency nowadays, even for countries that have large coal resources, such as Great Britain and the U.S.A., is to utilize their water power in order to conserve their resources of coal; in the latter country the use of large reservoirs derives added importance for flood control and navigation purposes, whilst the social benefits of large public works in times of wide unemployment are also important. Hydro-electric plants have become economic competitors with steam power plants in some places, and in the future they will acquire more economic advantage as the price of coal and oil rises steadily.

The energy available theoretically when a pound of water is at a height or head of h ft. is h ft.-lb. A rate of flow of 1 ft.^{3} per sec., called 1 *cusec*, carries 62·3 lb. of water per sec. so that the available power is 62·3 h ft.-lb. per sec.

$$= \frac{62{\cdot}3h}{550}\text{ h.p.} = \frac{h}{8{\cdot}83}\text{ h.p.} = \frac{h}{11{\cdot}8}\text{ kW.}$$

Thus the theoretical kilowatts available are

$$\text{cusecs} \times \text{head in feet} \div 11{\cdot}8.$$

Assuming an 80 per cent efficiency in the mechanical part of the installation, when there is a direct drive from the waterwheel to the generator, and 75 per cent if the drive is indirect, and a 95 per cent generator efficiency, the available power in kilowatts is

$$(\text{cusecs} \times \text{head in feet})\left(\frac{1}{11{\cdot}8} \times \frac{80}{100} \times \frac{95}{100}\right)$$

$$= (\text{cusecs} \times \text{head in feet}) \div 15{\cdot}5 \text{ for direct drive}$$

and $(\text{cusecs} \times \text{head in feet}) \div 16{\cdot}5$ for indirect drive.

A *catchment* area is the area, bounded by watersheds, which drains into a river so that the water flows past a given point; a point on the river nearer the sea has a larger catchment area. When estimating the water in an area it is useful to remember that one acre is 43 560 ft.^{2} and one square mile is 27·9 million ft.^{2}

$$\text{Also } 1 \text{ cusec} = 31{\cdot}5 \text{ million ft.}^3 \text{ per year}$$

$$= 1{\cdot}13 \text{ sq. mile-ft. per year.}$$

Evaporation losses would raise the requirement to 40 million ft.^{3} per year for 1 cusec.

EXAMPLE. The catchment area is 10 sq. miles, the available head 100 ft., the rainfall 60 in. per annum, and 70 per cent of the total rainfall can be used. Find the available electrical power.

The available flow is

$$10 \times \frac{60}{12} \times \frac{70}{100} \text{ sq. mile-ft. per year}$$

$$= 10 \times \frac{60}{12} \times \frac{70}{100} \times \frac{1}{1{\cdot}13} \text{ cusecs}$$

$$= 31 \text{ cusecs.}$$

For direct drive the available electrical power is thus

$$\frac{31}{15{\cdot}5} \times 100 \text{ kW.} = \underline{200 \text{ kW.}},$$

whilst for indirect drive the power is

$$\frac{31}{16{\cdot}5} \times 100 = \underline{188 \text{ kW.}}$$

EXAMPLE. Find the available electrical energy stored by a million ft. of water at a head of 1 ft., assuming direct drive.

The potential energy available is 62·3 million ft.-lb., of which the plant delivers

$$62{\cdot}3 \times 0{\cdot}8 \times 0{\cdot}95 \text{ million ft.-lb.}$$

as electrical energy. This is

$$\frac{62{\cdot}3 \times 0{\cdot}8 \times 0{\cdot}95 \times 1\,000\,000 \times 746}{550 \times 3\,600 \times 1\,000} \text{ kWh.}$$

$$= \underline{17{\cdot}9} \text{ kWh.}$$

The amount of rain falling in a catchment area that is available for use depends largely upon the nature of the country; in flat

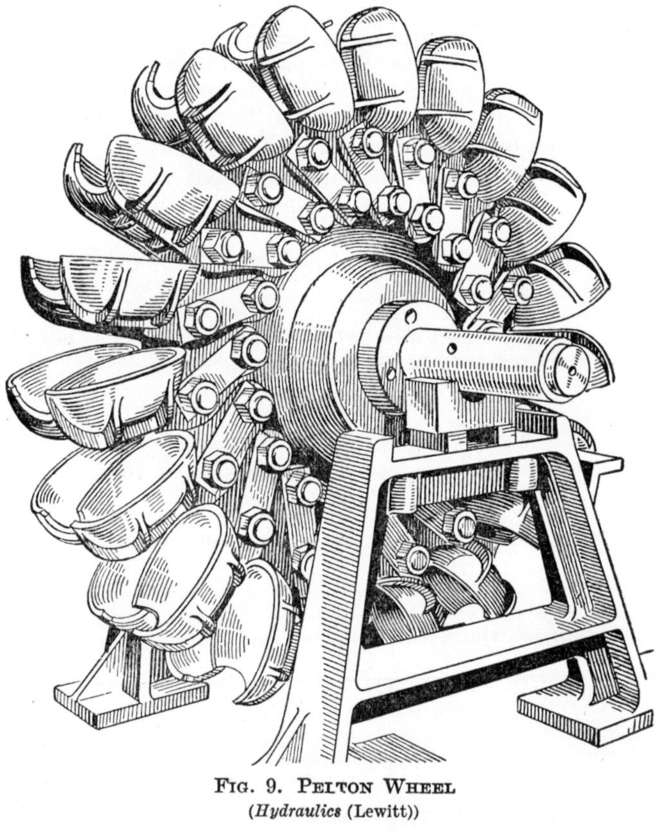

FIG. 9. PELTON WHEEL
(*Hydraulics* (Lewitt))

regions covered with vegetation much of the rain is absorbed, whilst in steep and rocky places most of it is available for use.

Choice of Site. The site of a hydro-electric plant is determined by natural conditions. The water must be available at a usable head and in sufficient quantity. If the flow is not regular enough for continuous supply, there must be convenient accommodation for a reservoir or dam at a reasonable cost; the lower the head the larger must be the reservoir. Other considerations may help in the fixing of the site; for instance, the erratic flow of the Colorado River has created dangerous conditions resulting in damage to life and property, so that the building of the Boulder Dam was welcome for flood control as well as for the construction of the hydro-electric station. If water is stored by a dam and reservoir, it must be carefully arranged that the heavy flows during peak loads do not cause floods downstream.

Prime Movers. Low heads up to 150 ft. use the vertical shaft Francis turbine in which the action is similar to that in the reaction type steam turbine; the water glides over the blades with a small and fairly constant velocity, and exerts a pressure which varies from a maximum at the top to a small value near the bottom.

Modified forms of the Francis turbine are used for medium heads up to 500 or 600 ft., the shaft being horizontal for the larger heads.

High heads, above 500 ft., use the Pelton wheel in which the action is similar to that in the de Laval or impulse type steam turbine. The water is allowed to run down from the head or height and attains a considerable velocity. It is then directed on to the blades, or buckets, which absorb the kinetic energy of the moving water. Fig. 9 shows a Pelton wheel; the bucket is of the Doble form, in which a part is cut away to allow the jet access to the place where it can act most efficiently. The theoretical velocity of the jet is $\sqrt{(2gh)}$ ft. per sec., where $g = 32{\cdot}2$ and h is the net head in feet, viz. the actual head less the head lost by friction in the pipes. The loss in head due to the pipes is given by

$$4mlv^2/2g\mathrm{D},$$

where v is the velocity in ft. per sec., l the length of the pipe in ft., D = diameter in ft., and m is a coefficient depending upon the diameter and inner surface of the pipe. Some values of m are as follows—

Diameter (in.)	Clean Pipes	Slightly Pitted Pipes	Foul Pipes
6	0·0067	0·0077	0·0110
18	0·0052	0·0062	0·0073
36	0·0042	0·0049	0·0055

On a 1 000 ft. head the actual issuing velocity is roughly 240 ft. per sec. The peripheral speed of the Pelton wheel is a little less than half the water velocity.

In the reaction type of wheel the speed can be varied within wide limits by design of the blades, as in the Kaplan and propeller turbines. The propeller type runner is designed to maintain axial flow as nearly as possible, whereas in the Francis turbine there is considerable loss at part loads due to the rotary motion of the water. This kind of wheel has a peaked efficiency curve, so that the efficiency on part load is very low. This drawback is avoided in the Kaplan turbine, in which the inclination of the blades is adjusted by a governor. Thus in a propeller type the efficiency drops from 92 per cent at full load to 65 per cent at half load, while the Kaplan turbine maintains a uniform efficiency of about 90 per cent.

In the case of low heads it is necessary to achieve as high an angular speed of the wheel as possible, in order to avoid gearing to the generator or to reduce the size of the latter. The angular velocity can be increased by reducing the diameter of the wheel, and several wheels can be used to produce a desired output. The term *specific speed* has been introduced so as to enable a comparison to be made between different sizes of machines as regards angular velocity. The specific speed is the number of revolutions per minute that a geometrically similar machine would make if it were to deliver 1 h.p. under a head of 1 ft. It can be shown that the specific speed is given by

$$n_s = (\mathrm{N} \times \sqrt{\mathrm{h.p.}})/h^{5/4},$$

where n_s = specific speed;

N = revolutions per min.;

h.p. = output in horse-power,

and h = head in ft.

With a given head the velocity is constant; the water flow is proportional to the square of the diameter of the wheel, as also is h.p., whereas the speed N is inversely proportional to the diameter and thus is inversely proportional to $\sqrt{\mathrm{h.p.}}$ The value of n_s given above is in British units; the metric unit is given by expressing h in metres and h.p. in metric horse-power (= 0·986 British h.p.); thus metric specific speed = 4·38 × British specific speed.

With the same head the specific speed of a Kaplan or propeller turbine is twice or three times that of a Francis turbine. The advantage can be used to increase the diameter of the wheels and decrease their number for a given output of energy, or to avoid or decrease the gearing, or to decrease the size of the generator.

In a certain case, Kaplan and Francis turbines give the same output under the same head, the former driving direct at 125 r.p.m. while the latter runs at 41·7 r.p.m. and requires gearing.

Approximate values of the specific speed for water turbines are 20 to 120 for the Francis type, 120 to 200 for the propeller type, and 10 to 20 for the impulse type.

Arrangement of Hydro-electric Installation. The layout depends upon geographical conditions and the head.

When the head is very low, say 8 ft., a dam is constructed which raises the level and provides storage. The water passes through sluices to the penstock and thence into the head race. Then the water flows through racks into the forebay, then through the turbines, which are of the "open" construction, the flow being controlled and regulated by gates. The turbines discharge into the tail race downstream. The Linton Lock station has a maximum head of 11 ft. and an average of 8·8; the output is 750 kW., and there are three Francis turbines, two of 430 h.p. and one of 330 h.p. Special precautions were taken in the shape of the retaining walls, etc., to protect the river bank from floods.

Another interesting example of a plant using a very low head, average less than 7 ft., is at Chester; the output is 635 kW. and the installation has proved decidedly profitable.

With a medium head of 100 to 500 ft. the arrangement is quite different, since the flow for a given output of energy is much lower. The Grampian Electricity Supply Co. have recently constructed hydro-electric installations at Loch Rannoch (head 465 ft. maximum) and on the River Tummel (160 ft.). Fig. 10 shows the geography of the scheme. The two stations act in parallel, Loch Ericht acting as the reservoir for the whole watershed. "The guiding principle was that in times of flood the Tummel station would take the base load, and storage would be effected at Loch Ericht; and vice versa during low water periods."*

Fig. 11 shows the Civil Engineering Works of the Rannoch plant. A concrete-lined tunnel, 18 ft. in diameter, runs 14 080 ft. to the valve house; the tunnel crosses the River Ericht, at which it is in the form of a riveted steel pipe-line, 11 ft. 10 in. diameter. There are two vertically arranged Francis turbines enclosed in a spiral casing. Fig. 12 shows a cross-section of the Rannoch power station. The capacity of this station is 32 000 kW. and of Tummel 34 000 kW.

The hydro-electric plant in Scotland is 660 MW., 350 MW. having been installed since 1943 by the North of Scotland Hydro-electric Board at a cost of £114 per kW. The output in 1955 is 1·62 milliard, or equivalent to 960 000 tons of coal. Plans are in hand for a further 400 MW., and it is hoped to achieve a pumped storage of 2 000 MW. between 1961 and 1975.

The Boulder Dam power station in California is the largest in the world. The reservoir created by the dam is 227 sq. miles and has a capacity of 30 million acre-feet; the top 9·5 million is allocated to flood control, the bottom 5 million for storage of silt, and the middle 16 million is active storage for the regulation of flow to suit irrigation and power production. There will be available a firm continuous output of 663 000 h.p. The power plant is designed for an ultimate

* "Hydro-electric development in Great Britain, . . ." by A. S. Valentine and E. M. Bergstrom, *Journal of the I.E.E.*, Vol. 76, page 125, describes the Grampian Electricity Supply Co.'s schemes in full.

installation of 15 generating units of 82 500 kVA. each. Fig. 13 shows the general scheme. There is a spillway on each side of the river; each is an open channel 650 ft. long, 150 ft. wide, and 120 ft. deep, lined with concrete with an ogee-shaped crest on the river side. The turbines are of the vertical-shaft single-runner Francis type with spiral casings of cast steel. The generators have each a capacity of 82 500 kVA., 180 r.p.m., 60 cycles, 16·5 kV.; they have a very large moment of inertia, so that the short-circuit ratio is the very low figure of 2·74. The power is transmitted 266

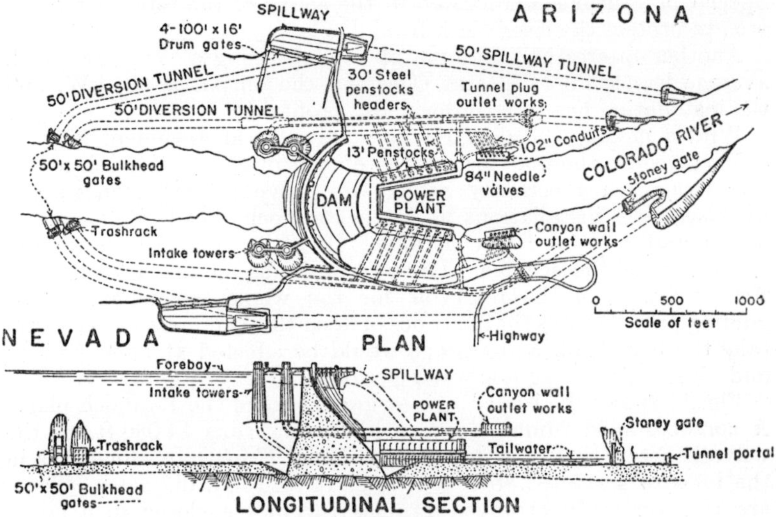

FIG. 13. BOULDER DAM
(*Electrician*)

miles to Los Angeles at a voltage of 275 kV., the conductors being hollow copper tubes.

Several schemes have been put forward for the utilization of the water power in tides. For this purpose it is necessary that conditions be favourable for the construction of an enormous reservoir. One of the most widely discussed projects is that of the Severn barrage, which was estimated to be capable of producing 500 000 h.p. during a 10-hour day with a peak load capacity of 1 000 000 h.p. In order to overcome the disadvantage of ebb and flow there were to be two reservoirs, and the effects of uneven flow were to be avoided by pumping water up to an elevated reservoir during high level periods. The large cost of the necessary constructional work has prevented as yet the exploitation of the scheme.

Operating Costs of a Hydro-electric Station. The main characteristics of a hydro-electric installation are the high capital cost and the very low running costs. The installation cost may vary up to £200 per kW., and running costs may be as low as 0·1d. per kWh. If we assume an interest rate of 5 per cent on a capital cost of £200 per kW., the annual cost is

$$\pounds(10{\cdot}0 \times \text{kW.} + 0{\cdot}0004 \times \text{kWh.}),$$

which compares with

$$\pounds(5{\cdot}0 \times \text{kW.} + 0{\cdot}0020 \times \text{kWh.})$$

the cost of a large steam power station. The importance of a high load factor is greater for the hydro-electric plant than for the steam power station. Values of load factor and cost in pence per unit for the steam station and the hydro-electric plant are given below.

Load Factor (%)	20	40	60	80	100
Steam station . . .	1·15	0·82	0·71	0·65	0·62
Hydro-electric . . .	1·47	0·78	0·56	0·44	0·37

It is seen how important it is to keep the plant running at as high a load factor as possible. To raise the load factor, arrangements have been made in the Pyrenees district for electro-chemical firms to buy blocks of surplus power at certain seasons and at a few hours' notice at very low rates. In some countries domestic consumers are encouraged to buy power for hot water storage at times of low load, the supply being controlled by a time-clock. One way to obtain a high load factor is to use a hydro-electric station for the base load, which is continuous, and to supply the peak loads by steam or internal combustion engine stations.

NUCLEAR POWER STATIONS

The physical principles involved in the production of heat by nuclear fission are described in Appendix VIII. Briefly the facts are as follows. Natural uranium consists of 99·3 per cent of ^{238}U and 0·7 per cent (1 part in 140) ^{235}U. ^{235}U can absorb fast or slow (*thermal*) neutrons; in doing so it becomes unstable, splits in about 10^{-14} sec. into two approximately equal fragments with large kinetic energy and with an emission of 2 or 3 neutrons with an energy of about 2 MeV. ^{238}U requires neutrons with energy greater than 1·1 MeV. for fission, but absorbs neutrons with less energy without splitting: in the latter case it is converted after several transformations to plutonium, ^{239}Pu. ^{235}U has a very large capture cross-section for thermal neutrons, i.e. it has a great attraction for slow neutrons, so that even though ^{235}U is present as 1 part in 140 in natural uranium the thermal neutrons are more likely to be captured by the ^{235}U than by the ^{238}U.

The action of a thermal (i.e. slow) reactor is thus as follows. Natural uranium is used. An atom of ^{235}U absorbs a neutron, splits and produces 2 or 3 neutrons: the kinetic energy of the fragments produces heat. The emitted neutrons would be slowed down below 1·1 MeV. by collisions and then absorbed without fission by ^{238}U, because of its abundance: the action would then stop. It is necessary to ensure that at least one of the 2 or 3 neutrons will be absorbed by ^{235}U, and this is achieved by a *moderator*. A moderator is an element of small atomic number, e.g. hydrogen, carbon, beryllium, which slows down neutrons by collision without absorbing them: heavy water and pure graphite are the most commonly used moderators. By arranging the uranium rods and the moderator, pure graphite say, in a suitable configuration, the fast neutrons are quickly slowed down to thermal velocity of the order of 1 eV. before they can be absorbed by ^{238}U. At this low speed of neutrons, the preference of ^{235}U for neutrons overrides the abundance of ^{238}U, and at least one of the neutrons is absorbed by the ^{235}U with a consequent fission. The process thus continues.

In order that the production of heat shall proceed at a desired rate, it is necessary that exactly one neutron shall be absorbed by ^{235}U from each earlier fission. To do this the configuration is designed to give rather more neutrons than are necessary, and to absorb the excess in *control rods:* these are usually made of cadmium or boron carbide, since cadmium and boron have a great absorptive capacity for neutrons. The control rods can then be moved in or out of the reactor to decrease or increase the rate of fission and heat. Finally, the time constant of the control required is rendered practicable by the production of *delayed neutrons*. The neutrons are produced by the fission in about 10^{-14} sec., and therefore are too rapid to control. Fortunately there is another source of neutron production, viz. by the disintegration of the fission fragments, and these are produced up to a minute or so later than the fission. The time constant of the control can then be of the order of minutes, and this is practicable.

Calder Hall power station has reactors of the sort described above, one of the main purposes being the production of plutonium from the ^{238}U. This Pu can be separated chemically from the ^{238}U and used later to enrich the material of a reactor. The amount of Pu produced can exceed the ^{235}U consumed.

If the proportion of ^{235}U is increased, to, say, 25 per cent, or if Pu is added, then even fast neutrons will cause fission because there is so much more chance of their being absorbed by ^{235}U or Pu. Such a system is called a *fast reactor:* note that the control time constant is still determined by delayed neutrons. The advantage of a fast

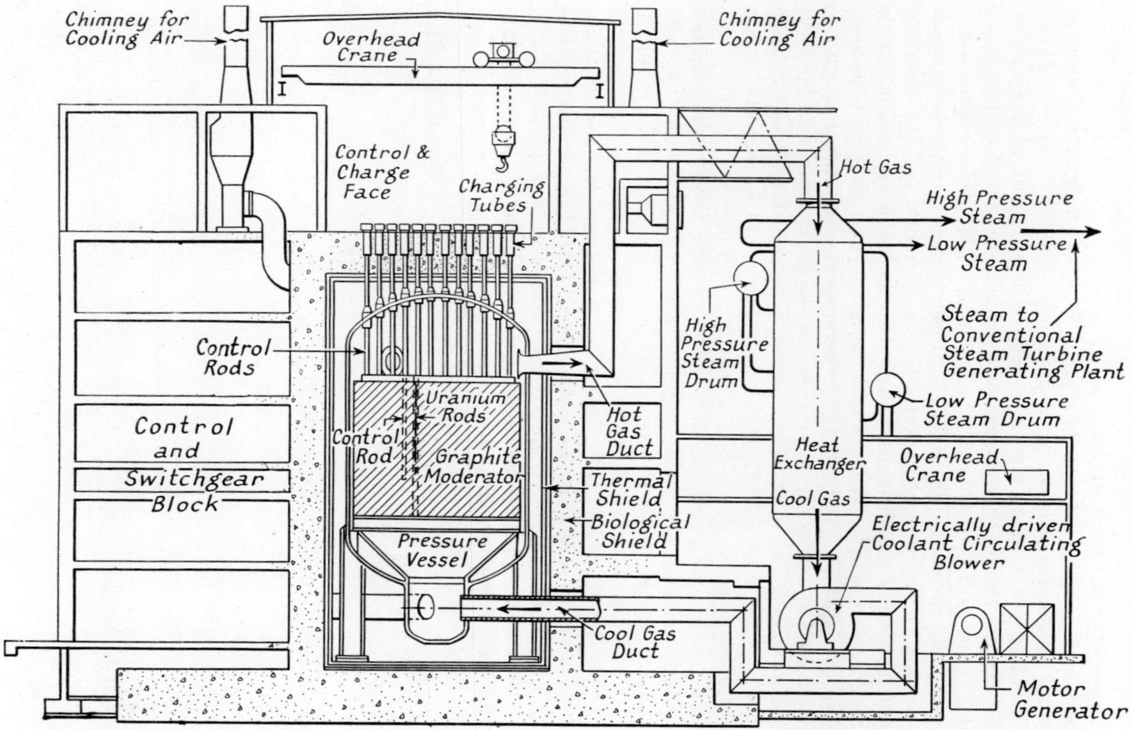

Fig. 14. Calder Hall Reactor

reactor is that it can be made much smaller than a thermal reactor.

The conversion of ^{238}U to Pu is called *breeding*, and is an important process as it leads to the consumption of the whole of the natural uranium for the production of energy.

Fig. 14 shows a reactor and associated equipment of the Calder Hall station: there are two reactors, which are stated to give over 50 MW., but will probably give nearer 100 MW. The reactors are enclosed in a pressure drum 37 ft. diameter and about 70 ft. high of 2 in. steel. The fuel elements are 1 in. diameter rods of uranium sheathed in an aluminium alloy, and are placed in vertical channels in the graphite bricks: control rods hang in other vertical channels. The cooled gas CO_2 from the heat exchanger passes up through the graphite core at 100 p.s.i. and receives a temperature rise of 200° C.; and then passes via four 54 in. ducts to the heat exchanger. The reactor is shielded by 6 in. mild steel plates and a concrete screen 7 ft. thick.

Each reactor produces steam to drive two 23 MW. turbo-alternators; at rated load the steam supplies are 185 p.s.i. at 590° F. and 40 p.s.i. at 340° F. which enters at the fifteenth stage of the turbine.

Four such stations are being built and they should be contributing about 300 MW. to the grid by 1960; by 1965 the C.E.A. are to build twelve nuclear power stations to develop 2 000 MW., and thereafter the annual construction is hoped to be at the rate of 2 000 MW. The costs of a graphite-moderated, gas-cooled reactor power station for 150 MW. is estimated as follows—

Capital Costs	Capital Cost (£ million)	Annual Cost (£ million)	Cost (d./kWh.) at 80 per cent load factor
Reactor plant . . .	7·5	0·68	
Other plant . . .	11·3	0·69	
Fuel at £20 000 per ton uranium . . .	5·0	0·20	
Total	23·8	1·57	0·36
Operating Costs			
Site operating . . .		0·26	
Cartridge replacements .		1·72	0·40
			0·76

From this cost of 0·76d./kWh. has to be deducted the value of the plutonium produced. If the Pu is credited at £5/g. there is an annual allowance of £0·17 million, or 0·17d./kWh., so that the net cost is 0·59d./kWh.; if the Pu is credited at £10/g., the allowance is 0·33d./kWh., and the net cost is 0·42d./kWh. It is seen that these figures are comparable with the figure of 0·565d./kWh. for the C.E.A. stations in 1954.

THE CENTRAL ELECTRICITY AUTHORITY AND THE GRID

The partial failure of Ferranti in 1890 to supply high voltage a.c. from Deptford to London had the effect of concentrating the attention of supply engineers upon d.c. Distribution was at 110 volts d.c. and could be used only in small densely populated areas. The generation and supply was in the hands of private companies or municipalities who had a monopoly for their district. Moreover, the generation was local and unco-ordinated, so that each station required a large amount of spare plant. Tariffs varied widely in neighbouring districts. The Coal Conservation Committee reported during the war that a great development of electricity supply was required to conserve coal supplies and to maintain the manufacturing position of the country.

The Weir Committee, formed in 1921, issued its report in 1926 which was a favourable political year for its proposals. It recommended that there be constructed a Grid of high voltage transmission lines interconnecting selected large generating stations. The principal advantages of the interconnection are the use of the most economical stations situated where fuel is cheap and the site is convenient, the relegation of the less efficient stations to peak load and stand-by use, and the reduction of spare plant.

Interconnection, however, is not enough; there must also be control by a central organization which will specify the load to be dispatched by each station. It was expected that in this way electric supply would be made available throughout the country at a greatly reduced cost. The Electricity (Supply) Act of 1926 embodied these suggestions and established the *Central Electricity Board* to supply electricity to authorized undertakers, but not to generate (except with a special Order from the Electricity Commissioners).

After the war it was decided to nationalize the generation as well as the supply of electricity, and the *British Electricity Authority* was set up on 1st April, 1948, to operate the nationalized industry. It runs fourteen Area Electricity Boards in Great Britain. A slight reorganization took place in 1955, and the name was changed to *Central Electricity Authority* on 1st April, 1955. This body has issued a report "Seven Year Record—1948 to 1955," reviewing the outstanding developments of electricity supply, of which the following are some points of interest.

The grid, operating at 132 kV., was completed in 1936, and had the effect of reducing excess plant from 80 to 20 per cent. Average generating costs have been progressively reduced and much coal saved. Since before the Second World War the grid has been successfully operated as a solidly interlinked system.

Generator units are larger (up to 200 MW.) and operate up to 2 350 lb./in.^{2} and 1 050° F. The average overall thermal efficiency of operation of all power stations has been increased from 20·87 per cent in 1947–8 to 23·85 per cent in 1954–5. A supergrid at 275 kV. has been installed to transmit power in bulk over the country and feed the 132 kV. grid. The C.E.A. is now planning nuclear power stations. It is also planning a cross-Channel 132 kV. cable link to carry 100 MW.; it expects that it will thus be able to import "spill-over" power from the vast French hydro-electric developments. By 1965 some sixteen stations using oil will be installed, at a cost equivalent of 10 million tons per year.

EXAMPLES I

1. A power station burns about 1·8 lb. of coal per kWh. generated. Assuming the coal to have an average calorific value of 13 500 B.Th.U. per lb. find the overall efficiency, and state approximately how the losses would be distributed between the boilers, engines and electrical generators. (*Lond. Univ.*, 1927.)

2. Find the ideal efficiency of an engine using steam (Rankine cycle) at 875° F. and 650 lb. per in.^{2}, the condenser exhausting at 0·5 lb. per in.^{2} Find the ideal efficiency if (*a*) the condenser exhausts at 1 lb. per in.^{2}, (*b*) the temperature is 750° F.

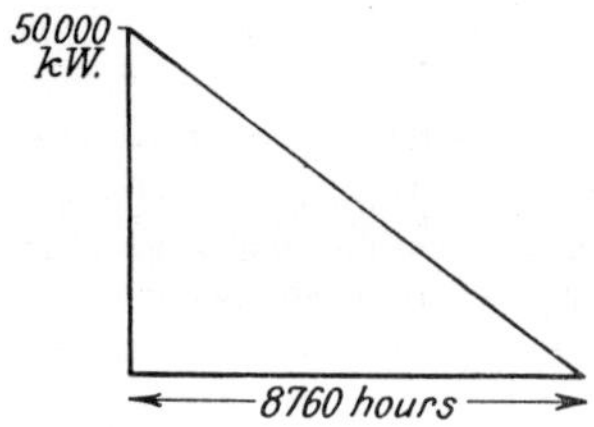

FIG. 15

3. The estimated total annual operating costs and capital charges for two proposed power stations are given by the following expressions—

Annual cost for station A

$$= £(6\,000 + 0{\cdot}3 \times \text{kW.} + 0{\cdot}0015 \times \text{kWh.}).$$

Annual cost for station B

$$= £(9\,000 + 0{\cdot}5 \times \text{kW.} + 0{\cdot}0014 \times \text{kWh.})$$

where kW. represents the capacity of the station and kWh. represents its total annual energy output.

The stations are to be used to supply a load having a "load duration" curve as shown in Fig. 15. The ordinate of a point on this curve represents a certain load on the station and its abscissa represents the number of hours per year during which the load is equal to or exceeds this amount. Which station should be used to supply the peak load, what should be its installed capacity and for how many hours per year should it be in operation to give the minimum total cost per unit generated? Calculate the total cost per unit generated under these conditions. (*Lond. Univ.*, 1932.)

4. Explain the principle and applications of the water storage system for improving the load factor of a generating station.

A load having a maximum peak of 100 000 kW. and a load factor of 30% is to be supplied by one of the following schemes—

(*a*) a steam power station capable of supplying the whole load;

(*b*) a steam power station in conjunction with a water storage station, the latter supplying 100 × 10^6 kWh. per annum with maximum output of 40 000 kW.

The costs are as follows—

Capital cost of steam station	£20 per kW. installed.
Capital cost of storage station	£14 per kW. installed.
Operating cost of steam station	0·1d. per kWh.
Operating cost of storage station	0·01d. per kWh.
Efficiency of storage station	60%
Interest on capital cost	15%

Assume no spare plant required.

Calculate the average overall generating cost per unit in the two cases. (*Lond. Univ.*, 1933.)

5. A 30 000 kW. turbo-alternator has an overall efficiency of 24·1 per cent. If the coal burned has a calorific value of 12 500 B.Th.U. per lb. find the coal consumption per kWh. and per day of 24 hours if the load factor is 27%. (*Faraday House Diploma, March*, 1935.)

6. A Diesel generator is found on test to have a specific fuel consumption of 0·56 lb. per kWh., the oil having a calorific value of 19 200 B.Th.U. per lb. If the generator has an efficiency of 95% find the overall efficiency of the engine. What would you expect the thermal and the mechanical efficiency to be? (*Faraday House Diploma, July*, 1935.)

7. A proposed hydro-electric station has an available head of 100 ft., a catchment area of 20 sq. miles, the rainfall for which is 48 in. per annum. If 70% of the total rainfall can be collected, calculate the power that could be generated. Suggest a suitable size for the machines. Assume the following efficiencies: penstock 95%, turbine 80%, generator 85%. (*Nat. Cert.*, 1935.)

8. Explain carefully the various factors which govern the relative amounts of load which should be supplied by the steam and hydro-electric stations respectively in order to secure the most economical operation of a system supplied by both types of station.

Under what circumstances is a pumped-water-storage system likely to prove an economic proposition when used in conjunction with a power system? (*Lond. Univ.*, 1934.)

9. An undertaking owns a steam driven station equipped with three 5 000 kW. generating sets. It is estimated that the maximum load will reach 14 000 kW. in two years' time. To meet this with a spare set, the undertaking can either install a further 5 000 kW. set in the station at a cost of £16 per kW. or purchase a bulk supply from the C.E.B. at a two-part tariff of £3 15s. per annum per kW. of maximum demand plus 0·18d. per unit. The present generating costs of the steam-driven station are £3 10s. per annum per kW. of maximum load plus 0·30d. per unit. Annual charges on new plant may be taken as 15% on capital cost and the total station output in two years at 50 million units. Which alternative should the undertaking adopt? (*Nat. Cert.*, 1934.)

10. Give a brief account of the influence of diversity factor and of load factor upon the total cost of supply of electric energy.

From the following data, estimate the generating cost per unit delivered at the station: capacity of the generating plant installed, 50 000 kW.; annual load factor, 40%; capital costs, £600 000; annual cost of fuel, oil, taxation, wages and salaries, £80 000: rate of interest, 5%; rate of depreciation, 5% of initial value (*Lond. Univ.*, 1932.)

11. Enumerate the principal auxiliary plant (excluding boiler and coal-handling plant) required for a large turbo-alternator in a generating station. Discuss the type of drive which would be employed for these auxiliaries, giving reasons for the choice. Discuss also the system of power supply to any electrically-driven units in order to ensure continuity of service under all operating conditions.

An electrically-driven boiler-feed pump delivers 300 000 lb. of water per hour at a pressure of 500 lb. per in.^{2} Calculate the current input to the 3 000 V., 3-phase driving motor assuming the efficiencies of pump and motor to be 85%, and 92% respectively, the power factor 0·93, and the weight of 1 ft.^{3} of hot water 59 lb. (*Lond. Univ.*, 1947.)

12. Discuss the factors which influence the choice of site for a steam generating station.

A supply is required for a new project having a maximum demand of 50 000 kW. and an estimated load factor of 50%. The energy can be supplied by either

(*a*) a steam power station having a capital cost of £50 per kW. and a maintenance cost of 0·5d. per kWh., using coal of calorific value 11 000 B.Th.U. per lb. The overall thermal efficiency of the station is estimated at 25% (1 kWh. is equivalent to 3 409 B.Th.U.); or

(*b*) a hydro-electric station having a capital cost, including transmission lines, etc., of £210 per kW. and a running cost of 0·1d. per kWh.

Allow 12% for the steam station, and 10% for the hydro-electric station, for interest and depreciation. Calculate the price of coal above which the steam station is uneconomical. (*Lond. Univ.*, 1950.)

13. Explain why a closed-circuit system of circulating the cooling air or gas is advisable for a large turbo-alternator. Sketch the general arrangement of such a system and explain how the temperature rise of the machine is maintained within the specified limits. State the advantages of hydrogen, compared with air, as a cooling agent and mention some of the difficulties of its application to turbo-alternators.

Calculate the volume of air required per minute at full load for a 50 MVA., 0·8 power factor, turbo-alternator assuming that 85% of the total losses is dissipated by the internally circulating air, the inlet and outlet temperatures of which are 40° C. and 70° C. respectively, and that the full load efficiency is 97·5%. Assume specific heat of air = 0·24: weight of 1 cubic foot of air = (1/13) lb. (*Lond. Univ.*, 1947.)

CHAPTER II

TRANSMISSION OF ELECTRICAL ENERGY

Transmission by Low Voltage Direct Current. The Radial System. The early supply of electrical energy was by means of low voltage direct current, and a number of such systems still exist. A populated area is served by one power station, and the system is as shown in Fig. 16. *Feeder mains*, F, which are cables of large current-carrying capacity, carry the current in bulk to *feeding points*, where *distributors*, D, tap off the current to the *service mains*, S; the latter are small cables which lead the current to the consumers'

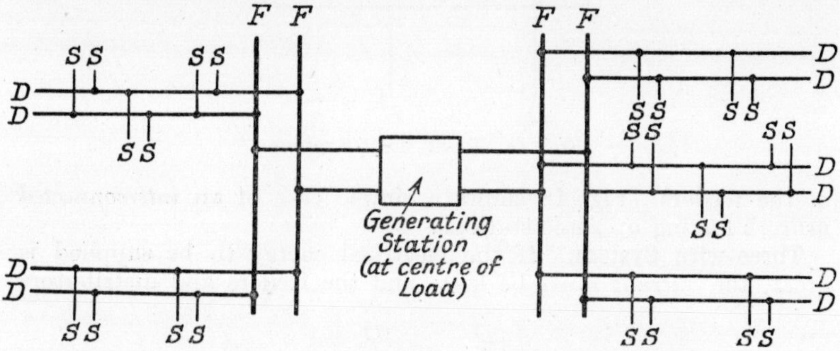

FIG. 16. EARLY DISTRIBUTION SCHEME, RADIAL SYSTEM

premises. Such a system is called a *radial system*, as the feeders, distributors, and service mains radiate outwards from the generator. The size of the feeders is determined mainly by the current to be transmitted, as the volt-drop along them can be allowed for by regulation or compounding. The size of the distributors, however, is determined by the fact that the voltage fluctuation at the consumers' terminals must not exceed the Regulation limit of $\pm$ 6 per cent. The disadvantage of the radial system is that the consumer is dependent on a single feeder, so that a fault on any feeder or distributor cuts off the supply from all consumers who are on the side of the fault away from the station.

The Ring System. This disadvantage is removed by the *ring system*, in which each consumer is supplied via two feeders. A simple example of the ring system is shown in Fig. 17; for simplicity the two (or three) wires of the supply lines are represented by a single line. If there is a fault on a feeder at a point A, the section between B and C can be switched out without interrupting the supply to any consumers.

When the ring main is employed, the electrical energy can be supplied by two or more generators at the same or different points

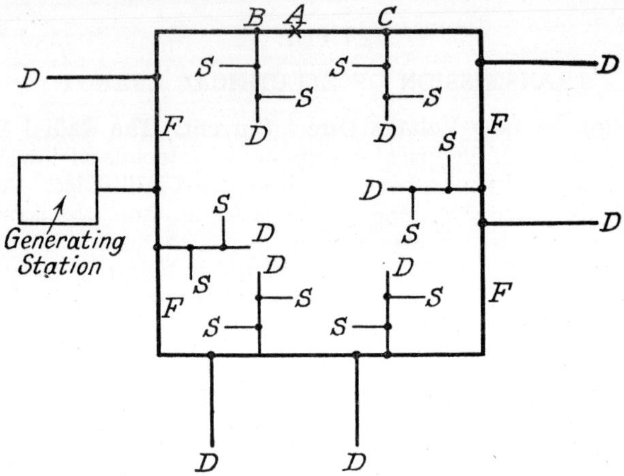

Fig. 17. Ring Main System

of the feeders. Fig. 18 shows a simple case of an *interconnected network* linking up three stations.

Three-wire System. If the electrical energy to be supplied is great, the current must be large and the feeders and distributors

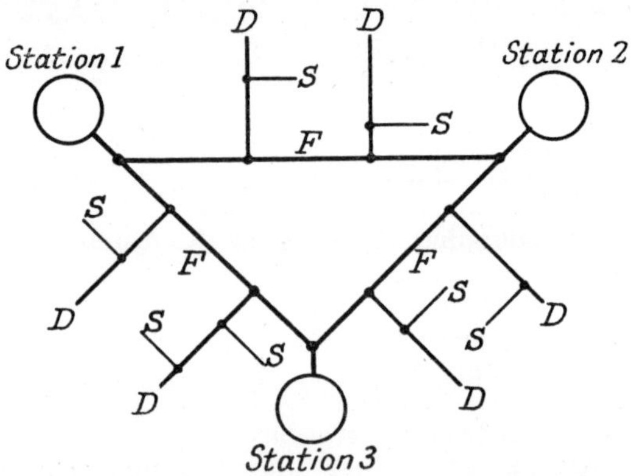

Fig. 18. Simple Interconnected System

must have a large cross-section. A considerable economy is effected by the use of a 3-wire system, which is shown in Fig. 19. One generator supplies current via the wires *1* and *N*, and the other via *N* and 2; *N* is the neutral and is earthed. If the generators supply equal loads, the currents I_1 and I_2 are equal and the neutral carries no current. In practice, I_1 and I_2 are not quite equal and the neutral carries a current which is small compared with either. The neutral wire cannot be made of very small cross-section or the unbalance current $I_1 - I_2$ will cause a large voltage drop between different points and vary the potential from the earth value. It is usual for the neutral to have a cross-section equal to half that of the outer wires. The saving in copper is found in the following way. Let V be the voltage between the wires in the 2-wire case and between the outers and the neutral in the 3-wire case, P the power to be transmitted, R_1 the resistance per cm. per wire in the former

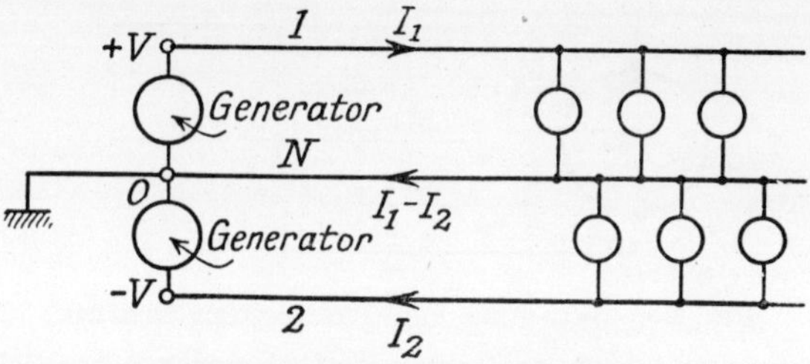

FIG. 19. THREE-WIRE SYSTEM

case, and R_2 the resistance per cm. per outer wire in the latter; the resistance of the neutral is $2R_2$ per cm., as it has half the cross-section of the outers. We have to find the relation between R_1 and R_2 so that there are the same I^2R losses in both cases. In the 2-wire case the current per wire is P/V, and the losses per cm. for the two wires are

$$2(P/V)^2R_1.$$

In the 3-wire case, if we assume balanced loads, the current in the outers is $(P/2V)$ and zero in the neutral. The losses per cm. are

$$2(P/2V)^2R_2.$$

We get therefore

$$2(P/V)^2R_1 = 2(P/2V)^2R_2,$$

or

$$R_2 = 4R_1.$$

The cross-section of the outer is thus one-fourth of that in the 2-wire case, so that the copper ratio is

$$\frac{\text{3-wire}}{\text{2-wire}} = \frac{(2 \times \frac{1}{4}) + (\frac{1}{2} \times \frac{1}{4})}{2 \times 1} = \frac{5}{16} = 31{\cdot}25 \text{ per cent.}$$

If the neutral has the same cross-section as the outer, the ratio is

$$\frac{\text{3-wire}}{\text{2-wire}} = \frac{(2 \times \frac{1}{4}) + (1 \times \frac{1}{4})}{2 \times 1} = \frac{3}{8} = 37{\cdot}5 \text{ per cent.}$$

If the currents in the outers are not equal, the main consideration is that there shall not be an excessive voltage drop along the neutral due to the unbalance. In this case the 3-wire system must be made of larger gauge cable and the saving is much reduced. It is therefore essential that the loads be balanced as far as possible.

Balancers. The 3-wire system shown in Fig. 19 has two generators which can be regulated independently, so that unbalanced loads can be allowed for. It is more usual to connect a single generator across the outers with a great saving in cost. In this case a *balancer* must be used to deal with the unbalance in the loads. The balancer consists of two shunt-wound dynamos coupled mechanically, the fields

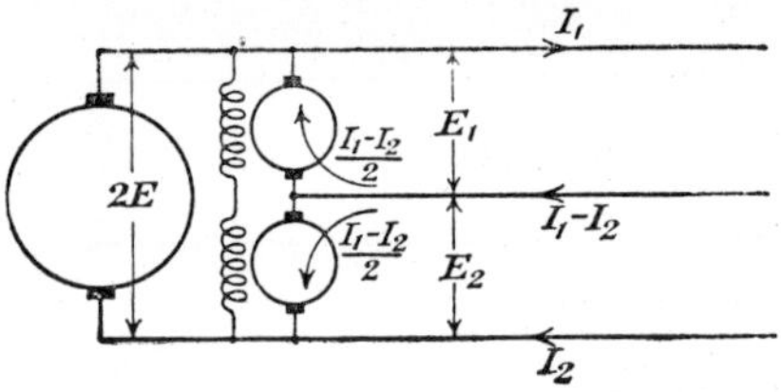

FIG. 20. BALANCER FOR THREE-WIRE SYSTEM

and armatures of which are connected in series across the outers, whilst the middle wire is connected to the junction of the armatures, as shown in Fig. 20. If the loads are balanced the balancer dynamos run as unloaded motors and take merely their no-load currents. Suppose now the load on the upper half is greater than that on the lower, so that I_1 is greater than I_2. The voltage across dynamo 1 drops and that across dynamo 2 rises, with the result that the latter acts as a motor and drives the former which then generates and tends to equalize the voltages between the outers and the neutral.

If the armature resistances were zero and the windage and friction losses negligible, the balancing action would be perfect. For then the voltages across the machines would be equal to their e.m.f.'s, which are the same since the fields are the same and the machines run at the same speed (as they are coupled mechanically). Moreover, one-half of the unbalance current, $\frac{1}{2}(I_1 - I_2)$, goes up through one machine and down through the other as shown in Fig. 20; this follows from the fact that the dynamo 1 acts as a generator with an output equal to its e.m.f. times its current, whilst the dynamo 2 acts as a motor to drive it and must therefore *receive* this same value of current.

If the armature resistance and losses are not negligible, the balancing action is not perfect and may be calculated as follows. Let us consider the case when I_1 is greater than I_2, suppose that dynamo 1 gives a current i_1 upwards and dynamo 2 i_2 downwards as in Fig. 21. Then

$$i_1 + i_2 = I_1 - I_2 \quad . \quad . \quad . \quad . \quad . \quad \text{(i)}$$

Let the e.m.f.'s of the dynamos be e (equal, since they have the same field and speed). The voltages across the machines, which are the voltages between the outers and neutral, are

$$E_1 = e - ri_1 \text{ and } E_2 = e + ri_2 \quad . \quad . \quad . \quad \text{(ii)}$$

where r is the resistance of each armature. Let w be the windage and friction losses. Then

$$w = ei_2 - ei_1 = e(i_2 - i_1) \quad . \quad . \quad . \quad . \quad \text{(iii)}$$

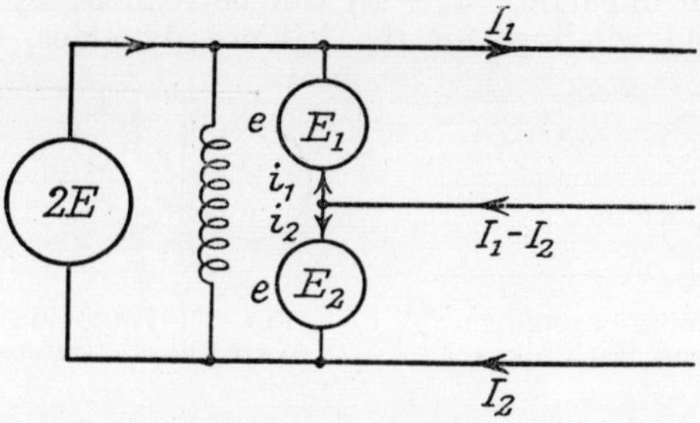

FIG. 21. BALANCER FOR THREE-WIRE SYSTEM

Equations (i), (ii), and (iii) are sufficient to determine E_1, E_2, i_1 and i_2. Eliminating e from equations (ii) we get the voltage unbalance as

$$E_2 - E_1 = r(i_2 + i_1) = r(I_1 - I_2).$$

To find i_1 and i_2 we proceed as follows. Adding equations (ii) we get

$$E_1 + E_2 = 2E = 2e + r(i_2 - i_1).$$

Substituting from this equation in (iii) for e we get

$$w = (i_2 - i_1)\,[E - \tfrac{1}{2}r(i_2 - i_1)],$$

which is a quadratic equation for $i_2 - i_1$, the solution of which is

$$i_2 - i_1 = \frac{E}{r} - \sqrt{\frac{E^2}{r^2} - \frac{2w}{r}}$$

$$\simeq \frac{E}{r} - \frac{E}{r}\left(1 - \frac{wr}{E^2}\right)$$

$$= \frac{w}{E}.$$

With the help of equation (i) we get

$$i_1 = \tfrac{1}{2}(I_1 - I_2) - \frac{w}{2E},$$

$$i_2 = \tfrac{1}{2}(I_1 - I_2) + \frac{w}{2E}.$$

It is clear that the terms $\frac{w}{2E}$ represent the current drawn from the mains of voltage $2E$ to provide the friction and windage losses w.

The voltage unbalance $E_2 - E_1$ can be reduced by using cross-connected field windings for the balance dynamos, as shown in

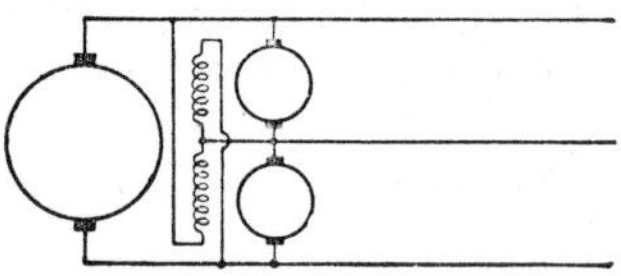

FIG. 22. CROSS-CONNECTED FIELDS FOR BALANCERS

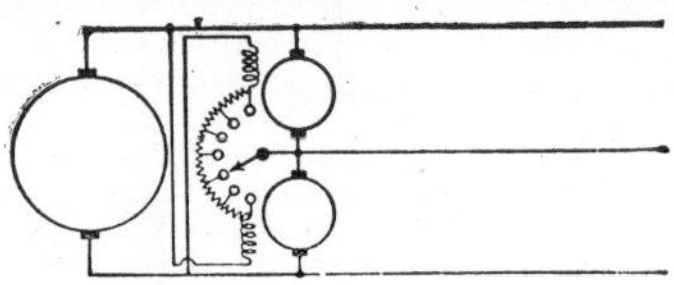

FIG. 23. RHEOSTATIC CONTROL AND CROSS-CONNECTED FIELDS

Fig. 22. It can be shown in the same way as used above, that if the dynamo fields have a linear characteristic the voltage unbalance with this system is

$$\tfrac{1}{2}r(I_1 - I_2)$$

instead of $r(I_1 - I_2)$ for the straight through fields. Since E_2 is greater than E_1, the field of dynamo 1 is increased and that of dynamo 2 is decreased, so that the e.m.f. of dynamo 1 is increased and that of dynamo 2 is decreased.

Hand regulation of the balancer may be employed in addition to the cross-connecting of the fields, as shown in Fig. 23. If the rheostat is set to give equal voltages E_1 and E_2 on half the maximum unbalance current, the unbalance voltage from zero to maximum unbalance current is then about $\pm \tfrac{1}{4}r(I_1 - I_2)$.

Boosters. These are generators inserted into a circuit to compensate for a variable voltage drop. For instance, if the current in a feeder varies, the voltage supplied to the distributors may vary more than the legal amount. This difficulty may be overcome by using very large gauge feeders, but this is costly. A more economical method is to insert a *feeder booster* in the feeder. This booster is a series generator in which the e.m.f. is proportional to the field current, which is here the feeder current. By proper choice of the constants of the booster, the e.m.f. can exactly neutralize the voltage drop in the feeder. Fig. 24 shows the method adopted in practice. The booster is clearly a low-voltage, heavy current machine.

The effect of voltage drop in the feeders can be overcome by using compound d.c. generators, but the use of boosters is more convenient when there are feeders of different lengths.

In a tramway system it may be desirable to raise the voltage of the line at a distant point. This can be achieved by running a feeder from the generator to the point and inserting a booster in series with the feeder.

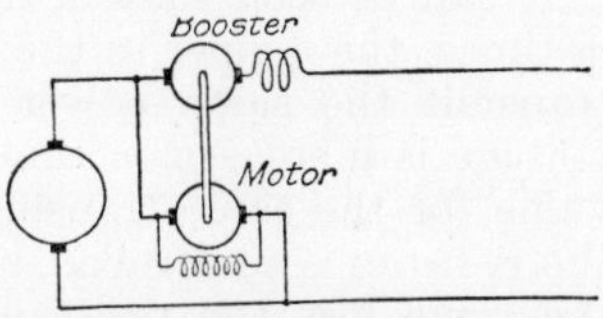

FIG. 24. FEEDER BOOSTER

NEGATIVE BOOSTERS subtract from the voltage, and are used in earth return systems in order to keep the potential of all points of the return rail within the Board of Trade regulation limit of 4·2 volts (to avoid the troubles of electrolysis). Fig. 25 shows how the negative booster is used; in this case a known fraction of the feeder current, which is shunted by R, is used for the field winding.

If a d.c. system is subjected to violent fluctuations of load, e.g. a traction system, it is usual to "float" a battery between the bus-bars, as shown in Fig. 26. If the resistance of the battery were zero, its

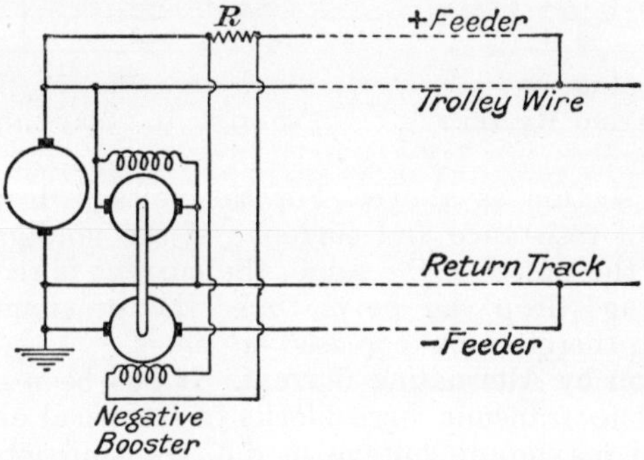

FIG. 25. NEGATIVE BOOSTER CONNECTIONS

potential difference would be constant and equal to its e.m.f., which would then be the voltage between the bus-bars. In times of small load the generator will send charge into the battery as well as into the line; but when the load is heavy the voltage across the generator terminals will drop, the battery will cease to receive current and will send it out on to the line. But the battery has an appreciable resistance, r; if its e.m.f. is E, its voltage on charge is $E + rI$ and on discharge it is $E - rI$. The voltage fluctuation is thus $2rI$, which may be neutralized by means of a *battery booster* as shown in Fig. 27. There is an additional voltage fluctuation due to the fact that E varies on charge and discharge. There are various boosters, e.g. the *Entz* booster, which attempt to allow for this.

Advantage of High Voltage. It was shown in the last section that the use of a 3-wire system causes a saving in the amount of copper required, the reason being that the voltage of transmission is effectively doubled.

It can be seen that if the voltage of transmission is multiplied m times, the copper in the conductors can be reduced $1/m^2$ times to transmit the same power with the same ohmic loss. For if the voltage is increased m times, the current is $1/m$ times the previous value for the same transmitted power. The ohmic loss is equal to the resistance multiplied by the square of the current, so that for the same loss the resistance can be m^2 times the previous value and thus the copper in the conductors need be only $1/m^2$ as much as before.

It can be seen that if the criterion is that the voltage drop be the same percentage in both cases, the conductors can be made $1/m^2$ of

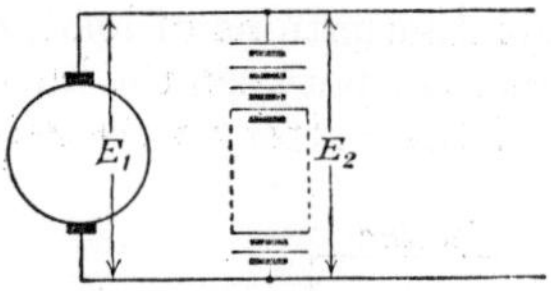

FIG. 26
FLOATING BATTERY

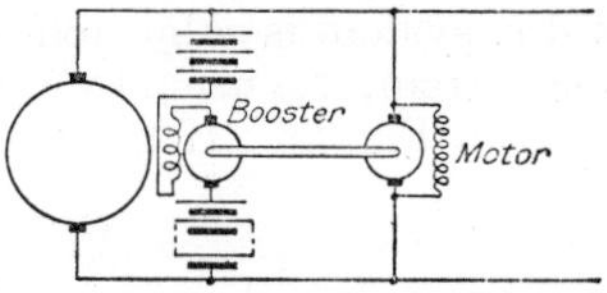

FIG. 27
FLOATING BATTERY AND BOOSTER

the previous section as above. For the voltage drop is equal to the product of resistance and current; if the voltage is increased m times and the power is the same, the current is reduced to $1/m$, and the voltage drop can be m times the previous value. The resistance can therefore be increased m^2 times.

Transmission by Alternating Current. It has been shown that it is economical to transmit large blocks of electrical energy at high voltage. The maximum voltage in d.c. transmission was limited by the voltage that was considered safe for the consumer, about 200 volts; the 3-wire system was a method of doubling the effective voltage of transmission. As there were no convenient means of transforming d.c. from one voltage to another, the main trend in the last forty years has been in the direction of high voltage, alternating current transmission.

For transmission and primary distribution the three-phase, 3-wire system of high voltage has been universally adopted. For secondary distribution the three-phase, 4-wire system has been standardized, as it gives 400 volts three-phase for large motors and 230 volts single-phase (between one line and the neutral) for small consumers. Radial and ring mains are used. There is still a fair amount of d.c. 3-wire distribution, and a main obstacle to the conversion to the three-phase, 4-wire system is that new 4-core cables would be required. A single-phase 3-wire system 460/230 volts could easily be supplied, but is not welcome because of the comparatively poor working of single-phase motors.

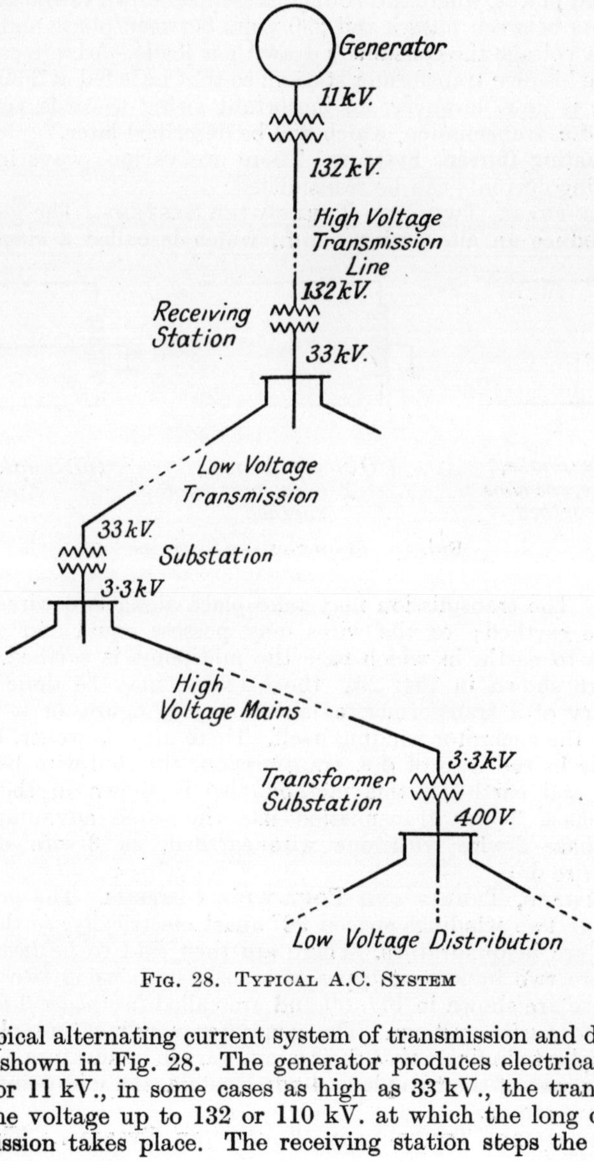

FIG. 28. TYPICAL A.C. SYSTEM

A typical alternating current system of transmission and distribution is shown in Fig. 28. The generator produces electrical power at 6·6 or 11 kV., in some cases as high as 33 kV., the transformer steps the voltage up to 132 or 110 kV. at which the long distance transmission takes place. The receiving station steps the voltage down to 33 kV. and feeds the substations, which step the voltage down to 3·3 kV. (3 300 volts) and radiate out in a system of high voltage distribution mains. There are transformer stations at various populated places, where the voltage is stepped down to 400/230 volts (400 volts between phases and 230 volts between phase and earth), at which voltage the consumers draw their loads. A large consumer will have his own transformer station, so that he is fed at 3 300 volts.

There is now, however, an important swing towards very high voltage d.c. transmission, which will be described later.

Alternating Current Systems. There are various ways in which alternating currents can be transmitted.

SINGLE-PHASE, TWO- AND THREE-WIRE SYSTEMS. The generator may produce an alternating e.m.f., which is called a *single-phase*

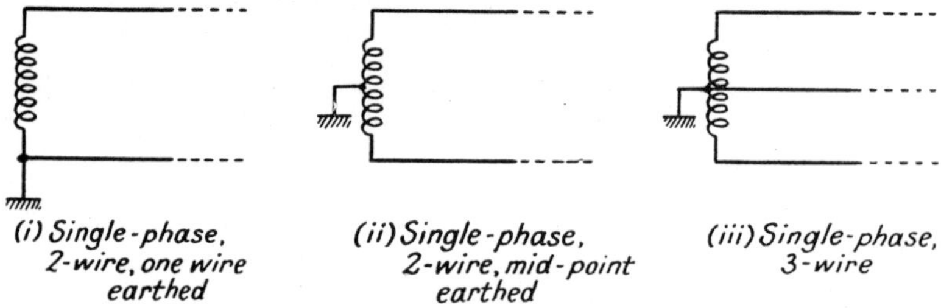

FIG. 29. SINGLE-PHASE SYSTEMS

voltage. The transmission may take place along two wires one of which is earthed; or the wires may possess equal and opposite voltages to earth, in which case the mid-point is earthed. These cases are shown in Fig. 29; the earthing may be done on the secondary of a transformer, as shown in the figure, or it may be done on the generator winding itself. There may, however, be three wires, as in the case of d.c. transmission, the mid-wire being the neutral and earthed; this method also is shown in the figure. Single-phase 3-wire transmission has the same advantage over single-phase 2-wire with one wire earthed, as 3-wire d.c. has over 2-wire d.c.

TWO-PHASE, THREE- AND FOUR-WIRE SYSTEMS. The generator may have two windings spaced 90° apart electrically, so that their e.m.f.'s are in quadrature. There are then said to be *two phases*. There are two important types of transmission using two phases; and these are shown in Fig. 30 and are called *two-phase* 3-*wire* and *two-phase* 4-*wire* systems. The generator windings are placed at right angles to indicate that their e.m.f.'s are in quadrature.

In the case of the two-phase 4-wire system, the mid-points of the phases are joined.

THREE-PHASE, THREE- AND FOUR-WIRE SYSTEMS. The most

common method of alternating current transmission is by the three-phase system. In this case the generator has three windings spaced 120° apart electrically, so that the e.m.f.'s are equal in magnitude but 120° apart in phase.

Fig. 31 shows the three-phase 3-wire systems that are used in practice. They are called the *star* or ⅄ and the *delta* or Δ connec-

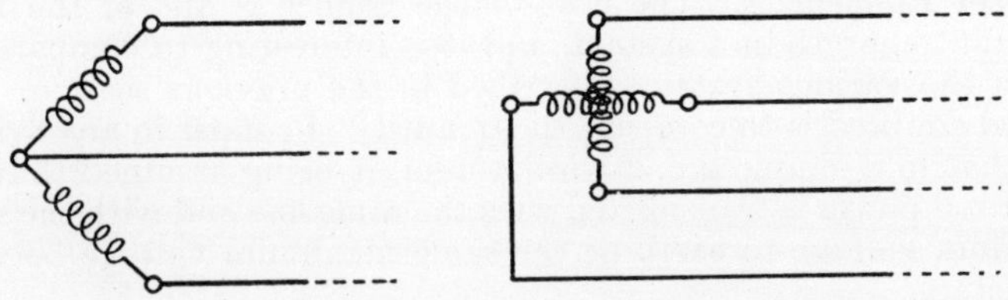

FIG. 30. TWO-PHASE SYSTEMS

tions. The common point in the star connection is the neutral point O and is generally earthed, either directly or through a resistance or an inductance coil (Petersen coil).

In the star system the line currents are equal to the phase currents, but the voltages between lines are $\sqrt{3}$ times the phase voltages and lag behind them by 90°. In the delta system the line voltages are

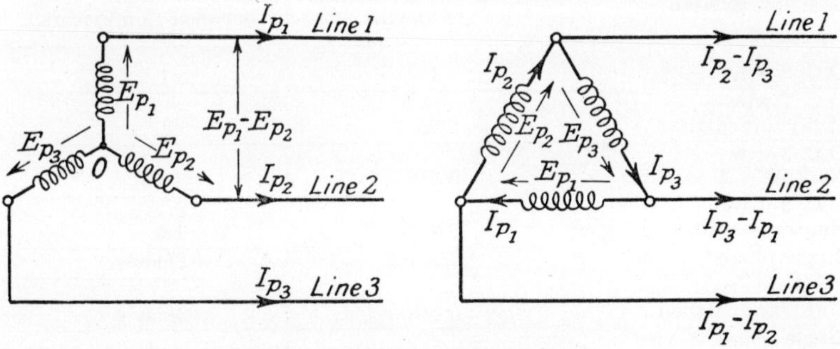

FIG. 31. THREE-PHASE SYSTEMS

equal to the phase voltages, but the line currents are $\sqrt{3}$ times the phase currents and lag behind them by 90°.

In both the star and delta arrangements the power supplied is given by $E_p'I_p' \cos \phi$ per phase, where E_p' is the r.m.s. phase voltage, I_p' the r.m.s. phase current, and ϕ the phase difference between them; the total power is $3E_p'I_p' \cos \phi$ if the loads are equal. It is seen that in both cases the power is equal to $(\sqrt{3})E_l'I_l' \cos \phi$, where E_l' and I_l' are the r.m.s. values of the line voltages and currents. For in the star system $I_p' = I_l'$ and $(\sqrt{3})E_p' = E_l'$, whilst in the delta system $(\sqrt{3})I_p' = I_l'$ and $E_p' = E_l'$. Thus the power is

$$P = 3E_p'I_p' \cos \phi = (\sqrt{3})E_l'I_l' \cos \phi$$

In low-voltage distribution a neutral wire is used, so that household loads can be taken between one line and neutral. We have thus a three-phase, 4-wire system.

Copper Efficiencies. The cost of the copper is one of the most important charges in a system, and it is interesting to compare the cost in the various systems described in the previous section. The method adopted is to compare the quantity of copper in any system with that in a simple d.c. 2-wire system, it being assumed that the same total power is transmitted with the same loss and with the same maximum voltage to earth or the same maximum voltage between conductors.

Both of the last conditions are important, the maximum voltage to earth being the quantity of importance in overhead lines and in single core cables, the maximum voltage between conductors being important in multi-core cables.

Table VII gives the ratio of copper in any system compared with that in the corresponding d.c. 2-wire system; $\cos \phi$ is the power factor in an a.c. system.

TABLE VII

Copper Efficiencies

System	Same Maximum Voltage to Earth	Same Maximum Voltage between Conductors
D.C. 2-wire . . .	1	1
D.C. 2-wire Mid-point earthed . .	0·25	1
D.C. 3-wire Neutral = ½ × outer .	0·3125	1·25
D.C. 3-wire Neutral = outer . .	0·375	1·5
Single-phase, 2-wire .	$2/\cos^2\phi$	$2/\cos^2\phi$
Single-phase, 2-wire Mid-point earthed . .	$0{\cdot}5/\cos^2\phi$	$2/\cos^2\phi$
Single-phase, 3-wire Neutral = ½ × outer .	$0{\cdot}625/\cos^2\phi$	$2{\cdot}5/\cos^2\phi$
Two-phase, 4-wire .	$0{\cdot}5/\cos^2\phi$	$2{\cdot}0/\cos^2\phi$
Two-phase, 3-wire .	$1{\cdot}46/\cos^2\phi$	$2{\cdot}91/\cos^2\phi$
Three-phase, 3-wire .	$0{\cdot}5/\cos^2\phi$	$1{\cdot}5/\cos^2\phi$
Three-phase, 4-wire Neutral = outer . .	$0{\cdot}67/\cos^2\phi$	$2/\cos^2\phi$

Two of the cases of the d.c. 3-wire system have been worked out on page 33, and two cases of three-phase systems will be discussed to show the method.

To compare the copper required in a d.c. 2-wire system and a three-phase, 3-wire system having the same maximum voltage to earth.

Let E be the voltage between conductors in the d.c. system and I the current. The power is EI and the loss is $2RI^2$, R being the resistance of each wire per unit length.

We shall consider a star-connected, three-phase system. The maximum phase voltage is E and the r.m.s. voltage is $E/\sqrt{2}$. If I' is the r.m.s. value of the line currents, the power is $(3/\sqrt{2})EI' \cos \phi$ and the loss is $3R'I'^2$, where R' is the resistance per wire. We have therefore

$$EI = (3/\sqrt{2})EI' \cos \phi$$

and

$$2RI^2 = 3R'I'^2,$$

so that

$$R = R'/3 \cos^2 \phi.$$

Each wire in the three-phase system has thus a cross-section $1/3 \cos^2 \phi$ of that in the d.c. system. As there are two conductors in the d.c. system and three in the a.c., the d.c. requires

$$(3/2) \times (1/3 \cos^2 \phi) = 0{\cdot}5/\cos^2 \phi$$

as much copper. The delta system will require the same amount of copper as the star system, since the power and loss are the same in both systems.

To compare the copper required in the d.c. 2-wire system and a three-phase, 4-wire system having the same maximum voltage between lines and a neutral equal to the outers.

We consider again a star-connected system. If E is the maximum voltage between lines, the maximum phase voltage is $E/\sqrt{3}$ and the r.m.s. is $E/\sqrt{6}$. If the line current is I', the power is $(3/\sqrt{6})EI' \cos \phi$ and the loss is $3R'I'^2$. Therefore

$$EI = (3/\sqrt{6})EI' \cos \phi$$

and

$$2RI^2 = 3R'I'^2,$$

so that

$$R = R'/\cos^2 \phi.$$

The cross-section in the a.c. system is $1/\cos \phi$ times that in the d.c. system, and as there are two conductors in the latter and four in the former, the copper ratio is $(4/\cos^2 \phi) \div 2 = 2/\cos^2 \phi$.

The effect of raising the voltage has already been discussed. Multiplying the voltage m times reduces the copper to $1/m^2$ of its previous value.

High Voltage, Direct Current. There are several considerable advantages to be gained by the use of high voltage, direct current transmission. If we consider overhead lines or single-core cables, which alone are used with extra high voltage so that the maximum voltage to earth is the criterion, the ratio of copper in the d.c., 2-wire system with earthed mid-point and the three-phase, 3-wire system is $(0{\cdot}25) \div (0{\cdot}5/\cos^2 \phi) = 0{\cdot}5 \cos^2 \phi$. Thus if the power factor is 0·85, the d.c. system requires only 0·36 as much copper as the a.c. system. Furthermore in a.c. systems the charging currents contribute to a continuous loss even when there is no load, whilst the d.c. system will have losses only when the load is on. It is held that the losses due to the charging current are a determining factor in the economics of long distance transmission.

Furthermore the transmission of a.c. for great distances is attended with instability, i.e. a synchronous machine will not be pulled back into phase if it departs from its correct position; and it is necessary to inject reactive power at intervals of about 100 miles to limit the reactive drop which is the cause of instability. D.C. transmission is not affected by instability and the lines may have any length.

Insulation difficulties are much greater with a.c. than with d.c., and the adoption of d.c. will raise the permissible transmission voltage without extra cost or trouble. Thus the current-carrying capacity of buried cables used for very high voltages (100 kV. and above) is determined partly by thermal instability due to the rise of the power factor of the dielectric with temperature. As there are no appreciable dielectric losses with d.c. the current-carrying capacity can be increased considerably.

Until recently there was no adequate method of transforming electrical energy from low voltage a.c. or d.c. to high voltage d.c., and from the latter to the former. The only method of utilizing high voltage d.c. was the Thury method. In this system series-wound generators are connected in series, the current is kept constant and the power is varied by varying the voltage of transmission. This is done by varying the speed of the generators or by inserting more generators into the circuit. The Thury system between Moutiers and Lyons, 112 miles apart, has a constant line current of 75 amperes and a maximum voltage of 60 000 volts, so that the maximum output is 4 500 kW. The generators have to be insulated from earth for the maximum voltage.

Power is taken from the circuit by motor-generators, the motors being series-wound and connected in series with the main circuit. The generators driven by them can give d.c. or a.c. at any desired voltage. The motors have to be insulated from earth for the maximum voltage, and are short-circuited when they are not required.

The main disadvantages of the Thury system are the facts that the line losses are constant at all loads so that the efficiency at low loads is very poor, and an increase of power of the system necessitates fresh insulation of the line for higher voltage since the current is constant.

There are now available various rectifiers, of the mercury vapour and atmospheric arc types, which can handle 30 000 kW. at 400 kV. Energy can be transformed from a.c. to d.c. and from d.c. to a.c. with a very high efficiency and at reasonable cost.

EXAMPLES II

1. Discuss the advantages, possibilities and difficulties of transmitting large amounts of power over long distances by means of high-voltage direct current as compared with the usual three-phase system of transmission. (*Lond. Univ.*, 1932.)

2. Discuss the advantages of a 3-wire system as compared with a 2-wire system for a d.c. distribution network and explain carefully how you would maintain approximately equal voltages across the two sides of a 3-wire system at the end of a long radial feeder with a heavily unbalanced load. (*Lond. Univ.*, 1934.)

3. A power of 150 000 kW. is to be transmitted over a distance of 200 miles. Discuss the relative advantages of (*a*) three-phase and (*b*) direct current, transmission. Explain why, at the present time, nearly all transmission systems are three-phase. (*Nat. Cert.*, 1935.)

CHAPTER III

OVERHEAD LINES: MECHANICAL DESIGN

Introduction. Transmission by overhead lines is very much cheaper than by cables. In fact, it is held that the determining factor in the Grid scheme is their cheapness compared with that of cables. In thinly populated rural areas the cost of distribution can be kept reasonably low only by the use of overhead lines.

The advantages of overhead lines are cheapness, the comparative ease with which the highest voltages can be employed, and the accessibility for extensions and repairs. The disadvantages are exposure, especially to lightning, smoke, ice, and the opposition to their presence in the country-side. (The necessity of overhead lines for long distance, high voltage transmission is such that way-leave is now legally obligatory.) Another disadvantage is the interference with communication circuits, and the Post Office must be consulted whenever an overhead line is proposed.

Regulations. There are regulations for securing the safety of the public, and these have to be observed in the design. A brief summary will be given of those regulations which affect the design directly. All line conductors must comply as regards elongation, breaking load and elasticity with the specification of the British Standards Institution. The minimum permissible size for copper and other lines (other than service lines) must be such as to have an actual breaking load of not less than 1 237 lb., the minimum cross-sectional area for copper being 0·0201 in.^{2} (No. 8 S.W.G.) and the weight per mile 409 lb. For service lines the minimum values are 816 lb., 0·0129 in.^{2} (No. 10 S.W.G.) and 262 lb. per mile. The lines must be inaccessible except by a ladder or other special appliance. The line conductors must be attached to suitable insulators carried on supports of wood, iron, steel, or reinforced concrete.

All wooden supports, other than oak or hard-wood cross-arms must, unless otherwise specified by the Electricity Commissioners, be of red fir impregnated with creosote. Special precautions must be taken to prevent the corrosion of all metal work at or below the surface of the ground. The supports must withstand the loads due to ice and wind pressure specified without damage or movement in the ground. The strength of a support in the direction of the overhead line must never be less than one-quarter the required strength in the transverse direction. The factor of safety must be 2·5 for iron or steel and 3·5 for wood or reinforced concrete, and is calculated on the assumption that all line conductors, cables, and wires carried by the supports are at 22° F., have a covering of ice to a radial thickness of $\frac{3}{16}$ in. for voltages not exceeding 650 volts d.c. or 325 volts a.c. or $\frac{3}{8}$ in. for higher voltages, and a wind pressure of 8 lb. per ft.^{2} calculated on the whole of the projected area.

Service lines may be connected to line conductors at a point of support only and must be fixed to insulators on consumers' premises.

The factor of safety of line conductors must be 2, and based on the breaking load under the conditions of temperature, ice, and wind given above. The minimum heights of conductors are given in Table VIII; the temperature is taken as 122° F.

TABLE VIII
MINIMUM HEIGHTS OF CONDUCTORS

Voltage	Minimum Height (feet)	Remarks
Less than 650 V., d.c.	19	Across public road
or 325 V., a.c.	17	Other positions
	15	Inaccessible to vehicles
Less than 66 kV.	20	
Between 66 and 110 kV.	21	
Between 110 and 165 kV.	22	
Exceeding 165 kV.	23	

To prevent danger from a broken line conductor, an earthed conductor must be carried on wooden pole supports below the line conductors so that a broken conductor will be earthed immediately. When the support is of metal, the earthed conductor is usually above the line conductors in order to protect them from lightning, and then the metal support must be efficiently earthed; for this purpose a continuous earth wire must be provided and connected with earth at four points in every mile, or the metal work must be connected to an efficient earthing device at each individual support. In this last case, the design and construction of the system of earth connections must be such that when a line conductor makes contact with an earthed metal, the leakage current must be not less than twice the leakage current required to operate the protective devices which make the line dead. When lines of voltage exceeding 650 volts, d.c., or 325 volts, a.c., run along or across public roads or canals, duplicate insulators must be used to support the conductors and a device must be included to prevent a live wire from falling to the ground; when the high voltage line crosses telephone wires, the Post Office requires that a cradle guard shall protect the telephone wires.

Stranded Conductors. Conductors for overhead lines are usually stranded, unless the section is small, for the continuous swinging and vibration would produce mechanical fatigue and finally fracture at the point of connection of a large solid wire to an insulator. Trolley wires are solid and are specially suspended so as to avoid fracture, and some very high voltage wires are made in the form of hollow tubes in order to avoid corona losses (see page 94).

Stranded conductors usually have a central wire around which are successive layers of 6, 12, 18, . . . wires. If there are n layers, the number of strands is $3n(n + 1) + 1$. If the diameter of each strand is d, the diameter of the cable or stranded conductor is $(2n + 1)d$. Adjacent layers are spiralled in opposite directions with the result that the layers are bound to one another; the binding is rendered more effective by spiralling the conductors before they are applied to their respective layers. The axial length of a spiral of a wire in a layer is called the *lay*, and is often expressed as a multiple of the mean diameter of the layer containing the wire. The lays in a three-layer conductor may be 20, $17\frac{1}{2}$, 15. If the lay ratio is r, the length of a wire is $\sqrt{[1 + (\pi/r)^2]}$ times its axial length. For if a complete spiral is unwound it will form a line rd along the axis and πd perpendicular to the axis, so that its length is

$$\sqrt{[(rd)^2 + (\pi d)^2]} = rd\sqrt{[1 + (\pi/r)^2]}.$$

If r is 15 or 20, the factor $\sqrt{[1 + (\pi/r)^2]}$ is 1·0217 or 1·01227. In practice it is assumed that stranding causes a 2 per cent increase in the resistance, which corresponds to a lay of $15\frac{1}{2}$; it is assumed that the current flows through the strands along their spirals and does not flow axially from strand to strand because of dirt and oxide films on the strand surfaces.

Conductor Materials. The most important material is copper because of its high conductivity and great tensile strength. Aluminium is used to a large extent, especially with a steel core, for high voltage lines. Among the other materials used are copperweld, cadmium copper, phosphor bronze, and galvanized steel. The choice of material depends upon the cost, the required electrical and mechanical properties, and on local conditions. The conductivities of copper and aluminium are decreased greatly by very small quantities of impurities: their alloys, with the exception of cadmium copper, have too low a conductivity to be of use. For this reason composite conductors are used, which consist of pure copper or aluminium with a galvanized steel core for mechanical strength.

COPPER. Hard-drawn copper has a high conductivity and a great tensile strength; cold working decreases the conductivity slightly but increases the strength considerably. Within the range of 20 to 30 tons per in.^{2} tensile strength, the conductivity is lowered by T/10 per cent where T is the tensile strength in tons per in.^{2}; thus at 20 tons the conductivity is 98 per cent, at 30 it is 97 per cent. The following list gives the properties of copper.

Coefficient of linear expansion . .	$16{\cdot}6 \times 10^{-6}$ per ° C.
Specific gravity	8·890
Weight of 1 $ft.^3$ (60° F.)	554·98 lb.
Modulus of elasticity	18×10^6 lb. per $in.^2$
Resistivity of annealed copper at 60° F. .	1·694 $\mu\Omega$. per cm. cube
Temperature coefficient of resistivity .	0·00400 per ° C.

The resistance of a solid conductor of annealed copper at 60° F., 1 000 yd. long and 1 $in.^2$ cross-section is 0·024008 ohms. Hard-drawn copper at 20 tons per $in.^2$ has 98 per cent conductivity and has a resistance of 0·024498 ohms. The temperature coefficient for hard-drawn copper must be multiplied by the conductivity ratio; thus 96 per cent conductivity copper has a temperature coefficient of 0·002133 per ° F. as compared with 0·002222 per ° F. for annealed copper.

Table IX gives the properties of standard copper conductors, conductivity 97 per cent.

TABLE IX

STANDARD COPPER CONDUCTORS

Equivalent Cross-section of Solid Conductor ($in.^2$)	Resistance (Ω./mile)	Strands and Diameter (in.)	Overall Diameter (in.)	Ultimate Strength (lb./$in.^2$)	Calculated Strength of Strand (lb.)	Weight per Mile (lb.)
0·025	1·74	3/·104	0·220	64 700	1 520	529
·05	0·870	3/·147	·317	62 250	2 910	1 057
·075	·580	3/·180	·388	60 500	4 250	1 585
·100	·434	7/·136	·408	62 750	5 870	2 105
·125	·348	7/·152	·456	61 750	7 240	2 629
·150	·291	7/·166	·498	61 250	8 520	3 136
·175	·248	7/·180	·540	60 500	9 920	3 687
·200	·215	7/·193	·579	59 750	11 270	4 239
·25	·174	7/·215	·645	59 000	13 800	5 260
·30	·143	19/·144	·720	62 500	17 380	6 416
·40	·108	19/·166	·830	61 250	22 680	8 526

CADMIUM COPPER. An addition of about 1 per cent of cadmium increases the tensile strength 50 per cent and reduces the conductivity by 15 per cent only. The effects of cold working are the same as for pure copper. Thus 0·9 per cent cadmium produces an alloy of tensile strength 45 tons per $in.^2$, 85 per cent conductivity, with the same modulus of elasticity and coefficient of linear expansion as for copper. Cadmium copper is relatively expensive, and is most useful for long spans with a line of small cross-section.

ALUMINIUM. The effects of cold working are as for copper. The properties are as follows.

Coefficient of linear expansion . . .	23×10^{-6} per ° C.
Specific gravity	2·71
Weight of 1 ft.^{3} (normal purity) . . .	169·18 lb.
Modulus of elasticity	$9{\cdot}9 \times 10^6$ lb. per in.^{2}
Tensile strength (min.) (for hard drawn) .	9 tons per in.^{2}
Standard resistivity at 20° C. (for hard drawn)	2·8735 $\mu\Omega$. per cm. cube
Temperature coefficient of resistivity .	0·00407 per ° C.

As the resistivity of 95 per cent conductivity copper is 1·7585 and the specific gravity is 8·89, an aluminium conductor of the same resistance per unit length has a cross-section $2{\cdot}87/1{\cdot}76 = 1{\cdot}63$ times that of a copper conductor, but the weight is $(1{\cdot}63 \times 2{\cdot}7)/8{\cdot}89 = 0{\cdot}496$ that of the copper. The tensile strength is about one-third that of copper, but as the cross-section is 1·63 times that of the equivalent copper conductor, the aluminium conductor has about 0·6 of the ultimate strength; since the weight of the aluminium conductor is only half of the copper conductor, there is a slight balance in favour of the former. This small advantage is more than neutralized by the increased loading due to wind and ice, so that copper is definitely superior to aluminium.

STEEL-CORED ALUMINIUM. Steel-cored aluminium conductors have a core of galvanized steel strands and a layer or layers of aluminium wires outside. Usually the aluminium and steel wires are of the same diameter. There may be one wire of steel surrounded by six wires of aluminium, or seven of steel by twelve or thirty of aluminium. There are many types of such composite conductors.

The conductivity of the steel-cored aluminium conductor is taken as that of the aluminium portion alone, since the steel wires have a high resistance to alternating currents.

The strength is taken as 85 per cent of the sum of the steel wires plus 95 per cent of the sum of the strengths of the aluminium wires. The factors 85 per cent and 95 per cent allow for the stranding. The strength of aluminium varies from 23 000 lb. per in.^{2} (for large wires) to 28 000 (for small), and of steel from 179 000 to 200 000. The total strength of a steel-cored aluminium conductor is normally 50 per cent greater than that of the equivalent copper conductor, and the weight only three-quarters as much (one-half due to aluminium and a quarter to steel). It is claimed that the result is a conductor with a smaller ratio of loading to strength than any other conductor, even allowing for the increased wind and ice loads due to the increased diameter as compared with that of the equivalent copper conductor. The sag is therefore the least so that the supporting towers may be shorter, or the span length greater for a given sag, than for any other conductor. The larger diameter is useful in very high voltage lines, as the corona losses are then less.

The virtual modulus of elasticity is given by

$$E = \frac{9{\cdot}9m + 28}{m + 1} \times 10^6 \text{ lb. per in.}^2,$$

where m is the ratio of aluminium section to steel section. The virtual coefficient of linear expansion is

$$\alpha = \frac{12{\cdot}78m + 18{\cdot}1}{m + 2{\cdot}83} \times 10^{-6} \text{ per } ^\circ \text{F}.$$

The weight per ft. per in.² of total section is

$$\delta = \frac{1{\cdot}21m + 3{\cdot}31}{m + 1} \text{ lb.}$$

EXAMPLE. Find the virtual modulus of elasticity, the coefficient of linear expansion, and the weight per ft. per in.² of total section of a steel-cored aluminium conductor having one steel and six aluminium wires of equal diameter.

Here $m = 6$, so that

$$E = \frac{(9{\cdot}9 \times 6) + 28}{6 + 1} \times 10^6 \text{ lb. per in.}^2 = \underline{\underline{12{\cdot}49 \times 10^6}} \text{ lb. per in.}^2,$$

$$\alpha = \frac{(12{\cdot}78 \times 6) + 18{\cdot}1}{6 + 2{\cdot}83} \times 10^{-6} = \underline{\underline{10{\cdot}74 \times 10^{-6}}} \text{ per } ^\circ \text{F.},$$

$$\delta = \frac{(1{\cdot}21 \times 6) + 3{\cdot}31}{6 + 1} = \underline{\underline{1{\cdot}51}} \text{ lb.}$$

STEEL-CORED COPPER. This is made in the same way as steel-cored aluminium, but there is an insulating tape over the steel wires to prevent corrosion of the copper. The mechanical characteristics are given by

$$E = \frac{18m + 28}{m + 1} \times 10^6 \text{ lb. per in.}^2,$$

$$\alpha = \frac{9{\cdot}21m + 9{\cdot}95}{m + 1{\cdot}55} \times 10^{-6} \text{ per } ^\circ \text{F.},$$

and

$$\delta = \frac{1{\cdot}21m + 3{\cdot}97}{m + 1} \text{ lb.}$$

Steel-cored copper has a smaller diameter than steel-cored aluminium for the same resistance per foot. For the same ultimate strength it requires less steel, since copper has a much higher strength than aluminium, so that it weighs less. It will probably be a strong competitor to steel-cored aluminium in the near future.

COPPERWELD. Copper is welded on to steel rods and the composite bars are rolled to form copperweld wires, which are stranded into conductors. It is well suited for very long spans. The conductivity is 30 or 40 per cent of that of a copper conductor with equal diameter.

PHOSPHOR BRONZE. When an exceptionally long span is required and the atmosphere contains harmful gases such as ammonia, phosphor bronze is a useful material for an overhead line conductor. The 132 kV. overhead crossing of the River Thames is 3 060 ft. wide, and the towers are 487 ft. high. The conductor, which weighs just over 2 lb. per ft., consists of a core of seven cadmium-copper

FIG. 32. THAMES CROSSING
(*Callender's Cable & Construction Co.*)

strands, to raise the conductivity, and 84 strands of phosphor bronze; thus there are four layers of the latter. The overall diameter is 0·942 in. and the breaking strength is 57 600 lb. The equivalent copper section is 0·12 in.^{2} Aluminium bronze is liable to chemical attack under high stress and on prolonged exposure to ammonia present in the atmosphere. Fig. 32 shows a picture of the crossing.

GALVANIZED STEEL AND IRON. These have been used on very long spans, but are most useful in rural areas where cheapness is the main consideration. Owing to the magnetic properties of iron and steel the resistance and inductance are very high, and the voltage drop is equally high.

Calculation of Sag. Let a line of weight w per unit length be suspended from the points A and B, and let O be the lowest point, i.e. where the curve is horizontal. (See Fig. 33.) We choose for the x-axis the horizontal line through O and for the y-axis the vertical. Let the point P have co-ordinates x and y, the length of the curve from O to P be s, and let ψ be the angle that the tangent at P makes with the horizontal. The forces which act on the portion OP are the horizontal tension H at O, the tensions T_x and T_y at P, and the weight ws vertically downwards through the centre of gravity of OP. (See Fig. 33.) Equating horizontal and vertical components we get

$$H = T_x,$$
$$T_y = ws.$$

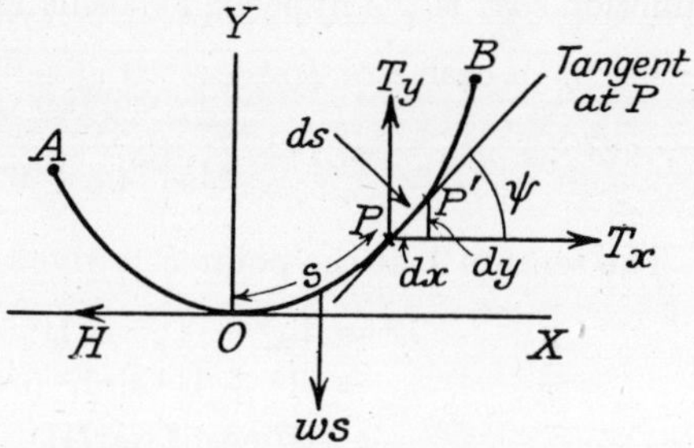

FIG. 33. SAGGING OVERHEAD LINE

The first equation shows that the horizontal component of the tension is constant throughout the length of the line. The resultant of T_x and T_y must be along the tangent at P, so that

$$T_y/T_x = \tan\psi.$$

But $\tan\psi = dy/dx$, so that

$$dy/dx = T_y/T_x = ws/H \quad . \quad . \quad . \quad . \quad (1)$$

It is seen from the elementary piece PP' of the line that

$$ds = \sqrt{(dx^2 + dy^2)},$$

so that $$ds/dx = \sqrt{[1 + (dy/dx)^2]} = \sqrt{[1 + (w^2s^2/H^2)]}$$

or $$dx = \frac{ds}{\sqrt{[1 + (w^2s^2/H^2)]}}$$

Integrating both sides we get

$$x + C = (H/w)\sinh^{-1}(ws/H),$$

where C is the constant of integration. As we have chosen the origin at O, $x = 0$ when $s = 0$, so that $C = 0$. We get therefore

$$s = (H/w)\sinh(wx/H) \quad . \quad . \quad . \quad (2)$$

Substituting from this equation in (1) we get

$$dy/dx = \sinh(wx/H)$$

so that $$y = \int \sinh\frac{wx}{H}dx$$
$$= (H/w)\cosh(wx/H) + D.$$

where D is another constant of integration. At O, $y = 0$ when $x = 0$, so that

$$0 = (\mathrm{H}/w) \cosh 0 + D = (\mathrm{H}/w) + D,$$

so that $$D = -(\mathrm{H}/w).$$

The equation of the line is thus

$$y = (\mathrm{H}/w)\,[\cosh(wx/\mathrm{H}) - 1] \quad . \quad . \quad . \quad (3)$$

(3) is the equation of the curve, which is called the *catenary*. The function *cosh* is the hyperbolic cosine and is such that

$$\cosh z = \tfrac{1}{2}(\varepsilon^{z} + \varepsilon^{-z})$$
$$= 1 + \frac{z^2}{2!} + \frac{z^4}{4!} + \frac{z^6}{6!} + \ldots \quad . \quad . \quad (4)$$

The tension T at the point P is given by

$$\mathrm{T}^2 = \mathrm{T}_x{}^2 + \mathrm{T}_y{}^2 = \mathrm{H}^2 + w^2 s^2$$
$$= \mathrm{H}^2 + \mathrm{H}^2 \sinh^2(wx/\mathrm{H})$$
$$= \mathrm{H}^2 \cosh^2(wx/\mathrm{H}),$$

so that $$\mathrm{T} = \mathrm{H} \cosh(wx/\mathrm{H}) \quad . \quad . \quad . \quad . \quad . \quad (5)$$

If the line is supported between points A and B at the same level and half the span length is l, then at the supports $x = \pm\, l$, and

$$\mathrm{T} = \mathrm{H} \cosh(wl/\mathrm{H}) \quad . \quad . \quad . \quad . \quad . \quad (6)$$

The sag d is the value of y at A or B and is given by

$$d = (\mathrm{H}/w)\,[\cosh(wl/\mathrm{H}) - 1] \quad . \quad . \quad . \quad (7)$$

The length of the line in a half span is

$$s = (\mathrm{H}/w) \sinh(wl/\mathrm{H}) \quad . \quad . \quad . \quad . \quad (8)$$

A table of the hyperbolic cosine is given in Appendix 1.

EXAMPLE. Calculate the maximum sag of a conductor 650 ft. span, 0·48 lb. weight per ft., the maximum tension allowable being 3 000 lb.

Here $l = 325$ ft. and $w = 0{\cdot}48$. Equation (6) gives the maximum tension so that

$$3\,000 = \mathrm{H} \cosh \frac{325 \times 0{\cdot}48}{\mathrm{H}} = \mathrm{H} \cosh \frac{156}{\mathrm{H}}.$$

H can be found by trial and error from the table of cosh z. A quicker method is obtained by observing that H is very nearly equal to T and is very much greater than wl. Expanding equation (6) we obtain

$$\frac{T}{H} = \cosh\frac{wl}{H} = 1 + \frac{w^2l^2}{2H^2} + \frac{w^4l^4}{24H^4} + \ldots$$

$$\simeq 1 + \frac{w^2l^2}{2H^2} \quad . \quad . \quad . \quad . \quad . \quad . \quad (9)$$

since terms higher than the inverse square power are very small. This gives a quadratic equation for H, viz.

$$H^2 - HT + \tfrac{1}{2}w^2l^2 = 0,$$

so that

$$\begin{aligned} H &= \tfrac{1}{2}T + \tfrac{1}{2}\sqrt{[T^2 - (4w^2l^2/2)]} \\ &= \tfrac{1}{2}T[1 + \sqrt{(1 - 2w^2l^2/T^2)}] \\ &\simeq \tfrac{1}{2}T[1 + 1 - w^2l^2/T^2] \\ &= T - w^2l^2/2T. \end{aligned}$$

In this case

$$H = 3\,000 - \frac{156^2}{2 \times 3\,000} = 3\,000 - 4{\cdot}05 = 2\,995{\cdot}95.$$

A check is obtained by seeing whether equation (6) is satisfied.

$$T = H\cosh\frac{156}{H} = 2\,995{\cdot}95\cosh\frac{156}{2\,995{\cdot}95} = 2\,995{\cdot}95\cosh 0{\cdot}052$$

$$= 2\,995{\cdot}95 \times 1{\cdot}00136 = 2\,999{\cdot}9.$$

The sag is given by

$$d = \frac{2\,996}{0{\cdot}48}(1{\cdot}00136 - 1) = \underline{\underline{8{\cdot}64\text{ ft.}}} = 8\text{ ft. }8\text{ in.}$$

APPROXIMATE FORMULAE. If $(w^2l^2/2H^2)$ is small compared with unity, T is very nearly equal to H. Equation (3) of the catenary can be expanded as

$$y = \frac{H}{w}\left(1 + \frac{w^2x^2}{2H^2} + \frac{w^4x^4}{24H^2} \ldots - 1\right)$$

$$\simeq wx^2/2H \simeq wx^2/2T.$$

The sag is thus $\quad d = wl^2/2T.$

The arc length

$$s = \frac{H}{w}\sinh\frac{wx}{H}$$

$$= \frac{H}{w}\left(\frac{wx}{H} + \frac{w^3x^3}{6H^3} + \ldots\right)$$

$$\simeq x + \frac{w^2x^3}{6T^2},$$

so that the cable length in half the span is

$$s = l + w^2l^3/6T^2$$

$$\left.\begin{array}{lr} \text{Thus the tension} & T = H, \\ \text{the curve is the parabola} & y = wx^2/2T, \\ \text{the sag is} & d = wl^2/2T \\ \text{and the half-span length is} & s = l + w^2l^3/6T^2. \end{array}\right\} \quad . \quad . \quad (10)$$

These formulae are sufficiently accurate in most practical cases, in which the sag is less than 10 per cent of the span.

EXAMPLE. Calculate the sag in the previous problem by means of the approximate formula.

The sag is

$$d = \frac{wl^2}{2T} = \frac{0{\cdot}48 \times 325^2}{2 \times 3\,000} = \underline{\underline{8{\cdot}45 \text{ ft.}}} = 8 \text{ ft. } 5 \text{ in.};$$

which is 3 in. less than the value given by the exact method. The error is only 2·2%.

Effects of Wind and Ice. When the temperature drops to about 22° F., conditions are favourable for the formation of sleet on the wire, and appreciable thicknesses of ice can be formed. The Regulations insist that the effect of a thickness of $\frac{3}{8}$ in. for high voltage and $\frac{3}{16}$ in. for low voltage lines be considered. In addition there may be a high wind which acts on the increased surface; a 50 m.p.h. wind is considered to exert a pressure of 8 lb. per ft.² of projected area, and this must be allowed for in the calculation of sag. If the weight of the ice is w_i per ft., the total weight of the conductor is $w + w_i$. It is assumed that the wind pressure, w_w, acts horizontally. The total force on 1 ft. of conductor is

$$W = \sqrt{[(w + w_i)^2 + w_w{}^2]} \quad . \quad . \quad . \quad (11)$$

and acts at an angle to the vertical. The conductor sets itself in a plane at this angle to the vertical, but keeps the shape of a catenary with the constant W in place of w.

If the overall diameter of the conductor is D in. and the radial thickness of ice is R in., the weight of ice per foot is

$$\begin{aligned} w_i &= 57 \times \pi[(\tfrac{1}{2}D + R)^2 - (\tfrac{1}{2}D)^2] \div 144 \text{ lb.} \\ &= 1{\cdot}25\, R(D + R) \text{ lb.}, \end{aligned}$$

since 1 ft.³ of ice weighs 57 lb. The maximum wind pressure per foot run is

$$w_w = 8 \times (D + 2R) \div 12 = \tfrac{2}{3}(D + 2R) \text{ lb.}$$

EXAMPLE. A line conductor has an effective diameter of 0·77 in. and weighs 571 lb. per 1 000 ft. Find the additional loading due to ice of $\frac{1}{2}$ in. radial thickness and a horizontal wind pressure of 8 lb. per ft.²; find also the total load per foot run.

$$\begin{aligned} D &= 0{\cdot}77,\ R = 0{\cdot}5. \\ w_i &= 1{\cdot}25\,(0{\cdot}5)\,(1{\cdot}27) = 0{\cdot}794. \\ w_r &= \tfrac{2}{3}(0{\cdot}77 + 1{\cdot}0) = 1{\cdot}18. \end{aligned}$$

The total load is

$$\begin{aligned} W &= \sqrt{[(0{\cdot}571 + 0{\cdot}794)^2 + (1{\cdot}18)^2]} \\ &= \sqrt{[1{\cdot}365^2 + 1{\cdot}18^2]} \\ &= \underline{\underline{1{\cdot}804 \text{ lb. per ft.}}} \end{aligned}$$

EXAMPLE. In the transmission lines of The National Grid Scheme the line conductor has an effective diameter of 0·77 in., weighs 571 lb. per 1 000 ft., and has an ultimate strength of 17 722 lb. Calculate the height above ground at which a conductor with a span of 900 ft. must be supported in order that the total tension shall not exceed half the ultimate strength with a $\frac{1}{2}$ in. radial coating of ice and a horizontal wind pressure of 8 lb. per ft.^{2} of projected area. The ground clearance required is 22 ft. Weight of 1 ft.^{3} of ice = 57 lb. (*B.Sc. Lond. Univ.*, 1932.)

The total loading is W = 1·804 lb. per ft.
The maximum permissible tension is $T = \frac{1}{2} \times 17\,722 = 8\,861$ lb.
Half the span length is $l = 450$ ft.
The approximate formula gives a sag

$$d = \frac{Wl^2}{2T} = \frac{1{\cdot}804 \times 450^2}{17\,722} = 20{\cdot}6 \text{ ft.} = 20 \text{ ft. } 7 \text{ in.}$$

The more exact method gives

$$H = T - w^2l^2/2T = 8\,861 - 37{\cdot}3 = 8\,824.$$

$$\frac{Wl}{H} = \frac{1{\cdot}804 \times 450}{8\,824} = 0{\cdot}092.$$

The sag is

$$\begin{aligned} d &= (H/W)\,[\cosh\,(Wl/H) - 1] \\ &= (8\,824/1{\cdot}804)\,(\cosh 0{\cdot}092 - 1) \\ &= (8\,824/1{\cdot}804) \times 0{\cdot}00424 = 20{\cdot}8 \text{ ft.} = 20 \text{ ft. } 10 \text{ in.} \end{aligned}$$

The height of the supports must therefore be greater than 42 ft. 10 in., say $\underline{\underline{43 \text{ ft.}}}$ It has been assumed that the safety factor has the regulation value of 2.

Tension and Sag at Erection. Since the line will never be erected under the conditions of ice and wind which are postulated by the regulations, it is necessary to calculate the tension and sag under the conditions occurring at the time of erection. In practice, the conductors are hung over pulleys or snatch-blocks, pulled up to the correct sag or tension, and then transferred to the insulators. Usually the sag is measured, as the friction in the snatch-blocks makes it difficult to measure the tension accurately.

There are two factors which vary the sag and tension, viz. elasticity and temperature. The load at erection will be less than that of the regulation conditions; this removal of load will lessen the tension so that the sag will be reduced because of elastic contraction. An increase of temperature will cause the length of the conductor to increase, so that the sag will increase. If the temperature increases sufficiently, the sag may reach the value corresponding to the regulation conditions for ice and wind at 22° F.; this high temperature is called the *critical temperature* and must be higher than 122° F., which is considered a temperature that may occur in England in summer.

The combined effect of elasticity and temperature is found in the following way, assuming that the approximate formulae of equations (10) are sufficiently accurate. Let w_1, f_1, s_1, d_1 and θ_1 be the load per unit length, the stress (i.e. tension per in.^{2}), the half-span, sag and temperature at the maximum load conditions; w_2, f_2, s_2, d_2 and θ_2 the values under stringing conditions; A the area of cross-section of the conductor, α the coefficient of linear expansion, and E the Young's modulus. The length of half the span under any conditions is given by equations (10) as

$$s = l + w^2l^3/6T^2 = l + w^2l^3/6A^2f^2 = l + 2d^2/3l.$$

The half-span length at maximum load conditions is

$$s_1 = l + w_1{}^2l^3/6f_1{}^2A^2.$$

The temperature rise from θ_1 to θ_2 causes an increase in the half-span length of $(\theta_2 - \theta_1)\alpha s_1$, which is practically equal to $(\theta_2 - \theta_1)\alpha l$. The fall in stress from f_1 to f_2 causes a decrease in the length of

$$\frac{f_1 - f_2}{E} s_1 \simeq \frac{f_1 - f_2}{E} l.$$

The new length, which is s_2, is thus given by

$$s_2 = s_1 - \frac{f_1 - f_2}{E} l + (\theta_2 - \theta_1)\alpha l.$$

But $$s_2 = l + w_2{}^2l^3/6f_2{}^2A^2,$$

so that $$l + \frac{w_2{}^2l^3}{6f_2{}^2A^2} = l + \frac{w_1{}^2l^3}{6f_1{}^2A^2} - \frac{f_1 - f_2}{E} l + (\theta_2 - \theta_1)\alpha l.$$

Cancelling l from both sides and rearranging we get

$$f_2{}^2\left[f_2 - f_1 + \frac{w_1{}^2l^2E}{6f_1{}^2A^2} + (\theta_2 - \theta_1)\alpha E\right] = \frac{w_2{}^2l^2E}{6A^2},$$

or $$\left.\begin{aligned} &f_2{}^2[f_2 - F] = G, \\ \text{where}\quad &F = f_1 - w_1{}^2l^2E/6T_1{}^2 - (\theta_2 - \theta_1)\alpha E \\ \text{and}\quad &G = w_2{}^2l^2E/6A^2 \end{aligned}\right\} \quad . \quad (12)$$

The sag d_2 is then given by

$$d_2 = w_2l^2/2T_2 = w_2l^2/2f_2A \quad . \quad . \quad . \quad (13)$$

A *stringing chart* can be drawn by calculating f_2 and d_2 from equations (12) and (13) and plotting them against the temperature, θ_2, since θ_1 is constant (22° F.). Fig. 34 shows a stringing chart calculated for the example on pages 56–7.

It is possible to obtain a relation between the sag and temperature directly in the following way.

$$f_1 = T_1/A = w_1l^2/2Ad_1 \text{ and } f_2 = w_2l^2/2Ad_2.$$

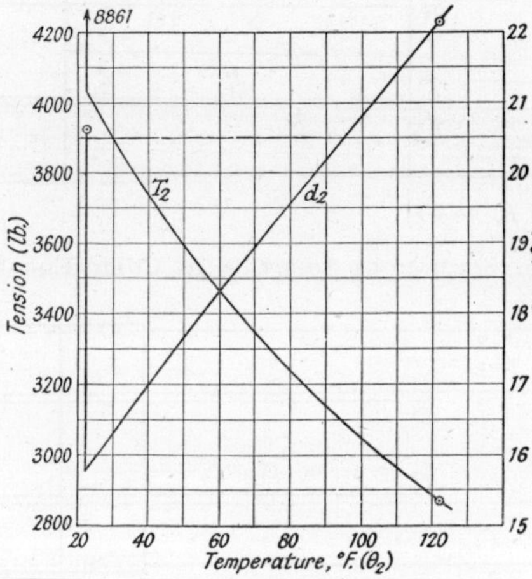

FIG. 34. STRINGING CHART

Substituting for f_1 and f_2 in equation (12) and rearranging we get

$$d_2^2 - d_1^2 = (3l^4/4AE)(w_2/d_2 - w_1/d_1) + \tfrac{3}{2}(\theta_2 - \theta_1)\alpha l^2$$

$$\left.\begin{aligned} &\text{or} && d_2(d_2^2 - K) = L, \\ &\text{where} && K = d_1^2 + \tfrac{3}{2}(\theta_2 - \theta_1)\alpha l^2 - \tfrac{3}{4}(w_1l^4/4AEd_1) \\ &\text{and} && L = 3w_2l^4/4AE. \end{aligned}\right\} \quad (14)$$

It is possible to calculate f_2 from equation (12) and d_2 from equation (14) by a slide-rule manipulation.

Fig. 35 shows the method of calculating f_2. *A* is the fixed scale of squares and *B* the sliding scale; *C* and *D* are the sliding and fixed scales of ordinary numbers. The cursor is moved so that the hair-line is at *G* on scale *A*. The sliding scale is moved until the zero of the sliding scale indicates on *D* a number F in excess of that indicated by the hair-line on *B*. Then the number indicated on *D* is f_2. For if this is so, the zero of the sliding scale is opposite f_2^2 on the scale *A*, the number on *B* is $f_2 - F$, and the product $f_2^2(f_2 - F)$ is G, as is shown on scale *A*.

Fig. 36 shows the method of calculating d_2. The cursor is moved so that the hair-line is opposite L on scale *D*. The sliding scale is moved until the hair-line meets scale *C* at a number which is K less than the number on scale *A* at the zero of the sliding scale.

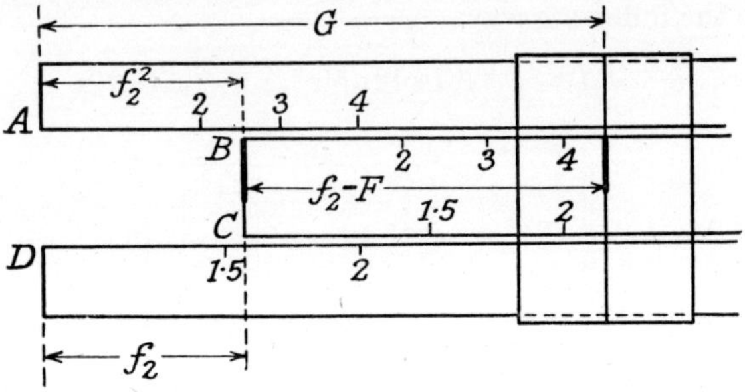

FIG. 35. SLIDE-RULE SOLUTION OF CUBIC EQUATIONS

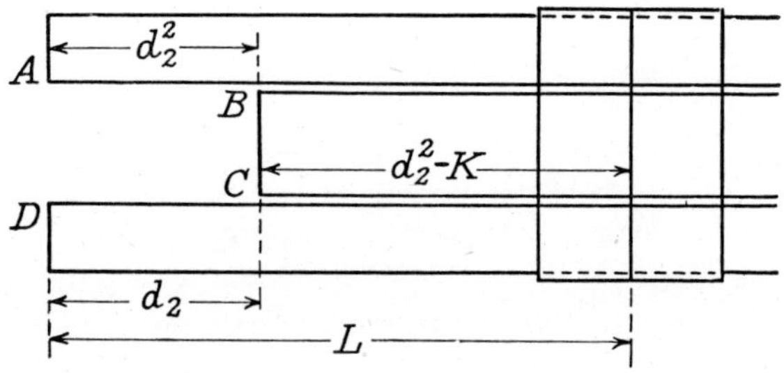

FIG. 36. SLIDE-RULE SOLUTION OF CUBIC EQUATIONS

EXAMPLE. Construct the stringing chart for the line in the example on page 57, assuming a wind pressure of 4 lb. per ft.²

We have already found that $w_1 = 1{\cdot}804$, $T_1 = 8\ 861$.

Now
$$A = (\pi/4)\,(0{\cdot}77)^2 \div 144 = 0{\cdot}00324 \text{ ft.}^2,$$
$$f_1 = T_1 \div A = 2{\cdot}74 \times 10^6 \text{ lb. per ft.}^2,$$
$$E = 18 \times 10^6 \text{ lb. per in.}^2 = 2{\cdot}592 \times 10^9 \text{ lb. per ft.}^2,$$
$$\alpha = 9{\cdot}22 \times 10^{-6} \text{ per °F.}$$
$$w_2 = \sqrt{(0{\cdot}571^2 + 0{\cdot}257^2)} = 0{\cdot}626,$$

since the wind pressure is 0·257.

Therefore

$$G = \frac{w_2{}^2 l^2 E}{6A^2} = \frac{(0{\cdot}626)^2\,(450)^2\,(2{\cdot}592)10^9}{6(0{\cdot}00324)^2}$$
$$= 3{\cdot}26 \times 10^{18}.$$

$$F = f_1 - \frac{w_1^2 l^2 E}{6T_1^2} - (\theta_2 - \theta_1)\alpha E$$

$$= 2{\cdot}74 \times 10^6 - \frac{(1{\cdot}804)^2\,(450)^2\,(2{\cdot}592)10^9}{6 \times 8861^2}$$

$$- (\theta_2 - \theta_1)\,(9{\cdot}22)\,(2{\cdot}592) \times 10^3$$

$$= -\,0{\cdot}89 \times 10^6 - 2{\cdot}39 \times 10^4\,(\theta_2 - \theta_1).$$

θ_1 is 22° F. and F is found for different values of θ_2 by simple substitution. The values of f_2, T_2, and d_2 are then found and given in the following table.

θ_2 (° F.)	22	42	62	82	102	122
$-F \times 10^{-6}$ . . .	0·88	1·359	1·838	2·314	2·798	3·277
$f_2 \times 10^{-6}$ . . .	1·245	1·145	1·060	0·995	0·936	0·885
T_2 (lb.)	4 030	3 700	3 440	3 220	3 035	2 870
d_2 (ft.)	15·75	17·15	18·47	19·74	20·95	22·17

Fig. 34 shows the graphs of T_2 and d_2 against temperature, and is the stringing chart for this line. The maximum sag occurs at 122° F. and is 22·17 ft., so that the conductors must be supported from insulators which are at least 44·17 ft. above ground, say 45 ft.

It is seen that F is negative in these calculations, but the slide-rule manipulation described in Fig. 35 still applies, the number opposite

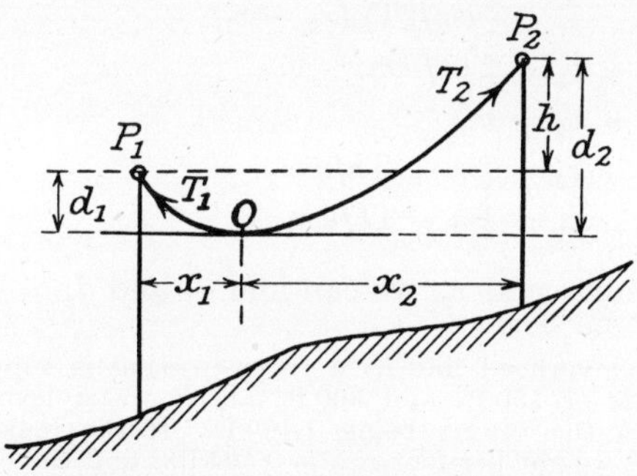

FIG. 37. SPAN IN HILLY COUNTRY

the hair-line on scale *B* being greater than that opposite the zero of the sliding scale on scale *D*.

Spans in Hilly Country. Fig. 37 shows a line suspended at two points P_1 and P_2 at different levels, the vertical distance between the points being h. If the ground is not too steep there will be a point O at which the line is horizontal. Then OP_1 is a catenary of half-span x_1 and OP_2 of half-span x_2. Equations (6), (7), and (8) will apply to each part, so that

$$d_1 = (\mathrm{H}/w)\,[\cosh\,(wx_1/\mathrm{H}) - 1],$$
$$d_2 = (\mathrm{H}/w)\,[\cosh\,(wx_2/\mathrm{H}) - 1],$$
$$\mathrm{T}_1 = \mathrm{H}\cosh\,(wx_1/\mathrm{H}), \text{ and } \mathrm{T}_2 = \mathrm{H}\cosh\,(wx_2/\mathrm{H}).$$

It is clear that the maximum tension occurs at the higher support, P_2 in the figure, since x_2 is greater than x_1.

The quantities that are given are w, h, and the horizontal distance between the supports, viz. $x_1 + x_2$. Thus we have w given,

$$\left.\begin{aligned} d_2 - d_1 &= (\mathrm{H}/w)\,[\cosh\,(wx_2/\mathrm{H}) - \cosh\,(wx_1/\mathrm{H})] = h \\ \text{and } x_2 + x_1 &= m, \text{ say.} \\ \text{Also}\quad T_2 &= \mathrm{H}\cosh\,(wx_2/\mathrm{H}) = \text{maximum permissible value} \end{aligned}\right\} \quad (15)$$

These three equations are sufficient to determine the unknown quantities x_1, x_2, and H, from which the sag and tension can be found. The equations can be solved by the method of successive approximations, and it is useful to obtain an approximate solution by assuming that the curve is a parabola. The approximate solution is obtained as follows.

$$d_1 = wx_1^2/2\mathrm{T},\ d_2 = wx_2^2/2\mathrm{T},\ \mathrm{T} = \mathrm{H},$$
$$\begin{aligned} d_2 - d_1 = h &= (w/2\mathrm{T})\,(x_2^2 - x_1^2) \\ &= (w/2\mathrm{T})\,(x_2 + x_1)\,(x_2 - x_1) \\ &= (wm/2\mathrm{T})\,(x_2 - x_1), \end{aligned}$$

so that $\quad x_2 - x_1 = 2\mathrm{T}h/wm.$

But $\quad x_2 + x_1 = m,$

and therefore $\quad x_2 = \tfrac{1}{2}m + \mathrm{T}h/wm$

$$x_1 = \tfrac{1}{2}m - \mathrm{T}h/wm.$$

Having found x_1 and x_2, we can find d_1 and d_2.

EXAMPLE. An overhead line at a river crossing is supported from two towers at heights of 150 ft. and 300 ft. above water level, the horizontal distance between the towers being 1 100 ft. If the maximum tension is 4 260 lb. and the weight per foot run is 0·594 lb., find the clearance between the conductor and the water at a point mid-way between the towers.

First we find the approximate solution.

$$x_2 = \frac{1\,100}{2} + \frac{4\,260 \times 150}{0{\cdot}594 \times 1\,100} = 550 + 976 = 1\,526,$$
$$x_1 = 550 - 976 = -\,426.$$

The negative sign of x_1 shows that P_1 lies on the same side of O as P_2, as shown in Fig. 38. The mid-point P has a value of $x = \frac{1}{2}(1\,526 + 426) = 976$, so that its height above O is given by

$$d = \frac{wx^2}{2T} = \frac{0{\cdot}594 \times 976^2}{2 \times 4\,260} = 66{\cdot}4 \text{ ft.}$$

The height of P_2 above O is given by

$$d_2 = \frac{wx_2{}^2}{2T} = \frac{0{\cdot}594 \times 1\,526^2}{2 \times 4\,260} = 162 \text{ ft.}$$

The mid-point is thus 96 ft. below P_2 and is therefore 204 ft. above the water. As a check we find that $d_1 = 12{\cdot}6$ ft., so that P_1 is 150 ft. below P_2 as it should be.

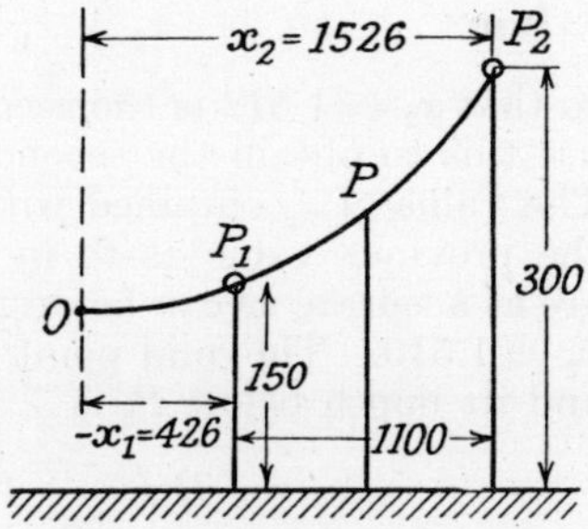

FIG. 38
SPAN IN HILLY COUNTRY

In order to obtain the exact solution it is preferable to express the first of equations (15) in the form

$$d_2 - d_1 = h = \frac{H}{w}\left(\cosh\frac{wx_2}{H} - \cosh\frac{wx_1}{H}\right)$$

$$= \frac{2H}{w}\sinh\frac{w(x_2 - x_1)}{2H}\sinh\frac{w(x_2 + x_1)}{2H},$$

i.e.

$$h = \frac{2H}{w}\sinh\frac{wm}{2H}\sinh\frac{w(x_2 - x_1)}{2H} \quad . \quad . \quad . \quad (16)$$

We now apply the method of successive approximations in the following way. Insert the value of x_2 obtained by the approximate method in the third equation of equations (15) to obtain the first approximation for H.

Thus $4\,260 = H\cosh\dfrac{0{\cdot}594 \times 1\,526}{H} = H\cosh\dfrac{906}{H}$. This is of the form $\cosh z = 4{\cdot}70\,z$, where $z = 906/H$, and from the tables of $\cosh x$ we see that $z = 0{\cdot}2175$, so that $H = 906 \div 0{\cdot}2175 = 4\,170$, which is the first approximation for H. We now insert this value of H in equation (16) so that

$$\sinh\left[\frac{w(x_2 - x_1)}{2H}\right] = \sinh\frac{0{\cdot}594(x_2 - x_1)}{2 \times 4\,170}$$

$$= \frac{wh}{2H\sinh\dfrac{wm}{2H}} = \frac{0{\cdot}594 \times 150}{2 \times 4\,170 \times \sinh\dfrac{0{\cdot}594 \times 1\,100}{2 \times 4\,170}}$$

$$= \frac{0{\cdot}594 \times 150}{2 \times 4\,170 \times 0{\cdot}0785} = 0{\cdot}1372.$$

Therefore $w(x_2 - x_1)/2\text{H} = 0{\cdot}1368,$

$$x_2 - x_1 = \frac{2 \times 4\,170 \times 0{\cdot}1368}{0{\cdot}594} = 1\,923.$$

But $$x_2 + x_1 = 1\,100,$$

so that $x_2 = 1\,512$ is the second approximation for x_2. We can now use this to obtain the second approximation for H, which is 4 160. The value of x_2 obtained with value of H is 1 510, which is so near the previous value as to indicate that the second approximations are as accurate as can be desired. We therefore take H = 4 160 and $x_2 = 1\,510$. The mid-point P has $x = 1\,510 - \frac{1}{2} \times 1\,100 = 860$, and its depth below P_2 is

$$\frac{H}{w}\left(\cosh\frac{1\,510w}{\text{H}} - \cosh\frac{860w}{\text{H}}\right)$$

$$= \frac{2\text{H}}{w}\sinh\frac{650w}{2\text{H}}\sinh\frac{2\,370w}{2\text{H}} = 107{\cdot}6.$$

The clearance is thus 192 ft. It is seen that the approximate method over-estimates the clearance by 12 ft.

Trolley Wires. These must possess a very small sag since the collectors should run at high speeds without much oscillation. To

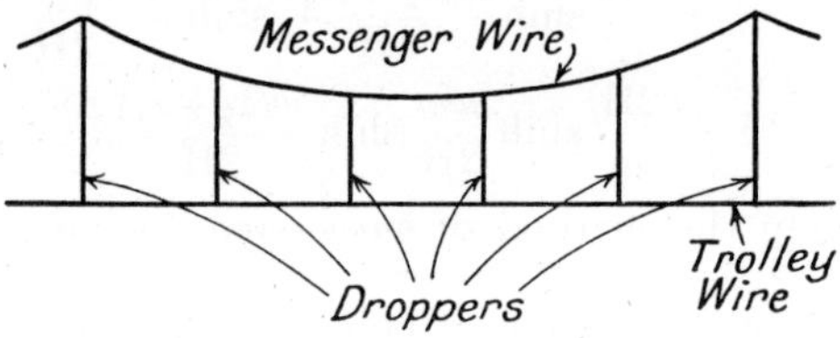

FIG. 39. TROLLEY WIRE SUSPENSION

achieve very small sags the trolley wire is suspended from a messenger wire by means of droppers. The trolley wire can be tensioned by counterweights or fixed rigidly at the ends. (See Fig. 39.)

Poles and Towers. Small lines for rural supply at low voltage use wood or steel poles and have short spans, i.e. less than 250 ft. Creosoted wood poles have been found more suitable in this country. Pin-type insulators are used, because they are cheap for low voltages. Fig. 40 shows an "A"-type wood pole. Copper is used for the conductor.

A pole which acts as a terminal, or carries switchgear or a transformer, or is used at a tapping point, may be of the "H"-type shown in Fig. 41.

In the case of single unguyed poles, it is found satisfactory to set the pole in the earth for one-fifth of its overall length, provided the soil is firm. If the soil is not firm, it is necessary to provide baulks, as shown in Fig. 42.

A pole cap of zinc, aluminium or cement is usually fitted on wood poles to prevent exposure of the end grain. The cross-arms are generally of steel, and two arrangements are shown in Fig. 43 (note end caps).

Reinforced-concrete poles are used in large numbers on the Continent. (See Fig. 44.)

Steel towers are invariably used for long spans, the steel being galvanized or painted. The initial cost is higher than that of wood

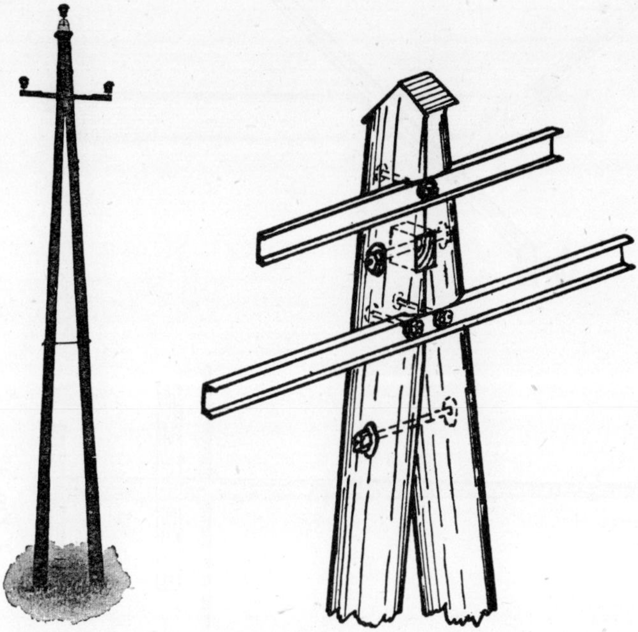

Fig. 40. "A"-type Wood Pole

poles in small and medium sizes, but for tall supports and long spans they are more economical. Fig. 45 shows a narrow-base, lattice-type support; the insulators are the suspension type. When the span is very great, it is better to have a broad base tower, as shown in Fig. 46, as it is then much easier to make an adequate foundation. Notice the cradle guard in the figure.

Fig. 47 shows a standard double-circuit tower used in the Grid scheme. The clearance of the lowest line conductor to earth at 50° C. is 22 ft., over buildings the minimum clearance is 12 ft., the vertical distance between conductors is 12 ft., and the projected

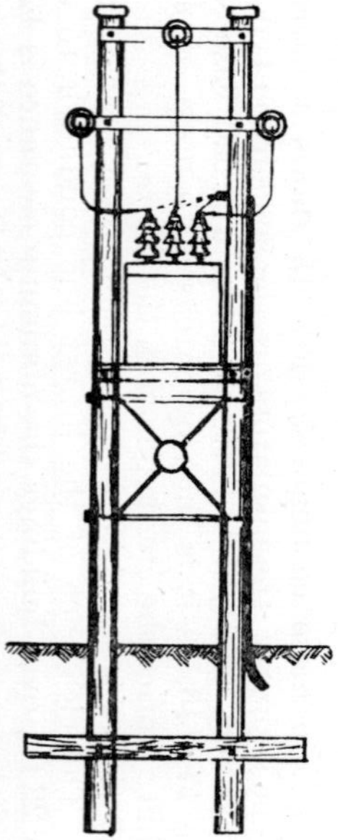

FIG. 41
"H"-TYPE WOOD POLE

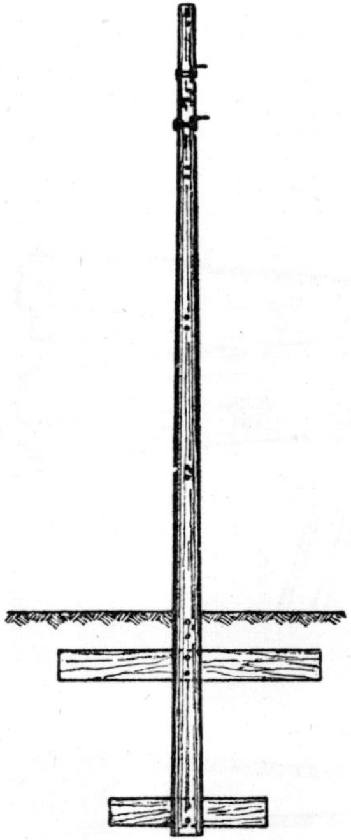

FIG. 42
BAULK FOUNDATION FOR POLE

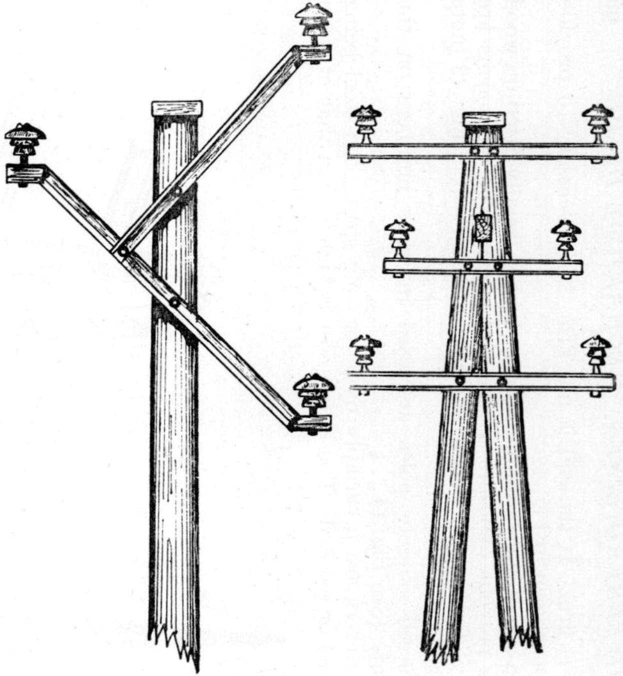

FIG. 43
CROSS-ARM FORMATIONS

horizontal distance is 3 ft. The minimum normal clearance between live metal and the tower steelwork is 3 ft. 6 in. under the worst conditions of line displacement. The foundations are excavated earth, concrete ball or ordinary concrete, depending on the soil and the cost. In the second method a cylindrical hole is drilled for each tower leg, and the lower end of the hole is expanded to a spherical shape by exploding a charge of dynamite in it. The tower leg stubs are fitted into position by templets, and the cavities are filled with concrete.

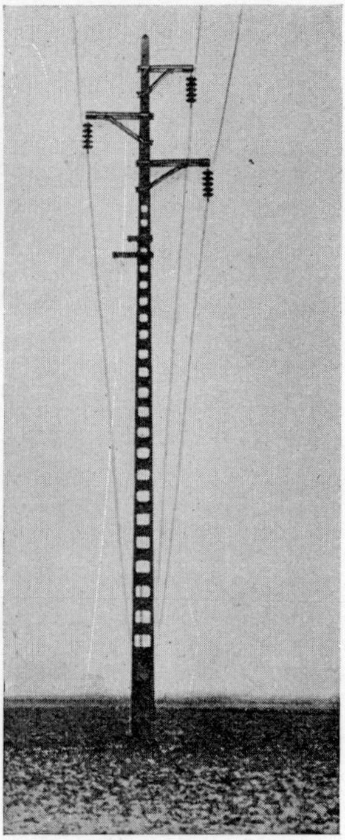

FIG. 44. CONCRETE POLE

PULL ON SUPPORTS. The supports must stand the normal load, which is transverse and subject to the conditions of ice and wind specified above, with a factor of safety of 2·5 (steel) or 3·5 (wood or reinforced concrete). If one or more conductors break there is in addition a longitudinal pull on the supports. If the supports are flexible they bend away from the broken conductor or conductors, so that the sags in the undamaged spans are increased. This results in a decrease in the pull. After three or four spans the longitudinal pull is very small. The practice is to have rigid anchoring towers at intervals of a mile or so, these towers being strong enough to withstand the breaking of one wire in three on one side. The narrow-base mast of Fig. 45 is designed with the decrease of pull due to flexibility in view.

Light flexible supports are not favoured for long spans, because of the twisting action which takes place when conductors on one side of the mast break. The British Regulations therefore specify that the strength of a support in the longitudinal direction shall not be less than one-quarter of the strength in the transverse direction. For very long spans and large-size conductors, the broad-base towers of Figs. 46 and 47 are employed.

Insulators. These are made of hard porcelain which contains two parts of kaolin (china clay), one of quartz, and one of felspar. Fig. 48 shows a *pin-type* insulator; the conductor passes through the groove, which is on the upper end of the insulator and is bound by annealed wire of the same material as the conductor. Overhead line conditions in Great Britain are particularly severe because of

Fig. 45. Narrow-base, Lattice-type Mast

FIG. 46. BROAD-BASE TOWER AND CRADLE GUARD

fog, and the atmospheric pollution caused by the intense industrialization, whilst in Wales and the West of England the westerly winds bring a great deal of brine which is deposited on the insulators. Design has therefore taken place along the lines of producing an insulator with long sheds with surfaces easily washable by rain.

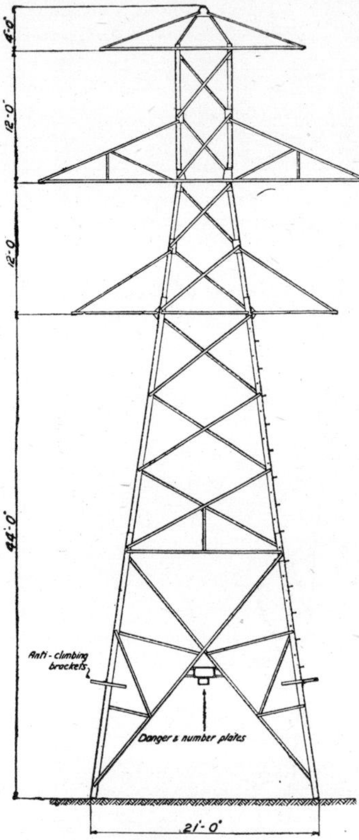

FIG. 47
STANDARD DOUBLE-CIRCUIT TOWER OF GRID SCHEME
(I.E.E. Journal)

FIG. 48. PIN-TYPE INSULATOR

An insulator is most easily washed by rain when the axis is not vertical.

At high voltages (above 33 kV.) it is convenient to make the insulators in units which can be connected in series. Fig. 49 shows a *suspension-type* insulator, of the cap and pin type, of nine units with grading ring and arcing horns. Fig. 50 shows the Hewlett

FIG. 49. SUSPENSION-TYPE INSULATOR WITH GRADING RINGS AND ARCING HORNS

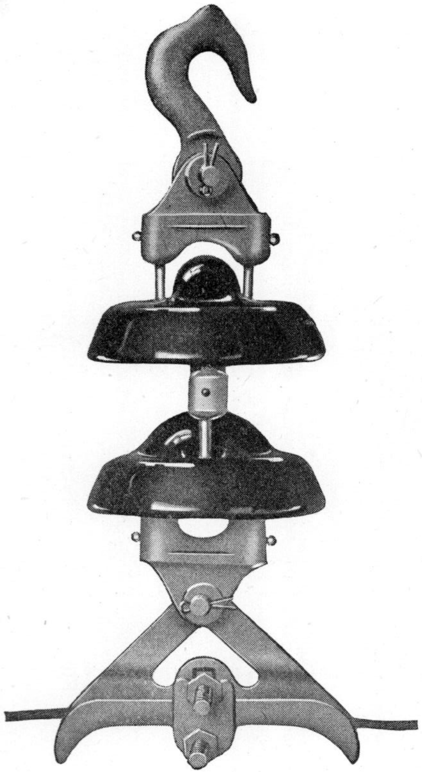

Fig. 50. Hewlett Disc Insulator
(*Bullers. Ltd.*)

Fig. 51. Strain Insulator
(*Bullers, Ltd.*)

disc insulator. When the insulator is to stand a very large tension, two strings are put in parallel, as shown in Fig. 51.

Pin-type insulators are tested for transverse strength. The insulator is mounted on its spindle, the shank being held in a metal

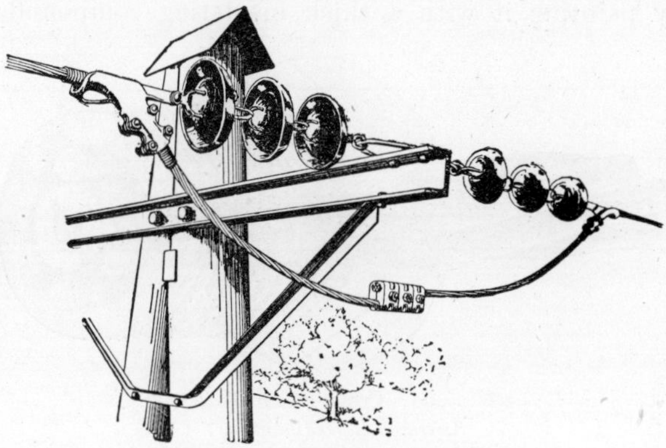

FIG. 52. OVERHEAD JOINT
(*Overhead Lines* (Painton))

bracket on the testing machine. A flexible wire rope loop is placed round the neck to apply a load transverse to the axis of the insulator, and the load is increased to the specified test value, which is $2\frac{1}{2}$ times

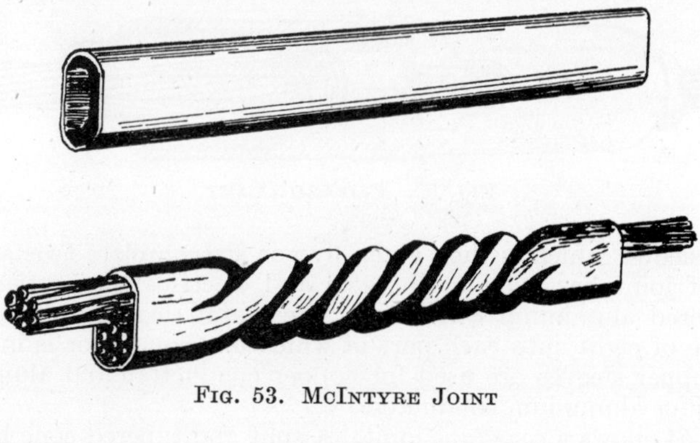

FIG. 53. McINTYRE JOINT

the specified maximum working load. The period of application is 1 min.

Suspension-type insulators are tested in pure tension, since they cannot and do not maintain a transverse load.

Joints and Clamps. Whenever possible joints are made at anchor supports, so that the joint need have no mechanical strength and may be constructed as shown in Fig. 52. In this method the conductors terminate through strain clamps, which take the stress, and are bolted in a metal clamp which ensures good electrical conductivity. It is important to prevent the entrance of water into the clamp by painting it with a thick insulating compound. If the

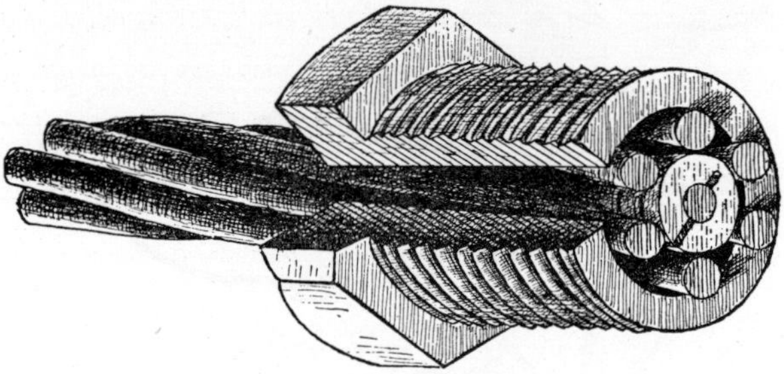

FIG. 54. CONE-TYPE JOINT
(*Overhead Lines* (Painton))

conductors are of copper they may simply be soldered together and then painted to avoid chemical action at the joint of dissimilar metals.

If the joint is in a span it must take the full tension, and soldering is avoided. The *McIntyre joint*, Fig. 53, is suitable for copper and aluminium conductors. The conductors are inserted into an oval

FIG. 55. PAIRARD CLAMP

metal sleeve, which is then given five or six complete twists. This type of joint has good mechanical and electrical properties. For steel-cored aluminium conductors, the metal sleeve is shaped like a figure of eight, into each part of which one conductor is inserted. Soft copper sleeves are used for copper conductors and aluminium sleeves for aluminium conductors.

Fig. 54 shows a *cone-type* joint. A split and tapered cone has one wire through its centre whilst the cone is pushed in among the other wires. A connecting piece screws the two conductors together. Tension of the conductors wedges the cones more tightly among the strands. The *Pairard clamp*, shown in Fig. 55, is of the cone type and may be used to make a joint or to serve as a terminal

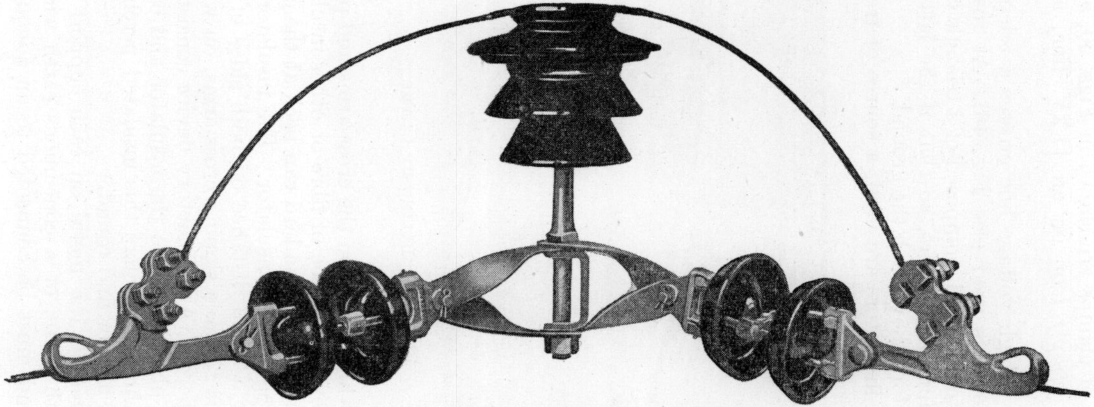

FIG. 56. JUMPER LOOP ON 11 kV. LINE
(*Bullers, Ltd.*)

clamp. It is specially used with aluminium and steel-cored aluminium conductors.

Different types of clamps are shown in Figs. 50, 51, 55, and 56. Fig. 56 shows a *jumper loop* on an 11 kV. line, and Fig. 57 a *pulling-up clamp*.

Guards. When a high voltage line crosses a telephone wire an earthed cradle guard is used (see Fig. 46). At road crossings a simpler guard system is used, composed of a triangle of earthed wire surrounding the line conductors, so that if the latter break they will be earthed before they reach the ground.

When pin-type insulators are used, a source of trouble is arcing

FIG. 57. PULLING-UP CLAMP

caused by birds, which perch on the cross-arms and peck at the live conductors. Sometimes a tube of fibre or other insulator is slipped over the cross-arm, so that the bird can peck at the conductor only when it is standing on the insulator. The insulator may be over the conductor for a small distance on both sides of the insulator. Another remedy is to have sloping cross-arms, on which the birds do not prefer to perch. A satisfactory device consists of a stick of ebonite, which is placed near to the insulator and carries a perch. It is found that the bird prefers this insulated perch to the cross-arm, because it is narrow and round.

Earthing. Either all the metal on each support must be connected to an earth plate, or a continuous earth wire is run from tower to tower and must be connected to an adequate earth four times per mile.

On wood pole lines, the regulations are met by using stranded galvanized steel wire attached to each pole below the conductors, all metal work on the pole being bonded to the earth wire by a single copper wire, usually No. 8 gauge. The earth wire is earthed four times per mile by running the bonding wire down the pole and connecting it to an earth plate. A good earth, i.e. one with a very low resistance, is of the utmost importance. The best practice is to bury a plate or pipe about 5 ft. below the surface, this being surrounded by about 6 in. of screened coke. The fear of corrosion has led to the use of copper earth plates. This is expensive and seems unnecessary as good results are obtained with galvanized cast-iron plates 2 to 3 ft. square and not less than $\frac{1}{2}$ in. thick. The bonding wire is best connected to the earth plate by a *tramway bond.* Whenever possible the earth pipe or plate should be buried well below the surface and connected by an insulated lead to the earth wire. This procedure diminishes considerably the potential gradient in the neighbourhood of the earth connection. Fig. 58 shows such an arrangement, which is, of course, applicable only when wood poles are used.

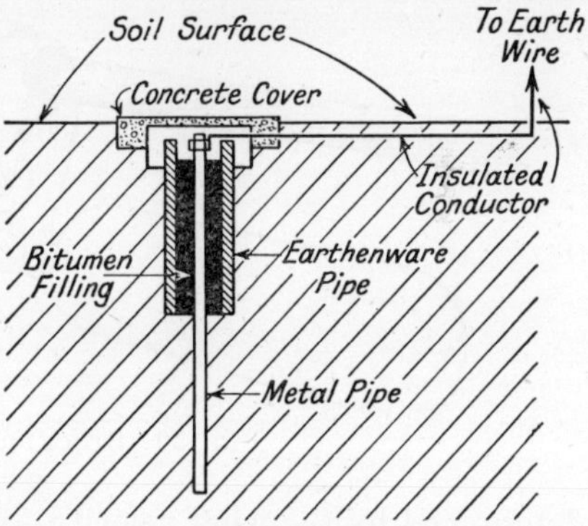

FIG. 58

On steel supports, and on wood poles where lightning storms are frequent, the earth wire is run above the line conductors. In the former case the earth wire is bonded to the mast at its point of support and a stout bonding wire runs from the base of the mast to the earthing plate, the steelwork being relied upon for good conductivity. The earth wire is of galvanized steel, and its size is chosen from considerations of mechanical factors and the span length, so that the maximum sag is not greater than that of the line conductors. Galvanized steel has a comparatively short life, so that the earth wire is sometimes made of the same material as the conductors. This involves extra expense and increased spacing, since the sag is greater than for a thick steel wire. Copper-clad steel offers an excellent compromise, as it has great strength and lasts long.

Vibration and Dampers. There are three kinds of vibrations of overhead line conductors. There is firstly the simple swinging of the conductors in the wind; this is harmless provided the clearance is sufficiently large so that the conductors do not approach within sparking distance of each other.

Then there is a low frequency vibration, about one cycle per sec., that occurs during sleet storms with a strong wind. The amplitude is very large, about 20 ft. or more, and the conductors are said to "dance." Operation is almost impossible for the conductors touch one another, since the "dancing" takes place horizontally and vertically. It is thought that the phenomenon is due to the fact that the line conductor receives an irregular coating of sleet. If we consider that the cross-section of the coating is an ellipse, the line will experience a drag; furthermore, if the major axis of the ellipse makes an angle of 45° to the wind, there will be a considerable aerodynamic lift (or downward thrust) such as is experienced by an aerofoil. The conductor will thus dance horizontally and vertically with a large amplitude, and in an irregular manner due to the irregular deposition of sleet.

Finally there is a high frequency oscillation, 5 to 100 cycles per sec., of amplitude about 1 in. This vibration is due to the formation of eddies on the leeward side of the conductor. The frequency is

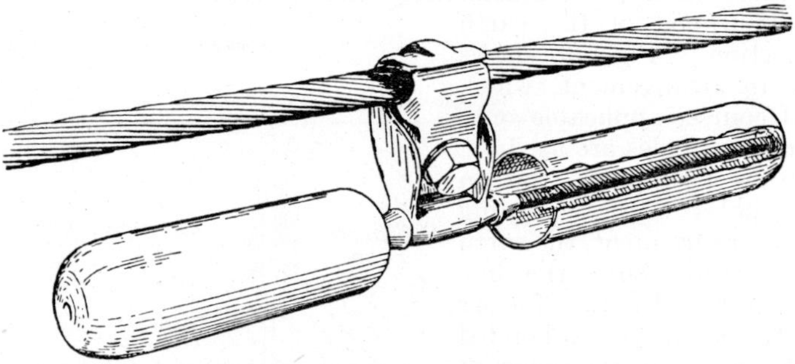

FIG. 59. STOCKBRIDGE DAMPER

determined by the wind velocity—which must be fairly steady and between 2 and 10 m.p.h. for the maintenance of the vibrations—and the diameter of the wire, and is given by

$$\text{frequency} = 3{\cdot}26 \times (V/d) \text{ cycles per sec.},$$

where V is the wind velocity in miles per hour and d the diameter in inches. The line conductor will vibrate in a number of loops, depending on its length, mass per unit length, and tension. The harmful effects of this vibration take place at the clamps or supports, where the conductor suffers fatigue and breaks eventually. The conductors are protected by dampers, which prevent the vibration from reaching the conductors at the clamps or supports. The *Stockbridge damper*, shown in Fig. 59, consists of two weights attached to a piece of stranded cable one or two feet long, which is clamped to the line conductor. The energy of vibration is absorbed by the stranded cable, and the vibration is rapidly damped out. The Stockbridge damper is made in various sizes, which weigh from 2 to 16 lb. Another successful damper consists of a box, which contains a weight resting on a spring. In this case the spring absorbs the energy of vibration. The fatigue may be lessened by reinforcing the conductor for a few feet on either side of the clamp by binding metal rods or a length of the same conductor to the main conductor outside the clamp.

EXAMPLES III

1. State the relation between the conductivity of copper and the temperature and tensile strength. Assuming that the resistance of 1 000 yd. of annealed copper of 1 in.^{2} cross-section at 60° F. is 0·0240 Ω., find the resistance of 2 500 yd. of 0·15 in.^{2} hard-drawn copper of tensile strength 27 tons per in.^{2} at 75° F. ($\alpha = 0{\cdot}002222$ per ° F.).

2. Find the tensile strength and Young's modulus of a steel-cored aluminium conductor consisting of 7 strands steel and 54 strands aluminium of 0·105 in. dia. Calculate also the weight per foot. Assume tensile strengths of 26 000 lb. per in.^{2} for aluminium and 190 000 for steel.

3. Develop an expression for the sag of a transmission line conductor suspended between supports at equal heights, assuming the curve taken up by the conductor to be a parabola.

A copper conductor is suspended between two points at the same height and is strung to a tension of 3 000 lb. The load on the conductor (including its weight) is 1·5 lb. per ft. and its cross-sectional area is 0·39 in.^{2} If the load on the conductor is increased to 2 lb. per ft., calculate the tension in the conductor and the sag under these new conditions. The span is 600 ft., and the modulus of elasticity of copper is 18×10^6 lb. per in.^{2} Use the approximate method, assuming the curve to be a parabola.

(*B.Sc., Lond. Univ.*, 1933.)

4. Calculate the sag allowable for a 500 ft. span of copper wire 0·4 in. dia. under the following conditions. Ice $\frac{3}{8}$ in. thick, wind pressure 8 lb. per ft.^{2} of projected area. Ice weighs 57 lb. per ft.^{3}, copper 555 lb. per ft.^{3} Tensile strength of copper is 50 000 lb. per in.^{2}, and the factor of safety is 2.

If the conductor is supported by an insulator string 3 ft. long, calculate also (1) the height of the tower to the lowest cross-arm if the ground clearance is to be 25 ft., and (2) the length of the cross-arm if the width of the tower is 2 ft. and the minimum conductor clearance is to be 18 in.

(*Nat. Cert.*, 1934.)

5. Determine accurately the maximum sag of a transmission line conductor having a span of 2 200 ft., the supports being on the same level, the total load on the conductor (under maximum ice and wind loads) being 2·0 lb. per foot run and the maximum allowable tension being 8 000 lb. Values of cosh x may be taken from the following table—

x	0·23	0·24	0·25	0·26	0·27
cosh x	1·0265	1·0289	1·0314	1·0339	1·0367

Describe briefly how you would proceed to construct a stringing chart for the above conductor showing the tension to which it must be strung if erected under other conditions of loading and temperature.

(*B.Sc. Lond. Univ.*, 1932.)

6. A transmission line conductor at a river crossing is supported from two towers at heights of 100 ft. and 300 ft. above water level. The horizontal distance between the towers is 900 ft. If the tension in the conductors is 3 000 lb., find the clearance between the conductor and water at a point mid-way between the towers. Wt. of conductor per ft. run = 0·57 lb.

Explain the necessity for a stringing chart in erecting a transmission line and describe how you would proceed to construct such a chart.

(*B.Sc. Lond. Univ.*, 1934.)

7. Construct the stringing chart for the previous example.

8. A 132 kV. grid line conductor has an effective diameter of 0·77 in., weighs 571 lb. per 1 000 ft., and is under a tension of 3 200 lb. If the span length is 950 ft. and the wind velocity is 6 m.p.h., calculate the frequency of the oscillation and the number of loops in the span. The formula connecting frequency, tension, mass, and length of loop is

$$f = (1/2l)\sqrt{(\mathrm{T}g/w)},$$

where f = frequency in cycles per sec.,
l = length of loop in ft.,
T = tension in lb.,
w = mass in lb. per ft.,
and g = 32·2.

9. Explain how to determine the sag and stress of an overhead line and indicate how the effects of wind, ice loading, and temperature variation are taken into account. Show how to deal with any peculiarities which may occur with steel-cored aluminium conductors. (*Lond. Univ.*, 1949.)

10. Explain the meaning of the terms "virtual modulus of elasticity" and "virtual coefficient of expansion."

Calculate the values of these quantities for a steel-cored aluminium conductor with 7 steel strands and 30 aluminium strands, all of the same diameter. The modulus of elasticity for steel is 30×10^6 lb. per in.^{2} and for aluminium is $9{\cdot}5 \times 10^6$ lb. per in.^{2}, the coefficient of expansion for steel is $6{\cdot}4 \times 10^{-6}$ and for aluminium is $12{\cdot}8 \times 10^{-6}$ per ° F.

Explain under what conditions the virtual values do not apply for a steel-cored aluminium conductor. (*Lond. Univ.*, 1948.)

11. A steel-cored aluminium overhead conductor has 7 steel and 30 aluminium strands each 0·11 in. dia. The modulus of elasticity for steel is 30×10^6 lb. per in.^{2} and for aluminium $9{\cdot}5 \times 10^6$ lb. per in.^{2} and the temperature coefficient of linear expansion for steel is $6{\cdot}4 \times 10^{-6}$ and for aluminium $12{\cdot}8 \times 10^{-6}$ per °F. Determine the virtual modulus of elasticity and virtual coefficient of expansion for the composite conductor.

Explain why it is possible to use these values and state under what conditions these values would not apply. (*Lond. Univ.*, 1953.)

CHAPTER IV

OVERHEAD LINES : ELECTRICAL DESIGN

Resistance of Overhead Lines. The resistance of a conductor to direct current is given by the formula

$$R = \rho l/A,$$

where ρ = the resistivity, l = the length, and A = the area of cross-section. The values of resistivity of the various materials used are given in Chapter III, and also their temperature coefficients. There is a tendency for alternating currents to concentrate near the surface of a conductor. This phenomenon is called the *skin effect* and results in an increased resistance. At 50 cycles the effect is noticeable only with conductors of large diameter. Thus the resistance to alternating current at 50 cycles of a copper conductor of diameter 0·752 in., of 97 per cent conductivity, is only 1·02 times the d.c. resistance; for the same ratio of resistances to a.c. and d.c. a 61 per cent conductivity aluminium conductor can have a diameter of 0·948 in. The skin effect is thus of small importance in overhead lines. The skin effect of a stranded conductor is the same as for a solid conductor of the same cross-section, provided the stranded conductor has no core. If it has a core, the effect is the same as for a hollow conductor and is still very small.

The effect of stranding is to increase the resistance in the way described in Chapter III. In practice it is taken that stranding increases the resistance by 2 per cent, which corresponds to a lay of $15\frac{1}{2}$.

The increase in length of a line due to sag is approximately $(2d^2/3l)$. For a sag ratio, d/l, of 10 per cent this gives an extra length of only $\frac{2}{3}$ per cent, so that it may be neglected.

Inductance of an Overhead Line. A conductor which carries current is surrounded by a magnetic field, which reacts upon the currents in this and other conductors. A long straight conductor is surrounded by a field in which the lines of magnetic force are concentric circles with the axis of the conductor through their centre. Fig. 60 shows a plane section of the magnetic field surrounding a cylindrical conductor carrying a current of I e.m. units; the direction of the lines is for the case of a current passing into the paper. At a point P distant x cm. from the centre O the force is

$$H = 2I/x.$$

The force at a point P' inside the conductor depends upon the current distribution inside the conductor. In overhead lines it may be assumed without appreciable error that the current is uniformly distributed. Then the force at P' is due only to the currents inside a circle through P' with O as centre, so that at P' the magnetic force is

$$H' = (2I/x) \times (x^2/r^2) = 2Ix/r^2,$$

where r = the radius of the conductor.

If the currents can be considered as flowing along geometrical lines (i.e. of no thickness) it is sufficient to calculate the total flux and divide it by the current in order to get the inductance. But when the currents flow along wires of finite thickness it is necessary to calculate the linkages and divide by the square of the current, a

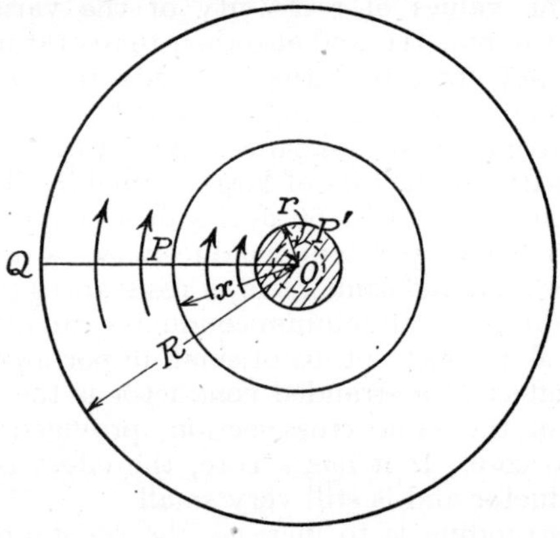

FIG. 60. MAGNETIC FIELD SURROUNDING A CONDUCTOR

linkage being the product of a current element and a flux. The linkages inside the conductor are for unit length along the axis

$$\int_0^r H' \times I\frac{x^2}{r^2}dx = \frac{2I^2}{r^4}\int_0^r x^3dx = \tfrac{1}{2}I^2,$$

since the flux H' at P' links only a current $I \times (x^2/r^2)$.

The linkages outside up to a distance R (considered large) are

$$\int_r^{\mathrm{R}} H \times I dx = 2I^2 \int_r^{\mathrm{R}} \frac{1}{x}dx = 2I^2 \operatorname{logh}^* \frac{\mathrm{R}}{r}.$$

It is seen that the inductance *per unit length* of a long conductor is infinite, since R should be taken as infinite. In practice there is always a return conductor at a finite distance; at very great distances from both wires the flux due to this return wire neutralizes

* logh (R/r) = natural or hyperbolic or napierian logarithm of (R/r), viz. to the base ε = 2·71828.

the flux due to the first wire, so that the linkages become finite. Fig. 61 shows the magnetic field due to the wire and its return conductor at a distance D from it. The linkages that conductor 1 creates for itself between O and Q have been found to be

$$\tfrac{1}{2}I^2 + 2I^2 \operatorname{logh} (\mathrm{R}/r).$$

The linkages created by conductor 2 round conductor 1 are

$$I\int_{\mathrm{D}}^{\mathrm{R}}\left(-\frac{2I}{x}\right)dx = -2I^2 \operatorname{logh}\frac{\mathrm{R}}{\mathrm{D}},$$

so that the total linkages round conductor 1 are

$$\tfrac{1}{2}I^2 + 2I^2 \operatorname{logh} (\mathrm{R}/r) - 2I^2 \operatorname{logh} (\mathrm{R}/\mathrm{D})$$
$$= \tfrac{1}{2}I^2 + 2I^2 \operatorname{logh} (\mathrm{D}/r)$$

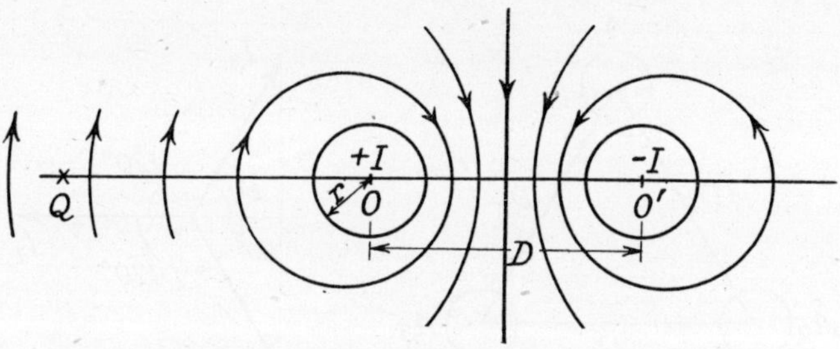

FIG. 61. MAGNETIC FIELD SURROUNDING TWO PARALLEL CONDUCTORS

which we see is independent of R and is finite. This gives for the inductance per unit length of conductor 1

$$L_1 = \tfrac{1}{2} + 2 \operatorname{logh} (\mathrm{D}/r).$$

The inductance of conductor 2 is the same. The inductance of the loop per unit length is

$$L_1 + L_2 = 1 + 4 \operatorname{logh} (\mathrm{D}/r) \quad . \quad . \quad . \quad (17)$$

If the wires are magnetic and have a permeability μ, the inductance of the loop per cm. length is

$$L_1 + L_2 = \mu + 4 \operatorname{logh} (\mathrm{D}/r) \quad . \quad . \quad . \quad (17a)$$

The unit of inductance in equations (17) and (17a) is the electromagnetic unit, which is 10^{-9} henry. The inductance per mile of the loop is

$$L = L_1 + L_2 = [0{\cdot}161 + 1{\cdot}482 \log^* (\mathrm{D}/r)]\ \mathrm{mH.} \quad (18)$$

whilst $$L_1 = L_2 = [0{\cdot}080 + 0{\cdot}741 \log (\mathrm{D}/r)]\ \mathrm{mH.} \quad . \quad . \quad (18a)$$

* Log (D/r) is the logarithm of (D/r) to the base 10.

Stranded conductors have a slightly higher inductance, the constant term being somewhat higher. Thus the constant for 3-strand is 0·125, 7-strand 0·103, 19-strand 0·089, 37-strand 0·085, 61-strand 0·083, solid conductor 0·080.

EXAMPLE. Calculate the inductance per mile of a conductor 0·4 in. dia. at a spacing of 18 in.

$$L_1 = [0{\cdot}080 + 0{\cdot}741 \log (18/0{\cdot}2)] \text{ mH.}$$
$$= (0{\cdot}080 + 0{\cdot}741 \log 90)$$
$$= 0{\cdot}080 + 1{\cdot}457 = 1{\cdot}537 \text{ mH.}$$

Inductance of Three-phase Lines. Fig. 62 shows a cross-section of a three-phase system, consisting of three conductors A_1, A_2,

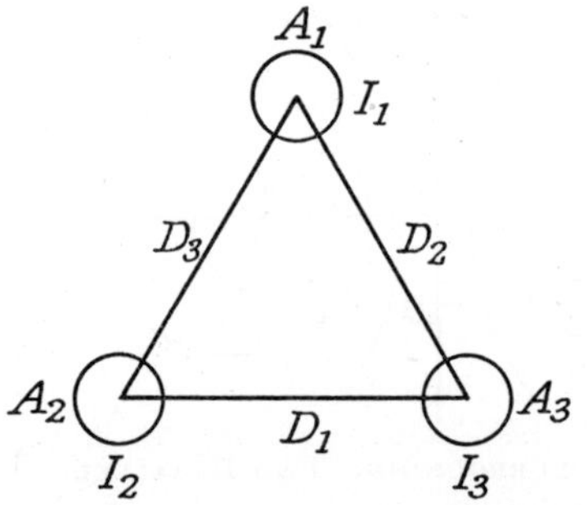

FIG. 62
CROSS-SECTION OF THREE-PHASE SYSTEM

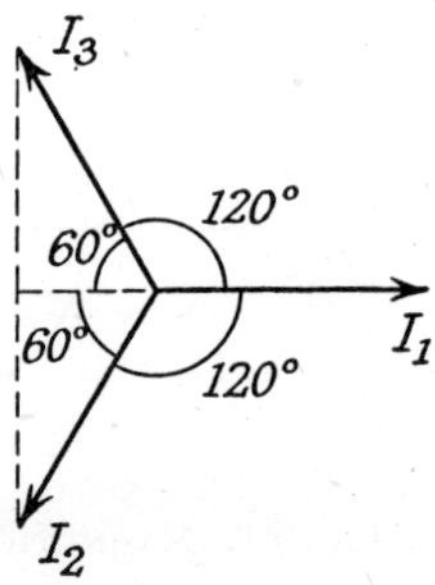

FIG. 63
CURRENTS IN BALANCED, THREE-PHASE SYSTEM

and A_3 of radius r and spacings D_1, D_2, and D_3. The currents are I_1, I_2, and I_3, which must satisfy the relation $I_1 + I_2 + I_3 = 0$. If the loads and voltages are balanced, the currents can be represented by the three vectors shown in Fig. 63. The real components of I_2 and I_3 are $I \cos 120° = -\frac{1}{2}I_1$, whilst their imaginary components are $I_1 \sin(-120)$ and $I_1 \sin(+120°)$, viz. $-(\frac{1}{2}\sqrt{3})I_1$ and $+(\frac{1}{2}\sqrt{3})I_1$. Thus

$$\left.\begin{aligned} I_1 &= I_1 \\ I_2 &= [-\tfrac{1}{2} - j(\tfrac{1}{2}\sqrt{3})]I_1 = \lambda I_1, \text{ say} \\ \text{and} \quad I_3 &= [-\tfrac{1}{2} + j(\tfrac{1}{2}\sqrt{3})]I_1 = \lambda^2 I_1, \end{aligned}\right\} \quad . \quad . \quad (19)$$

$$\left.\begin{aligned} &\text{where} & j &= \sqrt{-1}, \text{ and} \\ &\text{where we have put} & \lambda &= -\tfrac{1}{2} - j(\tfrac{1}{2}\sqrt{3}). \\ &\text{It is easy to prove that} & \lambda^2 &= -\tfrac{1}{2} + j(\tfrac{1}{2}\sqrt{3}), \\ &\text{and} & 1 + \lambda &+ \lambda^2 = 0. \end{aligned}\right\} \quad . \quad . \quad (20)$$

1, λ, and λ^2 are the cube roots of unity, for

$$\sqrt[3]{1} = 1^{\frac{1}{3}} = (\cos 0 + j \sin 0)^{\frac{1}{3}} = \cos 0 + j \sin 0 = 1$$

or
$$(\cos 360 + j \sin 360)^{\frac{1}{3}} = \cos 120 + j \sin 120 = \lambda^2,$$

or
$$(\cos 720 + j \sin 720)^{\frac{1}{3}} = \cos 240 + j \sin 240 = \lambda.$$

This fact is also obvious from the vector diagram of Fig. 63.

The *flux* linking conductor 1 up to some large distance R due to its own current is, as shown on page 82,

$$[\tfrac{1}{2} + 2 \text{ logh } (\text{R}/r)]I_1 \text{ per cm. length.}$$

The flux linking conductor 1 up to the same distance due to current I_2 is, as shown on page 82,

$$[2 \text{ logh } (\text{R}/\text{D}_3)]I_2,$$

whilst the flux due to I_3 is

$$[2 \text{ logh } (\text{R}/\text{D}_2)]I_3.$$

The total flux linking conductor A_1 is thus

$$[\tfrac{1}{2} + 2 \text{ logh } (\text{R}/r)]I_1 + [2 \text{ logh } (\text{R}/\text{D}_3)]I_2 + [2 \text{ logh } (\text{R}/\text{D}_2)]I_3 \quad (21a)$$
$$= [\tfrac{1}{2} + 2 \text{ logh } (1/r)]I_1 + [2 \text{ logh } (1/\text{D}_3)]I_2 + [2 \text{ logh } (1/\text{D}_2)]I_3 + (2 \text{ logh R})(I_1 + I_2 + I_3)$$
$$= [\tfrac{1}{2} + 2 \text{ logh } (1/r)]I_1 + [2 \text{ logh } (1/\text{D}_3)]I_2 + [2 \text{ logh } (1/\text{D}_2)]I_3 \quad (21b)$$

since $I_1 + I_2 + I_3 = 0$, even when the system is not balanced. We see that this flux is independent of R and is thus finite, even when we take the flux up to infinity (at which $\text{R} = \infty$). To find the inductance of conductor A_1 in the presence of A_2 and A_3 we must divide the preceding expression by I_1 and we get

$$L_1 = \tfrac{1}{2} + 2 \text{ logh } (1/r) + [2 \text{ logh } (1/\text{D}_3)](I_2/I_1) + [2 \text{ logh } (1/\text{D}_2)](I_3/I_1),$$

which depends upon the currents if they are unbalanced.

This means that we cannot represent the system by three simple inductances for unbalanced currents. It is necessary to include mutual inductances between pairs. If the currents are balanced so that equations (19) hold, we have

$$\left.\begin{aligned} L_1 &= \tfrac{1}{2} + 2 \text{ logh } (1/r) + 2\lambda \text{ logh } (1/\text{D}_3) + 2\lambda^2 \text{ logh } (1/\text{D}_2) \\ &= \tfrac{1}{2} + 2 \text{ logh } \frac{\sqrt{(\text{D}_2\text{D}_3)}}{r} + j\sqrt{3} \text{ logh } \frac{\text{D}_3}{\text{D}_2} \\ &\text{Similarly} \\ L_2 &= \tfrac{1}{2} + 2 \text{ logh } \frac{\sqrt{(\text{D}_3\text{D}_1)}}{r} + j\sqrt{3} \text{ logh } \frac{\text{D}_1}{\text{D}_3} \\ \text{and} \quad L_3 &= \tfrac{1}{2} + 2 \text{ logh } \frac{\sqrt{(\text{D}_1\text{D}_2)}}{r} + j\sqrt{3} \text{ logh } \frac{\text{D}_2}{\text{D}_1}. \end{aligned}\right\} \quad (22)$$

The imaginary terms in L_1, L_2, and L_3 represent the transfer of power between phases due to mutual inductance, and it is seen that they add up to zero.

If the line is transposed along its length, i.e. the positions of the conductors are interchanged so that each occupies all three positions for an equal length, the inductance of each wire becomes the mean of the expressions in equations (22), so that

$$L_1 = L_2 = L_3 = \tfrac{1}{2} + \tfrac{1}{3}\left[2 \operatorname{logh} \frac{\sqrt{(D_2D_3)}}{r} + 2 \operatorname{logh} \frac{\sqrt{(D_3D_1)}}{r} + 2 \operatorname{logh} \frac{\sqrt{(D_1D_2)}}{r}\right] = \tfrac{1}{2} + 2 \operatorname{logh} \frac{\sqrt[3]{(D_1D_2D_3)}}{r} \quad . \quad . \quad (23)$$

or

$$0{\cdot}080 + 0{\cdot}741 \log \frac{\sqrt[3]{(D_1D_2D_3)}}{r} \text{ mH. per mile} \quad . \quad . \quad (23a)$$

If the spacings are equal to D, equations (21*a*) *et seq*. show that

$$\left.\begin{aligned} L_1 = L_2 = L_3 &= \tfrac{1}{2} + 2 \operatorname{logh} (D/r) \\ \text{or} \quad &0{\cdot}080 + 0{\cdot}741 \log (D/r) \text{ mH. per mile} \end{aligned}\right\} \quad . \quad . \quad . \quad (24)$$

even for unbalanced currents.

Skin effect decreases the inductance but only by a very small amount. The fact that the current follows spiral paths in a stranded conductor causes a small increase of inductance, but the increase is negligible at normal supply frequencies for non-magnetic wires.

Capacitance of Overhead Lines. When two cylindrical conductors have a potential difference V, they acquire charges $+ Q$ and $- Q$ per cm. length, and we say that they have a capacitance C per cm. to each other of

$$C = Q/V.$$

The charges are not spread uniformly over the surfaces but are concentrated at the inner parts of the cylinders. The exact calculation of the capacitance between two parallel, circular cylinders is known to give the value

$$\frac{1}{4 \operatorname{logh}\left[\dfrac{D + \sqrt{(D^2 - 4r^2)}}{2r}\right]} \text{ e.s.u. per cm. length} \quad . \quad (25)$$

The e.s.u. (electrostatic unit) of capacitance is the centimetre and is 10/9 $\mu\mu$F.

It is interesting and important to find the error caused by assuming that the charges are distributed uniformly over the cylinders. Fig. 64 shows a uniformly charged cylinder, $+ Q$ per cm. length. The electric force at every point P is radial and is the same at all points distant x from the axis. If a cylinder of radius x is drawn about the axis and has a length of 1 cm., the total electric flux crossing the cylinder is $F \times$ the area of the curved part of the cylinder, viz. $F \times 2\pi x$. But by Gauss's theorem this flux is $4\pi \times$ charge $= 4\pi Q$.

$$\therefore \quad F \times 2\pi x = 4\pi Q,$$

$$\text{or} \qquad F = 2Q/x.$$

Fig. 65 shows two cylinders of radius r spaced D cm. apart and having charges $+Q$ and $-Q$. At the point P the force is

$$2Q/x + 2Q/(\mathrm{D} - x).$$

The difference in potential between P_1 and P_2 is found by integrating this force between P_1 and P_2: this should be the potential

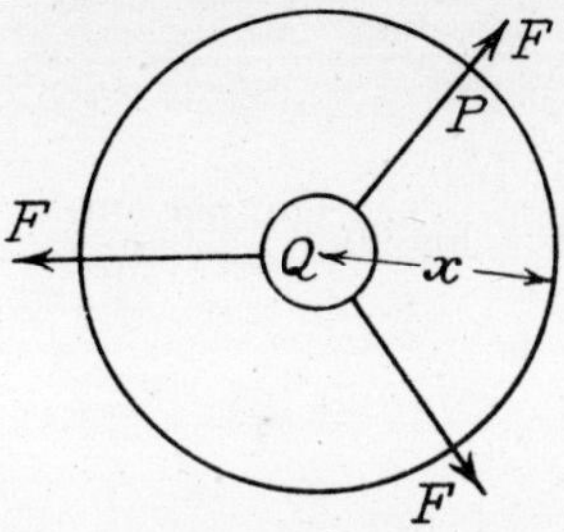

FIG. 64

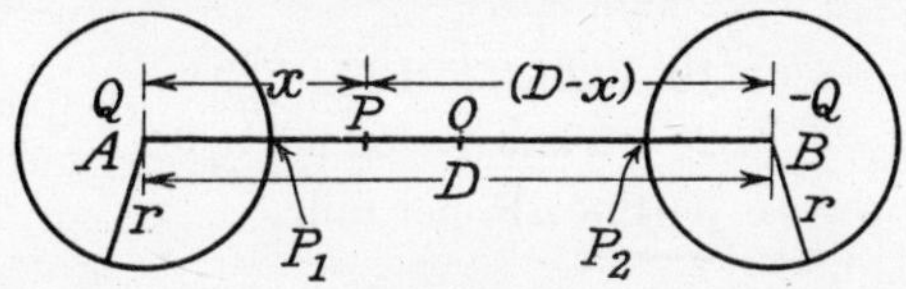

FIG. 65

difference between the cylinders, but actually the cylinders will not be equipotentials for the two line charges. It is more usual to integrate from P_1 to the centre of the other cylinder, B, for the force $2Q/x$, and from A to P_2 for the force $2Q/(\mathrm{D} - x)$. Then

$$\begin{aligned} V &= \int_r^{\mathrm{D}} \frac{2Q}{x} dx + \int_0^{\mathrm{D}-r} \frac{2Q}{\mathrm{D}-x} dx \\ &= 2Q \operatorname{logh} (\mathrm{D}/r) + 2Q \operatorname{logh} (\mathrm{D}/r) \\ &= 4Q \operatorname{logh} (\mathrm{D}/r). \end{aligned}$$

This gives

$$C = \frac{Q}{V} = \frac{1}{4 \operatorname{logh} (\mathrm{D}/r)} \text{ cm. per cm. length} \quad . \quad (26)$$

If $\mathrm{D} = 8r$, the capacitance given by the exact formula of equation (25) is

$$1/4 \operatorname{logh} 7{\cdot}87 = 0{\cdot}1212 \text{ cm. per cm. length,}$$

whilst the approximate value given by equation (26) is

$$1/4 \operatorname{logh} 8 = 0{\cdot}1203 \text{ cm. per cm. length.}$$

The error is thus only $\frac{3}{4}$ per cent. In overhead lines the approximate method may thus be used without much error.

The point O which is mid-way between the axes is at zero potential and is called the *neutral*. The potentials of the conductors are $\frac{1}{2}V$ and $-\frac{1}{2}V$ with respect to O, so that the capacitances C_0 to the neutral, are $2C$. Thus

$$\left.\begin{aligned} C_0 &= \frac{1}{2 \operatorname{logh} (D/r)} \text{ cm. per cm. length} \\ \text{or} \qquad &\frac{0{\cdot}0388}{\log (D/r)} \mu\text{F. per mile} \end{aligned}\right\} \quad . \quad . \quad . \quad (27)$$

EXAMPLE. Find the capacitance to neutral of the conductors of the previous example.

Here D $= 18$ in., $r = 0{\cdot}2$ in.

Therefore

$$C_0 = \frac{1}{2 \operatorname{logh} 90} \text{ cm. per cm.} = \frac{1}{2 \times 2{\cdot}303 \times \log 90} \text{ cm. per cm.}$$

$$= \frac{10 \times 2{\cdot}54 \times 12 \times 5\,280}{2 \times 2{\cdot}303 \times \log 90 \times 9} \mu\text{F. per mile}$$

$$= \underline{\underline{0{\cdot}0198}} \ \mu\text{F. per mile.}$$

It may be noted that

$$1/\sqrt{(LC)} = 1/\sqrt{(L_1C_0)} = 181\,000,$$

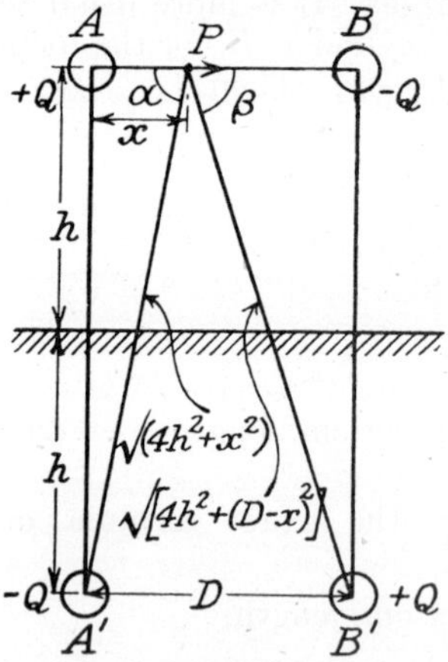

FIG. 66

which is the velocity of electro-magnetic waves along these conductors in miles per sec. The quantity

$$\sqrt{(L/C)} = \sqrt{(2L_1/\tfrac{1}{2}C_0)} = 2\sqrt{(L/C_0)} = 556\ \Omega,$$

is the *surge impedance* of the line.

Effect of Earth. The effect of the earth may be allowed for by the method of images. Fig. 66 shows a single phase line, conductors A and B carrying charges $+Q$ and $-Q$ per cm. length, and an earth distance h below. Let A' and B' be the images of A and B in the earth. Then if conductor A' is considered to have a charge $-Q$ per cm. and B', $+Q$ per cm., the potential at any point on the earth is zero, as it should be. The electric field at any point above the earth may be considered as due to the conductors A, B, A', and B'; the field below the earth is, however, zero, since the earth is considered to be a perfect conductor. The electric force at P, distance x from A, along AB is

$$\frac{2Q}{x}+\frac{2Q}{\mathrm{D}-x}-\frac{2Q\cos\alpha}{\sqrt{(4h^2+x^2)}}-\frac{2Q\cos\beta}{\sqrt{[4h^2+(\mathrm{D}-x)^2]}}$$

$$=\frac{2Q}{x}+\frac{2Q}{\mathrm{D}-x}-\frac{2Qx}{4h^2+x^2}-\frac{2Q+(\mathrm{D}-x)}{4h^2+(\mathrm{D}-x)^2}.$$

The potential V is the integral of this force from A to B and is

$$4Q\ \mathrm{logh}\frac{\mathrm{D}}{r}+4Q\ \mathrm{logh}\frac{2h}{\sqrt{(4h^2+\mathrm{D}^2)}}\ *$$

$$=4Q\ \mathrm{logh}\frac{\mathrm{D}2h}{r\sqrt{(4h^2+\mathrm{D}^2)}}=4Q\ \mathrm{logh}\frac{\mathrm{D}}{r\sqrt{(1+(\mathrm{D}^2/4h^2)}}.$$

The capacitance to neutral is thus

$$\left.\begin{aligned}C_0&=\frac{1}{2\ \mathrm{logh}\dfrac{\mathrm{D}}{r\sqrt{(1+\mathrm{D}^2/4h^2)}}}\text{ cm. per cm. length}\\ \text{or}\quad&\frac{0{\cdot}0388}{\log\dfrac{\mathrm{D}}{r\sqrt{(1+\mathrm{D}^2/4h^2)}}}\ \mu\text{F. per mile}\end{aligned}\right\}\qquad(27a)$$

Example. Find the capacitance to neutral of the conductors in the previous example if they are 10 ft. above ground.

$$C_0=\frac{0{\cdot}0388}{\log\dfrac{90}{\sqrt{\left(1+\dfrac{1{\cdot}5^2}{4\times10^2}\right)}}}=\frac{0{\cdot}0388}{\log 90{\cdot}25}=\frac{0{\cdot}0388}{1{\cdot}959}$$

$$=\underline{\underline{0{\cdot}0199}}\ \mu\text{F. per mile.}$$

In the previous case $C_0=\dfrac{0{\cdot}0388}{\log 90}=\dfrac{0{\cdot}0388}{1{\cdot}954}$, which is only $\frac{1}{4}$ per cent less.

Generally the effect of the earth on overhead lines is negligible.

Capacitance of Three-phase Line. Suppose that the line is as shown in Fig. 62, and that voltages V_1, V_2, V_3 are applied to the conductors with the result that the charges per cm. length are Q_1, Q_2, Q_3. The electric force due to a charged cylinder is $2Q/x$, where Q is the charge per cm. length and x the distance from the axis. In order to calculate the potential of A, we imagine a unit charge brought from infinity, which is assumed at zero potential, up to A_1.

* $\displaystyle\int\frac{2x}{4h^2+x^2}dx=\int\frac{d(x^2)}{4h^2+x^2}=\mathrm{logh}\ (4h^2+x^2).$

If the unit charge is brought from a large distance R to within a distance r of A_1, D_3 of A_2, and D_2 of A_3, the work done is

$$\int_r^R (2Q_1/x)\,dx + \int_{D_3}^R (2Q_2/x)dx + \int_{D_2}^R (2Q_3/x)dx$$

$$= 2Q_1 \operatorname{logh}(R/r) + 2Q_2 \operatorname{logh}(R/D_3) + 2Q_3 \operatorname{logh}(R/D_2)$$

$$= 2Q_1 \operatorname{logh}(1/r) + 2Q_2 \operatorname{logh}(1/D_3) + 2Q_3 \operatorname{logh}(1/D_2) + 2(Q_1 + Q_2 + Q_3) \operatorname{logh} R.$$

When R is infinite, this work is to be the potential V_1. It follows that a necessary condition is $Q_1 + Q_2 + Q_3 = 0$. Also then

$$\left.\begin{aligned} V_1 &= 2Q_1 \operatorname{logh}(1/r) + 2Q_2 \operatorname{logh}(1/D_3) + 2Q_3 \operatorname{logh}(1/D_2) \\ &\text{Similarly} \\ V_2 &= 2Q_1 \operatorname{logh}(1/D_3) + 2Q_2 \operatorname{logh}(1/r) + 2Q_3 \operatorname{logh}(1/D_1) \\ &\text{and} \\ V_3 &= 2Q_1 \operatorname{logh}(1/D_2) + 2Q_2 \operatorname{logh}(1/D_1) + 2Q_3 \operatorname{logh}(1/r) \end{aligned}\right\} \qquad (28)$$

Q_1 can be found from the equations for V_1 and V_2 by replacing Q_3 by $-(Q_1 + Q_2)$ and solving. Then

$$Q_1 = \tfrac{1}{2}V_1 \frac{\operatorname{logh}(D_1/r) - (V_2/V_1)\operatorname{logh}(D_2/D_3)}{\operatorname{logh}(D_2/r)\operatorname{logh}(D_1/r) - \operatorname{logh}(D_1/D_3)\operatorname{logh}(D_2/D_3)} \qquad (28a)$$

There are similar expressions for Q_2 and Q_3. The charging currents are

$$I_1 = j\omega Q_1,\ I_2 = j\omega Q_2,\ I_3 = j\omega Q_3.$$

As the charge in equation (28*a*) contains a term (V_2/V_1), it is clear that the currents in a line are determined by all the line voltages.

The representation of the charging currents by capacitances is complicated except in certain cases.

Equilateral Spacing. Here $D_1 = D_2 = D_3 = D$. Then $Q_1 = V_1/2 \operatorname{logh}(D/r)$, $Q_2 = V_2/2 \operatorname{logh}(D/r)$ and $Q_3 = V_3/2 \operatorname{logh}(D/r)$, independent of the values of the voltages. Then $Q_1/V_1 = Q_2/V_2 = Q_3/V_3 = 1/2 \operatorname{logh}(D/r)$, and this is called the capacitance of each line to neutral, i.e.

$$\left.\begin{aligned} C_0 &= 1/2 \operatorname{logh}(D/r) \text{ e.s.u. per cm. length} \\ \text{or} \quad &\quad 0{\cdot}0388/\log(D/r)\ \mu\text{F. per mile} \end{aligned}\right\} \quad . \quad . \quad (29)$$

Transposed Line with Balanced Voltages. In this case

$$V_1 = V,\ V_2 = \lambda V,\ V_3 = \lambda^2 V, \text{ where } \lambda = -\tfrac{1}{2}(1 + j\sqrt{3}).$$

Also the conductors are transposed so as to occupy all three positions for equal distances. It can then be shown that $Q_1 = C_0V_1$, $Q_2 = C_0V_2$ and $Q_3 = C_0V_3$, where

$$\left.\begin{aligned} C_0 &= \left[1/2 \text{ logh } \frac{\sqrt[3]{(D_1D_2D_3)}}{r}\right] \text{ cm. per cm. length} \\ \text{or} \quad &\left[0{\cdot}0388/\log \frac{\sqrt[3]{(D_1D_2D_3)}}{r}\right] \mu\text{F. per mile} \end{aligned}\right\} \quad . \quad (30)$$

Here again we call C_0 the capacitance to neutral.

The effect of the earth can be found by the method of images described above, but the effect is generally negligible. The effect of an earth wire is found by assuming an induced charge Q_0 on it. Then $Q_1 + Q_2 + Q_3 + Q_0 = 0$. Furthermore, by equating the potential of the earth wire to zero we obtain another equation in addition to those obtained from the potentials of the lines. These equations are sufficient to solve the problem. The charging currents are increased by about 5 or 7 per cent in practical cases by the presence of the earth wire.

Electric Stress. The electric force or stress in the single-phase system is

$$2Q/x + 2Q/(D - x).$$

It is seen that this has a minimum value at $x = \frac{1}{2}D$ and a maximum value for the smallest possible value of x, viz. r. The maximum stress is thus

$$g = 2Q/r + 2Q/(D - r) \simeq 2Q/r.$$

But $V = 4Q$ logh (D/r), so that

$$g = \frac{V}{2r \text{ logh } (D/r)}.$$

If E_0 is the voltage to neutral, $E_0 = \frac{1}{2}V$, so that

$$g = \frac{E_0}{r \text{ logh } (D/r)} \quad . \quad . \quad . \quad . \quad (31)$$

This formula holds for the symmetrically spaced three-phase system also.

In an actual case the gradient at the surface of a conductor is a maximum at the inner point, and varies round the circumference. Thus if $D/r = 20$ and we represent the true maximum gradient by 1, the surface gradient at the outer point is 0·81 and the approximate formula of equation (31) gives a value of 0·91. Thus the approximate formula gives a correct value for the average surface stress. For a value of $D/r = 100$ or more, the various gradients differ by less than 2 per cent.

Stranding increases the maximum stress, but the effect is best included in the formula for the corona voltage.

Corona Discharge. If the voltage is high the surface stress may reach a value at which the air breaks down, and becomes a conductor. The conducting layer of air forms part of the conductor, so that r increases and the maximum stress decreases. If the spacing is small enough, the *corona* may bridge the conductors and cause flash-over. Generally the spacing is large enough for the corona to cease spreading long before it bridges the conductors; values of r and g are reached such that the stress is insufficient to ionize any more air.

The phenomenon of corona is accompanied by a faint glow and a hissing noise. There is also an energy loss.

Disruptive Critical Voltage. The breakdown strength of air at 76 cm. pressure and 25° C. is 30 kV. per cm. or 21·1 kV. (r.m.s.) per cm. This value is called g_0. At a barometric pressure of b cm. of mercury and τ° C., the breakdown strength is δg_0, where

$$\delta = \frac{3{\cdot}92b}{273 + \tau} \quad . \quad . \quad . \quad . \quad (32)$$

If E_0 is the voltage to neutral that causes this breakdown equation (31) gives

$$E_0 = \delta g_0 r \text{ logh } (\mathrm{D}/r).$$

In practice it is necessary to allow for the condition of the surface of the wire, so that

$$\left.\begin{aligned} E_0 &= m\delta g_0 r \text{ logh } (\mathrm{D}/r) \\ &= 21{\cdot}1 m\delta r \text{ logh } (\mathrm{D}/r) \text{ kV. (r.m.s.) to neutral,} \end{aligned}\right\} \quad (33)$$

where m = 1·0 for clean smooth wires;

$$\left.\begin{aligned} &0{\cdot}98 \text{ to } 0{\cdot}93 \text{ for roughened or weathered wires;} \\ &0{\cdot}87 \text{ to } 0{\cdot}80 \text{ for stranded wires.} \end{aligned}\right\} \quad (34)$$

The value of E_0 given by equations (33) and (34) is the *disruptive critical voltage*. Bad atmospheric conditions, such as fog, rain, or sleet, may reduce E_0 to 0·8 of the value given above.

Visual Critical Voltage. When the voltage of the line is the disruptive critical value, there is no visible corona. This is explained as being due to the fact that the charged ions in the air must be able to receive a finite energy before they can cause further ionization by collision, which is necessary for the corona discharge. Peek states that the disruptive critical voltage must be so exceeded that the stress is greater than the breakdown value up to a distance of $0{\cdot}3\sqrt{(\delta r)}$ cm. from the conductor. Thus visual corona will occur when the breakdown value is attained at the distance $r + 0{\cdot}3\sqrt{(\delta r)}$ from the axis, instead of at the distance r. This requires that the voltage to neutral be $(1 + 0{\cdot}3/\sqrt{(\delta r)})$ times the disruptive critical voltage. Thus the visual critical voltage is

$$E_v = 21{\cdot}1 m_v \delta r \left(1 + \frac{0{\cdot}3}{\sqrt{(\delta r)}}\right) \text{logh } \frac{\mathrm{D}}{r} \text{ kV. (r.m.s.) to neutral} \quad (35)$$

m_v is a roughness factor, which is unity for smooth conductors. When the wire is stranded or rough, m_v is less than unity. Peek states that m_v may be taken as 0·72 for stranded wires, when corona is likely to occur in local places only; for decided corona along the length of the conductor he takes the value 0·82.

Corona Power Loss. If the visual corona voltage is exceeded the power loss due to corona is given by

$$P = (390/\delta)\,(f + 25)\,[\sqrt{(r/\mathrm{D})}]\,(E - E_0)^2\,10^{-5}\ \text{kW. per mile per phase,} \qquad (36)$$

where f = the frequency in cycles per sec.,
and E = voltage to neutral.

High voltage lines are seldom designed to work with a decided visual corona: what corona there is is local, and the formula of equation (36) will not hold accurately.

Example. Find the disruptive critical and visual corona voltages of a Grid line operating at 132 kV., the conductors being 37/0·11 in. diameter steel-cored aluminium at a minimum spacing of 12 ft. 6 in. Temperature 60° F., barometer 29·0 in.

Here D = 150 in. and $r = 7 \times 0{\cdot}11$ in. $\div\ 2 = 0{\cdot}98$ cm.

$$\tau = 60°\,\text{F.} = 15{\cdot}6°\,\text{C.};\ b = 73{\cdot}7\ \text{cm.}$$

$$\delta = \frac{3{\cdot}92 \times 73{\cdot}7}{273 + 15{\cdot}6} = 1{\cdot}00.$$

$$E_0 = 21{\cdot}1m \times 0{\cdot}98\ \text{logh}\,\frac{150}{0{\cdot}385}$$

$$= m \times 21{\cdot}1 \times 0{\cdot}98 \times 2{\cdot}303 \times 2{\cdot}60$$

$$= 126m\ \text{kV. (r.m.s.) to neutral.}$$

As the conductors are stranded, we take $m = 0{\cdot}83$ for fine weather and $m = 0{\cdot}83 \times 0{\cdot}8$ for rough weather. We then get

$$E_0 = \begin{cases} \underline{\underline{104\ \text{kV.}}}\ \text{in fine weather} \\ \underline{\underline{83\ \text{kV.}}}\ \text{in rough weather.} \end{cases}$$

$$1 + \frac{0{\cdot}3}{\sqrt{(\delta r)}} = 1 + \frac{0{\cdot}3}{\sqrt{0{\cdot}98}} = 1{\cdot}30.$$

$$E_v = 1{\cdot}30 \times 126m_v$$

$$= 164m_v.$$

For local visual corona $m_v = 0{\cdot}72$
so that $E_v = \underline{\underline{118\ \text{kV.}}}$,

whilst for decided corona $m_v = 0{\cdot}82$
and $E_v = \underline{\underline{134\ \text{kV.}}}$

The actual voltage to neutral is $132/\sqrt{3} = 76$ kV. Thus there is no corona under normal circumstances.

Avoidance of Corona. The critical voltage can be raised either by increasing the spacing or the diameter of the conductors. The spacing cannot be increased greatly or the cost of the supports will be very high. The diameter of the conductors can be increased by using hollow conductors with a hemp core. Steel-cored aluminium conductors have a large diameter for a given conductivity and weight, and are thus good from the point of view of corona.

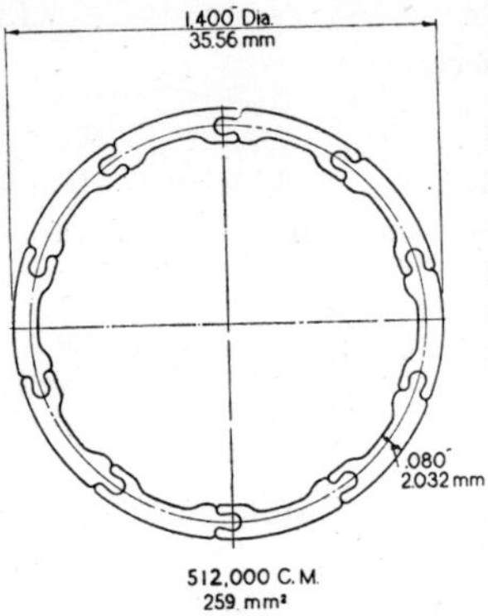

FIG. 67. HOLLOW CONDUCTOR FOR HIGH VOLTAGE (*Electrician*)

For very high voltages, such as the 275 kV. of the Boulder Dam transmission scheme, it has been found economical to use the hollow conductor shown in section in Fig. 67. The conductor consists of tongued and grooved rectangular strips of copper, spiralled along the length of the conductor.

A line is usually designed to work at a voltage just below the disruptive critical voltage for fair weather (taking $\delta = 1$). It is economical to have a small corona loss in bad weather, rather than have larger conductors to avoid corona entirely. Moreover, corona acts as a safety valve for surges.

Current Effects of Corona. Corona forms when the voltage of a conductor passes the disruptive critical voltage, and disappears when the voltage descends through the same value. This occurs on each conductor every half-cycle and contributes a triple harmonic to the charging current, since the effective capacitance of the conductor increases when the corona is present. The triple harmonic currents pass through the neutral to earth in an earthed system; in a non-earthed system the neutral has a voltage to earth of triple frequency.

Insulators. Pin-type, suspension and strain insulators are shown in Figs. 48 and 49. Single-piece, pin-type insulators are used up to 26 kV., and multi-part types up to 80 kV. Above 50 kV. it is more economical to use suspension-type insulators, as they are cheaper and lighter. Moreover, the insulation of the line can be co-ordinated by using insulators with more units at some points of the line where breakdown or flash-over is to be avoided, and less units at other points where flash-over can do no costly damage. If it is desired to raise the working voltage at some later date, suspension insulators are very useful, as extra units can be added at small cost. The insulators on the Boulder Dam line are cap- and pin-type suspension insulators and have 24 units. The suspension type is preferable from the mechanical point of view, as it allows the conductor to take up a position in which the insulator experiences a pure tension only. Furthermore the earthed cross-arm is above the conductor and protects the line to some extent from lightning.

If a flash-over creates an arc from the conductor to the earthed cross-arm, the porcelain is liable to be shattered. Fig. 49 shows arcing horns and ring, which draw the arc away from the insulator, lengthen the path and help to extinguish the arc. The device lowers the flash-over voltage of the insulator, but the protective value outweighs this slight disadvantage.

Suspension-type Insulators. The voltage applied between the cross-arm and conductor is not shared equally between the units of a suspension-type insulator because of the earth capacitances of the units. In consequence the flash-over voltage of a string of insulators is less than the flash-over voltage of a unit multiplied by the number of units. The *string efficiency* is defined as

$$(E/ne) \times 100 \text{ per cent} \quad . \quad (37)$$

where E is the flash-over voltage of the string, e that of a unit, and n is the number of units.

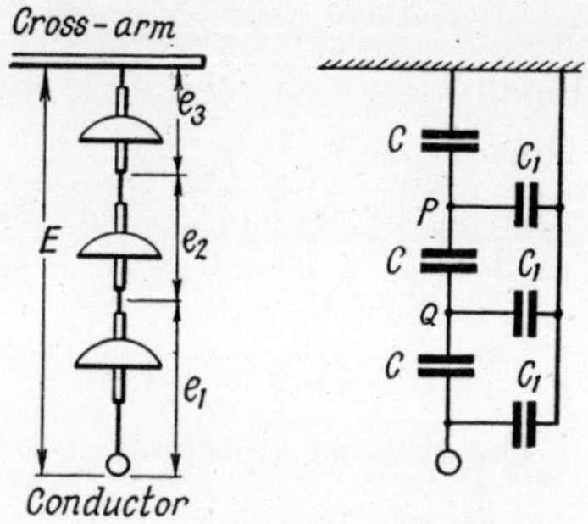

FIG. 68. STRING EFFICIENCY OF SUSPENSION INSULATOR

When the insulator is wet the direct capacitance between units is increased, whilst the earth capacitances are not increased (except for the unit nearest the cross-arm). The result is a more uniform distribution of potential than occurs when the unit is dry, and the string efficiency is higher.

The string efficiency can be increased in several ways. One method is to design the units such that the direct capacitance between them is much greater than the capacitance to earth: a ratio of 6 to 10 can be achieved. Another method is to increase the capacitances between the lower units: this is very inconvenient and not used, as it is desirable that units be interchangeable. A third method is to use a grading ring placed near the lower units and connected to the line, as shown in Figs. 49 and 51. This ring screens the lower units, thus decreasing their earth capacitances; and it introduces a number of capacitances between the line and the insulator caps, these capacitances being greater for the lower units and thus reducing the voltages across them.

EXAMPLE. Sketch the construction of a suspension-type insulator. Show how the voltage distribution over a string of insulators may be improved, and explain what is meant by "string efficiency."

A string of suspension insulators consists of three units. The capacitance between each link pin and earth is one-sixth of the self-capacitance of the unit. If the maximum peak voltage per unit is not to exceed 35 kV., determine the greatest working voltage and the string efficiency.

(*B.Sc., Lond. Univ.*, 1938.)

Let the capacitances and potentials be as shown in Fig. 68. The simplest method is to calculate the induced charges at the junctions P and Q, at which, since they are insulated, no net charge can gather. At P there are in effect three condenser plates joined together, their charges being Ce_3, C_1e_3, $-Ce_2$. We have therefore

$$Ce_3 + C_1e_3 - Ce_2 = 0.$$

If we put $C_1 = kC$, this reduces to

$$e_3(1+k) - e_2 = 0.$$

The equation for junction Q is

$$Ce_2 + C_1(e_2 + e_3) - Ce_1 = 0$$

or

$$e_3k + e_2(1+k) - e_1 = 0.$$

Finally

$$e_3 + e_2 + e_1 = E$$

Thus

$$\frac{e_2}{1+k} = \frac{e_3}{1} = \frac{e_1}{k + (1+k)^2} = \frac{E}{(1+k) + 1 + k + (1+k)^2}$$

$$= \frac{E}{(1+k)(3+k)}$$

The greatest voltage is clearly e_1, which is equal to

$$e_1 = E\frac{1 + 3k + k^2}{(1+k)(3+k)} = 35 \text{ kV.},$$

so that

$$E = \frac{(1+k)(3+k)}{1+3k+k^2} \times 35 \text{ kV. peak}$$

$$= \frac{(1{\cdot}167)(3{\cdot}167) \times 35}{1{\cdot}528} = 84{\cdot}5 \text{ kV. peak.}$$

The string efficiency is

$$\frac{(1+k)(3+k)}{3(1+3k+k^2)} = 80{\cdot}5 \text{ per cent.}$$

As a check we note that $(84{\cdot}5) \div (3 \times 35) = 0{\cdot}805$.

General Case. When there are many insulator units the method is laborious by simple methods, but a reasonable form of solution has been derived by the method of difference equations. Let the voltages across the units from the line upwards be $e_1, e_2, \ldots e_m, \ldots e_n$, there being n units. Then

$$e_m = E\frac{2\sinh[\tfrac{1}{2}\sqrt{k}]\cosh[(n - m + \tfrac{1}{2})\sqrt{k}]}{\sinh(n\sqrt{k})}$$

The largest voltage is

$$e_1 = E\frac{2\sinh[\tfrac{1}{2}\sqrt{k}]\cosh[n - \tfrac{1}{2})\sqrt{k}]}{\sinh(n\sqrt{k})},$$

so that the string efficiency is

$$\frac{E}{ne_1} = \frac{\sinh (n\sqrt{k})}{2n \sinh (\frac{1}{2}\sqrt{k}) \cosh [(n - \frac{1}{2})\sqrt{k}]}.$$

Thus for the previous example, $n = 3$, $\sqrt{k} = 0{\cdot}408$, and the string efficiency is $(1{\cdot}555)/(6 \times 0{\cdot}205 \times 1{\cdot}568) = 80{\cdot}5$ per cent.

EXAMPLES IV

1. Four suspension-type insulators are connected in series on a 33-kV., 3-phase overhead line. If the capacitance to earth is 20% of the capacitance of each unit, determine the voltage distribution across the four units under clean dry conditions.

Explain how the voltage distribution might change under damp deposit conditions and describe one method of obtaining a more uniform distribution. (*Lond. Univ.*, 1949.)

2. Explain the factors which determine the formation of corona on overhead lines. A single-phase overhead line has two conductors 0·5 in. diameter with 3 ft. spacing. Deduce expressions for the potential and potential gradient at any point on a line joining the centres of the two conductors if the r.m.s. voltage between them is 33 kV. Assuming the breakdown strength of air to be 21 kV. (r.m.s.) per cm., calculate the voltage at which corona occurs. (*Lond. Univ.*, 1953.)

3. A three-phase transmission line 30 miles long has its conductors of 0·5 cm. dia. spaced at the corners of an equilateral triangle of 120 cm. side at an average height from the ground of 1 000 cm.

The line is fed from a star-connected transformer with neutral point earthed at 44 000 V. between lines, 50 cyc. Find the charging current. (*Lond. Univ.*, 1932.)

4. A single overhead conductor 0·77 in. diameter is mounted 22 ft. above the ground. Deduce expressions for the inductance and capacitance and calculate the values of these per mile.

State and discuss any assumptions made. (*Lond. Univ.*, 1947.)

5. A three-phase line has spacings 10, 12 and 14 in. The line is transposed, and the conductor diameter is 0·2 in. Find the capacitance and inductance to neutral.

6. Explain the factors which determine the formation of corona on overhead lines. A single-phase overhead line has two conductors 0·5 in. diameter with a spacing of 3 ft. between them. Deduce expressions for the potential and potential gradient at any point between the conductors if the r.m.s. voltage between them is 33 kV. Assuming the breakdown strength of air at 76 cm. pressure and 25° C. is 21·1 kV. (r.m.s.) per cm., calculate the voltage on the above line at which corona occurs. (*Lond. Univ.*, 1948.)

7. The following figures give the percentage voltage distribution across the individual units of 3 identical strings of 4 insulators erected on the same tower (*a*) when clean and dry and (*b*) under deposit conditions at a certain instant with a humidity of 80%. In each case unit number 1 is connected to the line.

Unit number	1	2	3	4
Each string, dry . . .	36	25	17	22
String A wet	37	12	20	31
,, B ,, ...	17	33	15	35
,, C ,, ...	24	22	30	24

Explain the reasons for the figures obtained and describe a method of improving the voltage distribution. (*Lond. Univ.*, 1953.)

CHAPTER V

UNDERGROUND CABLES

Introduction. In towns and densely populated areas overhead lines are clearly impossible. In the very early days of distribution the principles of overhead lines were employed in underground work. One very successful system was devised by Crompton, who stretched bare conductors between glass insulators contained in underground ducts or culverts; one such system was in use in 1926 after thirty years' service. The method was successful for low voltage systems, except that explosions sometimes occurred due to pressure of gas produced by the metallic sodium which was formed by electrolysis, due to slow leakage at the insulators. The method is unsuitable for high voltages because of the flash-over of the insulators. It was found advisable to insulate the conductor before it was laid, and the combination of conductor and its insulation is called a *cable*.

Vulcanized rubber-insulated cables were then used, and because of their failure vulcanized bitumen cables. The latter are still used for low-voltage distributors and in mines.

One of the earliest attempts to make a high voltage, paper-insulated cable was due to Ferranti, who wrapped oil-impregnated paper round copper tubes and pushed the insulated conductor into lengths of wrought-iron pipes which he filled with compound.

Types of Cable. There are many types of cable used at the present time; the type for a particular service is determined by the mechanical properties required and the voltage of transmission, mainly the latter.

LOW-TENSION CABLES (BELOW 1 000 VOLT). The insulating material may be impregnated paper, varnished cambric, vulcanized rubber, or vulcanized bitumen.

Paper is the most important insulating material, and is made of wood pulp, manila fibre or rag. The impregnating compound is thin or thick oil, with or without resin. The resin is added to increase the viscosity at working temperatures so that drainage should not be excessive, whilst it does not increase the viscosity greatly at the temperature of impregnation. The thickness of dielectric at 660 volts is between 0·08 and 0·11 in.

Varnished cambric is coated with petroleum jelly to provide lubrication between layers so that bending of the cable is possible without damage. An advantage is that there is no drainage from the ends of the cable.

In *vulcanized bitumen* cables the refined bitumen is melted and sulphur and vegetable oils are added. On cooling, the bitumen hardens and acts as the waterproof envelope of the cable. No elaborate sealing is required. A major disadvantage is that the bitumen becomes soft under the action of alkaline solutions, and the conductor becomes decentralized. Vulcanized bitumen cables were once favoured for shaft and roadway cables because they are easily jointed, do not require elaborate sealing, and are light compared with lead-covered, paper-insulated cables. They are not now made because of the softening and decentralization, and are being replaced by the paper cables, which are often cheaper and as suitable, and have a greater current-carrying capacity.

FIG. 69. FIVE-CORE TRAILING CABLE
(*Henley's Telegraph Works Co.*)

Lead-sheathed, rubber-insulated cables have a vast field of use in house wiring, large buildings, and ships.

Cables in mines are subjected to specially rough handling and they are constructed accordingly. Trailing cables must be flexible and must be capable of being dragged along without being damaged. Fig. 69 shows a 5-core trailing cable, with a tough rubber central cradle, insulated pilot, tough rubber sheath, and braided wire screen. One of the conductors acts as the earth conductor, and all the five cores are rubber-insulated flexibles. The central cradle prevents a short circuit from occurring when a fall of coal or stone crushes the cable. The braided wire screen is of tinned copper and is embedded in the tough rubber sheathing; the screen is earthed but is not used as the earth conductor. The pilot wire serves for the remote control of coal cutter and other circuits.

Shaft cables should be drained and not contain free compound, otherwise the compound will settle and force its way out at the lower termination. The installation of a shaft cable is best performed in the following way, in order that the cable may not suffer a large tension due to its weight. The cleats are first fixed. Then the cable is run from a winch over pulleys into the shaft, where a short length is allowed to swing free. It is fastened to a winch rope with a spun-yarn lashing and then lowered a short distance, say 30 ft. It is lashed again, and lowered a further distance. This is repeated until the cable reaches the bottom of the shaft. The cable is then fastened to the cleats starting from the bottom.

Fig. 70 shows various types of paper-insulated power cables. In the concentric double-wire armoured cable the conductors are a solid (stranded) conductor and a concentric layer of strands. The armouring consists of two insulated and concentric layers of steel wire.

HIGH-TENSION CABLES (UP TO AND INCLUDING 11 kV.). These are usually paper-insulated, although some cambric cables are used

FIG. 70. PAPER-INSULATED POWER CABLES

for running up poles to overhead lines. The minimum thickness of paper dielectric for various voltages is given in the following table for 3-core cables of 0·4 in.^{2} cross-section.

Voltage (volts) . .	3 300	6 600	11 000
Thickness (in.) . .	0·10 — 0·12	0·13 — 0·16	0·16 — 0·22

The smaller thicknesses are for centre point earthed, and are between conductors only. The larger thicknesses are for centre point not earthed, and are the minimum values between conductors and between a conductor and the sheath.

In the 3-core belted-type cable, the conductors are each lapped with paper and the cores are wormed together, the interstices being filled with packing to form the cable to a circular section; a further insulating belt of impregnated paper of the same nature surrounds the three cores. A lead sheath encloses the whole to prevent the ingress of moisture. If the cable is likely to experience mechanical forces, it has steel armouring which has an outer serving of hessian or jute dressed with preservative compound.

In the case of the 11 kV. 3-core belted-type cable, the separate walls and the belt are each 0·15 in. thick, so that there is a dielectric thickness of 0·3 in. between conductors and between any conductor and the sheath.

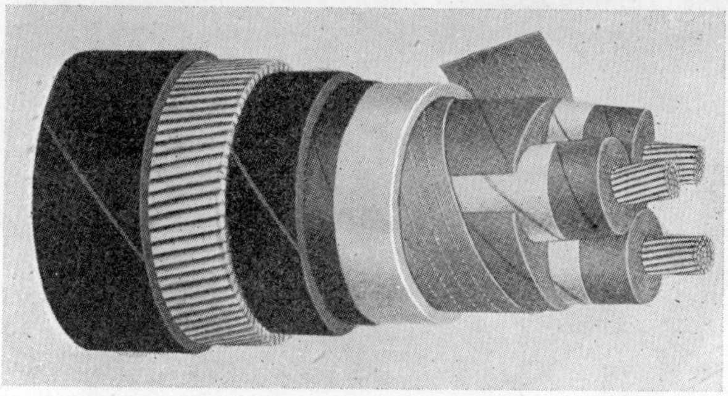

FIG. 71. "H"-TYPE CABLE
(Henley's Telegraph Works Co.)

SUPER TENSION CABLE (22 AND 33 kV.). The 22 kV. cable is of the 3-core belted type, the insulation thickness being 0·45 in. between conductors. A single-core cable has an insulation thickness of 0·24–0·28 in.

At first the same construction was used for 33 kV. with a dielectric thickness of 0·6 in. Trouble was, however, experienced, and the cables broke down. It appeared that the tangential stresses caused deterioration. This difficulty was overcome in the "H"-type cable, shown in Fig. 71, and which is due to M. Höchstädter; in this cable each core is surrounded by a metallized and perforated paper which is kept at earth potential. There is thus no stress in the wormings; in fact the electrical condition is that of three single-core cables. In some cables, "S.L." type, each core has its own lead sheath and a fourth lead sheath encloses the three cores.

In the "H.S.L." type there is a perforated screen, and a separate lead sheath for each core (see Fig. 72). Sometimes the metallized paper is replaced by a copper tape.

EXTRA HIGH VOLTAGE CABLES. For voltages above 66 kV., 3-core cables are too large and single-core cables are preferred. The disadvantages of single-core cables is that steel armouring causes sheath losses equal to or greater than the copper losses; they are generally not armoured. Fig. 73 shows a 66 kV. "H"-type single-core cable, 0·25 in.^{2} section. The conductor is 37/·093 plain copper of diameter 0·651 in. There are 129 layers of paper of 5 mils

FIG. 72. "H.S.L."-TYPE CABLE
(*Siemens Bros.*)

thickness giving a dielectric thickness of 0·65 in. Then there is one layer of metallized paper of 7·5 mils, a 40 mil clearance, and then the sheath, which is of antimonial lead and 80 mils thick; the outside sheath diameter is 2·30 in. Then there are two layers of waterproof paper and two of hessian; the overall diameter is 2·50 in. The treatment is as follows: there is steam or electrical drying for 120 hours at 240° F. at a pressure of 1 mm. of mercury or less, 96 hours impregnation at 280° F., and 15 days cooling to 65° or 70° F. to prevent the formation of cavities.

It has been found that the effect of variation of load and the temperature of the cable is to cause the circular lead sheath to expand when warm and not to recover when cool. Successive cycles result in an increased distension of the sheath, the formation of voids in the dielectric and ultimate breakdown. In an attempt to avoid this phenomenon, cables have been made with an oval conductor and an approximately circular sheath, the core being a loose fit in the sheath. The expansion of the oil can be taken up by an elastic deformation of the core, which contracts along the major axis and expands along the minor axis of the ellipse.

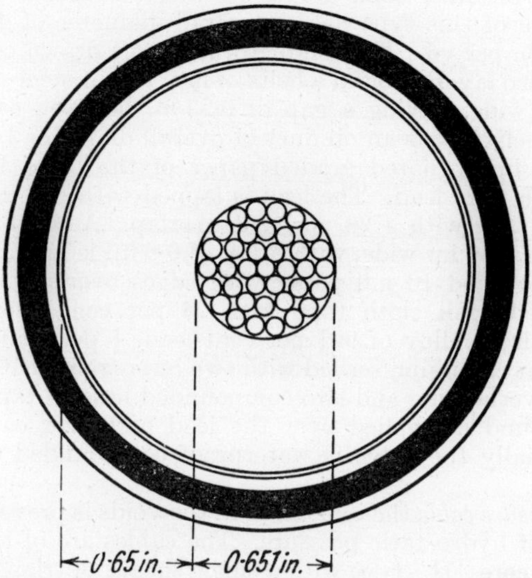

FIG. 73. 66 kV. "H"-TYPE SINGLE-CORE CABLE

FIG. 74. 220 kV. SINGLE-CORE, OIL-FILLED CABLE
(Pirelli General)

The solid-type cable described above is successful for voltages up to and including 66 kV., but becomes unreliable for higher voltages. If the ionization of voids could be prevented, the cable could be used at higher voltages. There are two ways in which ionization can be obviated. In one way the voids are prevented from forming by using a very thin oil under pressure. In this type of cable, which is termed *oil-filled cable*, the conductor is hollow and is fed with oil from reservoirs placed at intervals along the line. Fig. 74 shows a single-core oil-filled cable for a working voltage of 220 kV. A 0·30 in.^{2} cable of this type has an overall diameter of 4·05 in. and weighs 60·2 lb. per yd. The conductor is of soft-drawn copper wires stranded in two layers around a helix of flat, bright steel strip 40 mils thick, $\frac{5}{16}$ in. wide, having a gap of 0·23 in. between each turn of spiral. The helix forms an oil duct of overall diameter 1·18 in. The insulation is impregnated graded paper of thickness 1 in. and is sheathed with pure lead. The lead is lapped with one impregnated 10 mil cloth tape with a 28 per cent overlap. A plain brass tape, 16 mils thick, 0·39 in. wide, with a lay of 0·6 in. is applied, and also one of impregnated 10 mil paper with edges overlapping, and one impregnated 10 mil cloth tape with 28 per cent overlap. It is sheathed with an alloy of 99$\frac{1}{4}$ per cent lead, $\frac{1}{2}$ per cent antimony, and $\frac{1}{4}$ per cent cadmium, served with two impregnated 10 mil papers with edges overlapping and two compounded hessian tapes. Waterproof compound is applied over the lead and over each layer of tape, and finally the cable is waterproof compounded overall and whitewashed.

In the *pressure cable* the ionization in the voids is prevented by the application of hydrostatic pressure. The cables are of the ordinary solid type, 3-core "H" type with a single sheath of triangular shape. The cable is inserted into a welded steel pipe which is gas-tight and filled with nitrogen at 180 to 200 lb. per in.^{2} The effect of pressure is to keep the voids small by compression and also by making the lead contract when the oil cools; furthermore the pressure in the void has the effect of diminishing the ionization considerably.

Recently there has appeared the *gas cable*, which has dry or oil-impregnated paper for insulation and has nitrogen under a pressure of 200 lb. per in.^{2} inside the lead sheath to stop ionization. Here there is no question of void formation, and the cables work well at the highest voltages.

Manufacture of Paper-insulated Cables. The operations in their natural sequence are the following: stranding the conductors, applying the dielectric, laying up the cores, drying and impregnating, lead sheathing, armouring, sealing and battening.

The conductors consist of untinned and very pure stranded copper wires. The wires are pre-spiralled and formed into the correct conductor by a stranding machine. In the slow speed, cage type of stranding machine the bobbins of wire are mounted in carriages disposed around the inner circumference of the cage. The wires from the bobbins pass through a revolving die plate just before arriving at the point where the revolution of the cage causes them to be stranded. Successive cages revolve in opposite directions. To obtain a definite lay the speed of drawing off the strand must bear

FIG. 75. HIGH-SPEED, TUBULAR-TYPE STRANDING MACHINE
(*Larmuth*)

a certain relation to the speed of revolution of the cages. In the high-speed or tubular type of stranding machine the bobbins are mounted in tandem along the centre line of a steel tube of comparatively small diameter. Fig. 75 shows a high-speed machine for seven

FIG. 76. THREE-ARMED PAPER-LAPPING HEAD
(Johnson & Phillips)

strands. The seven bobbins are visible in the seven cut-away portions of the tube.

Fig. 76 shows a three-armed paper-lapping head applying paper tapes to a conductor. Laying up the cores is similar to stranding, but on a large scale.

Drying and impregnation takes place in four stages, preliminary heating, drying under vacuum, impregnating with compound, and cooling. Heating may be carried out in the vacuum chamber, or in a preheating chamber so that the vacuum chamber can be used with preheated cables. Extreme care in drying under vacuum is essential for high voltage cables. The cable is then immersed in hot oil for 24 hours or more and allowed to cool whilst still under oil. The cable then passes through the lead press, which extrudes a lead sheath on to the core (Fig. 77).

Steel armouring cannot be applied directly to the lead sheath for it would bite into it, particularly when the cable is bent. A bedding of hessian tape or jute is applied to the lead, and then the armouring of steel wire or tape is applied. The lead and the hessian

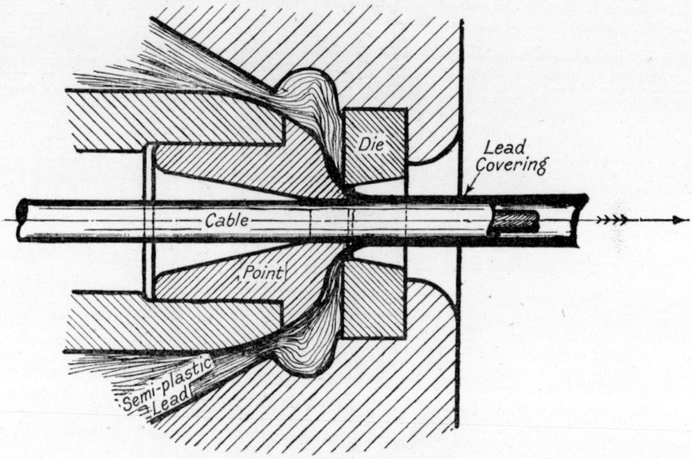

FIG. 77. LEAD PRESS

are usually compounded. The armouring is compounded, covered with hessian tapes, compounded again, and then whitewashed.

Sealing may be effected by soldering a lead cap over the end of the sheath. If it is desired to test the cable before laying, the conductor must be insulated from the lead sheath at one end. A long lead cap filled with compound must be soldered on to the sheath.

The cable is wound on to a large wooden or steel drum, and wooden battens or a steel cover enclose the cable round the circumference of the drum.

Laying. There are three main methods of laying cables: *direct laying*, *the draw-in* system, and the *solid* system.

In direct laying a trench is dug, in which the cable is laid and covered with soil. The cable may be protected by planks, bricks, tiles, or concrete slabs. It used to be the practice to armour such cables, but nowadays they are often laid with a bare sheath or with a serving of bituminized paper and hessian over the sheath. The former

FIG. 78. MACHINE LAYING OF CABLES
(*Weserhutte A.-G.*)

method is the only safe one in places where subsidence of the soil is likely to occur; then the cable should have steel wire armouring, so as to take a considerable tension. If the ground contains harmful chemicals, the serving must be adequate to protect the cable from corrosion and electrolysis. Bituminized paper is effective. It is clear that direct laying is cheap, but an extension of load is possible only by a completely new excavation which costs as much as the original work. In most cases, digging and trenching are done

FIG. 79. CABLES IN TRENCH PROTECTED BY CONCRETE COVERS

manually, but in places free from obstruction machine methods can be used. Fig. 78 shows a machine which dredges a trench 18 in. wide by 5 ft. deep, conveys the soil back along the top, and lays the cable. The trench is then filled and levelled. Fig. 79 shows a cable in a trench with concrete covers for protection.

In the draw-in system a line of conduits, ducts, or tubes is laid in a trench. The conduits or ducts are of glazed stoneware, cement, or concrete (Fig. 80). The tubes are of stoneware, fibre, steel, or wrought iron. The ducts being laid, the cables are pulled into position from manholes or brick-pits. It is unnecessary to armour the cable, but a serving of hessian tape of jute protects the cable when drawing in.

In the solid system the cable is laid in troughing in an open trench.

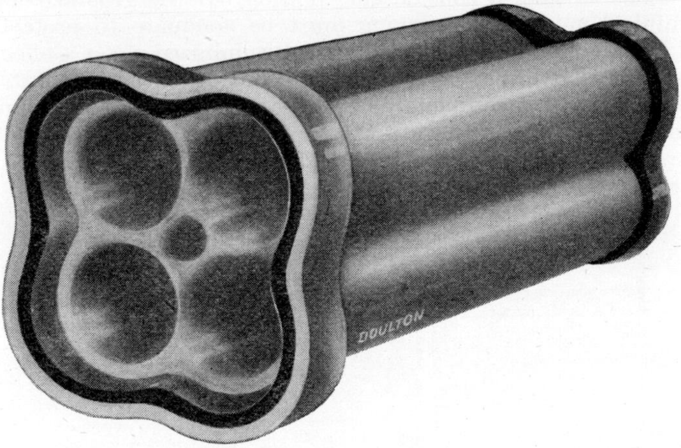

(*a*)

(*b*)

FIG. 80. STONEWARE CONDUITS

(*a*) Multiple duct. (*b*) Multiple butt-ended stoneware conduit.

(*Doulton*)

The troughing is of stoneware, cast iron, asphalt, or treated wood. After the cable is placed in position, the troughing is filled with a bituminous or asphaltic compound and covered over. The cable can be laid with a bare sheath, and is immune from electrolysis as the sheath is electrically insulated from earth. Fig. 81 shows a trough containing cables and covered with asphalt.

Jointing. The most common way of jointing the conductors is to insert the ends into a ferrule, which is a slotted metal tube, and solder the whole solid. With oval cables the ferrule is made in two halves which can be turned with respect to one another, in order that the cables need not be twisted to make their major axes parallel. Packing adaptors are also being used.

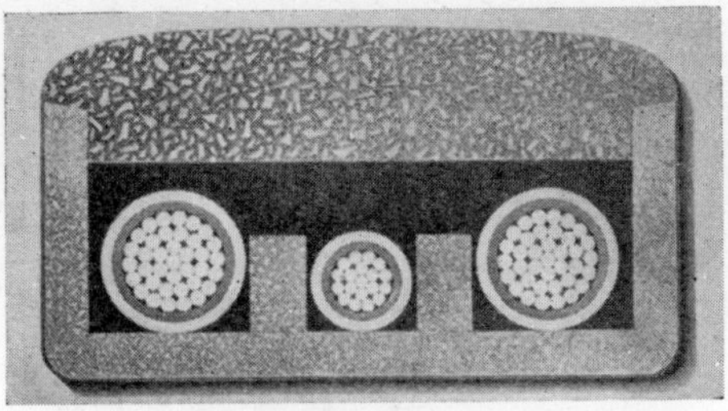

FIG. 81. ASPHALT TROUGHING CONTAINING CABLES
(*Howard*)

Fig. 82 shows a joint for a 66 kV., single-core, "H"-type cable with oval conductor. The following is a brief description of the procedure. The lead sheath is cut back $20\frac{5}{8}$ in. and the paper $2\frac{7}{8}$ in. The lead sleeve, lead flare and paper tube are passed back along the conductors. The ferrule is put on and the parts soldered together. The paper is pencilled back $1\frac{1}{4}$ in., and the metallized paper is cut back to within $1\frac{1}{2}$ in. of the lead. The paper tube is slipped into position and fixed by four narrow wedges, which are jammed in by layers of oiled silk tape. The paper tube is kept in a canted position by a kalanite spreader, which is made of two halves as shown in the cross-sectional diagram. This ensures that no air will be entrapped when the joint is filled with compound. Preshaped paper stress cones are lapped on, and oilproof (Kaleoilres) tape is wrapped over the end of the lead sheath to prevent oil from flowing from the joint into the cable. The lead flare is placed in position and $\frac{1}{8}$ in. diameter lead wire is wrapped on. The lead sleeve is put in position and plumbed at the ends and the middle. The bonding strands are fixed, and then the joint is filled with oil. It is seen that the upper part of the lead sleeve is shaped so that the sleeve cannot be filled completely with oil; the air dome is left to allow for the expansion of the oil. The black sectors shown in the figure represent steel reinforcing bands.

Fig. 83 shows a joint for a 66 kV. oil-filled cable, single core. It is essential that no moisture or dust shall enter the jointing tent.

Fig. 84 shows the recently designed styrene joint. The styrene is added as liquid and on being heated it polymerizes and sets into

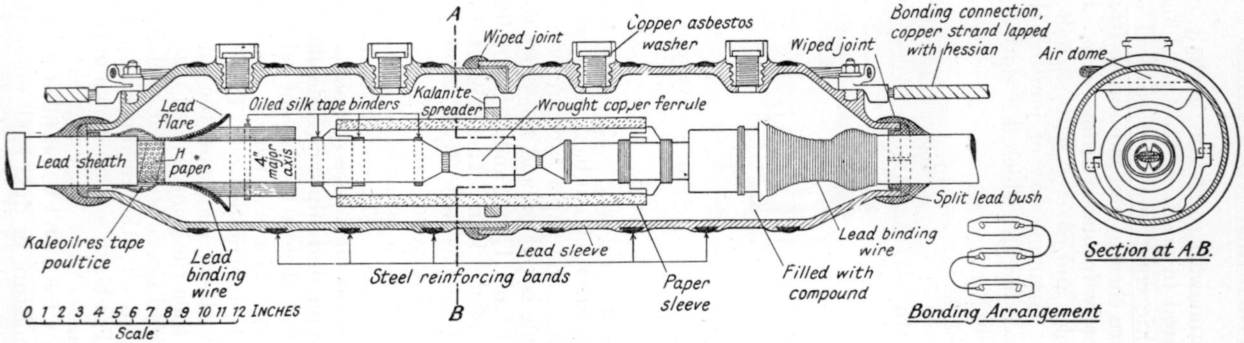

FIG. 82. 66 kV. CABLE JOINT
(Callender's Cable and Construction Co.)

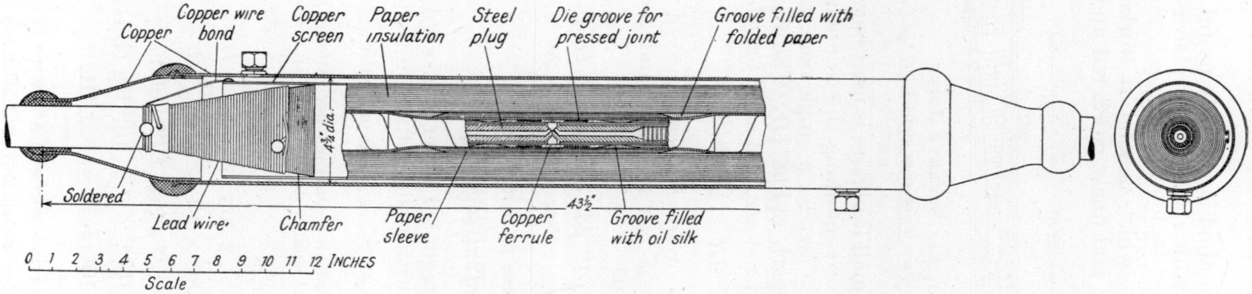

FIG. 83. 66 kV. OIL FILLED CABLE JOINT
(*Pirelli General*)

a very hard solid. The method is specially useful for 3-core cables, as the joint is mechanically rigid and displacement of the cores cannot occur.

A recent type of joint, which is solid at working temperatures, employs a mixture of oil and finely-powdered and cleaned sand in

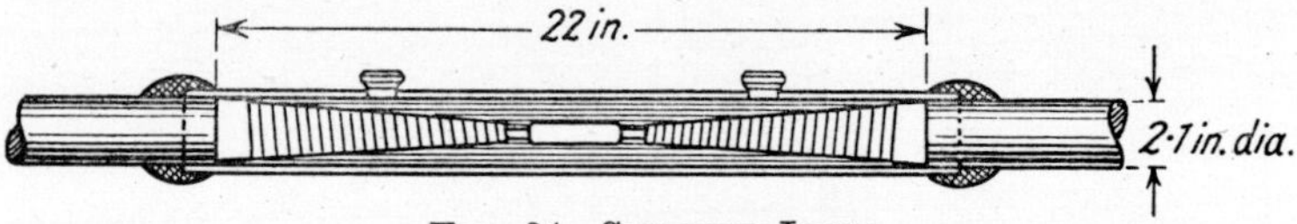

FIG. 84. STYRENE JOINT
(Standard Telephones and Cables)

place of the styrene. A small lead flare is used between the plumbed joint and the lead cylinder containing the compound.

Insulation Resistance. Consider a single-core cable of conductor radius r and internal sheath radius $\frac{1}{2}D$. The resistance of a thin shell, between radii x and $x + dx$, and axial length 1 cm. is

$$dR = \rho dx/2\pi x,$$

where ρ is the resistivity or specific resistance of the dielectric, for the area of cross-section is $2\pi x \times 1$ cm.^{2} and the thickness is dx. The total resistance is thus

$$R = \int_r^{\frac{1}{2}D} \frac{\rho dx}{2\pi x} = \frac{\rho}{2\pi} \operatorname{logh} \frac{D}{2r} \quad . \quad . \quad . \quad (39)$$

The insulation resistance of a length l is

$$R_l = (\rho/2\pi l) \operatorname{logh} (D/2r) \quad . \quad . \quad . \quad . \quad (39a)$$

The value of ρ for impregnated paper is about 5×10^{14} ohms, and decreases exponentially with temperature, so that

$$\rho_t = \rho_0 e^{-\alpha t} : \alpha \text{ is about } 0{\cdot}05.$$

EXAMPLE. Find the insulation resistance per mile of a cable of conductor diameter 0·4 in. and internal sheath diameter 0·7 in. $\rho = 6 \times 10^{14}$.

The resistance of 1 mile is

$$\frac{\rho \times 2{\cdot}303}{2\pi \times 5\,280 \times 12 \times 2{\cdot}54} \log \frac{D}{2r} \text{ ohms}$$
$$= 2{\cdot}28 \times 10^{-6} \rho \log (D/2r) \text{ ohms}$$
$$= 2{\cdot}28 \times 10^{-6} \times 6 \times 10^{14} \log (7/4)$$
$$= 3{\cdot}21 \times 10^{8} \text{ ohms} = \underline{\underline{321 \text{ M}\Omega}}.$$

If the positive and negative wires of a 2-wire system have leakages to earth, the values of the insulation resistances can be found by a voltmeter method described in the following example.

Example. Deduce a formula for the insulation resistance to earth of the positive and negative mains of a 2-wire system in terms of V, V_1, V_2 and r, where V is the voltage between the mains, and V_1 and V_2 are the respective readings on a voltmeter having a resistance r when connected between the positive main and earth and between the negative main and earth.

Find the insulation resistance of each main to earth when $V = 250$, $V_1 = 150$ and $V_2 = 30$ V., the voltmeter resistance r being 10 000 Ω.

(*Lond. Univ.*, 1932.)

Let the mains have earth resistances R_1 and R_2. It is clear that the resistance between the mains, R_3, does not affect the readings.

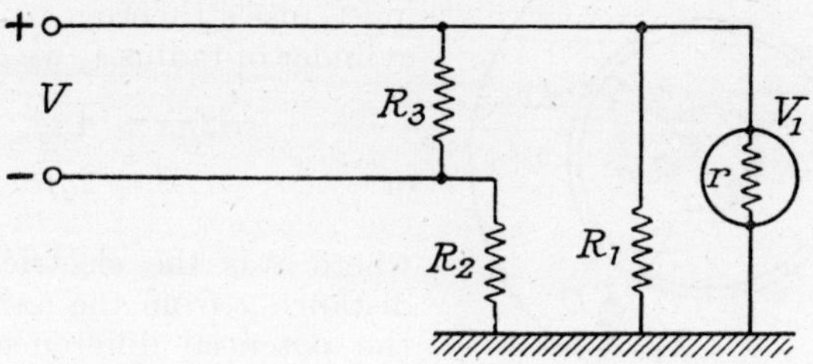

Fig. 85. Insulation Resistance of Mains

When the voltmeter is across the positive main and earth, the system is as shown in Fig. 85. The following currents then flow to and from the earth: (1) V_1/r downwards, (ii) V_1/R_1 downwards, and (iii) $(V - V_1)/R_2$ upwards.

Therefore

$$(V_1/r) + (V_1/R_1) = (V - V_1)/R_2,$$

giving

$$V_1\,(1/r + 1/R_1 + 1/R_2) = V/R_2.$$

Similarly in the other case

$$V_2\,(1/r + 1/R_1 + 1/R_2) = V/R_1.$$

Therefore, by division,

$$V_1/V_2 = R_1/R_2,$$

so that

$$\begin{aligned} V/V_2 &= R_1\,(1/r + 1/R_1 + 1/R_2) \\ &= R_1/r + 1 + R_1/R_2 \\ &= R_1/r + 1 + V_1/V_2, \end{aligned}$$

giving

$$R_1 = r\left(\frac{V}{V_2} - 1 - \frac{V_1}{V_2}\right) = r\frac{V - V_1 - V_2}{V_2}.$$

Similarly

$$R_2 = r\frac{V - V_1 - V_2}{V_1}.$$

In this case $R_1 = 10\,000\left(\frac{250 - 150 - 30}{30}\right) = \underline{\underline{23\,300\ \Omega}}.$

and $R_2 = 10\,000\left(\frac{250 - 150 - 30}{150}\right) = \underline{\underline{4\,660\ \Omega}}.$

Stress and Capacitance of Single-core Cable. Suppose that the cable of Fig. 86 has a dielectric of constant ε, no losses, and a charge q per cm. of axial length. Applying Gauss's theorem to a circular cylinder of radius x, we get

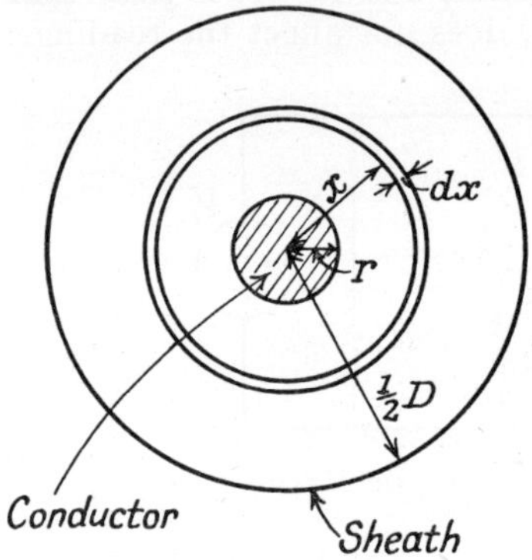

FIG. 86

$$\varepsilon S 2\pi x = 4\pi q,$$

or

$$S = 2q/\varepsilon x,$$

where S is the electric stress at distance x from the axis. If E is the potential difference between the conductor and sheath

$$E = \int_r^{\frac{1}{2}D} S dx = \frac{2q}{\varepsilon} \operatorname{logh} \frac{D}{2r},$$

so that the capacitance per cm. length is

$$C = \frac{q}{E} = \frac{\varepsilon}{2 \operatorname{logh}(D/2r)} \text{ electrostatic units (cm.)}$$

$$= \frac{\varepsilon}{2 \operatorname{logh}(D/d)} \text{ cm. per cm. length} \quad . \quad . \quad . \quad (40)$$

where d = the conductor diameter.

Remembering that

$$1 \text{ cm.} = 10/9\ \mu\mu\text{F}. \quad . \quad . \quad . \quad . \quad . \quad (41)$$

we find that

$$C = \frac{0{\cdot}0388\varepsilon}{(\log D/d)} \mu\text{F. per mile length} \quad . \quad . \quad (42)$$

Substituting for q in terms of E we see that the stress is

$$S = \frac{E}{x \operatorname{logh}(D/d)} \quad . \quad . \quad . \quad . \quad . \quad (43)$$

The stress is a maximum at the conductor where it is

$$S_{max} = \frac{E}{r \operatorname{logh}(D/d)} = \frac{2E}{d \operatorname{logh}(D/d)} \quad . \quad . \quad . \quad (44)$$

For a given voltage E and internal sheath diameter D, the stress S_{max} has a minimum value for variation of d when the differential coefficient of S_{max} with respect to d is zero. This occurs when

$$\text{logh}\,(D/d) = 1^{*} \text{ or } D = 2{\cdot}718d.$$

In low voltage cables the insulation is thin and d is greater than $D/2{\cdot}718$. In high voltage cables d may be less, and then it is advantageous to increase the diameter of the conductor to this value. There are two ways of doing this without using excessive copper; by making the conductor hollow or by building it up with a lead sheath. The latter method has the advantage of eliminating at the same time the increase of stress due to stranding which may be as high as 25 per cent. If d varies from $D/2$ to $D/4$, the maximum stress varies by only 6 per cent, so that no great care need be taken to fix the ratio D/d provided it is not too great.

The stress at the lead sheath is

$$\frac{2E}{D\,\text{logh}\,(D/d)},$$

so that the stress varies from a maximum at the conductor to a minimum (of d/D times the maximum) at the sheath. In some high voltage cables the dielectric is graded for strength by having a dense paper at the conductor and a less dense elsewhere. This does not appreciably alter the stress and voltage distribution unless the dielectric constants of the paper differ, but by using a paper of high electric strength near the conductor the total thickness of insulation can be reduced.

There are two main methods by which a more uniform distribution of stress may be achieved, by the introduction of intersheaths and by the use of layers of insulating material with different dielectric constants.

Effect of Intersheaths on Stress. Suppose that intersheaths of diameters d_1 and d_2 are inserted into the dielectric and maintained at potentials E_1 and E_2. The stress between any two metallic cylinders varies inversely as the distance from the axis; this is found by applying Gauss's theorem as on page 116. Thus between the conductor and the first intersheath the stress at a point distant x from the conductor is

$$S_1 = A_1/x,$$

* For
$$\frac{\partial S_{max}}{\partial d} = -\frac{2E}{[d\,\text{logh}\,(D/d)]^2}\cdot\frac{\partial[d\text{logh}(D/d)]}{\partial d}$$
$$= -\frac{2E}{[d\,\text{logh}\,(D/d)]^2}\left[\text{logh}\frac{D}{d} + \frac{d}{D/d}\left(-\frac{D}{d^2}\right)\right]$$
$$= -\frac{2E}{[d\,\text{logh}\,(D/d)]^2}\left[\text{logh}\frac{D}{d} - 1\right]$$

where A_1 is a constant which is found by integrating S_1 from the conductor to the intersheath as follows.

$$E - E_1 = \int_{\frac{1}{2}d}^{\frac{1}{2}d_1} S_1 dx = A_1 \operatorname{logh} \frac{d_1}{d},$$

so that $A_1 = (E - E_1)/\operatorname{logh} \frac{d_1}{d}$

and $$S_1 = \frac{E - E_1}{x \operatorname{logh} \frac{d_1}{d}}.$$

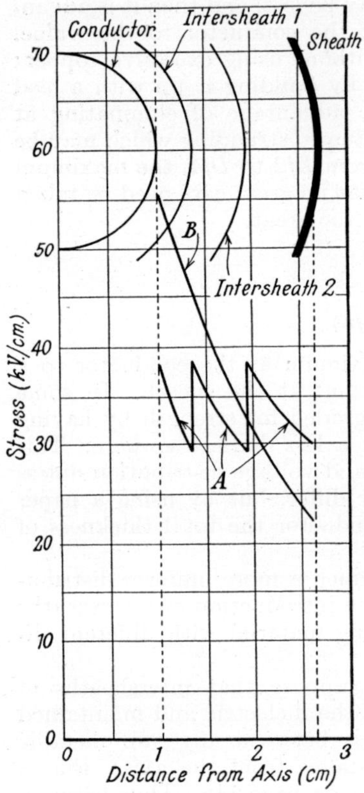

FIG. 87
EFFECT OF INTERSHEATHS ON STRESS

The maximum stress is

$$S_{1\,max} = \frac{E - E_1}{\frac{1}{2}d \operatorname{logh} \frac{d_1}{d}}.$$

Similarly the maximum stress between the first and second intersheaths is

$$S_{2\,max} = \frac{E_1 - E_2}{\frac{1}{2}d_1 \operatorname{logh} (d_2/d_1)},$$

whilst the maximum stress between the second intersheath and the sheath is

$$S_{3\,max} = \frac{E_2}{\frac{1}{2}d_2 \operatorname{logh} (D/d_2)}.$$

By choice of E_1 and E_2 the maximum stresses can be made equal, and the stress distribution is like that shown by curve A of Fig. 87, instead of curve B which represents the stress without intersheaths.

It is possible to choose d_1 and d_2 so that the stress varies between the same maximum and minimum in the three layers, by taking

$$d_1/d = d_2/d_1 = D/d_2 = \alpha,$$

for the minimum stresses are the maximum stresses divided by α. Equating the maximum stresses we get

$$E_2 = \frac{E}{1 + 1/\alpha + 1/\alpha^2}, \quad E_1 = \frac{E(1 + 1/\alpha)}{1 + 1/\alpha + 1/\alpha^2}$$

The maximum stress is then

$$\frac{E - E_1}{\frac{1}{2}d \operatorname{logh} \alpha} = \frac{E}{(1 + \alpha + \alpha^2)\frac{1}{2}d \operatorname{logh} \alpha} = \frac{E}{\frac{1}{3}(1 + \alpha + \alpha^2)\frac{1}{2}d \operatorname{logh} (D/d)},$$

since $\operatorname{logh} (D/d) = \operatorname{logh} \alpha^3 = 3 \operatorname{logh} \alpha$.

Without the use of intersheaths the maximum stress is

$$\frac{E}{\frac{1}{2}d \operatorname{logh} (D/d)},$$

so that the maximum stress has been reduced in the ratio

$$1 : \tfrac{1}{3}(1 + \alpha + \alpha^2).$$

EXAMPLE. A single-core 66 kV. cable has a conductor diameter of 2 cm. and a sheath of inside diameter 5·3 cm. Find the maximum stress. If two intersheaths are used, find the best positions, the maximum stress, and the voltages on the intersheaths.

Here $D/d = 5{\cdot}3/2 = \alpha^3$, so that $\alpha = 1{\cdot}384$.

Thus $d_1 = \underline{\underline{2{\cdot}77}}$ cm. and $d_2 = \underline{\underline{3{\cdot}84}}$ cm. are the diameters of the intersheaths. The peak voltage on the conductor is $66 \times \sqrt{2} \div \sqrt{3} = 53{\cdot}8$ kV., so that

$$E_2 = \frac{53{\cdot}8}{1 + \dfrac{1}{1{\cdot}384} + \dfrac{1}{1{\cdot}910}} = \underline{\underline{23{\cdot}9 \text{ kV.}}}$$

and

$$E_1 = \left(1 + \frac{1}{1{\cdot}384}\right)23{\cdot}9 = \underline{\underline{41{\cdot}1 \text{ kV.}}}$$

It should be remembered that in practice the system will be three-phase at this voltage, and the r.m.s. neutral to voltage is $(1/\sqrt{3})$ times these values. The maximum stress without the intersheaths is

$$\frac{53{\cdot}8}{1 \times \operatorname{logh} 2{\cdot}65} = \underline{\underline{55{\cdot}3}} \text{ kV. per cm.},$$

and the minimum stress is 20·8 kV. per cm. With the intersheaths the maximum stress is

$$\frac{55{\cdot}3}{\frac{1}{3}(1 + 1{\cdot}384 + 1{\cdot}91)} = \frac{55{\cdot}3}{1{\cdot}43} = \underline{\underline{38{\cdot}7}} \text{ kV. per cm.},$$

while the minimum stress is 27·9 kV. per cm. Fig. 87 shows the stress distribution in both cases. The maximum stress has been reduced by the ratio 1 : 1·43.

Example. Suppose that in the previous example the intersheaths are spaced at equal distances from each other, the conductor and the sheath. Find their voltages for the same maximum stresses in the layers, and find the maximum stress.

$D = 5{\cdot}3$ cm., $d = 2$ cm. $\frac{1}{3}(D - d) = 1{\cdot}1$ cm., so that $d_1 = 3{\cdot}1$ and $d_2 = 4{\cdot}2$.

$$S_{1max} = \frac{E - E_1}{\text{logh } 1{\cdot}55} = 2{\cdot}28(E - E_1),$$

$$S_{2max} = \frac{E_1 - E_2}{1{\cdot}55 \text{ logh } (4{\cdot}2/3{\cdot}1)} = 2{\cdot}12(E_1 - E_2),$$

and
$$S_{3max} = \frac{E_2}{2{\cdot}1 \text{ logh } (5{\cdot}3/4{\cdot}2)} = 2{\cdot}06E_2.$$

Equating these we get $E_1 = 45{\cdot}2$ kV. and $E_2 = 23{\cdot}0$ kV. The maximum stress is $47{\cdot}5$ kV. per cm. It is seen that the positions and voltages of the intersheaths are not very critical.

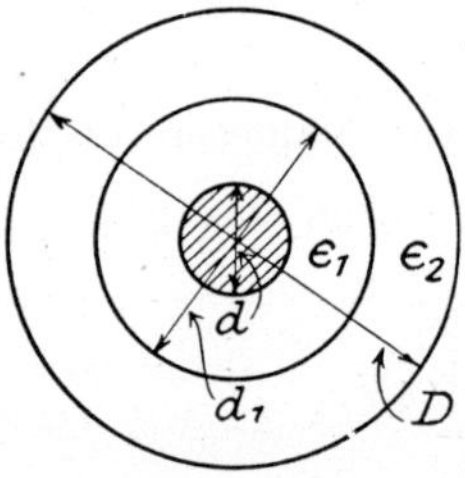

Fig. 88
Graded Dielectric

The use of intersheaths has not been general practice because of the complications involved. The sheaths must be supplied with the requisite potentials and must carry quite large charging currents. Jointing is made very difficult. Furthermore when there is a breakdown at one place, the stresses between the intersheaths containing healthy dielectric rise and breakdowns take place at other points of the cable, which is probably the greatest drawback.

Capacitance Grading. Suppose that the dielectric consists of two layers with a dividing diameter d_1, the dielectric constants being ε_1 and ε_2, as shown in Fig. 88. By Gauss's theorem the stress in the inner layer is

$$S_1 = 2q/\varepsilon_1 x,$$

whilst in the outer layer it is

$$S_2 = 2q/\varepsilon_2 x.$$

Then
$$E = \int_{\frac{1}{2}d}^{\frac{1}{2}d_1} S_1 dx + \int_{\frac{1}{2}d_1}^{\frac{1}{2}D} S_2\, dx$$

$$= 2q\left(\frac{1}{\varepsilon_1} \text{ logh } \frac{d_1}{d} + \frac{1}{\varepsilon_2} \text{ logh } \frac{D}{d_1}\right),$$

so that
$$C = \frac{q}{E} = \frac{1}{\dfrac{2}{\varepsilon_1} \text{ logh } \dfrac{d_1}{d} + \dfrac{2}{\varepsilon_2} \text{ logh } \dfrac{D}{d_1}}.$$

The maximum value of S_1 is

$$S_{1\,max} = \frac{4q}{\varepsilon_1 d} = \frac{2E}{d\left(\operatorname{logh}\frac{d_1}{d} + \frac{\varepsilon_1}{\varepsilon_2}\operatorname{logh}\frac{D}{d_1}\right)},$$

and $$S_{2\,max} = \frac{4q}{\varepsilon_2 d_1} = \frac{2E}{d_1\left(\frac{\varepsilon_2}{\varepsilon_1}\operatorname{logh}\frac{d_1}{d} + \operatorname{logh}\frac{D}{d_1}\right)}.$$

EXAMPLE. Suppose that the cable of the last two examples has an inner layer 1 cm. thick of rubber of dielectric constant 4·5 and the rest impregnated paper of constant 3·6. Find the maximum stress in the rubber and in the paper.

$$d = 2,\ d_1 = 4,\ D = 5{\cdot}3,\ \varepsilon_1 = 4{\cdot}5,\ \varepsilon_2 = 3{\cdot}6.$$

$$S_{1max} = \frac{2 \times 66}{2(\operatorname{logh} 2 + 1{\cdot}25 \operatorname{logh} 1{\cdot}325)} = \underline{\underline{63}}\ \text{kV. per cm.},$$

and $$S_{2max} = \frac{2 \times 66}{4(0{\cdot}8 \operatorname{logh} 2 + \operatorname{logh} 1{\cdot}325)} = \underline{\underline{39{\cdot}5}}\ \text{kV. per cm.},$$

both to be multiplied by $\sqrt{(2/3)}$ of course.

Thus the maximum stress has been reduced from 67·8 to 63. The reduction is hardly worth while, and in practice the only grading used is for strength, i.e. a better quality paper is put near the conductor than near the sheath.

This method of grading is quite practicable.

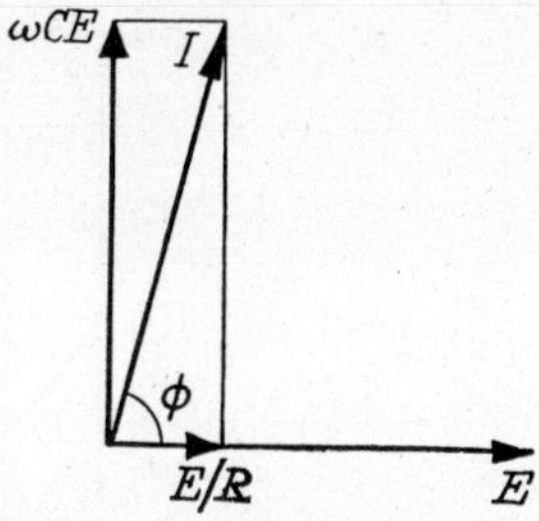

FIG. 89

Power Factor of Single-core Cable. Suppose that the dielectric has a resistivity ρ which is independent of the stress and may be considered as constant throughout the cable. Then upon the application of an alternating voltage E of frequency $\omega/2\pi$ there will be an in-phase current of E/R per cm. length, where R is given by equation (39).

The value of ρ from which G is to be calculated is generally very much less than that measured by direct current, and depends upon the frequency of the voltage. This is due to the fact that the losses with alternating currents are caused mainly by absorption phenomena.

There will be also a charging current ωCE, where C is given by equation (40), which leads the voltage by a right angle. Fig. 89 shows the vector diagram for this case. The total current I is the vector sum of E/R and ωCE, and leads the voltage by an angle ϕ where

$$\cot \phi = (E/R) \div \omega CE = 1/\omega CR.$$

It is usual to denote the reciprocal of R by G, which is the conductance of the cable per cm. length, so that

$$\cot \phi = G/\omega C$$

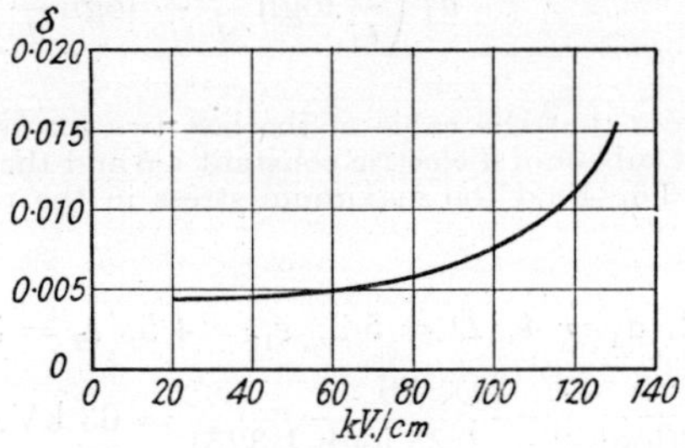

FIG. 90. VARIATION OF POWER FACTOR WITH STRESS

The power factor of the cable is given by

$$\text{P.F.} = \frac{\text{Watts}}{EI} = \frac{E^2/R}{EI} = \frac{E/R}{I} = \cos \phi.$$

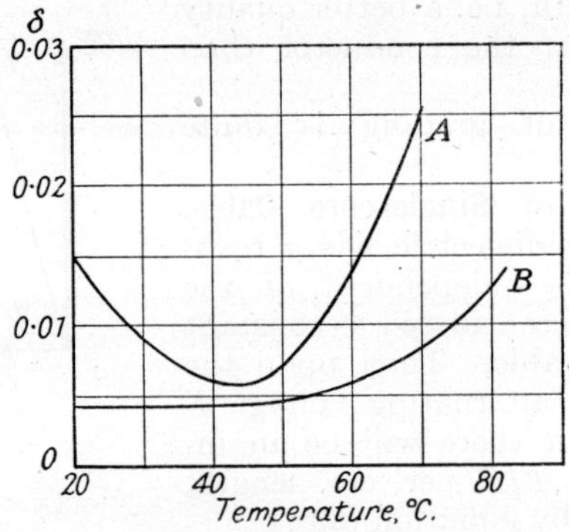

FIG. 91. VARIATION OF POWER FACTOR WITH TEMPERATURE

In well-made cables ϕ is so near to 90° that $\cos \phi$ and $\cot \phi$ are small and very nearly equal to each other and to δ, where δ is $(\pi/2) - \phi$ and is in radians.

We may thus put

$$\text{P.F.} = \delta = G/\omega C$$

The dielectric loss is

$$E^2/R = E^2G = \omega CE^2\delta \quad . \quad . \quad . \quad . \quad (45)$$

The power factor of impregnated paper varies with the electrical stress and the temperature. Fig. 90 shows how the power factor rises with the stress. At a stress of 60 kV. per cm. the power factor begins to rise, and above this stress it is said that the dielectric is *ionizing*. The term is unfortunate, as it implies that the gaseous

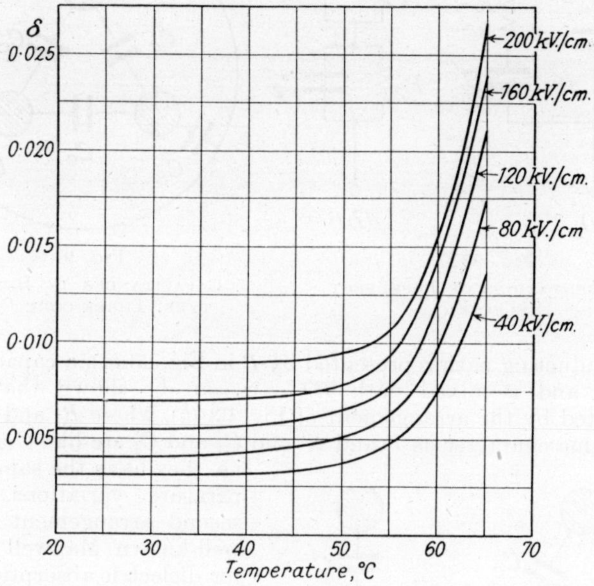

FIG. 92. POWER FACTOR VERSUS TEMPERATURE AND STRESS

voids are producing ions by collision and this may not be the case; for although ionization by collision will cause a rise of power factor, it is not true that a rise of power factor is necessarily caused by this phenomenon. The variation of power factor with temperature depends upon the paper and oil, and also upon the completeness of drying. It was once usual for the power factor-temperature curve to have a minimum at about 40° C., as shown in Fig. 91, curve A; this is the V-curve. With better drying and impregnation, the power factor-temperature curve is nowadays more like that of curve B, which is flat up to 50° C. or higher. Fig. 92 shows power factor-temperature curves for various stresses on high-grade impregnated paper insulation. The effect of resin is to make the power factor rise steeply at high temperatures, and the tendency in high-voltage cables is to omit the resin.

The existence of the V-curve can be explained by the presence of moisture or inhomogeneities in the dielectric. Thus if a dielectric

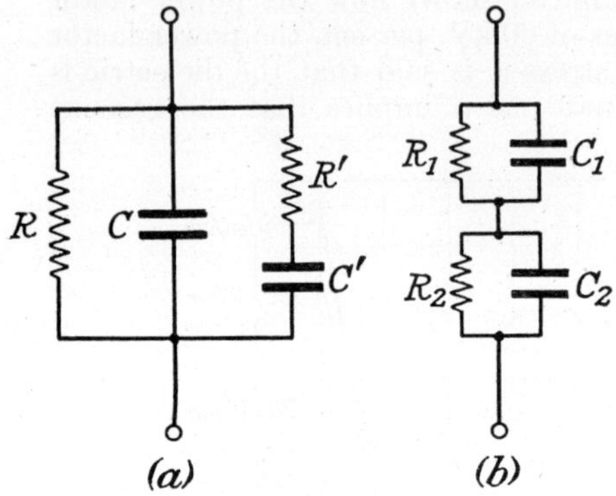

FIG. 93
REPRESENTATIONS OF IMPERFECT DIELECTRIC

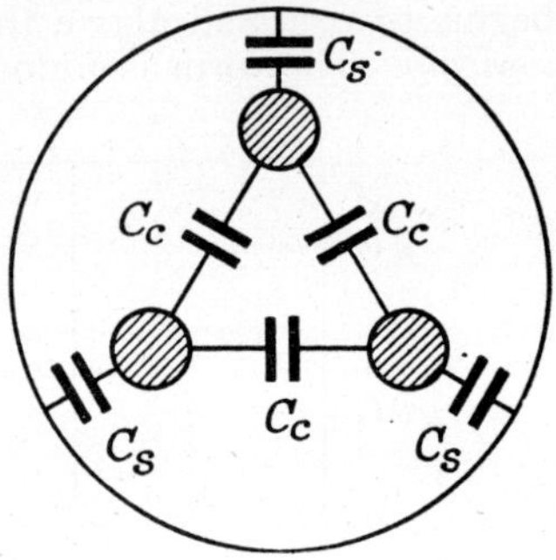

FIG. 94
CAPACITANCES IN BELTED-TYPE, THREE-CORE CABLE

has a conducting path represented by R in Fig. 93 (a), a capacitance path C, and a mixed path $R'C'$, it can be shown that it is represented by the arrangement of Fig. 93 (b), where R_1 and R_2 are of the same character as R and R' and C_1 and C_2 are like C and C', i.e. they have the same temperature variations. The second arrangement is the well-known Maxwell model for dielectric absorption and will exhibit the V-curve.

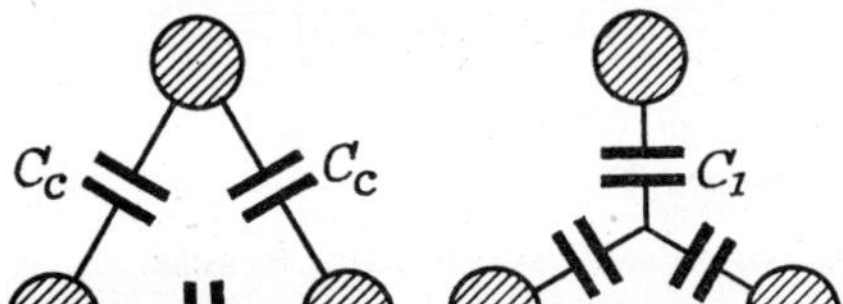

FIG. 95

Capacitance of Three-core, Belted-type Cable. The three-core S.L. and H types are equivalent, as far as capacitance and stress are concerned, to three separate single-core cables. In the belted-type cable the conductors have capacitance C_c to each other and C_s to the sheath, so that the system of capacitances is as shown in Fig. 94. The capacitances C_c form a delta of capacitances and can be replaced by a Y of capacitances C_1 as shown in Fig. 95. For this to be so, the capacitances between any two conductors in these arrangements must be the same, so that $C_c + \frac{1}{2}C_c = \frac{1}{2}C_1$, or $C_1 = 3C_c$. The centre point of the Y is the neutral, and as the sheath is at zero potential, we can consider that these capacitances act to the sheath, so that the neutral capacitance of each conductor is

$$C_0 = C_s + 3C_c.$$

It is very difficult to calculate the capacitance to neutral from the geometry of the table. The following empirical formula gives the capacitance with sufficient accuracy for design work.

$$C_0 = \frac{0{\cdot}048\varepsilon}{\log\left[1 + \frac{T+t}{d}\left\{3{\cdot}84 - 1{\cdot}70\frac{t}{T} + 0{\cdot}52\frac{t^2}{T^2}\right\}\right]} \mu\text{F. per mile} \quad (46)$$

where d = conductor diameter,

T = thickness of conductor insulation,

and t = thickness of belt insulation.

The capacitances C_c and C_s are found best by measurement, and the neutral capacitance calculated in the following way. Let con-

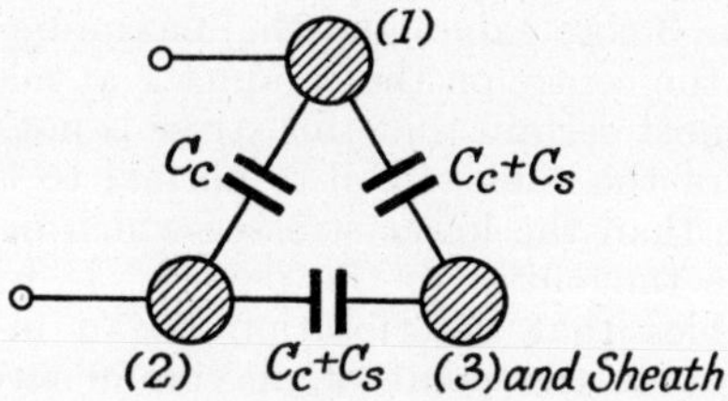

FIG. 96

ductors 2 and 3 be connected to the sheath and the capacitance be measured between conductor 1 and the rest. The value is

$$C_a = C_s + 2C_c.$$

Next let all the conductors be commoned and the capacitance, C_b, be measured between them and the sheath.

$$C_b = 3C_s.$$

Then $C_s = \frac{1}{3}C_b$

and $C_c = \frac{1}{2}C_a - \frac{1}{6}C_b.$

The capacitance to neutral is thus

$$C_0 = \tfrac{1}{3}C_b + \tfrac{3}{2}C_a - \tfrac{1}{2}C_b$$
$$= \tfrac{3}{2}C_a - \tfrac{1}{6}C_b \quad . \quad . \quad . \quad . \quad (47)$$

If C_b is not known it may be taken as $1{\cdot}8C_a$, so that

$$C_0 = 1{\cdot}5C_a - 0{\cdot}3C_a = 1{\cdot}2C_a \quad . \quad . \quad (48)$$

EXAMPLE. The capacitance of a length of three-phase cable is measured and the capacitance between two cores (the third being connected to the lead sheath) is found to be 3 μF. Find the charging current per core if the cable is connected to an 11 kV., 50 cyc., three-phase alternator. Prove each step. (*Nat. Cert.*, 1934.)

The capacitances are as shown in Fig. 96, so that the measured capacitance is

$$C_c + \tfrac{1}{2}(C_c + C_s) = \tfrac{1}{2}(3C_c + C_s),$$

which is half the capacitance to neutral. The neutral capacitance is therefore 6 μF. and the charging current per core is

$$\omega C_0 E = 2\pi \times 50 \times 6 \times 10^{-6} \times (11\,000 \div \sqrt{3})$$
$$= 11{\cdot}97 \text{ A.}$$

Stress in a Three-core Cable. Even when the dielectric is homogeneous the problem cannot be solved with accuracy, and as the dielectric is never homogeneous, because of the fillers, there is no point in quoting or working from formulae. There is a rotating electric field in the 3-core cable, and the maximum stress occurs at the point nearest the centre on the conductor at maximum voltage. It is, however, almost certain that this stress is not the determining factor in the life of the cable, for it is normal to the paper and is more easily borne than the lower stresses which occur in and near the fillers and are tangential to the papers. It was found in the 3-core 33 kV. cables that deterioration began in the fillers and wormings, and not at the point of maximum stress on the conductors; the H-type cable was designed to avoid these tangential stresses and solved at once the problem of the 3-core 33 kV. cable.

Inductance of Cables. The methods of calculating the inductance of overhead lines may be applied to underground cables, but the results will be in error because of the skin and proximity effects and the effect of the sheath. In low voltage cables the conductor spacings are small compared with the conductor diameters, so that the effects will not be negligible. It is then best to measure the inductances, if they are required, as calculation will be very laborious and inaccurate.

In high voltage cables the skin and proximity effects are negligible because of the increased thickness of insulation. In such cables the separate cores are often sheathed, or surrounded by metallized (H) paper which is connected to the sheath. The sheaths have mutual inductance to the conductors and influence considerably the resistance and inductance of the cores to neutral; the effects of the sheath will now be considered.

Sheath Effects. The currents induced in the sheaths are of two kinds: *sheath eddies*, whose paths lie in the sheath of a single cable and which flow even when the sheaths are isolated from each other, and *sheath-circuit eddies*, whose paths lie in the sheaths of separate cables and flow only when the sheaths are bonded.

Fig. 97 shows the formation of sheath eddies in the case of two single-core cables with separate and insulated lead sheaths. The

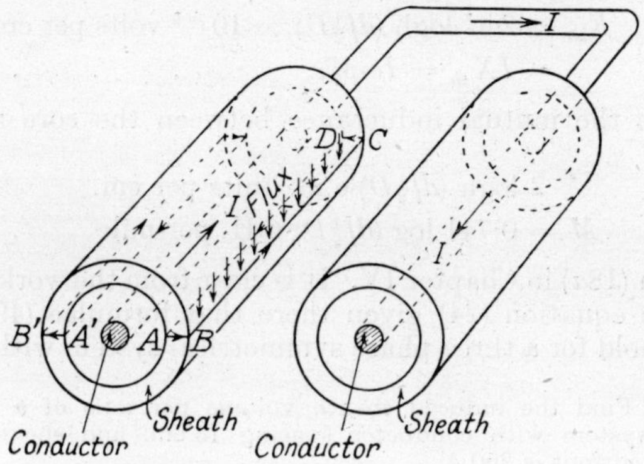

FIG. 97. SHEATH EDDY PATHS

conductor currents I produce a flux downwards through the sheath section $ABCD$. When I and the flux increase there is a sheath eddy round $ABCD$ from A to B to C to D to A. The sheath eddy at A' is outwards to B', along the outside of the sheath into the paper and back again inside the sheath to A'. The loss due to sheath

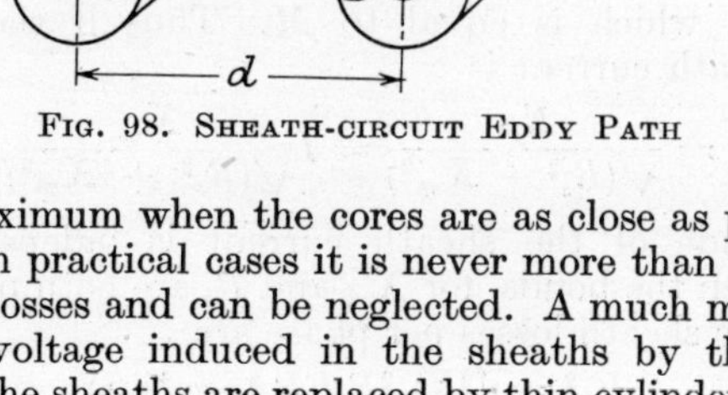

FIG. 98. SHEATH-CIRCUIT EDDY PATH

eddies is a maximum when the cores are as close as possible to one another, but in practical cases it is never more than a few per cent of the copper losses and can be neglected. A much more important effect is the voltage induced in the sheaths by the currents I. Suppose that the sheaths are replaced by thin cylinders of radius $\frac{1}{2}D$ and we consider a circuit $ABCD$ shown in Fig. 98. The flux through $ABCD$ is seen to be

$$0{\cdot}4I \text{ logh } (d/\tfrac{1}{2}D) \text{ per cm. length,}$$

where d is the conductor spacing. The induced e.m.f. is thus

$$4\omega I \text{ logh } (d/\tfrac{1}{2}D) \times 10^{-9} \text{ volts per cm. length.}$$

This is the e.m.f. along both sheaths and we may consider that each sheath has an induced e.m.f. of half this, viz.

$$E_{sh} = 2\omega I \text{ logh } (d/\tfrac{1}{2}D) \times 10^{-9} \text{ volts per cm.}$$
$$= IX_m = I\omega M, \quad . \quad . \quad . \quad . \quad . \quad (49)$$

where M is the mutual inductance between the core and sheath and is

$$2 \text{ logh } (d/\tfrac{1}{2}D) \text{ e.m. units per cm.}$$

or $$M = 0{\cdot}741 \log (d/\tfrac{1}{2}D) \text{ mH. per mile} \quad . \quad . \quad (50)$$

by equation (18*a*) in Chapter IV. It is clear from the work in Chapter IV and the equation (24) given there that formulae (49) and (50) just given hold for a three-phase symmetrical system with spacing d.

EXAMPLE. Find the induced sheath voltage per mile of a symmetrical three-phase system with conductor spacing 15 cm. and sheath diameters 5·5 cm. The current is 250 A.

$$M = 0{\cdot}741 \log (15/2{\cdot}75) = 0{\cdot}545 \text{ mH. per mile.}$$
$$E_{sh} = 250 \times 2\pi \times 50 \times 0{\cdot}545 \times 10^{-3}$$
$$= \underline{42{\cdot}8} \text{ V. per mile.}$$

If the sheaths are bonded at one end, the voltage between them at the other is

$$\sqrt{3} \times 42{\cdot}8 = \underline{74{\cdot}3} \text{ V. per mile.}$$

Currents in Bonded Sheaths. It is seen from the preceding example that large voltages are induced in the sheaths if they are open-circuited, and it is very probable that arcing will occur between them. It is therefore standard practice to bond the sheaths at both ends so as to avoid the high voltages. The impedance of the sheath current path is due to the sheath resistance R_s and the sheath self-inductance, which is equal to M. Thus if the sheaths are bonded the sheath current is

$$I_{sh} = \frac{E_{sh}}{\sqrt{(R_s^2 + X_m^2)}} = I\frac{X_m}{\sqrt{(R_s^2 + X_m^2)}} \quad . \quad . \quad (51)$$

The magnitude of the sheath current is independent of the distance between the bonds, for X_m and R_s are both proportional to the length. The sheath losses per phase are

$$R_s I_{sh}^2 = I^2 X_m^2 R_s/(R_s^2 + X_m^2) \quad . \quad . \quad (52)$$

so that the effective resistance per phase is the conductor resistance plus

$$X_m^2 R_s/(R_s^2 + X_m^2) \quad . \quad . \quad . \quad (53)$$

The ratio of sheath losses to copper losses is

$$\frac{I^2X_m{}^2R_s}{R_s{}^2+X_m{}^2} \div I^2R = \frac{X_m{}^2R_s}{R(R_s{}^2+X_m{}^2)}$$

where R is the resistance of the conductor. With large conductors and close spacing this ratio is approximately equal to

$$X_m{}^2R_s/RR_s{}^2 = X_m{}^2/RR_s,$$

as $R_s{}^2$ is large compared with $X_m{}^2$. If the conductor is made larger, R and R_s diminish whilst X_m remains fairly constant, so that the ratio increases rapidly. Thus for 66 kV. cables at a spacing of 6 in., the sheath losses exceed the conductor losses for conductor sections above 0·85 in.²

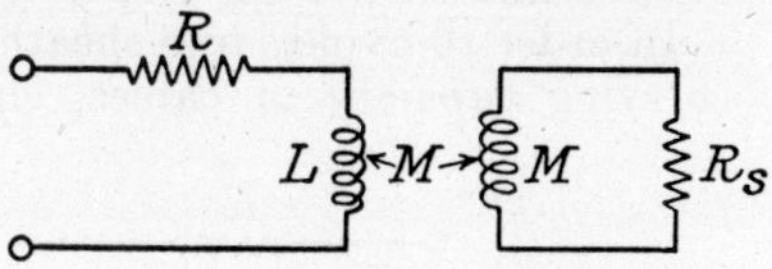

FIG. 99. ELECTRICAL EQUIVALENT OF SHEATH CIRCUIT

The effect of the sheath on the inductance of the cable may be found in the following way. The conductor has resistance R and self-inductance L, where

$$L = 2 \text{ logh } (d/r)$$

from the work in Chapter IV. This inductance is coupled to the sheath circuit by a mutual inductance M, where

$$M = 2 \text{ logh } (d/\tfrac{1}{2}D).$$

The sheath circuit itself has a resistance R_s and self-inductance M, so that the cable impedance is represented by the network of Fig. 99. By the well-known theorem of the equivalent network of the transformer (see page 471) the arrangement can be replaced by that shown in Fig. 100

$$\begin{aligned} L - M &= 2 \text{ logh } (d/r) - \text{logh } (d/\tfrac{1}{2}D) \\ &= 2 \text{ logh } (\tfrac{1}{2}D/r) = L_c, \end{aligned}$$

which is the leakage inductance of the core to the sheath. The total impedance of the conductor is thus

$$\begin{aligned} & R + j\omega L_c + \frac{R_s \,.\, j\omega M}{R_s + j\omega M} \\ &= R + j\omega L_c + \frac{R_s \,.\, j\omega M(R_s - j\omega M)}{R_s{}^2 + \omega^2M^2} \\ &= \left[R + \frac{R_s\omega^2M^2}{R_s{}^2+\omega^2M^2}\right] + j\omega\left[L_c + \frac{MR_s{}^2}{R_s{}^2+\omega^2M^2}\right] \\ &= \left[R + \frac{R_sX_m{}^2}{R_s{}^2+X_m{}^2}\right] + j\omega\left[L_c + \frac{MR_s{}^2}{R_s{}^2+X_m{}^2}\right]. \end{aligned}$$

The resistance is thus

$$R + R_s X_m^2/(R_s^2 + X_m^2) \quad . \quad . \quad . \quad . \quad (54)$$

which we have already found in equation (53), whilst the inductance is

$$L_c + MR_s^2/(R_s^2 + X_m^2) = L - M + MR_s^2/(R_s^2 + X_m^2)$$
$$= L - MX_m^2/(R_s^2 + X_m^2) \quad . \quad (55)$$

The decrease in inductance due to the sheath is thus

$$MX_m^2/(R_s^2 + X_m^2).$$

The resistivity of lead is $23{\cdot}2 \times 10^{-6}$ ohms per cm. cube at 30° C. Given the thickness of the sheath and its diameter, R_s can be calculated and thence the resistance and inductance of the cable.

In order to avoid large sheath currents, which lower the current-carrying capacity of cables, sheaths are sometimes *cross-bonded.*

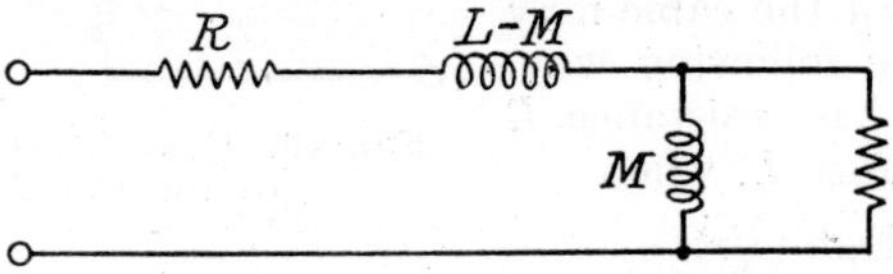

FIG. 100. ELECTRICAL EQUIVALENT OF SHEATH CIRCUIT

The sheath of conductor 1 is connected to that of conductor 2 and then to that of conductor 3 at equidistant points, and the induced sheath voltage is

$$I_1\omega M + I_2\omega M + I_3\omega M = (I_1 + I_2 + I_3)\omega M = 0.$$

(See Fig. 82.)

There will be no sheath current and yet the sheath voltage will never be greater than $I\omega M$ at any point. A combination of cross-bonding—except at every third joint, which is solidly bonded—and of simple reactances reduces sheath losses and voltages and also prevents the generation of third harmonic currents.

EXAMPLE. Find the resistance, inductance and capacitance per mile of a 3-core belted-type cable, in which the conductors are circular 37/0·093, conductor insulation is 0·20 in., belt thickness is 0·17 in., and the dielectric constant is 3·6.

From tables it is found that the resistance is 0·09933 Ω. per 1 000 yd. at 60° F. The resistance per mile at 55° C., which is taken as a normal working temperature, being 40° C. above the average temperature of 15° C., is

$$R = 0{\cdot}09933 \times 1{\cdot}760 \times [1 + 0{\cdot}004 \times 39{\cdot}5]$$
$$= \underline{\underline{0{\cdot}202\ \Omega.}},$$

stranding having been allowed for in the tables.

The formula for overhead lines is used in this case for inductance, and no great accuracy can be claimed.

$$L = 0{\cdot}085 + 0{\cdot}741 \log (d/r) \text{ mH. per mile.}$$

Here $r = 0{\cdot}325$ in. and $d = 2(0{\cdot}325 + 0{\cdot}20) = 1{\cdot}05$ in., so that

$$L = 0{\cdot}085 + 0{\cdot}741 \log (1{\cdot}05/0{\cdot}325) = \underline{\underline{0{\cdot}462}} \text{ mH. per mile.}$$

The capacitance is given by equation (46)

$$C = \frac{0{\cdot}048\varepsilon}{\log\left[1 + \dfrac{\mathrm{T}+t}{d}\left\{3{\cdot}84 - 1{\cdot}70\dfrac{t}{\mathrm{T}} + 0{\cdot}52\dfrac{t^2}{\mathrm{T}^2}\right\}\right]} \ \mu\text{F. per mile}$$

where d here is the conductor diameter, not the spacing.

$$C = \frac{0{\cdot}048 \times 3{\cdot}6}{\log\left[1 + \dfrac{0{\cdot}37}{0{\cdot}65}\left\{3{\cdot}84 - 1{\cdot}70 \times 0{\cdot}85 + 0{\cdot}52 \times 0{\cdot}85^2\right\}\right]}$$

$$= \frac{0{\cdot}048 \times 3{\cdot}6}{\log 2{\cdot}57} = \underline{\underline{0{\cdot}420}} \ \mu\text{F. per mile.}$$

EXAMPLE. Find the resistance, inductance and capacitance of a three-phase symmetrical arrangement of 66 kV. single-core cables, 61/0·103 (nominal 0·5 in.^{2}), insulation thickness 0·65 in., sheath thickness 0·15 in., serving thickness 0·15 in., dielectric constant 3·6; the cables are laid touching one another and the sheaths are bonded.

From the tables the resistance is 0·04913 Ω. per 1 000 yd. at 60° F., so that at 55° C. the resistance per mile per phase is

$$R = 0{\cdot}04913 \times 1{\cdot}760 \times [1 + 0{\cdot}004 \times 39{\cdot}5]$$
$$= 0{\cdot}1005 \ \Omega.$$

$$C = \frac{0{\cdot}0388\varepsilon}{\log (D/d)} \ \mu\text{F. per mile.}$$

Here $d = 9 \times 0{\cdot}103 = 0{\cdot}927$ in.,

and $D = 0{\cdot}927 + 2 \times 0{\cdot}65 = 2{\cdot}227$ in.

$$C = \frac{0{\cdot}0388 \times 3{\cdot}6}{\log 2{\cdot}40} = \underline{\underline{0{\cdot}378}} \ \mu\text{F. per mile.}$$

We will assume a sheath temperature of 30° C.

$$R_s = \frac{23{\cdot}2 \times 10^{-6} \times 5\,280 \times 12 \times 2{\cdot}54}{\pi[1{\cdot}263^2 - 1{\cdot}113^2] \times 2{\cdot}54^2} \ \Omega. \text{ per mile}$$
$$= 0{\cdot}516 \ \Omega. \text{ per mile.}$$

The spacing is $2\cdot227 + 4 \times 0\cdot15 = 2\cdot827$ in., so that

$$M = 0\cdot741 \log. (2\cdot827/1\cdot115)$$
$$= 0\cdot297 \text{ mH. per mile}$$
$$X_m = 2\pi \times 50 \times 0\cdot297 \times 10^{-3} = 0\cdot0931\ \Omega. \text{ per mile.}$$

The effective resistance per mile is therefore

$$0\cdot1005 + \frac{0\cdot516 \times 0\cdot0931^2}{0\cdot516^2 + 0\cdot0931^2}$$
$$= 0\cdot1005 + 0\cdot0167 = \underline{\underline{0\cdot1172}} \text{ ohms.}$$

The sheath loss is $0\cdot0167 \div 0\cdot1005 = 16\cdot6$ per cent of the conductor loss.

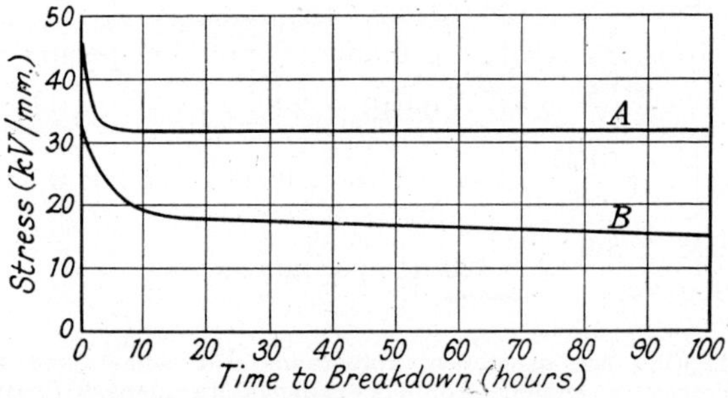

FIG. 101. V.T.B. CURVES OF WELL AND BADLY IMPREGNATED PAPER

The inductance is

$$L_e + MR_s^2/(R_s^2 + X_m^2) = L - MX_m^2/(R_s^2 + X_m^2)$$
$$= L - 0\cdot031M$$
$$\simeq L = 0\cdot741 \log 2\cdot827/0\cdot463 = 0\cdot583 \text{ mH. per mile.}$$

Actually the inductance is slightly lower by 0·0092 mH. due to the sheath bonding, so that it is 0·574 mH. per mile.

Breakdown Voltage and Mechanism of Breakdown. The voltage required to break down a certain insulation depends upon many factors such as time of application, shape of electrodes, temperature, pressure, the presence of moisture or gaseous spaces. The dependence of the voltage on time is very important, and tests are made to determine the curve relating the voltage and time of application; such a curve is called a V.T.B. curve, i.e. voltage-time-breakdown curve. Fig. 101, curve A, shows the V.T.B. curve of 1 mm. thickness of very well impregnated paper. The short-time breakdown voltage is about 45 kV. (stress 45 kV. per mm.), but the breakdown voltage reaches a steady value of 32 kV. in about 5 hours. If a voltage of 31·5 kV. is applied the insulation never breaks down. It is of the greatest advantage if the asymptotic value (i.e. final value) is reached in a short time, for then decisive tests may be short; presumably also the cable is stable and likely to give long service. Curve *B* shows the V.T.B. curve of badly impregnated paper which contains air spaces. The short-time breakdown voltage is not much less than for the good dielectric, but the asymptotic value is much lower and is not reached even in 100 hours. With slow deterioration of this kind it is difficult to say what voltage the dielectric can maintain indefinitely.

Moisture has the same effect as gaseous voids on the V.T.B. curve.

A carefully impregnated cable will initially contain no voids and so it will have a V.T.B. curve like curve *A*. When, however, this cable is subjected to fluctuating loads, the heating causes the oil to expand and the sheath is stretched; when the cable cools, the sheath does not recover and small voids are formed by cavitation. After a number of fluctuations the voids may be such that the V.T.B. curve is like curve *B*, and eventual failure will occur if the applied voltage is greater than the new asymptotic value. In order to ascertain whether the cable V.T.B. curve is stable, the cable is subjected to the working voltage or a higher voltage whilst the cable is alternately heated and allowed to cool. Such a test is called a *stability test* and is applied to all types of high voltage cables.

The formation of voids is accompanied by ionization by collision and a rise of the power factor of the cable.

The voids are eliminated in the oil-filled and pressure cables, whilst the gas cable prevents ionization by the application of hydrostatic pressure. All these cables have good V.T.B. curves.

There are two ways in which breakdown of cables usually occurs. One way is by a progressive coring and tracking, which always starts from the conductor or sheath, and ultimately bridges the electrodes. Another way is by thermal instability; this occurs when the power factor increases so rapidly with rise of temperature, that a small rise of temperature increases the dielectric losses by a greater amount than can be conducted away. This method will be considered later in detail. A marked difference between the methods of breakdown is that coring, once it commences, will continue until the cable breaks down, although the time may be considerable for the complete action. In thermal instability, however, no damage is done until just before breakdown, so that if the load is released before breakdown the cable will not have suffered any permanent change. A very common occurrence is for coring to start and then introduce thermal instability at the centre of coring.

Thermal Characteristics of Cables. MAXIMUM CURRENT CAPACITY. There are several reasons why cables should not be run too hot; differential expansion may create voids with resulting ionization; the expansion of the oil may burst the sheath; the oil may lose its viscosity and drain away from higher levels; thermal instability may arise due to the rapid increase of dielectric losses with temperature. The last phenomenon is not likely to occur in cables up to 33 kV., but it is being reached in cables above 66 kV. The calculation is difficult and will not be given.

In order not to incur the other harmful effects, a maximum conductor temperature of 65° C. has been adopted for cables impregnated with viscous oils in this country. The maximum current that a cable can carry with a conductor temperature of 65° C. is found in the following way.

Assume that the dielectric and sheath losses are negligible compared with the conductor losses, which are given by nRI^2, where R is the conductor resistance at 65° C., and n the number of phases. Let S be the thermal resistance of the cable, i.e. between the combined conductors and sheath, and G the thermal resistance from sheath to earth surface. The heat has to pass through the two thermal paths in series, so that the temperature difference between the conductors and ground is

$$nRI^2(S + G) = 65 - \theta, \qquad (56)$$

where θ is the ambient ground temperature. We may take $\theta = 18$, so that

$$nRI^2(S + G) = 43,$$

giving

$$I = \sqrt{[43/nR(S + G)]} \qquad (57)$$

If the dielectric and sheath losses are not negligible, we can replace equation (56) by

$$65 - \theta = nRI^2(S + G) + W(S + G) + R_o I^2 G,$$

where W is the dielectric losses and is conservatively taken as occurring all at the conductor, and R_o is an equivalent resistance due to sheath losses. Then the current capacity is

$$I = \sqrt{\left[\frac{65 - \theta - W(S + G)}{nR(S + G) + R_o G}\right]} \qquad (58)$$

Thermal Resistance. The unit is the *thermal ohm* and is that thermal resistance which requires a temperature difference of 1° C. to produce a heat flow of one watt (i.e. one joule per second). If the thermal resistivity of a cable dielectric is K, the thermal resistance of a single-core cable is

$$S_1 = (K/2\pi) \text{ logh } (D/d) \text{ thermal ohms per cm. length of cable} \qquad (59)$$

K is taken as 750 for cables up to and including 2 200 volts, and 550 for cables above 2 200 volts.

The thermal resistance of a 3-core belted-type cable is given by the empirical formula

$$S_1 = \frac{K}{6\pi}\left(0{\cdot}85 + \frac{0{\cdot}2t}{\mathrm{T}}\right) \text{ logh}\left[1 + \frac{2(\mathrm{T} + t)}{d}\left(4{\cdot}15 - \frac{1{\cdot}1t}{\mathrm{T}}\right)\right] \qquad (60)$$

The thermal resistance of the ground is

$$G = (g'/2\pi) \operatorname{logh} (2h/R_2), \quad . \quad . \quad . \quad . \quad (61)$$

where g' is the resistivity, h the depth of the cable below ground, and R_2 the overall radius of the cable. It is found that the thermal resistivity g as measured in the laboratory is too great for use in the above formula by about 50 per cent, so that $g' = \frac{2}{3}g$ and the thermal resistance is given by

$$G = (\tfrac{2}{3}g/2\pi) \operatorname{logh} (2h/R_2) = (g/3\pi) \operatorname{logh} (2h/R_2) \quad . \quad (62)$$

In practice g varies from 120 to 800 or 1 000 depending on the soil and its moistness.

EXAMPLE. Find the maximum current that a 3-core, 11 kV., 0·25 in.² cable can carry; $t = 0{\cdot}06$ in., $\mathrm{T} = 0{\cdot}15$ in., $d = 0{\cdot}65$ in., $K = 550$, buried 3 ft. deep, $g = 180$, ambient temperature 15° C.

$$S_1 = \frac{550}{6\pi}\left(0{\cdot}85 + \frac{0{\cdot}012}{0{\cdot}15}\right) \operatorname{logh}\left[1 + \frac{0{\cdot}42}{0{\cdot}65}\left(4{\cdot}15 - \frac{0{\cdot}066}{0{\cdot}15}\right)\right]$$

$$= 33{\cdot}2 \text{ thermal ohms per cm.}$$

The lead sheath has a very small thermal resistance, but there is a serving of thickness 0·31 in. of $K = 300$. This has a thermal resistance of

$$S_2 = (300/2\pi) \operatorname{logh} (1{\cdot}54/1{\cdot}22) = 11 \text{ thermal ohms,}$$

as the external radius of the lead sheath is 1·22 in., and that of the serving is 1·54 in. The ground resistance is

$$G = \frac{180}{3\pi} \operatorname{logh} \frac{2 \times 36}{1{\cdot}54} = 73{\cdot}4 \text{ thermal ohms.}$$

$$\therefore \quad S + G = S_1 + S_2 + G = 117{\cdot}6.$$

From tables, allowing for coring and stranding and temperature rise, $R = 1{\cdot}33 \times 10^{-6}\,\Omega$. per cm.

$$\therefore \quad I = \sqrt{\frac{65 - 15}{3 \times 1{\cdot}33 \times 10^{-6} \times 117{\cdot}6}}$$

$$= \underline{\underline{326 \text{ A.}}}$$

EXAMPLES V

1. The insulation resistance of a mile of cable having a conductor diameter of 1·5 cm. and an insulation thickness of 1·5 cm. is 500 MΩ. What would be the insulation resistance if the thickness of insulation were increased to 2·5 cm.? Prove any formula used. (*Lond. Univ.*, 1930.)

2. A voltmeter with a resistance of 4 000 Ω. reads 80 V. when connected between the positive terminal of a direct current 200-V. supply and earth, and 40 V. when connected between the negative terminal and earth. What is the insulation resistance of each main to earth? (*Lond. Univ.*, 1930.)

3. What is meant by "Dielectric constant"? Find the capacitance in μF. per mile of a single-core cable with conductor diameter 0·5 cm. and inner diameter of sheath 2·0 cm., when the dielectric constant of the insulation is 3·2. Prove any formula used. (*Lond. Univ.*, 1930.)

4. Explain briefly the principle of a condenser-type bushing.

A condenser bushing has an axial length of 30 cm., and the central conductor has a diameter of 2 cm. The outer earthed ring has an effective axial length of 6 cm., and a diameter of 10 cm. Two concentric cylindrical foils are interleaved with the insulation. Determine approximately (i.e. neglecting "end effects") a suitable diameter and length for each foil. (*Lond. Univ.*, 1948.)

5. Describe the mechanism of breakdown of a specimen of insulation due to electric stress. Discuss with the aid of voltage-time curves the effect of the duration of application on the behaviour of the specimen.

Describe with the aid of sketches and circuit diagrams the form and principle of operation of an equipment suitable for generating impulse voltages. (*Lond. Univ.*, 1953.)

6. A concentric cable has a conductor diameter of 1 cm. and an insulation thickness of 1·5 cm. Find the maximum field strength when the cable is subjected to a test pressure of 33 kV. (*Lond. Univ.*, 1931.)

7. A voltmeter with a resistance of 10 000 Ω. indicates 25 V. when connected between the positive main of a 2-wire distributor and earth, and 15 V. when connected between the negative main and earth. The voltage between the mains is 230 V. Find the insulation resistance of each main. (*Lond. Univ.*, 1932.)

8. A 66-kV., single-core, lead-covered cable laid 3 ft. below ground-level has the following dimensions—Conductor, 37/0·093 in.; dielectric thickness, 0·65 in.; lead sheath thickness, 0·15 in.; serving thickness, 0·15 in.

Calculate the current the cable can carry continuously if the thermal resistivity in ° C. per watt per cm. cube of the dielectric is 550, of the outer covering 300, and of the earth 160. The maximum temperature rise is not to exceed 50° C. above an air temperature of 15° C., and the specific resistance of the conductor is 0·8 microhm per in. cube. (*Lond. Univ.*, 1947.)

9. Explain the meaning of the terms ionization, power factor and stability used in connection with cables.

Show why the ordinary belted construction is not satisfactory for 3-core cables above 11 kV.

State the fundamental principles underlying the design and construction of high-voltage cables. (*Lond. Univ.*, 1949.)

10. Show on a diagram the effective capacitance relations of a belted cable having three symmetrical cores and an earthed sheath, and demonstrate that all necessary values may be obtained from two measurements.

A length of 1 000 ft. of such a cable takes a sinusoidal charging current of 0·36 A. from a 33-kV., 3-phase source at 50 c/s, and has a short-circuit impedance per core of $0{\cdot}030 + j\,0{\cdot}032\ \Omega$ at that frequency. Determine approximately the length of cable which would exhibit resonance with a superimposed 17th harmonic. (*Lond. Univ.*, 1954.)

11. Deduce an expression for the insulation resistance of a single-core cable in terms of the radii bounding the dielectric.

A single-core cable has a conductor of diameter 1·84 cm. and a dielectric of thickness 3·16 cm. Determine the resistivity of the dielectric, if the insulation resistance of 1 mile of the cable is 100 000 Ω. (*Lond. Univ.*, 1949.)

12. The inner diameter of the lead sheath of a single-core cable is 3·5 in., and the cable is to be used on a 3-phase, 66-kV. circuit. The permittivity of the dielectric is 4.

Develop an expression for the voltage gradient at the surface of the conductor and determine what value of conductor diameter will make this a minimum. Using this diameter, determine the maximum voltage gradient and the capacitance per mile. (*Lond. Univ.*, 1949.)

CHAPTER VI

TRANSMISSION CALCULATIONS

Short Lines. The effect of capacitance can be neglected in short lines; in overhead lines a length of 30 or 40 miles is short, but in cables the distance is considerably less before capacitance has an appreciable effect.

A single-phase line can then be represented by its resistance R and loop inductance L, where

$$L = [0{\cdot}161 + 1{\cdot}482 \log (D/r)] \text{ mH. per mile.}$$

In cables the sheath currents increase the effective resistance of the line, although the inductance is not much affected. The effective values of R and L can be calculated in the way developed in Chapter V.

A three-phase line can be represented by an impedance in each line, composed of the resistance and the inductance to neutral; the latter is

$$[0{\cdot}080 + 0{\cdot}741 \log (D/r)] \text{ mH. per mile}$$

for a line with symmetrical spacing. If the line is not symmetrically spaced, a simple treatment is possible only when the loads are balanced, and then $\sqrt[3]{(D_1D_2D_3)}$ replaces D. The problem of an unsymmetrically spaced line with unbalanced loads is best solved by the method of symmetrical components.

The effects of generators and transformers can be represented by adding to the series line impedance the synchronous impedance of the generators and the series impedance $(r + jx)$ of the transformers.

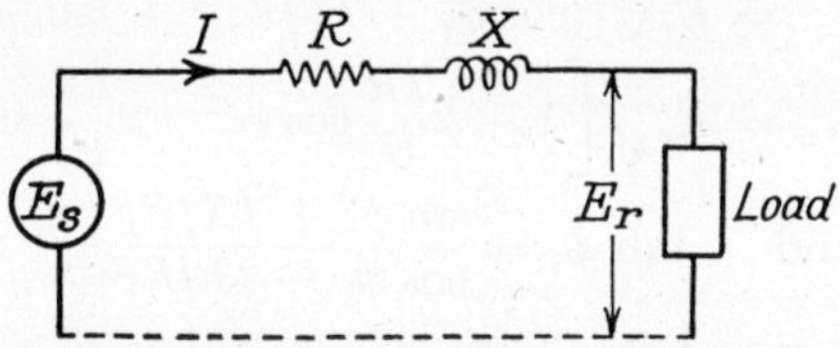

FIG. 102. ELECTRICAL EQUIVALENT OF SHORT LINE

In all cases the electrical condition is represented by the circuit of Fig. 102, where $R + jX$ is the sum of the generator, transformer, and line impedances. Fig. 102 represents a single-phase line if R and X are the resistance and reactance of the loop circuit. It represents a three-phase system if R and X are the resistance and reactance to the neutral, I is the current in one conductor, E_r is the receiving-end voltage to neutral, and the load is that in one phase; the voltage drop found in this way is the drop between a phase and the neutral. Fig. 103 shows the vector diagram of the current and voltages. The vector for I is drawn horizontally. If the received load is P at a power factor $\cos\phi_r$ and voltage E_r, the wattful component of the voltage is P/I and is equal to $E_r \cos\phi_r$. This is represented by OA. Then the line OB represents E_r, where $\angle OAB = 90°$ and $\angle AOB = \phi_r$. The voltage drop is IR horizontally and IX vertically, so that the sending-end voltage $E_s = OC$ and the sending-end power factor is $\cos\phi_s$ where $\angle AOC = \phi_s$. The power at the sending-end is $P + I^2R$. We see that

$$\left.\begin{aligned} OD &= E_s\cos\phi_s = E_r\cos\phi_r + IR \\ \text{and}\quad CD &= E_s\sin\phi_s = E_r\sin\phi_r + IX. \end{aligned}\right\} \quad (63)$$

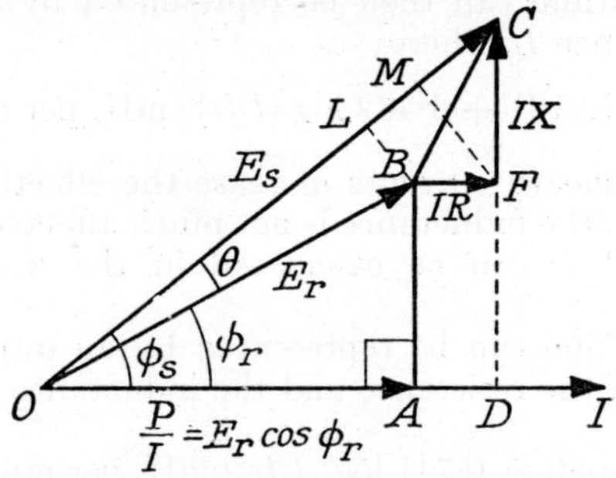

FIG. 103. VECTOR DIAGRAM FOR SHORT LINE

These give

$$\begin{aligned} E_s &= \sqrt{[(E_r\cos\phi_r + IR)^2 + (E_r\sin\phi_r + IX)^2]} \\ &= E_r\sqrt{[(\cos\phi_r + IR/E_r)^2 + (\sin\phi_r + IX/E_r)^2]} \\ &= E_r\sqrt{\left[1 + 2\frac{IR}{E_r}\cos\phi_r + 2\frac{IX}{E_r}\sin\phi_r + \frac{I^2(R^2+X^2)}{E_r^2}\right]}, \end{aligned} \quad (64)$$

$$\text{and}\quad \tan\phi_s = \frac{\sin\phi_r + IX/E_r}{\cos\phi_r + IR/E_r}. \quad (65)$$

Regulation. The *regulation* is defined as the percentage rise in voltage at the receiving-end when full load is thrown off, the sending-end voltage being unaltered. It is therefore

$$[(E_s - E_r)/E_r] \times 100 \text{ per cent.}$$

In equation (64), the last term is usually negligible, so that

$$\begin{aligned} E_s &= E_r\sqrt{\left(1 + \frac{2IR}{E_r}\cos\phi_r + \frac{2IX}{E_r}\sin\phi_r\right)} \\ &\doteqdot E_r\left(1 + \frac{IR}{E_r}\cos\phi_r + \frac{IX}{E_r}\sin\phi_r\right), \end{aligned}$$

$$\text{giving}\quad E_s - E_r = IR\cos\phi_r + IX\sin\phi_r \quad (66)$$

and the percentage regulation

$$= \frac{IR\cos\phi_r + IX\sin\phi_r}{E_r} \times 100 \text{ per cent} \quad . \quad (67)$$

This can be seen from the vector diagram, in which $LM = IR\cos\phi_r$ and $MC = IX\sin\phi_r$ approximately.

The regulation depends greatly upon the power factor of the load. Thus for unity power factor the regulation is IR; for zero power factor IX; for 0·8 power factor lagging it is $0{\cdot}8IR + 0{\cdot}6IX$, whilst for 0·8 power factor leading it is $0{\cdot}8IR - 0{\cdot}6IX$. In the latter case it is quite likely that the received voltage is greater than the sending voltage.

The regulation may be expressed in the more convenient form

$$IR\cos\phi_r + IX\sin\phi_r$$
$$= IZ\,[(R/Z)\cos\phi_r + (X/Z)\sin\phi_r]$$
$$\left.\begin{aligned} &= IZ\cos(\phi_r - \psi), \\ \text{where} \quad &Z = \sqrt{(R^2 + X^2)} \text{ and } \tan\psi = X/R \end{aligned}\right\} \quad . \quad . \quad (68)$$

ψ is the line impedance angle.

The regulation is zero when $\phi_r - \psi = -(\pi/2)$, and then $\cos\phi_r = \sin\psi = X/Z$. If $R + jX$ is the impedance of a line, generator, or transformer, ψ is positive since X is positive. It is then necessary that ϕ_r be negative for zero regulation. If it is the impedance of a synchronous condenser, ψ may be negative, so that ϕ_r may be positive for zero regulation. The formula just given, viz. $\phi_r = \psi - (\pi/2)$, for zero regulation is approximate, since we have neglected the last term in equation (64) and have made an approximation in taking the square root. The exact relation is found in the following way. Let $E_s = E_r$. Equation (64) then becomes

$$E_r^2 = (E_r\cos\phi_r + IR)^2 + (E_r\sin\phi_r + IX)^2$$
$$= E_r^2 + 2IE_r(R\cos\phi_r + X\sin\phi_r) + I^2(R^2 + X^2),$$

so that $$R\cos\phi_r + X\sin\phi_r = -IZ^2/2E_r \quad . \quad . \quad (69)$$

or $$\cos(\phi_r - \psi) = -IZ/2E_r,$$

so that $$\phi_r = \psi - \pi/2 - \sin^{-1}(IZ/2E_r). \quad . \quad . \quad . \quad (69a)$$

The regulation is negative, i.e. the receiving-end voltage is greater than the sending voltage, if ϕ_r is negative and numerically greater than the value given in this equation.

EXAMPLE. A three-phase transmission line has a resistance per phase of 5 Ω. and an inductive reactance per phase of 12 Ω., and the line voltage at the receiving-end is 33 kV.

(*a*) Determine the voltage at the sending-end when the load at the receiving-end is 20 000 kVA. at 0·8 power factor (lagging).

(*b*) The voltage at the sending-end is maintained constant at 36 kV. by means of a synchronous phase modifier at the receiving-end, which has the same rating at zero load at the receiving-end as for the full load of 16 000 kW. Determine the power factor of the full-load output and the rating of the synchronous phase modifier. (*Lond. Univ.*, 1947.)

(*a*) The VA. per phase is (20 000/3) kVA., the receiving voltage per phase is $E_r = 33$ kV.$/\sqrt{3} = 19{\cdot}05$ kV., so that $I = (20\,000/3) \div 19{\cdot}05 = 350$ A. Also $\cos\phi_r = 0{\cdot}8$ and $\sin\phi_r = 0{\cdot}6$, so that

$$E_s = E_r + IR\cos\phi_r + IX\sin\phi_r = (19{\cdot}05 + 1{\cdot}4 + 2{\cdot}4)\text{ kV.}$$
$$= 22{\cdot}85\text{ kV.},$$

which corresponds to a phase-to-phase voltage of 39·6 kV.

(*b*) Here $E_r = 19{\cdot}05$ kV. and $E_s = (36/\sqrt{3}) = 20{\cdot}8$ kV. Let the synchronous phase modifier have a phase current I_0, which is in quadrature with E_r. Then at zero load, $E_s = E_r + I_0X$, so that $I_0 = 150$ A.: this current lags E_r by 90°. The rating is $3 \times 150 \times 19{\cdot}05$ kVA. $= 8\,600$ kVA.

If the load has an in-phase current I, and quadrature current I_2, the sending-end voltage is given by

$$E_s = E_r + I_1R + I_2X + I_0X.$$

If E_s and E_r are fixed at the values given above, it follows that $I_1R + I_2X = 0$, i.e. $I_2/I_1 = \tan\phi_r = -R/X = -5/12$, $\phi_r = -22° 36'$, so that the power factor is 0·923 leading.

The effect of transformers is allowed for in the way shown in the next example.

Example. A three-phase load of 2 000 kVA., 0·8 power factor, is supplied at 6·6 kV., 50 cyc., by means of a 33 kV. transmission line 20 miles long and a 5 : 1 transformer. The resistance per mile of each conductor is 0·40 Ω. and reactance 0·5 Ω. The resistance and reactance of the transformer primary are 7·5 Ω. and 13·2 Ω., whilst the resistance of the secondary is 0·35 Ω. and reactance 0·65 Ω. Find the voltage necessary at the sending-end of the transmission line when 6·6 kV. is maintained at the load and find the sending-end power factor. Determine also the efficiency of the transmission.

(*Nat. Cert.*, 1935.)

The system is represented in Fig. 104. The resistance and reactance of the high voltage line are 8 Ω. and 10 Ω. per phase; to these we add the primary impedance of the transformer, giving $(15{\cdot}5 + j23{\cdot}2)$ Ω. in the high-tension side. This impedance can be transferred to the lower voltage side, giving

$$(15{\cdot}5 + j23{\cdot}2) \div 5^2 = 0{\cdot}62 + j0{\cdot}928.$$

We add to this the impedance of the secondary, so that the total impedance transferred to the low voltage side is

$$(0{\cdot}62 + j0{\cdot}928) + (0{\cdot}35 + j0{\cdot}65) = 0{\cdot}97 + j1{\cdot}578.$$

The receiving-end voltage is 6·6 kV. between phases, or $6{\cdot}6 \div \sqrt{3} = 3{\cdot}81$ kV. from phase to neutral. The load is 667 kVA. per phase, so that the line current is

$$I = 667/3{\cdot}81 = 175 \text{ A}.$$

The regulation per phase is

$$\begin{aligned} &IR \cos \phi_r + IX \sin \phi_r \\ &= 175(0{\cdot}97 \times 0{\cdot}8 + 1{\cdot}578 \times 0{\cdot}6) \\ &= 302 \text{ V}. \end{aligned}$$

The sending-end voltage (phase-to-neutral referred to the lower

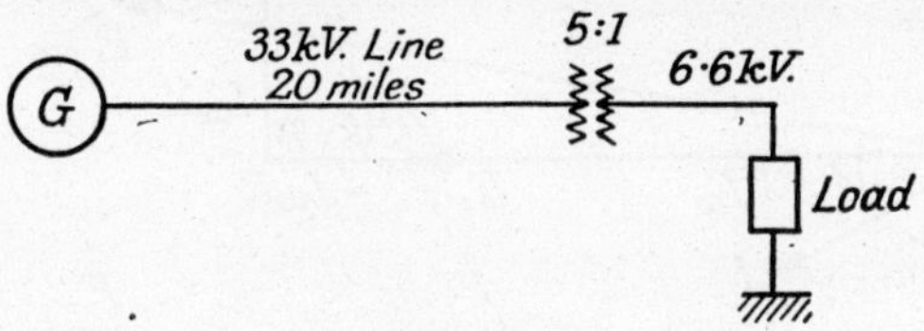

FIG. 104

voltage side) is thus 3 810 + 302 = 4 112 V. The sending-end voltage (between phases referred to the higher voltage side) is thus

$$4\ 112 \times \sqrt{3} \times 5 = \underline{\underline{35{\cdot}6}} \text{ kV}.$$

The sending-end power factor is $\cos \phi_s$, where

$$\tan \phi_s = \frac{\sin \phi_r + IX/E_r}{\cos \phi_r + IR/E_r}$$

$$= \frac{0{\cdot}6 + \dfrac{175 \times 1{\cdot}578}{3\ 810}}{0{\cdot}8 + \dfrac{175 \times 0{\cdot}97}{3\ 810}} = \frac{0{\cdot}672}{0{\cdot}844} = 0{\cdot}796.$$

This gives $\phi_s = 38° \ 31'$ and $\cos \phi_s = \underline{\underline{0{\cdot}782}}$.

The loss per phase is

$$I^2R = 175^2 \times 0{\cdot}97 = 29{\cdot}7 \text{ kW.},$$

whilst the received power per phase is

$$\tfrac{1}{3} \times 2\ 000 \times 0{\cdot}8 \text{ kW.} = 533{\cdot}3 \text{ kW}.$$

The efficiency is thus

$$\frac{533\cdot3}{533\cdot3 + 29\cdot7} \times 100 = \underline{\underline{94\cdot7\%}}.$$

Line and Machine Charts. In Fig. 103 the current was taken as the reference vector, but in Fig. 105 the receiving-end voltage is the reference vector. Let the current I be resolved into the power component I_w and the wattless component I_q. Let $OA = E_r$, $AB = I_wR$, $BC = I_wX$, $CD = I_qX$, and $DE = I_qR$. The sending-

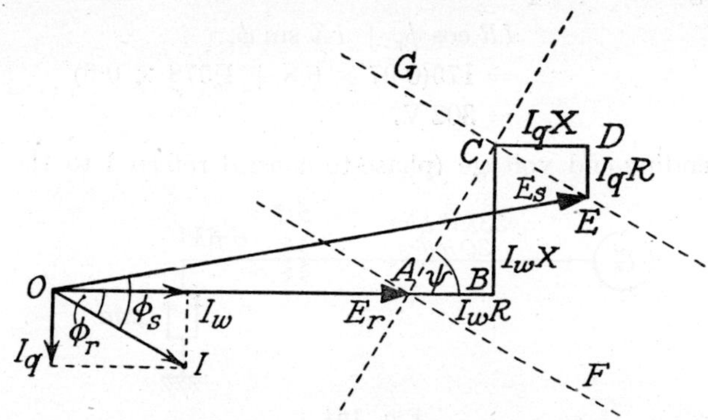

FIG. 105. VECTOR DIAGRAM FOR SHORT LINE

end voltage is $OE = E_s$ and makes an angle ϕ_s with the current vector I; the angle AOE is $(\phi_s - \phi_r)$.

The component of E_s along E_r is

$$\begin{aligned} E_s \cos(\phi_s - \phi_r) &= E_r + I_wR + I_qX \\ &= E_r + IR\cos\phi_r + IX\sin\phi_r. \end{aligned}$$

The component at right angles to E_r is

$$\begin{aligned} E_s \sin(\phi_s - \phi_r) &= I_wX - I_qR \\ &= IX\cos\phi_r - IR\sin\phi_r. \end{aligned}$$

Thus

$$\begin{aligned} E_s &= \sqrt{[(E_r + IR\cos\phi_r + IX\sin\phi_r)^2 + (IX\cos\phi_r - IR\sin\phi_r)^2]} \\ &\simeq E_r + IR\cos\phi_r + IX\sin\phi_r, \quad . \quad . \quad (70) \end{aligned}$$

since the second term is usually small compared with the first and *is in quadrature with it.* We have thus found the value of the regulation very quickly, but it is not easy to find the sending end power factor by this method.

It is seen that the power component of the current causes a voltage-drop along AC, whilst the wattless component causes a drop along CE which is at right angles to AC. We can use these facts as a basis for constructing a line (or machine) chart to determine the sending-end voltage for a given receiving-end voltage and various received loads and power factors. We take E_r as unity and draw circles with O as centre and radii 0·90, 1·00, 1·10, 1·20, etc. We draw rectangular axes AC and AF, where $\angle CAB = \psi = \tan^{-1} (X/R)$. The axis AC represents power; as unit distance represents a voltage of E_r, it represents a power component of current equal to (E_r/Z) or a power (E_r^2/Z) watts per phase, or $(3E_r^2/Z)$ watts total power. The axis AF represents VAR. and has the same scale. In practice it is more convenient to take AF and AC along the straight edges of the lined paper and draw OA to make an angle ψ with AC. The following example shows the construction and use of such a chart.

EXAMPLE. Power is to be transmitted over a short single-phase line, the output being at varying power factors and the voltage at the output end being 10 kV. The resistance and reactance of the line are 16 and 30 Ω. respectively. Construct a chart which enables the generator voltage to be determined readily for any load up to 1 000 kW. at any power factor. Determine from the chart the values of the generator voltage at 500 and 1 000 kW. when the power factors are 0·9, 0·7 and 0·5, leading and lagging respectively.
(*Lond. Univ.*, 1931.)

Here $\psi = \tan^{-1} (30/16) = 61° 56'$ and $Z = \sqrt{(16^2 + 30^2)} = 34{\cdot}0$.
Let $E_r = 10$ kV. be represented by x cm.; then x cm. will represent a power of

$$(10\,000^2/34) \text{ W.} = (100\,000/34) \text{ kW.}$$

The maximum power of 1 000 kW. is represented by a length

$$\left(1\,000 \div \frac{100\,000}{34}\right) \times x \text{ cm.} = 0{\cdot}34x \text{ cm.}$$

Before choosing a value of x it is necessary to find the maximum kVAR. involved. As

$$EI \cos \phi = W \text{ and } EI \sin \phi = \text{VAR.},$$

it follows that

$$\text{VAR.} = W \tan \phi,$$

so that the maximum VAR. is

$$1\,000 \times \tan (\cos^{-1} 0{\cdot}5) \text{ kVA.} = 1\,730 \text{ kVA.}$$

We will choose a length of 5 cm. to represent 2 000 kW. (or kVA.), giving

$$2 \times 0{\cdot}34 \times x = 5 \text{ cm., i.e. } x = 5/0{\cdot}68 = 7{\cdot}36 \text{ cm.}$$

The chart is shown in Fig. 106, and the following table gives the various quantities of importance,

Power	Power Factor	tan ϕ	kVAR.	Generator Voltage
500 kW.	0·9 lagging	0·484	242 kVA.	11·5 kV.
„	0·9 leading	— 0·484	— 242 „	10·3 „
„	0·7 lagging	1·02	510 „	12·4 „
„	0·7 leading	— 1·02	— 510 „	9·6 „
„	0·5 lagging	1·73	865 „	13·4 „
„	0·5 leading	— 1·73	— 865 „	8·7 „
1 000 kW.	0·9 lagging	0·484	484 „	13·3 „
„	0·9 leading	— 0·484	— 484 „	10·8 „
„	0·7 lagging	1·02	1 020 „	14·6 „
„	0·7 leading	— 1·02	— 1 020 „	9·7 „
„	0·5 lagging	1·73	1 730 „	16·8 „
„	0·5 leading	— 1·73	— 1 730 „	8·6 „

Let us check the sending-end voltage for 1 000 kW. and 0·5 power factor leading and lagging by means of the exact formula of equation (70).

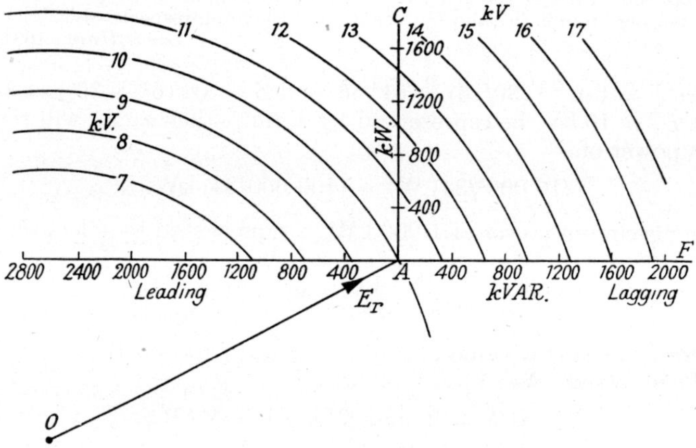

FIG. 106. REGULATION CHART FOR SHORT LINE

$$E_s = \sqrt{[(E_r + IR\cos\phi_r + IX\sin\phi_r)^2 + (IX\cos\phi_r - IR\sin\phi_r)^2]}.$$

We have $W = E_r I \cos\phi_r$,

so that $I = W/E_r \cos\phi_r = 1\,000/(10 \times 0{\cdot}5) = 200$ A

$$\cos\phi_r = 0{\cdot}5,\ \sin\phi_r = 0{\cdot}866,$$

$$IR\cos\phi_r = 1\,600,\ IX\sin\phi_r = 5\,200,$$

$$IR\sin\phi_r = 2\,780,\ IX\cos\phi_r = 3\,000.$$

Thus for lagging power factor

$$E_s = \sqrt{(16\ 800^2 + 220^2)} = 16{\cdot}8 \text{ kV.},$$

whilst for leading power factor

$$E_s = \sqrt{(6\ 400^2 + 5\ 780^2)} = 8{\cdot}62 \text{ kV.}$$

The results agree perfectly with those obtained from the chart, rather better than one would expect. We may note that the approximate formula

$$E_s = E_r + IR \cos\phi_r + IX \sin\phi_r$$

gives values of 16·8 kV. and 6·4 kV. It can be seen that the approximate formula will be substantially correct for lagging power factor, but gives values which are very low for leading power factor.

Maximum Power with a Given Regulation. The maximum power that can be transmitted with a given regulation is easily found from the line chart. Thus if the receiving- and sending-end voltages in the previous example are to be 10 kV., the maximum power is found by continuing the circle for 10 kV. until it shows a maximum kW. Thus in Fig. 106 the maximum power is 1 560 kW., when the reactive power is — 2 600 kVA. and the power factor is 0·515 leading.

For different sending- and receiving-end voltages, the appropriate circle for sending-end voltage is taken. The algebraic method is as follows. Equation (64) gives

$$E_s^2 = E_r^2 + 2IE_r(R\cos\phi_r + X\sin\phi_r) + I^2(R^2 + X^2).$$

The received power is

$$P = E_r I \cos\phi_r$$

and the received reactive voltamps

$$Q = E_r I \sin\phi_r,$$

so that the equation may be written

$$-E_s^2 + E_r^2 + 2PR + 2QX + (P^2 + Q^2)(R^2 + X^2)/E_r^2 = 0 \quad . \quad . \quad (70a)$$

For a given line, i.e. given values of R and X, and for given receiving- and sending-end voltages, this equation gives a relation between P and Q. If we vary Q we vary P, so that we consider that this equation gives P as an implicit function of Q. We find the maximum value of P by differentiating this equation with respect to Q and putting $(dP/dQ) = 0$. We thus get the condition for a maximum value of P as

$$2X + 2Q(R^2 + X^2)/E_r^2 = 0,$$

or

$$Q = -E_r^2X/(R^2 + X^2):$$

i.e. for maximum P the value of Q is a constant negative value. The truth of this is obvious from the line chart of Fig. 106, in which it is seen that the tops of the circles occur at the value of

$$Q = -2\,600 \text{ kVA.}$$

This value is

$$-\frac{(10\,000)^2 \times 30}{(16^2 + 30^2)} = -\frac{3 \times 10^9}{1\,156} = -2\,598 \text{ kVA.}$$

The maximum value of P is found by substituting this value of Q in the equation (70*a*), which then becomes

$$P^2(Z^2/E_r^2) + 2PR + E_r^2 - E_s^2 - 2E_r^2(X^2/Z^2) + (E_r^4X^2/Z^4)(Z^2/E_r^2) = 0$$

or
$$P^2(Z^2/E_r^2) + 2PR + E_r^2(R^2/Z^2) - E_s^2 = 0.$$

We find

$$P_{max} = (E_r^2/Z^2)\,[-R \pm \sqrt{(R^2 - R^2 + E_s^2Z^2/E_r^2)}] = (E_r^2/Z^2)\,[Z(E_s/E_r) - R],$$

since the positive sign of the square root must be taken.

Thus if we take $E_s = 11$ kV. and $E_r = 10$ kV. in the previous example, we find

$$P_{max} = (10^8/1\,156)\,[34 \times 1{\cdot}1 - 16] = 1\,850 \text{ kW.},$$

which is seen to be the value given by the line chart.

In the particular, when $E_s = E_r$, the value of Q for maximum power is the same and

$$P_{max} = (E_r^2/Z^2)\,(Z - R).$$

For the example of the chart this value is

$$(10^8/1\,156)\,(34 - 16) = 1\,556 \text{ kW.},$$

which again agrees with the value given by the chart.

If the reactance of the line can be varied, P_{max} will vary, and will have a final maximum when $(dP_{max}/dX) = 0$. As

$$P_{max} = (E_r^2/Z^2)\,[Z(E_s/E_r) - R] = (E_rE_s/Z) - E_r^2(R/Z^2) = E_rE_s/\sqrt{(R^2 + X^2)} - E_r^2R/(R^2 + X^2),$$

the condition is

$$E_rE_sX/(R^2 + X^2)^{\frac{3}{2}} = 2E_r^2RX/(R^2 + X^2)^2,$$

which yields

$$X = 0$$

or
$$(R^2 + X^2)^{\frac{1}{2}} = 2R(E_r/E_s), \text{ i.e. } X = R\sqrt{[(4E_r^2/E_s^2) - 1]}.$$

The value of P_{max} is then

$$P_{max} = E_s^2/4R.$$

For the particular case when $E_s = E_r$, the value of X is $R\sqrt{3}$ and P_{max} is $E_r^2/4R$.

Mixed Conditions. Suppose the sending-end voltage and the receiving-end power and power factor are given. It is required to find the receiving-end voltage and the sending-end power and power factor. We are given

$$E_r I \cos\phi_r = P, \text{ the received power,}$$

and

$$E_r I \sin\phi_r = Q, \text{ the VAR.}$$

We have

$$E_s^2 = E_r^2 + 2E_r I(R\cos\phi_r + X\sin\phi_r) + I^2(R^2 + X^2),$$

or

$$E_s^2 = E_r^2 + 2(PR + QX) + (R^2 + X^2)P^2/E_r^2\cos^2\phi_r.$$

This gives

$$E_r^4 - AE_r^2 + B = 0,$$

where

$$A = E_s^2 - 2(PR + QX)$$

and

$$B = (R^2 + X^2)P^2/\cos^2\phi_r.$$

Then

$$E_r = \sqrt{[\tfrac{1}{2}A \pm \tfrac{1}{2}\sqrt{(A^2 - 4B)}]}.$$

The usual value is that with the positive sign. The remaining quantities can then be found.

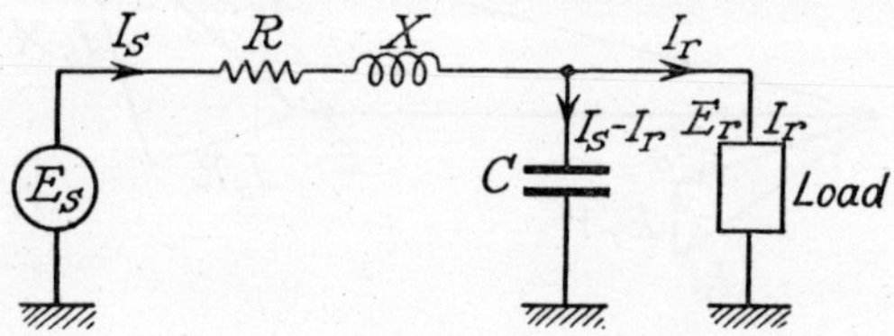

FIG. 107. MEDIUM LINE, LOCALIZED LOAD-END CAPACITANCE

Medium Lines. In overhead lines more than 50 miles long the charging currents cannot be neglected. If the line is less than 150 miles long, the line capacitance may be assumed as existing at one or more points, the more the points the closer the approximation. A first approximation is obtained by assuming that the line capacitance can be put across the load, as shown in Fig. 107. Putting $R + jX = Z$ and $Y = j\omega C$, we obtain

$$E_s = ZI_s + E_r \text{ and } I_s - I_r = E_r Y,$$

giving

$$\left.\begin{aligned} E_s &= E_r(1 + YZ) + I_r Z \\ I_s &= E_r Y + I_r. \end{aligned}\right\} \quad . \quad . \quad . \quad . \quad (71)$$

(and)

The vector diagram is shown in Fig. 108.

If we are given the receiving-end conditions, these equations give the sending-end conditions at once.

Example. A transmission line 100 miles long has the following constants—

Resistance per mile	0·25 Ω.
Reactance per mile	0·8 Ω.
Susceptance per mile	14×10^{-6} mho.
Receiving-end line voltage . . .	66 000 V.

Using a suitable approximate method, taking the susceptance into account, determine the sending-end voltage and the sending-end current when the line is delivering 15 000 kW. at 0·8 power factor lagging.

(*Lond. Univ.*, 1932.)

The system is clearly three-phase. Much needless arithmetic is avoided by the use of the following theorem. *A three-phase system of voltage between lines V, total power P, and impedance per line Z, has the same regulation and efficiency as a single-phase system of voltage V, power P, and loop impedance Z.*

For in the three-phase system the voltage to neutral is $V/\sqrt{3}$, the power per phase is $P/3$, and thus the line current is $(\frac{1}{3}P \div V/\sqrt{3}) = (P/V) \div \sqrt{3}$. The line equations are of the form $E_s/\sqrt{3} = (A\,E_r/\sqrt{3}) + BI_r = AE_r/\sqrt{3} + (BP_r/V_r\sqrt{3})$ or $E_s = AE_r + BP_r/V_r$, which is the line equation for a single-phase system with *loop* constants A and B. The only difference is that the line currents are the values for the single-phase circuit divided by $\sqrt{3}$.

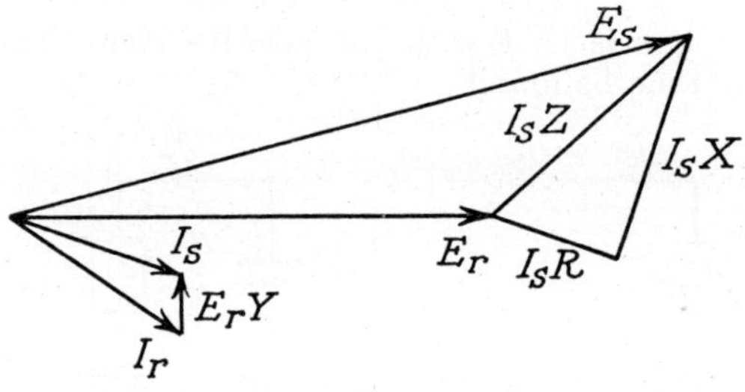

Fig. 108. Vector Diagram for Fig. 107

Let us assume that the susceptance may be concentrated at the load. Then

$$Z = 25 + j80 \text{ and } Y = j0{\cdot}0014,$$

giving
$$E_s = E_r(1 + j0{\cdot}035 - 0{\cdot}112) + I_r(25 + j80)$$
$$= E_r(0{\cdot}888 + j0{\cdot}035) + I_r(25 + j80)$$

and
$$I_s = E_r(j0{\cdot}0014) + I_r.$$

Let us take E_r as the basic vector, so that it is a real number 66 000. The power component of the received current is in phase with E_r and is $15\,000/66 = 227$ A.; the wattless component is $227 \tan \phi = 227 \times \frac{3}{4} = 170$ A. The current is thus

$$I_r = 227 - j170 = 284 \,\overline{|36° \; 52'}$$

since the power factor is lagging.

We have

$$\begin{aligned} E_s &= 66\,000\,(0{\cdot}888 + j0{\cdot}035) + (227 - j170)\,(25 + j80) \\ &= 58\,600 + j2300 + 19\,270 + j13\,900 \\ &= 77\,900 + j16\,200 = 79\,500\,\underline{|11^\circ\,44'} \\ I_s &= 66\,000\,(j0{\cdot}0014) + 227 - j170 \\ &= 227 - j78 = 240\,\overline{|18^\circ\,57'}. \end{aligned}$$

The vector power at the sending-end is

$$\begin{aligned} &E_s \times \text{conjugate of } I_s \\ &= 79\,500\,\underline{|11^\circ\,44'} \times 240\,\underline{|18^\circ\,57'} \\ &= 19\,100\,000\,\underline{|30^\circ\,41'} \\ &= 16\,400\text{ kW.} + j9\,750\text{ kVAR.} \end{aligned}$$

The sending-end voltage is 79·5 kV., current 138 A. (against the load current of 164 A.), power 16 400 kW., and power factor 0·86 lagging.

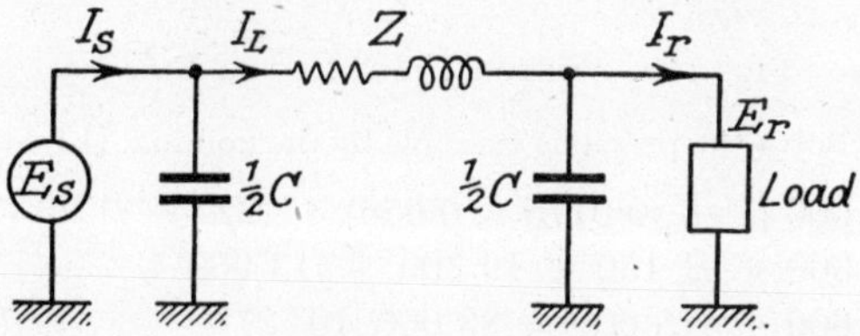

FIG. 109. MEDIUM LINE, NOMINAL Π-METHOD

If we ignore the line capacitance, the equations for the sending-end voltage and current are

$$\left.\begin{aligned} E_s &= E_r + I_r Z \\ \text{and} \quad I_s &= I_r \end{aligned}\right\} \qquad . \quad . \quad . \quad . \quad . \quad (71a)$$

giving

$$\begin{aligned} E_s &= 85\,300 + j13\,900 = 86\,400\,\underline{|9^\circ\,15^\circ}, \\ I_s &= 284\,\overline{|36^\circ\,52'}, \end{aligned}$$

and vector power $245\,000\,000\,\underline{|46^\circ\,7'}$

$$= 17\,000\text{ kW.} + j17\,700\text{ kVAR.}$$

The sending-end voltage is 86·4 kV., current 164 A., power 17 000 kW., and power factor 0·693 lagging.

We shall see later that the exact values lie between these two sets. The method of localizing the capacitance at the load-end overestimates the effect of the capacitance. An average of these two sets is much closer to the actual values.

NOMINAL Π-METHOD. In this method the line capacitance is assumed to be localized half at the receiving-end and half at the sending-end. The method is shown in Fig. 109, and the vector diagram in Fig. 110. We have

$$I_s - I_L = \tfrac{1}{2}E_sY,\ I_L - I_r = \tfrac{1}{2}E_rY$$

and $$E_s - E_r = I_LZ.$$

Eliminating I_L we obtain

$$\left.\begin{aligned} E_s &= E_r(1 + \tfrac{1}{2}YZ) + I_rZ \\ \text{and}\quad I_s &= E_rY(1 + \tfrac{1}{4}YZ) + I_r(1 + \tfrac{1}{2}YZ. \end{aligned}\right\} \quad . \quad (72)$$

The method of solution is as before.

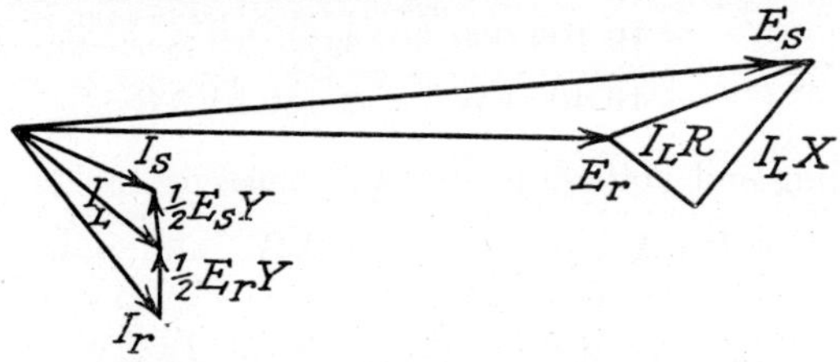

FIG. 110. VECTOR DIAGRAM FOR FIG. 109

EXAMPLE. Solve the previous example by the nominal Π-method.

$$\begin{aligned} E_s &= 66\ 000\ (1 + j0{\cdot}0175 - 0{\cdot}056) + (227 - j170)\ (25 + j80) \\ &= 62\ 300 + j1\ 150 + 19\ 300 + j13\ 900 \\ &= 81\ 600 + j15\ 000 = 83\ 000\ \underline{|10^\circ\ 27'} \end{aligned}$$

$$\begin{aligned} I_s &= 66\ 000\ (j0{\cdot}0014)\ (1 \times j0{\cdot}009 - 0{\cdot}028) \\ &\qquad + (227 - j170)\ (1 + j0{\cdot}0175 - 0{\cdot}056) \\ &= j90 - 0{\cdot}8 + 218 - j157 \\ &= 217 - j67 = 227\ \overline{|17^\circ\ 0'}. \end{aligned}$$

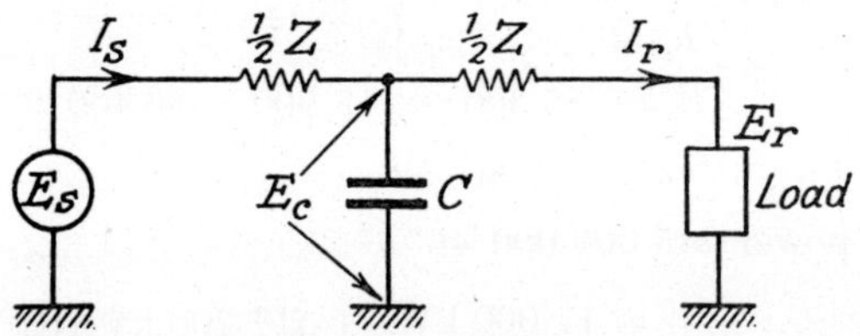

FIG. 111. MEDIUM LINE, NOMINAL T-METHOD

The vector power is 18 900 000 $\underline{|27^\circ\ 27'}$

$$= 16\ 800\ \text{kW.} + j8\ 800\ \text{kVAR.}$$

The sending-end power factor is 0·887 lagging.

NOMINAL T-METHOD. In this method the line capacitance is placed at the mid-point of the line. The method is shown in Fig. 111

and the vector diagram in Fig. 112. The equations can be shown to be

$$\left.\begin{aligned} E_s &= E_r(1 + \tfrac{1}{2}YZ) + I_rZ(1 + \tfrac{1}{4}YZ) \\ I_s &= E_rY + I_r(1 + \tfrac{1}{2}YZ) \end{aligned}\right\} \quad . \quad (73)$$

EXAMPLE. Solve the previous example by the nominal T-method.

From the preceding work

$$\begin{aligned} E_s &= 62\,300 + j1\,150 + (19\,300 + j13\,900)\,(1 + j0{\cdot}009 - 0{\cdot}028) \\ &= 81\,000 + j14\,800 = 82\,300\,\underline{|10° 22'}, \end{aligned}$$

and $I_s = j92 + 218 - j157$

$$= 218 - j65 = 229\,\overline{|16° 36'}.$$

The vector power is $189\,000\,000\,\underline{|26° 58'}$

$$= 16\,900\text{ kW.} + j8\,570\text{ kVAR.}$$

The sending-end power factor is 0·891 lagging. It is seen that the Π- and T-methods give substantially the same results.

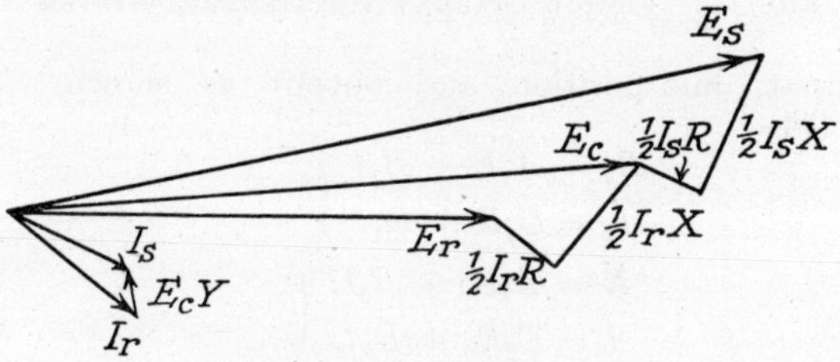

FIG. 112. VECTOR DIAGRAM FOR FIG. 111

Constants of the General Network. It is seen that the solutions of the network problems of Figs. 102, 107, 109, and 111 are given in equations (71*a*), (71), (72), and (73) in the form

$$\left.\begin{aligned} E_s &= AE_r + BI_r \\ I_s &= CE_r + DI_r, \end{aligned}\right\} \quad . \quad . \quad . \quad . \quad (74)$$

where A, B, C, and D are constants of the networks. It can be shown that the solution of the problem of any line or linear network with two input and two output terminals can be expressed in this form. A, B, C, and D are called the *T-constants* of the network, and they follow from the fact that the currents in the network are proportional to the voltages. When the line or network contains no internal e.m.f.s or sources (such as might be due to generators or synchronous condensers) it is called *passive*, and there is the following relation between the constants.

$$AD - BC = 1.$$

This is a useful check on the work. In symmetrical networks $A = D$. The vector diagram representing equations (74) is shown in Fig. 113.

Before the solution of the long transmission line is found, in the form of equations (74), we will find the T-constants of two networks in tandem.

Two Networks in Tandem. Fig. 114 shows two networks in tandem or series. The T-constants are A_1, B_1, C_1, D_1 and A_2, B_2, C_2, D_2. Let E_s and I_s, E and I, E_r and I_r be the voltages and currents

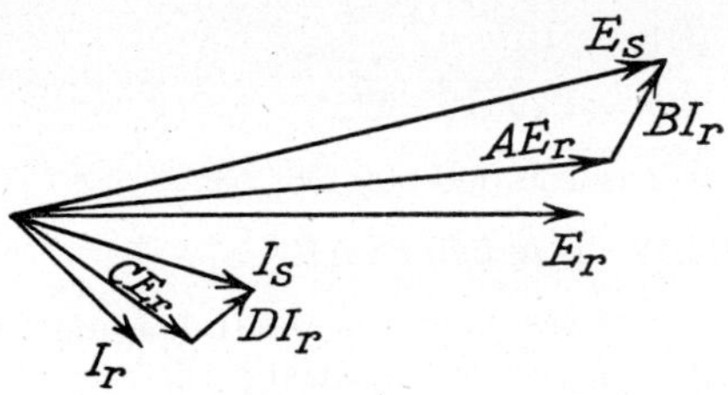

FIG. 113. VECTOR DIAGRAM FOR GENERAL SYSTEM

at the input, mid-position and output as shown. Then by equations (74)

$$\left.\begin{aligned} E_s &= A_1E + B_1I \\ I_s &= C_1E + D_1I \end{aligned}\right\}$$

and

$$\left.\begin{aligned} E &= A_2E_r + B_2I_r \\ I &= C_2E_r + D_2I_r \end{aligned}\right\}$$

Substituting for E and I we get

$$\begin{aligned} E_s &= A_1(A_2E_r + B_2I_r) + B_1(C_2E_r + D_2I_r) \\ &= (A_1A_2 + B_1C_2)E_r + (A_1B_2 + B_1D_2)I_r, \\ I_s &= (C_1A_2 + D_1C_2)E_r + (C_1B_2 + D_1D_2)I_r. \end{aligned}$$

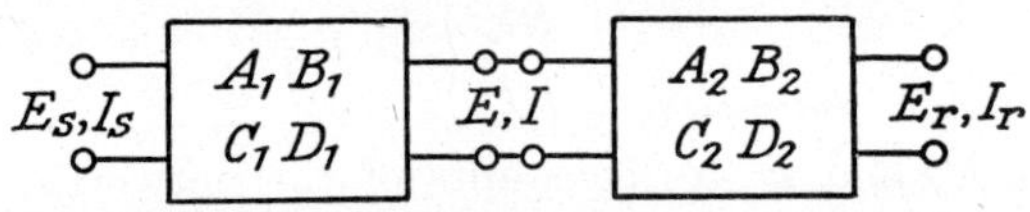

FIG. 114. TWO NETWORKS IN TANDEM

Thus the combination is a network whose T-constants are A, B, C, D where

$$\left.\begin{aligned} A &= A_1A_2 + B_1C_2, \\ B &= A_1B_2 + B_1D_2, \\ C &= C_1A_2 + D_1C_2, \\ D &= C_1B_2 + D_1D_2. \end{aligned}\right\} \quad . \quad . \quad . \quad . \quad (75)$$

Given the T-constants of the separate networks, we find the T-constants of the tandem combination by means of equations (75).

Rigorous Solution of Long Transmission Line. Let the line have impedance z and admittance y per unit length. At a distance x from an arbitrary zero (which later we choose to be the receiving end) let the voltage and current be E_x and I_x. Between the points at distances x and $x + dx$ there is a series impedance zdx and shunt admittance ydx, as shown in Fig. 115. There is thus a voltage drop from x to $x + dx$ of $I_x zdx$; but the voltage drop is

$$E_x - E_{x+dx} = E_x - \left[E_x + \frac{dE_x}{dx} dx \right] = - \frac{dE_x}{dx} dx,$$

since $E_{x+dx} = E_x + \dfrac{dE_x}{dx} dx$ to first order.

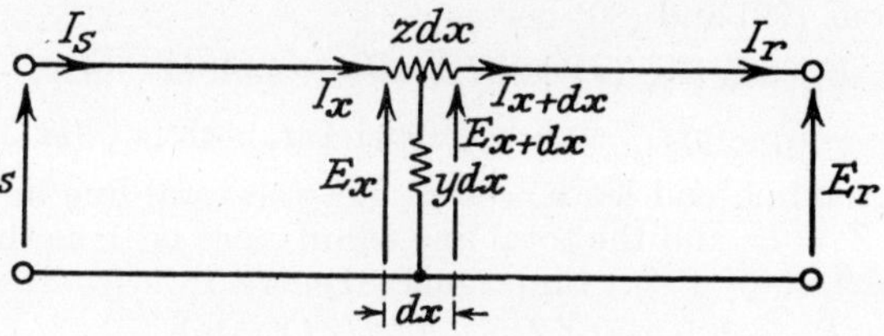

FIG. 115. LONG TRANSMISSION LINE

Therefore

$$I_x zdx = - \frac{dE_x}{dx} dx,$$

i.e.

$$\frac{dE_x}{dx} = - I_x z \quad . \quad . \quad . \quad . \quad . \quad . \quad (76)$$

The admittance ydx causes a current to flow between the lines (this is composed of charging and leakage currents) of amount $E_x ydx$; but this current must be equal to

$$I_x - I_{x+dx} = - \frac{dI_x}{dx} dx,$$

by the method used above. Equating we get

$$dI_x/dx = - E_x y \quad . \quad . \quad . \quad . \quad (77)$$

Equations (76) and (77) are the differential equations for the line voltage and current. They can be solved in the following way. Differentiating equation (76) and substituting for dI_x/dx from equation (77) we get

$$d^2E_x/dx^2 = -(dI_x/dx) = yzE_x \quad . \quad . \quad . \quad (78)$$

This is a well-known differential equation of which the solution is

$$E_x = E_1 \cosh [x\sqrt{(yz)}] + E_2 \sinh [x\sqrt{(yz)}] \quad . \quad (79)$$

where E_1 and E_2 are constants that are determined by the end conditions and sinh and cosh are the hyperbolic sine and cosine. For a brief treatment of the latter see Appendix I. From equation (76)

$$I_x = -\frac{1}{z}\frac{dE_x}{dx} = -E_1\sqrt{\left(\frac{y}{z}\right)} \,.\, \sinh\,[x\sqrt{(yz)}]$$
$$-E_2\sqrt{\left(\frac{y}{z}\right)} \,.\, \cosh\,[x\sqrt{(yz)}] \quad . \quad (80)$$

Let us choose the receiving end as the zero, so that $x = 0$ there. Then $E_0 = E_r$ and $I_0 = I_r$. Remembering that cosh $0 = 1$ and sinh $0 = 0$, we see that

$$E_r = E_1 \text{ and } I_r = -E_2\sqrt{(y/z)}.$$

Equations (79) and (80) become

$$\left.\begin{aligned} E_x &= E_r \cosh\,[x\sqrt{(yz)}] - I_r\sqrt{(z/y)} \,.\, \sinh\,[x\sqrt{(yz)}] \\ \text{and}\quad I_x &= -E_r\sqrt{(y/z)} \,.\, \sinh\,[x\sqrt{(yz)}] + I_r \cosh\,[x\sqrt{(yz)}] \end{aligned}\right\} \quad . \quad (81)$$

At the sending-end let $x = -l$. Let the total line impedance be Z so that $Z = lz$, and the total line admittance be Y so that $Y = ly$; then $l\sqrt{(yz)} = \sqrt{(YZ)}$. Equations (81) give

$$\begin{aligned} E_s &= E_r \cosh\,[-\sqrt{(YZ)}] - I_r\sqrt{(Z/Y)} \sinh\,[-\sqrt{(YZ)}] \\ &= E_r \cosh \sqrt{(YZ)} + I_r\sqrt{(Z/Y)} \sinh \sqrt{(YZ)} \\ \text{and}\quad I_s &= -E_r\sqrt{(Y/Z)} \sinh\,[-\sqrt{(YZ)}] + I_r \cosh\,[-\sqrt{(YZ)}] \\ &= E_r\sqrt{(Y/Z)} \sinh \sqrt{(YZ)} + I_r \cosh \sqrt{(YZ)}, \end{aligned}$$

since $\cosh(-x) = \cosh x$ and $\sinh(-x) = -\sinh x$. The equations can be written in the form

$$\left.\begin{aligned} E_s &= AE_r + BI_r \\ I_s &= CE_r + DI_r \end{aligned}\right\} \quad . \quad . \quad . \quad . \quad (74)$$

where

$$\left.\begin{aligned} A &= D = \cosh \sqrt{(YZ)}, \\ B &= \sqrt{(Z/Y)} \sinh \sqrt{(YZ)} \\ C &= \sqrt{(Y/Z)} \sinh \sqrt{(YZ)} \end{aligned}\right\} \quad . \quad . \quad . \quad (82)$$

The relation $AD - BC = 1$ holds, for

$$AD = BC = \cosh^2 \sqrt{(YZ)} - \sinh^2 \sqrt{(YZ)} = 1.$$

Methods of Applying the Rigorous Solution. There are three ways of applying the rigorous solution.

Method 1. In this method the hyperbolic sine and cosine are expressed in terms of their power series. The expansions are

$$\cosh x = 1 + \frac{x^2}{2!} + \frac{x^4}{4!} + \frac{x^6}{6!} + \ldots$$

and

$$\sinh x = x + \frac{x^3}{3!} + \frac{x^5}{5!} + \frac{x^7}{7!} + \ldots$$

It is never worth while going beyond the seventh power. We then get

$$\left.\begin{aligned} A = D &= 1 + \frac{YZ}{2} + \frac{Y^2Z^2}{24} + \frac{Y^3Z^3}{720}, \\ B &= Z\left(1 + \frac{YZ}{6} + \frac{Y^2Z^2}{120} + \frac{Y^3Z^3}{5\,040}\right), \\ C &= Y\left(1 + \frac{YZ}{6} + \frac{Y^2Z^2}{120} + \frac{Y^3Z^3}{5\,040}\right). \end{aligned}\right\} \quad . \quad . \quad (83)$$

It is instructive to note the values of A, B, C, and D in the approximate methods used. The simple series impedance method has

$$A = D = 1, B = Z, C = 0 \quad . \quad . \quad . \quad . \quad (71a)$$

the load-end capacitance method has

$$A = 1 + YZ, B = Z, C = Y, D = 1, \quad . \quad . \quad (71)$$

the nominal Π-method has

$$A = D = 1 + \tfrac{1}{2}YZ, B = Z, C = Y(1 + \tfrac{1}{4}YZ), . \quad (72)$$

and the nominal T-method has

$$A = D = 1 + \tfrac{1}{2}YZ, B = Z(1 + \tfrac{1}{4}YZ), C = Y \;. \quad (73)$$

Method 2. In this method the hyperbolic sines and cosines are expanded by the trigonometrical formulae given below, and tables of the trigonometric functions and hyperbolic functions of *real* numbers are used. Thus we express $\sqrt{(YZ)}$ in the form

$$\sqrt{(YZ)} = a + jb.$$

Then $\cosh\sqrt{(YZ)} = \cosh(a + jb)$

$$= \cosh a \,.\, \cosh jb + \sinh a \,.\, \sinh jb$$

so that

$$\left.\begin{aligned} \cosh\sqrt{(YZ)} &= \cosh a \,.\, \cos b + j \sinh a \,.\, \sin b. \\ \text{Similarly } \sinh\sqrt{(YZ)} &= \sinh a \,.\, \cos b + j \cosh a \,.\, \sin b. \end{aligned}\right\} \quad . \quad (84)$$

Method 3. In this way tables of the hyperbolic sine and cosine of complex numbers are used.

In general, the equations (83) are less laborious to use than equations (84); moreover the tables of the sinh and cosh of complex numbers are not easy to use, and are not given for small differences.

EXAMPLE. Solve the previous example by the rigorous method.

We will use the series method given in equations (83).

$$Z = 25 + j80 = 83{\cdot}8 \,\underline{|72^\circ\, 39'}$$

$$Y = j0{\cdot}0014 = 0{\cdot}0014 \,\underline{|90^\circ}$$

$$YZ = 0{\cdot}1164 \,\underline{|162^\circ\ 39'} = -\ 0{\cdot}1103 + j0{\cdot}0344$$

$$Y^2Z^2 = 0{\cdot}0138 \,\underline{|325^\circ\ 18'} = +\ 0{\cdot}0114 - j0{\cdot}00785$$

$$Y^3Z^3 = 0{\cdot}0016 \,\underline{|487^\circ\ 57'} = -\ 0{\cdot}00098 + j0{\cdot}00126$$

$$1 + \frac{YZ}{6} + \frac{Y^2Z^2}{120} + \frac{Y^3Z^3}{5\ 040}$$

$$= 1 - 0{\cdot}0184 + j0{\cdot}00573 + 0{\cdot}0001 - j0{\cdot}00006$$

$$= 0{\cdot}9817 + j0{\cdot}00567 = 0{\cdot}9817 \,\underline{|0^\circ\ 23'}$$

$$A = D = 1 - 0{\cdot}05515 + j0{\cdot}0172 + 0{\cdot}00047$$

$$- j0{\cdot}00033 - 0{\cdot}00001 + j0{\cdot}00001$$

$$= 0{\cdot}9453 + j0{\cdot}0169 = 0{\cdot}945 \,\underline{|1^\circ\ 1'}$$

$$B = 83{\cdot}8 \,\underline{|72^\circ\ 39'} \times 0{\cdot}9817 \,\underline{|0^\circ\ 23'}$$

$$= 82{\cdot}3 \,\underline{|73^\circ\ 2'}$$

$$C = 0{\cdot}0014 \,\underline{|90^\circ} \times 0{\cdot}9817 \,\underline{|0^\circ\ 23'}$$

$$= 0{\cdot}001376 \,\underline{|90^\circ\ 23'}$$

$$E_s = AE_r + BI_r$$

$$= 0{\cdot}945 \,\underline{|1^\circ\ 1'} \times 66\ 000 \,\underline{|0^\circ}$$

$$+ 82{\cdot}3 \,\underline{|73^\circ\ 2'} \times 284 \,\overline{|36^\circ\ 52'}$$

$$= 62\ 400 \,\underline{|1^\circ\ 1'} + 23\ 400 \,\underline{|36^\circ\ 10'}$$

$$= 62\ 400 + j1\ 100 + 18\ 900 + j13\ 800$$

$$= 81\ 300 + j14\ 900$$

$$= 82\ 600 \,\underline{|10^\circ\ 21'}$$

$$I_s = CE_r + DI_r$$

$$= 0{\cdot}001376 \,\underline{|90^\circ\ 23'} \times 66\ 000 \,\underline{|0^\circ}$$

$$+ 0{\cdot}945 \,\underline{|1^\circ\ 1'} \times 284 \,\overline{|36^\circ\ 52'}$$

$$= 91 \,\underline{|90^\circ\ 23'} + 268 \,\overline{|35^\circ\ 51'}$$

$$= -\ 0{\cdot}6 + j91 + 218 - j157$$

$$= 217 - j66$$

$$= 227 \,\overline{|16^\circ\ 55'}.$$

It is interesting to compare the various solutions of the problem.

Method	E_s	$I_s \times \sqrt{3}$
Simple series impedance . . .	$86\,400\ \underline{\vert 9^{\circ}\ 15'}$	$284\ \overline{\vert 36^{\circ}\ 52'}$
Load-end capacitance . . .	$79\,500\ \underline{\vert 11^{\circ}\ 44'}$	$240\ \overline{\vert 18^{\circ}\ 57'}$
Nominal Π-method . . .	$83\,000\ \underline{\vert 10^{\circ}\ 27'}$	$227\ \overline{\vert 17^{\circ}\ 0'}$
Nominal T-method . . .	$82\,300\ \underline{\vert 10^{\circ}\ 22'}$	$229\ \overline{\vert 16^{\circ}\ 36'}$
Exact method	$82\,600\ \underline{\vert 10^{\circ}\ 21'}$	$227\ \overline{\vert 16^{\circ}\ 55'}$

It is seen that the nominal Π- and T-methods are sufficiently accurate; the simple series impedance method is considerably in error, the sending-end voltage and current being much too high;

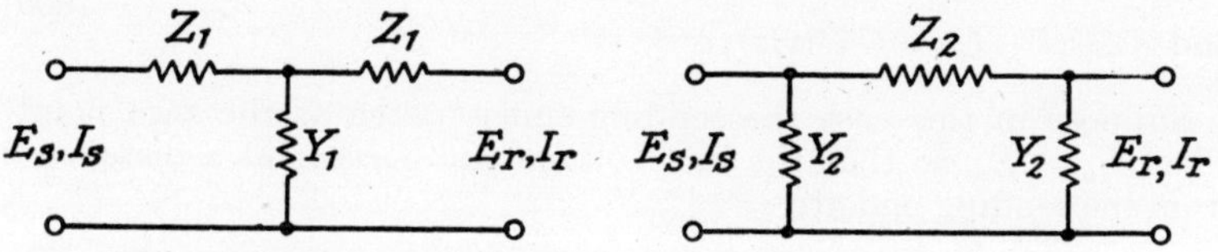

FIG. 116. EQUIVALENT T- AND Π-NETWORKS OF LONG LINE

the load-end capacitance method under-estimates the sending-end voltage, and over-estimates the current by a small amount.

Equivalent Network of a Long Line. Any symmetrical line can be represented by the simple T- and Π-networks of Fig. 116. In the T-network

$$\left.\begin{aligned} E_s &= E_r(1 + Y_1Z_1) + I_rZ_1(2 + Y_1Z_1) \\ \text{and} \quad I_s &= E_rY_1 + I_r(1 + Y_1Z_1). \end{aligned}\right\} \quad . \quad . \quad (85)$$

In the Π-network

$$\left.\begin{aligned} E_s &= E_r(1 + Y_2Z_2) + I_rZ_2, \\ \text{and} \quad I_s &= E_rY_2(2 + Y_2Z_2) + I_r(1 + Y_2Z_2). \end{aligned}\right\} \quad . \quad . \quad (86)$$

The general line has the equations (74). These agree with equations (85) and (86) provided

$$\left.\begin{aligned} Y_1 &= C \text{ and } Z_1 = (A - 1)/C \\ \text{or} \quad Z_2 &= B \text{ and } Y_2 = (A - 1)/B. \end{aligned}\right\} \quad . \quad . \quad . \quad (87)$$

Substituting from equations (82) we get for a continuous line

$$\left.\begin{aligned} Y_1 &= \sqrt{(Y/Z)} \sinh \sqrt{(YZ)} \\ \text{and} \quad Z_1 &= \sqrt{(Z/Y)} \tanh \tfrac{1}{2}\sqrt{(YZ)}, \\ \text{also} \quad Z_2 &= \sqrt{(Z/Y)} \sinh \sqrt{(YZ)} \\ \text{and} \quad Y_2 &= \sqrt{(Y/Z)} \tanh \tfrac{1}{2}\sqrt{(YZ)}. \end{aligned}\right\} \quad . \quad . \quad . \quad (88)$$

The Π-equivalent is the more convenient.

Very Long Line. The solution of equation (78) can be written in the form

$$E_x = F\varepsilon^{x\sqrt{(yz)}} + G\varepsilon^{-x\sqrt{(yz)}} \quad . \quad . \quad . \quad . \quad (79a)$$

and $$I_x = -F\sqrt{(y/z)}\varepsilon^{-x\sqrt{(yz)}} + G\sqrt{(y/z)}\varepsilon^{-x\sqrt{(yz)}} \quad . \quad (80a)$$

If x is great the term $\varepsilon^{x\sqrt{(yz)}}$ is great, and therefore the term in F must vanish or we should have infinitely increasing line voltage and current at distant points on a very long line. We therefore have

$$\left.\begin{aligned} E_x &= G\varepsilon^{-x\sqrt{(yz)}} \\ \text{and} \quad I_x &= G\sqrt{(y/z)}\,.\,\varepsilon^{-x\sqrt{(yz)}} \end{aligned}\right\} \quad . \quad . \quad . \quad . \quad . \quad (80b)$$

Suppose in this case the sending-end is taken as the zero point. Then $E_0 = E_s$, so that the line voltage and current at a distance x from the sending-end are

$$\left.\begin{aligned} E_x &= E_s\varepsilon^{-x\sqrt{(yz)}} \\ \text{and} \quad I_x &= E_s\sqrt{(y/z)}\varepsilon^{-x\sqrt{(yz)}} \end{aligned}\right\} \quad . \quad . \quad . \quad . \quad (80c)$$

We note first that $E_x/I_x = \sqrt{(z/y)} = Z_0$, say. Z_0 is called the *characteristic impedance* of the line, and we see that the ratio of the voltage to the current is constant at all points of the line and is equal to Z_0. The reciprocal of Z_0 is written Y_0 and is called the *characteristic admittance* of the line.

The product $\sqrt{(yz)}$ is called the *line angle per unit length,* and $x\sqrt{(yz)}$ is the angle for a length x. These may be written θ and $x\theta$ or Θ_x respectively. It is seen from equation (80*c*) that the voltage and current suffer a diminution by a factor $\varepsilon^{-\theta}$ for unit length or $\varepsilon^{-\Theta_x}$ for a length x. θ and Θ_x are complex, $a + jb$ say. Then the diminution factor is

$$\varepsilon^{-(a+jb)} = \varepsilon^{-a} \times \varepsilon^{-jb} = \varepsilon^{-a}\,\overline{|b},$$

i.e. there is an actual diminution of ε^{-a} in magnitude and a phase shift b.

Ferranti Effect. It has been found on page 139 that when the power factor is leading, the receiving-end voltage may be greater than the sending-end voltage. Ferranti noted that on a long line which is lightly loaded, the receiving-end voltage is greater than the sending-end voltage. Let us consider a single-phase unloaded line. The current is due to the capacitance and is

$$I = j\omega ClE_s,$$

where $\omega = 2\pi \times 50$, and Cl is the line capacitance. The voltage drop is

$$j\omega LlI = -\omega^2 LCl^2E_s,$$

so that the receiving-end voltage is

$$E_s + \omega^2 LCl^2 E_s = E_s(1 + \omega^2 LCl^2).$$

The product LC is nearly equal to the reciprocal of the square of the velocity of light, so that for a mile of line it is 3×10^{-11}. The voltage used is thus

$$(2\pi \times 50)^2 \times 3 \times 10^{-11} l^2 E_s$$
$$= 3 \times 10^{-6} l^2 E_s,$$

and l is in miles.

Thus in a 100-mile line at 132 kV. the rise is

$$3 \times 10^{-6} \times 10^4 \times 132 \text{ kV.}$$
$$= 3{\cdot}96 \text{ kV.} = 4 \text{ kV. approx.}$$

Losses in Open-circuited Line. Suppose a three-phase line of length l is on open circuit. The charging current per conductor at a distance x from the sending-end is

$$\omega C_0(l - x)E_s,$$

where C_0 is the capacitance to neutral per unit length. The loss per conductor is thus

$$\int_0^l \omega^2 C_0^2 (l - x)^2 E_s^2 R_0 dx$$
$$= \tfrac{1}{3}(\omega C_0 E_s)^2 R_0 l^3$$
$$= \tfrac{1}{3}(\omega C_0 l E_s)^2 R_0 l$$
$$= \tfrac{1}{3} I^2 R,$$

where I = total charging current per phase and R = total resistance per phase.

Tuned Power Lines. Equations (74) and (82) describe the behaviour of a transmission line. If we ignore the effects of resistance and leakage

$$Y = j\omega Cl \text{ and } Z = j\omega Ll,$$

where L and C are the inductance and capacitance per unit length and l the length. Thus

$$Z_0 = \sqrt{(L/C)},\ Y_0 = \sqrt{(C/L)},\ \text{and } \sqrt{(YZ)} = j\omega l\sqrt{(LC)}.$$

Since $\cosh [j\omega L\sqrt{(LC)}] = \cos [\omega l\sqrt{(LC)}]$
and $\sinh [j\omega l\sqrt{(LC)}] = j \sin [\omega l\sqrt{(LC)}]$,

we have
$$\left.\begin{aligned} E_s &= E_r \cos [\omega l\sqrt{(LC)}] + jZ_0I_r \sin [\omega l\sqrt{(LC)}] \\ I_s &= jY_0E_r \sin [\omega l\sqrt{(LC)}] + I_r \cos [\omega l\sqrt{(LC)}]. \end{aligned}\right\} \quad . \quad (90)$$

If there is no load, the receiving-end current I_r is zero. The charging current is

$$I_s = jY_0E_r \sin [\omega l\sqrt{(LC)}]$$

whilst
$$E_s = E_r \cos [\omega l\sqrt{(LC)}].$$

Thus I_s and E_s are in quadrature. If the relation

$$\omega l\sqrt{(LC)} = \pi \text{ or } 2\pi \quad . \quad . \quad . \quad . \quad . \quad . \quad (91)$$

is satisfied, there is no charging current on zero load. If the expression is equal to π, $\cos [\omega l\sqrt{(LC)}] = -1$ and then for *load conditions* the equations (90) become

$$E_s = -E_r \text{ and } I_s = -I_r.$$

If the expression is equal to 2π, equations (90) become

$$E_s = E_r \text{ and } I_s = I_r.$$

In both cases the receiving-end voltage and current are numerically equal to the sending-end values, so that there is no drop on load. The line is said to be *tuned*. We have already said that $1/\sqrt{(LC)}$ is the velocity of an electromagnetic wave along the line, and is approximately the velocity of light, i.e. 186 000 miles per sec. If the frequency is 50 cycles per sec., the length of line for tuning is

$$l = \frac{\pi \text{ or } 2\pi}{2\pi \times 50} \times 186\,000 \text{ miles}$$
$$= 1\,860 \text{ or } 3\,720 \text{ miles.}$$

If the length of line is fixed, tuning may be obtained by a proper choice of frequency. Thus if the line is 500 miles,

$$f = 50 \times (1\,860/500) \text{ or } 50 \times (3\,720/500)$$
$$= 186 \text{ or } 372 \text{ cycles per sec.}$$

It is undesirable to fix the frequency to suit the length of the line, as the length may vary.

If the length of the line and the frequency are fixed, the line may be tuned by putting capacitances at intervals between the line and neutral, as shown in Fig. 117. This has the effect of increasing C to the desired value given in equation (91). Alternatively, coils can be inserted in series with the line to raise L to the required value as shown in Fig. 117; this method is used in telephony but not in power transmission.

Another method is shown in Fig. 118, and is used in some systems. The capacitances in series with the line neutralize the voltage drop due to the line inductance, and the inductances across the line neutralize the charging current due to the line capacitance. This method is called the *method of tuning by the use of compensating*

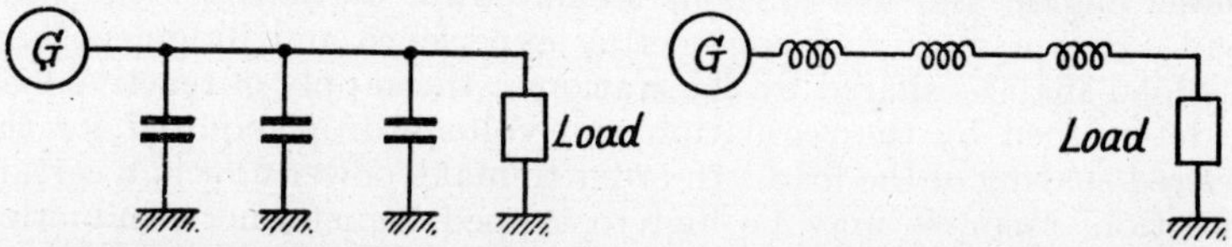

FIG. 117. TUNING BY SHUNT CAPACITANCE OR SERIES INDUCTANCE

sections. The values of L' and C' may be found with fair accuracy by replacing the line by its nominal-Π equivalent of Fig. 109. The two capacitances C' are equivalent to a single capacitance $\frac{1}{2}C'$, which will neutralize the inductance L (of Z) if $\omega^2 L(\frac{1}{2}C') = 1$. The inductances L' will neutralize the charging current provided $\omega^2 L'(\frac{1}{2}C) = 1$. The exact values required in long lines can be

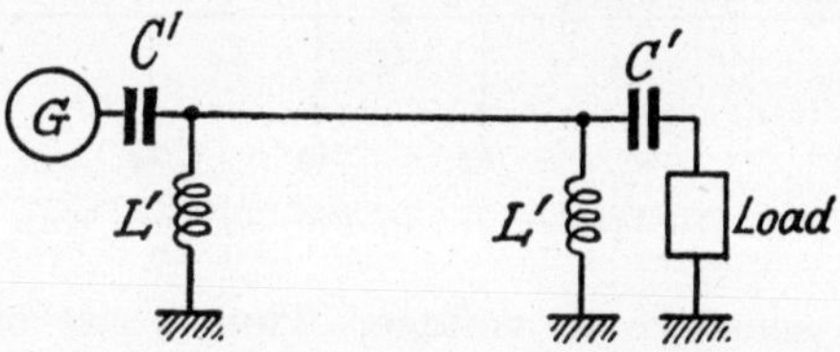

FIG. 118. TUNING BY COMPENSATING SECTIONS

obtained from the Π-equivalent of Fig. 116. Thus $\frac{1}{2}C'$ must neutralize Z_2, which is given by

$$Z_2 = \sqrt{(L/C)} \,.\, \sinh [j\omega\sqrt{(LC)}] = j\sqrt{(L/C)} \,.\, \sin [\omega\sqrt{(LC)}],$$

so that $1/j\omega C' + Z_2 = 0$, giving

$$C' = \frac{2}{\omega\sqrt{(L/C)} \,.\, \sin [\omega\sqrt{(LC)}]}$$

Similarly L' must neutralize Y_2, giving

$$L' = \frac{1}{\omega\sqrt{(C/L)} \,.\, \tan [\frac{1}{2}\omega\sqrt{(LC)}]}.$$

Interconnected Systems. It was shown in Fig. 17 that a ring main will give continuous service to the consumers even if one feeder is faulted, provided the faulty feeder is switched out of service. The ring main can thus give more secure service than a radial system. The interconnected system of Fig. 18 will give continuous service even if one of the stations is shut down. One of the main advantages of interconnection is increased security of service, another is the reduction in stand-by plant, and a third is the economy obtained by dividing the total load in such a way as to reduce the total capital cost and running costs to a minimum. The most economical arrangement is to use the very efficient generating stations, such as large hydro-electric and steam power stations of high capital cost and low running costs, for the base load, and the Diesel engine and less efficient steam-power stations for the peak load. It is a matter of engineering experience and judgment how the load shall be shared by the stations; the supply of reactive load is determined by the regulation and voltage drop required for the desired sharing of the load. In order to make power flow in a certain direction, recourse may be had to tapped transformers, induction regulators which introduce a boost of voltage, or synchronous reactors.

Voltage Drop in Distribution. It is essential that the voltage at a consumer's terminals be within narrow limits, for the consumer's

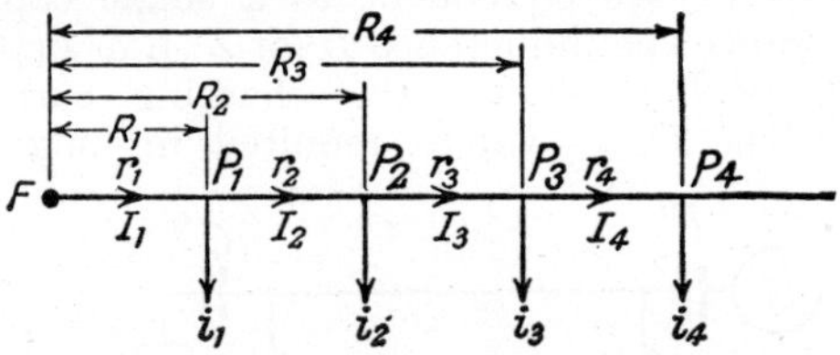

FIG. 119. DISTRIBUTOR FED AT ONE END

appliances are sensitive to voltage. Thus a rise of voltage may burn out lamps and heaters, and a drop will cause unsatisfactory operation. If the voltage drops very much, there will be an appreciable loss of revenue to the supply company. There is a legal limit of $\pm$ 4 per cent, which has been raised to $\pm$ 6 per cent in rural areas. The calculation of voltage drop is thus of importance.

We shall consider only systems with balanced loads; it is then convenient to find the voltage drop in each conductor, and we can represent a single or three-phase system by a single line. The total drop in a single-phase line is twice the drop in each conductor, whilst the voltage drop between phases in a three-phase system is $\sqrt{3}$ times the conductor voltage drop.

DISTRIBUTOR FED FROM ONE END. Let a distributor be fed at one end F and currents i_1, i_2, . . . be drawn off at P_1, P_2, . . . (Fig. 119). Let the resistance of $OP_1 = R_1$, $OP_2 = R_2$, . . . the current in FP_1 be I_1, in P_1P_2 be I_2, . . . The voltage drop at any point, say P_3, is

$$\begin{aligned} & I_1R_1 + I_2(R_2 - R_1) + I_3(R_3 - R_2) \\ = & R_1(I_1 - I_2) + R_2(I_2 - I_3) + R_3I_3 \\ = & R_1i_1 + R_2i_2 + R_3I_3. \end{aligned}$$

R_1i_1 is called the *moment* of i_1, R_2i_2 of i_2.

If the system transmits alternating current, the resistances must be replaced by impedances. The voltage drop is obtained by adding the products of the load currents multiplied by the resistances up to the tapping points, with the exception that the last term is the product of the resistance up to the last tapping point multiplied by the total current reaching the last tapping point.

A simple example of an alternating current feeder will illustrate the method.

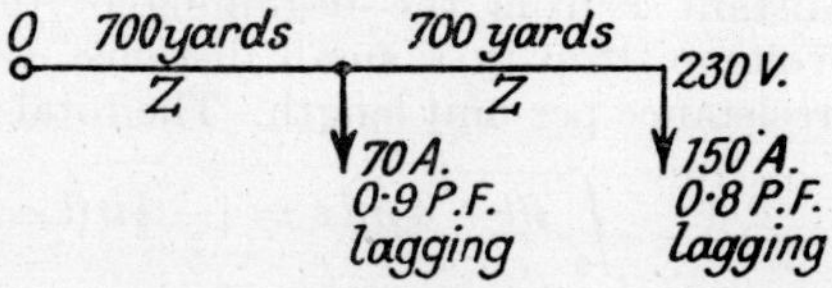

FIG. 120

EXAMPLE. A 2-wire feeder, 1 400 yd. long, supplies a load of 150 A. at 0·8 power factor at its far end and a load of 70 A. at 0·9 power factor at its mid-point. Both power factors are lagging and refer to the voltages at the respective load points. The resistance and reactance of the feeder per mile (go and return) are 0·1 Ω. and 0·15 Ω. respectively. If the voltage at the far end of the feeder is to be maintained at 230 V., calculate the voltage at the supply end. Calculate also the phase angle between the voltages at the two ends. (*Lond. Univ.*, 1932.)

Fig. 120 shows the system. The impedance per mile is $0{\cdot}1 + j0{\cdot}15$, so that for 700 yd. the impedance is

$$Z = (0{\cdot}1 + j0{\cdot}15) \times (700/1\,760) = 0{\cdot}0398 + j0{\cdot}0597.$$

Let us take as the datum vector the voltage at the far end. Then

$$i_2 = 150(0{\cdot}8 - j0{\cdot}6) = 120 - j90.$$

The voltage at the mid-point is thus

$$\begin{aligned} 230 + i_2Z &= 230 + (120 - j90)\,(0{\cdot}0398 + j0{\cdot}0597) \\ &= 240{\cdot}1 + j3{\cdot}57 = 240{\cdot}1\,\underline{|0^\circ\,51'}. \end{aligned}$$

The load at the mid-point lags behind this voltage by

$$\cos^{-1}(0{\cdot}9) = 25^\circ\,50',$$

so that it is

$$i_1 = 70\,\overline{|24^\circ\,59'}.$$

The voltage at the sending-end is thus

$$\begin{aligned} &240{\cdot}1 + j3{\cdot}6 + (i_1 + i_2)Z \\ &\quad = 240{\cdot}1 + j3{\cdot}6 + (70\,\overline{|24^\circ\,59'} \times 0{\cdot}0716\,\underline{|56^\circ\,18'}) + i_2Z \\ &\quad = 240{\cdot}1 + j3{\cdot}6 + 5{\cdot}01\,\underline{|21^\circ\,19'} + 10{\cdot}1 + j3{\cdot}6 \\ &\quad = 240{\cdot}1 + j3{\cdot}6 + 4{\cdot}7 + j1{\cdot}8 + 10{\cdot}1 + j3{\cdot}6 \\ &\quad = 254{\cdot}9 + j9{\cdot}0 = 255\,\underline{|2^\circ\,1'}, \end{aligned}$$

so that it has magnitude, 255 V. and leads the voltage at the far end by 2° 1′.

UNIFORMLY LOADED DISTRIBUTOR. Suppose that the current is tapped off uniformly along the length of the distributor at the rate of i amperes per unit length. Let the distributor have a length l at the far end of which there is no current. The current at a point distant x from the beginning of the current is $i(l-x)$, and the voltage drop in a small distance dx is $i(l-x)rdx$, where r is the resistance per unit length. The total voltage drop is

$$\int_0^l i(l-x)rdx = [-\tfrac{1}{2}ir(l-x)^2]_0^l$$
$$= \tfrac{1}{2}irl^2 = \tfrac{1}{2}(il)(rl)$$
$$= \tfrac{1}{2}IR,$$

where I is the total current and R the total resistance. If the uniformly loaded distributor is preceded by a length l' (resistance R') which takes the full current, the total drop is

$$IR' + \tfrac{1}{2}IR = I(R' + \tfrac{1}{2}R) = IR_m,$$

where R_m is the resistance of the distributor from the beginning of the unloaded length to the mid-point of the loaded part.

DISTRIBUTOR FED AT BOTH ENDS. The method is best explained by means of an example. Fig. 121 shows a distributor fed at both ends. It is required to find the point of lowest voltage, given that the resistance per yard is $50{\cdot}4 \times 10^{-5}$ ohms for both lines.

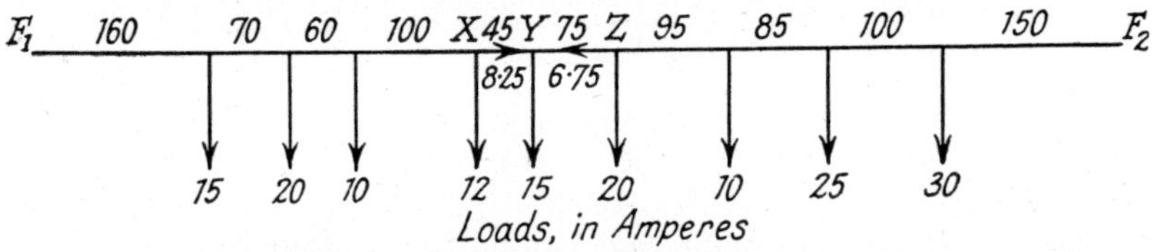

FIG. 121. DISTRIBUTOR FED AT TWO ENDS

We draw up a table (p. 165) giving the moments of the currents in ampere-yards from both ends, and add the moments. The point of lowest voltage is found by inspection of the table as that point at which the moments overlap.

The current thus divides at the sixth point from F_1. Let this point receive x amperes from F_1 and y amperes from F_2. Then

$$x + y = 20.$$

Moreover the moment about F_1 must be equal to the moment about F_2. Thus

$$21\,105 + 510x = 14\,100 + 430y.$$

Moment about F_1	Sum	Moment about F_2	Sum
$15 \times 160 = 2\,400$	2 400	$30 \times 150 = 4\,500$	4 500
$20 \times 230 = 4\,600$	7 000	$25 \times 250 = 6\,250$	10 750
$10 \times 290 = 2\,900$	9 900	$10 \times 335 = 3\,350$	14 100
$12 \times 390 = 4\,680$	14 580	$20 \times 430 = 8\,600$	22 700
$15 \times 435 = 6\,525$	21 105		
$20 \times 510 = 10\,200$	31 305		

These two equations give $x = 1{\cdot}7$ amperes and $y = 18{\cdot}3$ amperes. The voltage drop at the point is then

$$\begin{aligned} 21\,105 + 510 \times 1{\cdot}7 &= 21\,970 \text{ ampere-yards} \\ &= 21\,970 \times 50{\cdot}4 \times 10^{-5} \text{ volts} \\ &= 11{\cdot}1 \text{ volts.} \end{aligned}$$

Suppose F_2 is not at the same voltage as F_1, but is 4 volts above. There is then a circulating current of

$$\frac{4}{50{\cdot}4 + 10^{-5} \times 940} = 8{\cdot}45 \text{ amperes}$$

from F_2 to F_1. The current now divides at the fifth point, the current from Z to Y being $8{\cdot}45 - 1{\cdot}7 = 6{\cdot}75$ amperes, and from X to Y $8{\cdot}25$ amperes. The voltage drop at Y below F_1 is the value in the table, viz. 21 105 ampere-yards, *less* the drop of 8·45 amperes in the length of 435 yd., so that it is

$$\begin{aligned} 21\,105 - 8{\cdot}45 \times 435 & \\ = 17\,425 & \text{ ampere-yards} \\ = 8{\cdot}77 & \text{ volts.} \end{aligned}$$

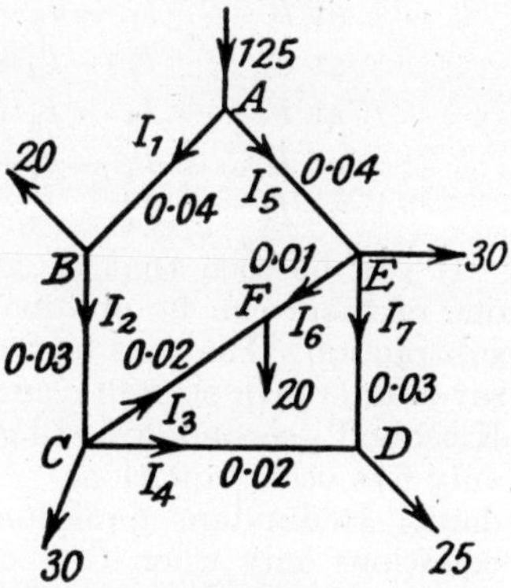

FIG. 122
PROBLEM OF INTERCONNECTION

The drop below F_2 is 12·77 volts.

Calculation of Interconnected Systems. The previous section dealt with the case of a distributor fed at both ends, which is a simple case of an interconnected system in which there are two generating stations. There are five main methods of solving the network problems in interconnected systems, in addition to a method of trial and error which is useful when a single operating condition is required. The methods are (1) direct application of Kirchhoff's laws, (2) superposition of currents, (3) network simplification, (4) circulating currents method, (5) use of Thévénin's theorem. The methods apply equally well to a.c. and d.c. systems.

(1) **Direct Application of Kirchhoff's Laws.** Currents are

written down as I_1, I_2, . . . in the various branches, and relations are obtained between them so that they satisfy Kirchhoff's first law, that the sum of the currents entering any junction is zero, and the second law, that the e.m.f. in any circuit is equal to the voltage drop in the circuit. The following example illustrates the method.

EXAMPLE. Four power loads B, C, D, and E are connected in this order to a 2-core distributor cable arranged as a ring main and take currents of 20, 30, 25, and 30 A. respectively. The ring is supplied from a substation at the point A between B and E. An interconnector cable joins the points C and E, and from a point F on this interconnector cable a current of 20 A. is taken. The total resistances of the cable between the load points are: $AB = 0{\cdot}04\ \Omega.$, $BC = 0{\cdot}03\ \Omega.$, $CD = 0{\cdot}02\ \Omega.$, $DE = 0{\cdot}03\ \Omega.$, $EA = 0{\cdot}04\ \Omega.$, $CF = 0{\cdot}02\ \Omega.$, and $FE = 0{\cdot}01\ \Omega$. Calculate the current in each section of the ring and interconnector. (*Lond. Univ.*, 1933.)

The network is shown in Fig. 122. Let the currents be I_1, I_2, . . . I_7 as shown. If we apply Kirchhoff's first law to the various junctions we get

$$
\begin{aligned}
&\text{at } A \quad & I_1 + I_5 &= 125 \text{ amperes,}\\
&\text{at } B \quad & I_1 - I_2 &= 20 \quad ,,\\
&\text{at } C \quad & I_2 - I_3 - I_4 &= 30 \quad ,,\\
&\text{at } F \quad & I_3 + I_6 &= 20 \quad ,,\\
&\text{at } E \quad & I_5 - I_6 - I_7 &= 30 \quad ,,\\
&\text{at } D \quad & I_4 + I_7 &= 25 \quad ,,
\end{aligned}
$$

It can be seen that these equations are not independent; any one of them can be obtained from the other five by addition and subtraction. This can be foreseen in the following way. There are seven unknown currents, and as there are two loops in the network, Kirchhoff's second law will give two equations, so that there remain only five other equations. Above we have six, so that one is redundant. Redundant equations may be avoided by writing down equations only when they contain a current which does not occur in preceding equations. Thus the last equation contains I_4 and I_7, both of which appear in the other equations, and thus it is redundant. A redundant equation is not harmful, provided it does not cause one to omit an essential equation. In the present problem there is no likelihood of such an omission, but when there are more than two loops the danger is present.

We apply Kirchhoff's second law to the two loops in the network. We get for the

loop $ABCFEA$

$$0{\cdot}04I_1 + 0{\cdot}03I_2 + 0{\cdot}02I_3 - 0{\cdot}01I_6 - 0{\cdot}04I_5 = 0,$$

loop $DCFED$

$$-\,0{\cdot}02I_4 + 0{\cdot}02I_3 - 0{\cdot}01I_6 + 0{\cdot}03I_7 = 0,$$

since there are no e.m.f.'s in the loops.

$I_2 = I_1 - 20$, $I_6 = 20 - I_3$, $I_4 = I_2 - I_3 - 30 = I_1 - I_3 - 50$, $I_7 = 25 - I_4 = 75 - I_1 + I_3$, and $I_5 = 125 - I_1$.

The equations for the loops become

$$0{\cdot}11I_1 + 0{\cdot}03I_3 = 5{\cdot}8$$

and

$$0{\cdot}05I_1 - 0{\cdot}08I_3 = 3{\cdot}05,$$

giving $I_1 = 53{\cdot}9$ and $I_3 = -4{\cdot}5$.

Then $I_2 = 33{\cdot}9$, $I_4 = 8{\cdot}4$, $I_5 = 71{\cdot}1$, $I_6 = 24{\cdot}5$, $I_7 = 16{\cdot}6$.

As a check we find the voltage drop between A and D by the paths AED and $ABCD$. These are

$$0{\cdot}04I_5 + 0{\cdot}03I_7$$
$$= 2{\cdot}844 + 0{\cdot}498 = 3{\cdot}342$$

and

$$0{\cdot}04I_1 + 0{\cdot}03I_2 + 0{\cdot}02I_4$$
$$= 2{\cdot}156 + 1{\cdot}017 + 0{\cdot}166 = 3{\cdot}339.$$

These are substantially the same, so that the solution is correct.

The fact that I_3 is negative means that the current flows from F to C, and is 4·5 amperes.

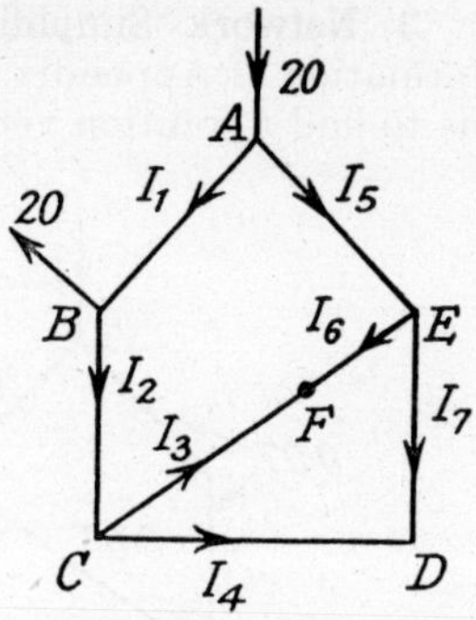

FIG. 123

(2) Superposition of Currents. In this method we find the currents in the branches due to one load. This is repeated for the various loads, and the final currents are the sums. As an example we will solve the preceding problem by this method.

When only load B acts, the currents are as in Fig. 123. It is clear that $I_3 = -I_6$ and $I_4 = -I_7$. The problem is most easily solved by compounding resistances in series and parallel. Resistance EDC is 0·05, EC is $(0{\cdot}03 \times 0{\cdot}05)/0{\cdot}08 = 0{\cdot}01875$, so that $AECB$ is 0·08875. Then

$$I_1 = \frac{0{\cdot}08875}{0{\cdot}04 + 0{\cdot}08875} \times 20 = 13{\cdot}8 \text{ amperes}$$

$$I_5 = -I_2 = \frac{0{\cdot}04}{0{\cdot}04 + 0{\cdot}08875} \times 20 = 6{\cdot}2 \text{ amperes}$$

$$I_6 = -I_3 = \frac{0{\cdot}05}{0{\cdot}3 + 0{\cdot}05} \times 6{\cdot}2 = 3{\cdot}88 \text{ amperes}$$

and

$$I_7 = -I_4 = 2{\cdot}32 \text{ amperes}.$$

We repeat the work for the loads C, D, E, F and construct the table shown on page 168.

The results agree exactly with those obtained by the application of Kirchhoff's laws.

Current	Load					
	B	*C*	*D*	*E*	*F*	All
I_1	13·8	13·7	9·9	9·3	7·2	53·9
I_2	− 6·2	13·7	9·9	9·3	7·2	33·9
I_3	− 3·88	− 10·2	− 3·2	5·8	7·0	− 4·5
I_4	− 2·32	− 6·1	13·1	3·5	0·2	8·4
I_5	6·2	16·3	15·1	20·7	12·8	71·1
I_6	3·88	10·2	3·2	− 5·8	13·0	24·5
I_7	2·32	6·1	11·9	− 3·5	− 0·2	16·6

(3) Network Simplification. The use of the star/delta transformation of Appendix II, page 467, equations (16) and (17), enables us to find a solution very quickly in some cases. For instance, let us

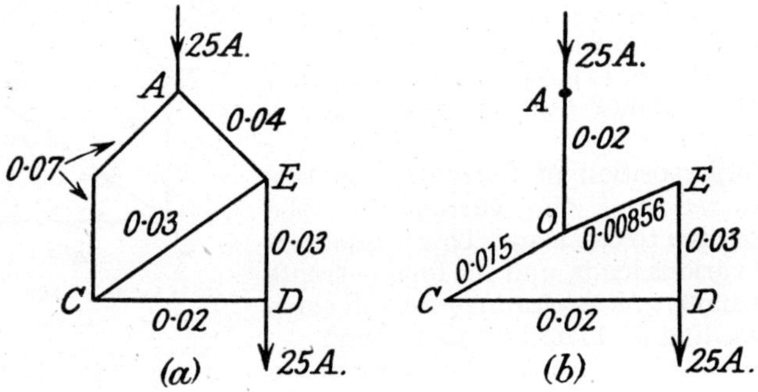

Fig. 124

find the solution of the previous problem when only the load *D* is acting. The network is then as shown in Fig. 124 (*a*). We apply the transformation to the delta *ACE* and we get the network of Fig. 124 (*b*) in which

$$AO = \frac{0{\cdot}04 \times 0{\cdot}07}{0{\cdot}04 + 0{\cdot}07 + 0{\cdot}03} = 0{\cdot}02,$$

$$CO = 0{\cdot}015, \text{ and } EO = 0{\cdot}00856.$$

The current along *AO* is 25 amperes, so that the current along *OCD* is

$$I_4 = \frac{0{\cdot}03856}{0{\cdot}03856 + 0{\cdot}035}\, 25 = 13{\cdot}1 \text{ amperes}$$

and $$I_7 = 25 - 13{\cdot}1 = 11{\cdot}9 \text{ amperes.}$$

The remaining currents are then found by Ohm's law applied to *ABCD* and *AED*. We get

$$0{\cdot}07 I_1 + (0{\cdot}02 \times 13{\cdot}1) = 0{\cdot}04\,(25 - I_1) + (0{\cdot}03 \times 11{\cdot}9),$$

giving $I_1 = 9{\cdot}9$ amperes.

The currents are thus

$$I_1 = I_2 = 9{\cdot}9,\ I_3 = -\,3{\cdot}2,\ I_4 = 13{\cdot}1,\ I_5 = 15{\cdot}1,\ I_6 = 3{\cdot}2,$$

and $I_7 = 11{\cdot}9$ amperes.

The case when the load F is acting is solved in a similar way.

(4) **Circulating Currents Method.** In this method certain branches of the network are removed so that any load is supplied from

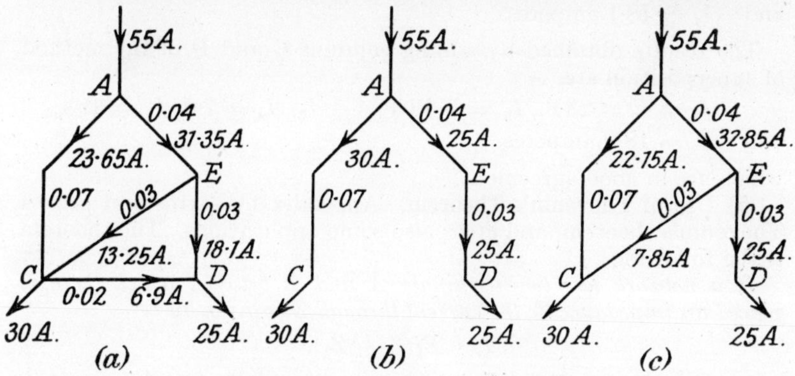

Fig. 125

the feeding point over routes having no connection with the routes of any other load. The voltage drops to the loads are calculated. One feeder is reinstated; it has to be put between points at different potentials, and thus potential difference drives a circulating current round the circuit just completed. This circulating current adds current to the previous currents. The work is repeated with each reinstated feeder until the network is made complete.

As an example let us find the currents in the previous problem due to the loads C and D. The network is in Fig. 125 (*a*). We remove the feeders *CD* and *CE* and obtain the very simple network of Fig. 125 (*b*). The voltage drop from A to C is $0{\cdot}07 \times 30 = 2{\cdot}1$ volts, whilst the drop from A to E is $0{\cdot}04 \times 25 = 1{\cdot}0$ volts. When we replace the feeder *CE*, there is a potential difference of 1·1 volts acting along *EC* producing a current of $1{\cdot}1 \div 0{\cdot}14 = 7{\cdot}85$ amperes round the circuit *ECAE*. The resulting network is shown in Fig. 125 (*c*), in which the circulating current has modified the branch currents to the values shown.

The voltage drop from A to C is now $0{\cdot}07 \times 22{\cdot}15 = 1{\cdot}55$ volts and the drop from A to D is $(0{\cdot}04 \times 32{\cdot}85) + (0{\cdot}03 \times 25) = 2{\cdot}06$ volts. When the feeder CD is replaced, the potential difference of 0·51 volts drives a circulating current from C to D via DE and EAC and EC in parallel. The resistance of EAC and EC in parallel is 0·0236, so that the circulating current has the value

$$\frac{0{\cdot}51}{0{\cdot}02 + 0{\cdot}03 + 0{\cdot}0236} = 6{\cdot}9 \text{ amperes.}$$

Of this the part that flows through EC is $0{\cdot}11/0{\cdot}14 \times 6{\cdot}9 = 5{\cdot}4$ amperes, and 1·5 amperes flows through EAC. The resulting system of currents is shown in Fig. 125 (*a*).

$$I_1 = I_2 = 23{\cdot}65,\ I_3 = -\,13{\cdot}25 = -\,I_6,\ I_4 = 6{\cdot}9,\ I_5 = 31{\cdot}35,$$

and $I_7 = 18{\cdot}1$ amperes.

The results obtained by adding columns C and D of the method of superposition are

$$I_1 = I_2 = 23{\cdot}6,\ I_3 = -\,13{\cdot}4 = -\,I_6,\ I_4 = 7{\cdot}0,\ I_5 = 31{\cdot}4,$$

and $I_7 = 18{\cdot}0$ amperes,

which are in good agreement.

(5) **Use of Thévénin's Theorem.** Appendix IV states and proves Thévénin's theorem, and gives also some applications. The theorem is the following:

If a network has two terminals A and B between which there is placed an impedance Z, the current through Z is given by

$$I = E/(Z + Z_1),$$

where E is the potential difference between A and B when Z is removed, and Z_1 is the impedance of the network between A and B calculated by assuming that generators are replaced by impedances equal to their internal impedances.

By the internal impedance of a generator is meant that impedance which causes a drop in the terminal voltage when current flows. Thus if Z_i is the internal impedance of a generator, the voltage when a current I is taken is the open-circuit e.m.f. less IZ_i (which is called the *internal drop*). In an alternator it is called the *synchronous impedance.*

EXAMPLE. Find the current I_1 in Fig. 122 by the use of Thévénin's theorem.

We assume that the feeder AB is removed and the network is then as shown in Fig. 126. The current along AE is 125 A.: let the current along ED be I. The remaining currents are then easily written down in terms of I, which is found by equating voltage drops along EFC and EDC. We get

$$0{\cdot}01(95 - I) + 0{\cdot}02(75 - I) = 0{\cdot}03I + 0{\cdot}02(I - 25),$$

giving $$I = 36{\cdot}9 \text{ A.}$$

The potential difference between A and B is

$$E = (0{\cdot}04 \times 125) + 0{\cdot}03I + 0{\cdot}02(I - 25) + (0{\cdot}03 \times 20)$$
$$= 5{\cdot}10 + 0{\cdot}05I = 6{\cdot}94 \text{ V}.$$

The resistance of the network between A and B is

$$Z_1 = 0{\cdot}04 + \frac{0{\cdot}03 \times 0{\cdot}05}{0{\cdot}03 + 0{\cdot}05} + 0{\cdot}03 = 0{\cdot}08875\ \Omega.$$

$$\text{giving } I_1 = \frac{\text{E}}{Z + Z_1} = \frac{6{\cdot}94}{0{\cdot}04 + 0{\cdot}08875} = \frac{6{\cdot}94}{0{\cdot}12875}$$
$$= \underline{\underline{53{\cdot}9 \text{ A}}}.$$

The method of circulating currents may be considered as an extension of Thévénin's theorem. Thus we have just found that the

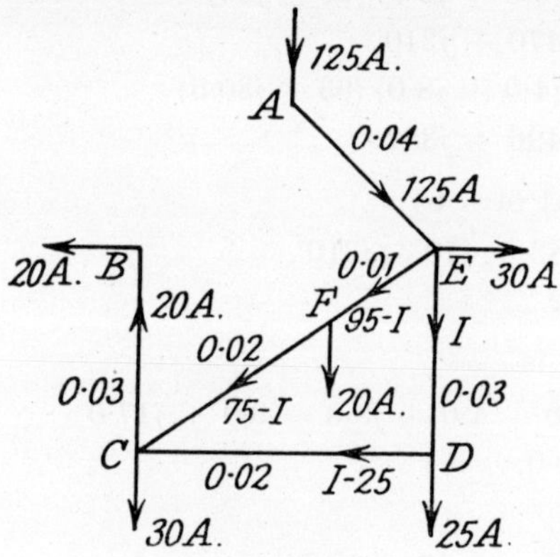

FIG. 126

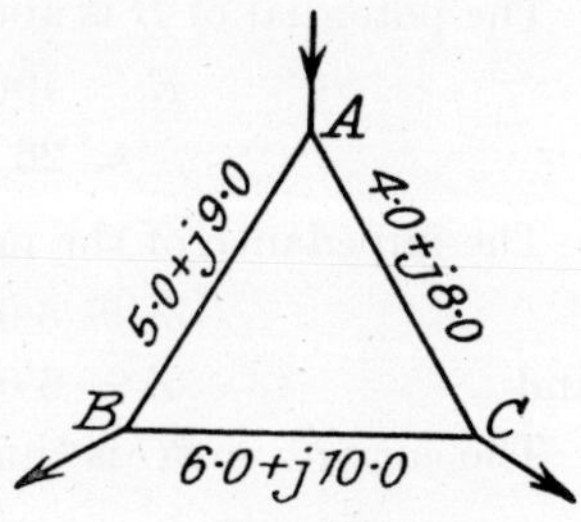

FIG. 127

insertion of the feeder AB causes a current of 53·9 A. to flow from A to B. This current must flow along B to C, from C to E by the paths through D and F, then from E to A. If we add the currents due to I_1 to those given in Fig. 126 we get the full solution.

A.C. Interconnected Systems. The method is the same as for d.c. but complex algebra is required. In systems that are not too complicated a vector diagram helps one to visualize what is happening. The following example illustrates the method.

EXAMPLE. A three-phase distribution system is as shown in Fig. 127. Power is supplied at A at 11 kV. (line voltage) and balanced loads of 50 A. per phase at 0·8 lagging power factor and 70 A. at 0·9 lagging power factor are taken at B and C respectively. The impedances of the feeders are $AB = (5{\cdot}0 + j9{\cdot}0)\ \Omega.$, $BC = (6{\cdot}0 + j10{\cdot}0)\ \Omega.$, $CA = (4{\cdot}0 + j8{\cdot}0)\ \Omega$. Calculate the voltage at B and C and the current in each branch. Power factors are assumed with respect to voltage at A. (*Lond. Univ.*, 1933.)

The simplest method of solution is by means of Thévénin's theorem or the method of circulating currents. We take the voltage at A as the basic vector. Then the current at B is

$$I_B = 50 \,\overline{|\cos^{-1} 0{\cdot}8} = 40 - j30$$

and the current at C is

$$I_C = 70 \,\overline{|\cos^{-1} 0{\cdot}9} = 63 - j30{\cdot}5.$$

The current at A is

$$I_A = I_B + I_C = 103 - j60{\cdot}5 = 119{\cdot}5 \,\overline{|30^\circ\, 9'},$$

so that A supplies at a power factor of 0·868.

If we assume that the feeder BC is removed, the current in AB is I_B and in AC it is I_C. The voltage drop per phase in AB is

$$(5{\cdot}0 + j9{\cdot}0)I_B = (5{\cdot}0 + j9{\cdot}0)\,(40 - j30)$$
$$= 470 + j210,$$

and in AC

$$(4{\cdot}0 + j8{\cdot}0)I_C = (4{\cdot}0 + j8{\cdot}0)\,(63 - j30{\cdot}5)$$
$$= 496 + j382.$$

The potential of B is above that of C by

$$E = 496 + j382 - 470 - j210$$
$$= 26 + j172.$$

The impedance of the network is

$$Z_1 = 5{\cdot}0 + j9{\cdot}0 + 4{\cdot}0 + j8{\cdot}0 = 9{\cdot}0 + j17{\cdot}0$$

and

$$Z = 6{\cdot}0 + j10{\cdot}0.$$

The current in BC is thus

$$\frac{E}{Z + Z_1} = \frac{26 + j172}{15{\cdot}0 + j27{\cdot}0} = \frac{174 \,\overline{|81^\circ\, 24'}}{30{\cdot}9 \,\overline{|61^\circ\, 0'}}$$
$$= \underline{\underline{5{\cdot}6 \,\overline{|20^\circ\, 24'} \text{ A.}}} = 5{\cdot}3 + j1{\cdot}96.$$

The current in AB is thus

$$40 - j30 + 5{\cdot}3 + j1{\cdot}96 = 45{\cdot}3 - j28{\cdot}0 = \underline{\underline{53{\cdot}3 \,\overline{|31^\circ\, 42'} \text{ A.}}}$$

and in AC

$$63 - j30{\cdot}5 - 5{\cdot}3 - j1{\cdot}96 = 57{\cdot}7 - j32{\cdot}5 = \underline{\underline{66{\cdot}2 \,\overline{|29^\circ\, 48'} \text{ A.}}}$$

The voltage drop between lines from A to B is

$$(\sqrt{3})(5{\cdot}0 + j9{\cdot}0)\,(45{\cdot}3 - j28{\cdot}0) = 846 + j463,$$

and from A to C

$$(\sqrt{3})(4{\cdot}0 + j8{\cdot}0)\,(57{\cdot}7 - j32{\cdot}5) = 850 + j474.$$

The voltage at B is therefore

$$11\,000 - 846 - j463 = 10\,154 - j463 = \underline{\underline{10\,164 \,\overline{|2° 36'}}}$$

and the voltage at C is

$$10\,150 - j474 = \underline{\underline{10\,160 \,\overline{|2° 40'}}}.$$

EXAMPLES VI

1. A 12·5 kV. transmission line connected with transformers at each end delivers 250 kVA. at a p.f. of 0·8 lagging to the low voltage bars in the sub-station. The line has a total resistance of 10 Ω. and an inductive reactance of 30 Ω.; each transformer has a ratio 2 000/11 000. The resistance and reactance on the low and high voltage sides are 0·04 Ω. and 0·125 Ω., 1·3 Ω. and 4·5 Ω. respectively.

Calculate the bus-bar voltage and p.f. at the generating end and the overall efficiency of transmission for a receiver pressure of 2 000 V.

2. A 3-phase overhead transmission line, in which the line voltage at each end is maintained constant at 33 kV., has a resistance per phase of 5 Ω. and a reactance per phase of 13 Ω.

Deduce an expression for the power delivered over the lines in terms of the angle between the sending-end voltage and the receiving-end voltage.

Plot to scale a curve of power to a base of angles and explain how to use this curve to determine the dynamic stability of the line.

(*Lond. Univ.*, 1948.)

3. Discuss the advantages of interconnecting generating stations (as in the National Grid) compared with the operation of stations as isolated plants. Explain, with the aid of convections and vector diagrams, the conditions governing the transference of power from one station to another, assuming the same bus-bar voltage at each station. Explain also how the magnitude, direction, and power factor of such power transference are controlled.

(*Lond. Univ.*, 1950.)

4. A 30-mile three-phase transmission line delivers 8 000 kW. at 33 kV. 0·8 p.f. lagging. The impedance of single conductors is $0{\cdot}60 + j0{\cdot}72$ Ω. per mile. Find the regulation, efficiency, generator voltage and power factor.

5. Construct a line chart for a three-phase line delivering a balanced load at 30 kV., the impedance of each conductor being $(1{\cdot}6 + j3{\cdot}3)$ Ω. Find the generator voltage at 1 000 kW. at power factors of 0·9 and 0·7 leading and lagging.

6. In the preceding example find the power that can be received at 0·8 p.f. lagging at 30 kV. if the generator voltage is 33 kV.

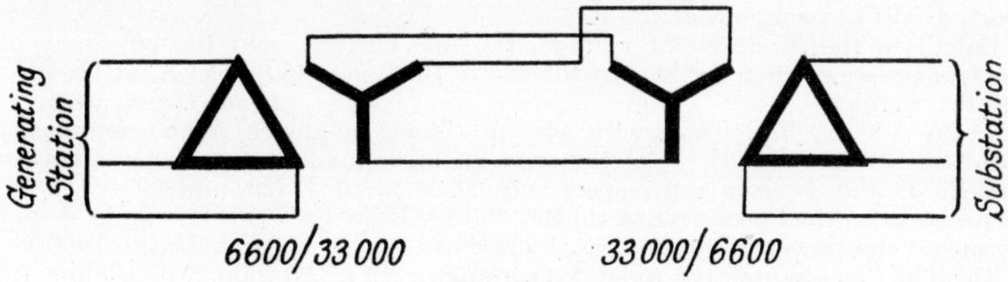

FIG. 128

7. A substation receives 6 000 kVA. at 6 kV. 0·8 p.f. lagging on the low voltage side of a transformer from a generating station through a three-phase cable system having a resistance of 7 Ω. and reactance of 2 Ω. per phase. Identical 6 600/33 000 transformers are installed at each end, the 6 600 V.

winding being delta- and the 33 000 V. star-connected (as shown in Fig. 128). The resistance and reactance of the star-connected windings are 0·6 and 4 Ω. respectively, and for the delta 0·06 and 0·36 Ω. Calculate the voltage at the generator bus-bars.

8. A building is supplied by a three-phase 4-wire service main, the resistance of each line wire being 0·25 Ω., and of the neutral 0·5 Ω. The main is fed with three-phase voltages at a line voltage of 420. Determine (*a*) the current in the neutral wire, and (*b*) the voltage, phase I to neutral, at the building end of the main, when the power *input* to the main comprises: 5 kVA. at 0·866 p.f. lagging, for phase I, 5 kVA. at 0·866 p.f., leading, for phase II, and 5 kVA. at 1·0 p.f. for phase III. (*Lond. Univ.*, 1933.)

9. A 3-phase transformer substation, *A*, supplies a load of 15 MW. at 0·8 power factor lagging, and another substation, *B*, supplies a load of 10 MW. at 0·7 power factor lagging, the bus-bar voltage of each station being 11 kV. and the frequency 50 c/s. By means of an interconnector 2 000 kW. are delivered to the bus-bars of *A* from station *B*, so that the loading of the transforming plant at *A* is reduced to 13 MW. and that of the plant at *B* is increased to 12 MW.

If the resistance and reactance per phase of the interconnector and regulating gear are 0·5 Ω 1·5 Ω respectively, determine, *either* graphically *or* analytically, the loading (MVA.) and power factor of the plant at each substation.

Draw a vector diagram. (*Lond. Univ.*, 1953.)

10. The resistance and reactance per phase of a three-phase transmission line are 23·5 Ω. and 67 Ω. respectively. Calculate the maximum steady-state power which can be transmitted over the line if the voltage at each end is maintained at 132 kV. Neglect the effect of the capacitance of the line.

Show that if the reactance, X, of the line could be varied, the resistance, R, remaining constant, the maximum steady-state power that could be transmitted over the line would be greatest when $X = \sqrt{3}.R$.
(*Lond. Univ.*, 1934.)

11. A two-wire distributor *ABCD*, 400 yd. long, each core of which has a resistance of 0·25 Ω per 1 000 yd., is fed at *A* and *D*, the voltage at *A* being 235 V. and that at *D* 232 V. There are concentrated loads of 100 A. at *B*, 50 yd. from *A*, and 70 A. at *C*, 150 yd. from *A*, and there is a uniformly distributed load of 1 A. per yd. over the 100-yd. length adjacent to *D*. Calculate (*a*) the minimum load voltage and (*b*) the load voltage at a point 50 yd. from *D*. (*Lond. Univ.*, 1953.)

12. Find the resistance and inductance to neutral of a three-phase line 15 miles long, the wires being 19/·083 copper conductors at a symmetrical spacing of 36 in. The temperature is 65° F.

13. A 50-frequency, three-phase transmission line has the following constants (line to neutral): resistance 11·0 Ω., reactance 38·0 Ω., susceptance $3{\cdot}0 \times 10^{-4}$ mho., leakance negligible. The capacitance can be assumed to be located half at each end of the line.

Calculate the sending-end voltage, the line current, and the efficiency of the transmission when the load at the end of the line is 40 000 kVA. at 110 kV., p.f. 0·7, lagging. (*Lond. Univ.*, 1933.)

14. A 400 V., 3-phase, 3-wire system (phase sequence *RYB*) supplies a star-connected load with isolated star point. The loads to the star point from the *R*, *Y* and *B* lines are respectively $25 + j0$, $0 + j20$, and $30 + j10$ Ω. Determine (*a*) the line currents, (*b*) the voltage between the *R* line and the star point, (*c*) the power dissipated by the circuit. (*Lond. Univ.*, 1950.)

15. The "constants" per mile, per conductor, of a 150-mile, 3-phase line are as follows: resistance, 0·25 Ω.; inductance, 2×10^{-3} H.; capacitance to neutral, 0·015 μF. A balanced 3-phase load of 40 MVA. at 0·8 power factor (lagging) is connected to the receiving end, and a synchronous condenser, operating at zero power factor (leading), is connected to the mid-point of the line. The frequency is 50 cycles per second.

If the voltage at the load is 120 kV., determine the kVA. rating of the synchronous condenser in order that the voltage at the sending-end may be equal (in magnitude) to that at the mid-point. The nominal T-circuit is to be used for the calculations. (*Lond. Univ.*, 1949.)

16. A transmission line consists of two circuits P and Q connected in series, the circuit P being at the sending end of the line. The circuits have the following auxiliary constants

Circuit P	*Circuit Q*
$A_P = 0{\cdot}982\underline{\vert 1{\cdot}2^\circ}$	$A_Q = 0{\cdot}808\underline{\vert 2{\cdot}0^\circ}$
$B_P = 77{\cdot}3\underline{\vert 80{\cdot}0^\circ}$	$B_Q = 30{\cdot}0\underline{\vert 45{\cdot}0^\circ}$
$C_P = 0{\cdot}000452\underline{\vert 91{\cdot}0^\circ}$	$C_Q = 0{\cdot}001\underline{\vert 92{\cdot}0^\circ}$
$D_P = 0{\cdot}982\underline{\vert 1{\cdot}2^\circ}$	$D_Q = 0{\cdot}808\underline{\vert 2{\cdot}0^\circ}$

Develop expressions for each of the four auxiliary constants A, B, C, and D of the whole line and calculate the numerical value of the constant A. (*Lond. Univ.*, 1934.)

17. The following data apply to a three-phase transmission line. The resistance per phase is 63·5 Ω., the reactance per phase 167·0 Ω., susceptance per phase $1{\cdot}1 \times 10^{-3}$ mho. Draw the vector diagram showing the sending- and receiving-end voltages and currents when the receiving-end voltage is 132 kV. and the line is supplying 40 000 kW. at a lagging power factor of 0·9. From the diagram determine the sending-end voltage and the efficiency of transmission under these conditions. The auxiliary line constants A, B, C are given by

$$A = 1 + ZY/2 + Z^2Y^2/24 + Z^3Y^3/720.$$
$$B = Z(1 + ZY/6 + Z^2Y^2120 + Z^3Y^3/5\,040).$$
$$C = Y(1 + ZY/6 + Z^2Y^2/120 + Z^3Y^3/5\,040).$$

(*Lond. Univ.*, 1933.)

18. Explain how (*a*) the power, (*b*) the reactive kVA., between two interconnected generating stations is controlled. Two 3-phase stations are linked by a 33 kV. cable. The conductors and one of the two similar transformers have a combined (line to neutral) reactance of 50 Ω. and a resistance of 4 Ω. Calculate the transformer voltage at one of the stations when a load of 10 000 kVA. at 33 kV. and unity p.f. is received from it by the other station. Find also the power factor of the load at the sending station. (*Lond. Univ.*, 1933.)

19. What are the necessary conditions under which power is transmitted through an interconnector between two power stations and how are these requirements met in practice?

Two power stations A and B are connected by a three-phase interconnector having a resistance of 15 Ω. per phase and a reactance of 50 Ω. per phase. If the voltage at B is 66 kV. and the voltage at A is 69 kV., calculate the power factor of the current in the interconnector when 10 000 kW. is being (*a*) received at B, (*b*) sent out from B. (*Lond. Univ.*, 1933.)

20. A total load of 12 000 kW. at a p.f. of 0·8 lagging is transmitted to a substation by two overhead three-phase lines connected in parallel. One line has a conductor resistance of 2 Ω. per conductor and a reactance (line to neutral) of 1·5 Ω., the corresponding values for the other line being 1·5 and 1·2 Ω. Calculate by an exact method the power transmitted by each overhead line. (*Lond. Univ.*, 1933.)

21. State (i) the essential, (ii) the desirable, conditions to be fulfilled in order that two 3-phase transformers may operate satisfactorily in parallel.

A 2 000 kVA. transformer (A) is connected in parallel with a 4 000 kVA. transformer (B) to supply a 3-phase load of 5 000 kVA. at 0·8 power factor (lagging). Determine the kVA. supplied by each transformer, assuming equal no-load voltages. The percentage voltage drops in the windings at the rated loads are as follows: transformer A, resistance 2%, reactance 8%; transformer B, resistance 1·6%, reactance 8%. (*Lond. Univ.*, 1948.)

22. What are the advantages of overhead lines as compared with underground cables for transmission at very high voltages?

A 3-phase, 50 c/s transmission line is 100 miles long and has the following constants—

Resistance per phase per mile = 0·2 Ω.;
Inductance per phase per mile = 2 mH.;
Capacitance (line to neutral) per mile = 0·015 μF.

If the line supplies a load of 50 MW. at 0·8 lagging power factor and 132 kV., determine, using the nominal T-method, the sending-end voltage, current and power factor and the line efficiency. (*Lond. Univ.*, 1950.)

23. A load of 8 000 kW. at a power factor of 0·8 lagging is to be transmitted over a three-phase line to a point 40 miles distant, the output voltage being 60 kV. at 50 cyc. An intermediate tapping point 25 miles from the input end of the line supplies a load of 6 000 kW. to a consumer at a power factor of 0·8

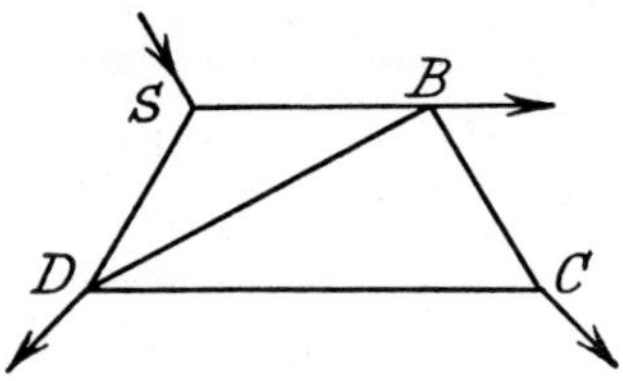

FIG. 129

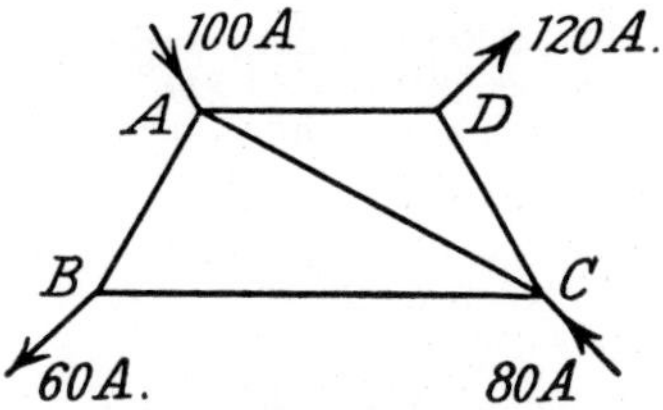

FIG. 130

lagging. The 10 000, 7 500, 17 500 kVA. transformers at the output end, intermediate point, and input end of the line respectively each have 1% resistance drop and 3% leakage reactance drop at their rated power and voltage of 66 kV. between lines. The resistance and reactance per mile of the line are 0·320 and 0·358 Ω. respectively. Find the voltage at the input end and at the intermediate tapping, assuming that the constants of the transformers may be added to those of the line and that the exciting currents and core losses are negligible. (*Lond. Univ.*, 1931.)

24. Discuss the advantages of a 3-wire system as compared with a 2-wire system for a d.c. distribution network and explain carefully how you would maintain approximately equal voltages across the two sides of a 3-wire system at the end of a long radial feeder with a heavily unbalanced load.

Fig. 129 represents a 2-wire d.c. network supplied at S and with loads of 50 A., 60 A., and 70 A. at B, C and D respectively. The resistances of the various sections are as follows (including go and return leads in each case)—

SB	0·05 Ω.
BC	0·15 Ω.
CD	0·10 Ω.
BD	0·10 Ω.
DS	0·05 Ω.

Calculate the voltage drop between the supply point S and the lead C when

(*a*) the section CD is not in circuit;
(*b*) the section CD is in circuit. (*Lond. Univ.*, 1934.)

25. Find the currents in the network of Fig. 130. The resistances are $AB = 0{\cdot}04$, $BC = 0{\cdot}06$, $CD = 0{\cdot}05$, $AD = 0{\cdot}02$, $AC = 0{\cdot}08$ Ω. Find the difference of potential between A and C.

CHAPTER VII

VOLTAGE REGULATION STABILITY

Voltage Control. All modern transmission systems, with the exception of the Thury constant-current system, operate at a constant voltage. It is essential for the satisfactory operation of the consumers' apparatus that the voltage be kept within narrow limits. Until recently there was a legal limit of $\pm$ 4 per cent, although the new regulations permit a variation of $\pm$ 6 per cent in order that rural lines may be erected at a reasonable cost.

In d.c. distribution the voltage can be kept fixed at the end of a feeder by using a compound generator. When there are feeders of different lengths supplied from one generator, it is clear that compounding is not a complete solution; the voltage at the end of each feeder can be kept constant by means of boosters, which are described briefly on pages 36–38.

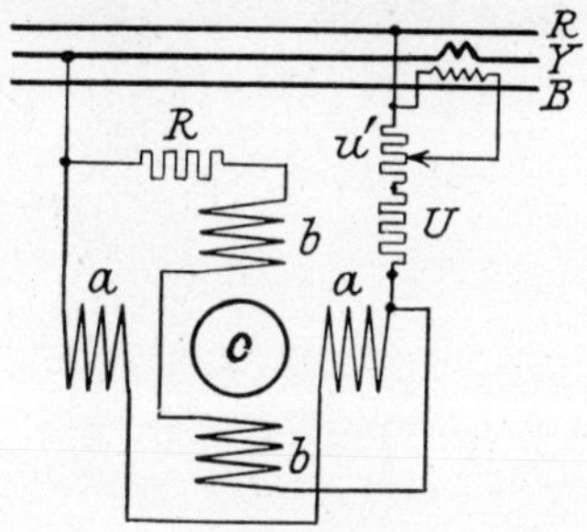

FIG. 131A
BROWN-BOVERI REGULATOR
(Brown-Boveri)

The regulation of an alternator is due mainly to the synchronous reactance of the armature. The synchronous reactance is intentionally not made as small as it can be, as a very small synchronous reactance renders the generator liable to damage when a short circuit occurs on the system at a point near to the generator. Modern practice is to have a synchronous reactance of 8 to 15 per cent, so that when a short circuit occurs near the generator the momentary peak value of the current is from 10 to 18 times the normal load value, and on continuous short circuit the current is only 2 to 6 times the normal. There are several excellent types of automatic voltage regulators; because of this and the protection afforded by a comparatively large synchronous reactance against short circuits, good inherent regulation is not sought in modern alternators. One very successful regulator is the *Tirrill* regulator, which is described in *Electrical Technology*.* Another is the *Brown-Boveri* regulator which is shown in Fig. 131A. The action of the latter is as follows.†

If the excitation voltage is raised to the required value, the field

* *Electrical Technology,* by Prof. H. Cotton (Pitman), Sixth Edition, pp. 363–4.

† The thanks of the author are due to British Brown-Boveri Limited for this information.

current will not attain the steady value corresponding to it for some seconds, because of the fairly large time-constant of the field windings. For this reason the Brown-Boveri regulator acts by overshooting the mark considerably and then falling back to the proper value: in this way the excitation current reaches the required steady value much more quickly.

Figs. 131B and C show the moving part of the regulator, which acts on the Ferraris principle. An annular core of laminated sheeting

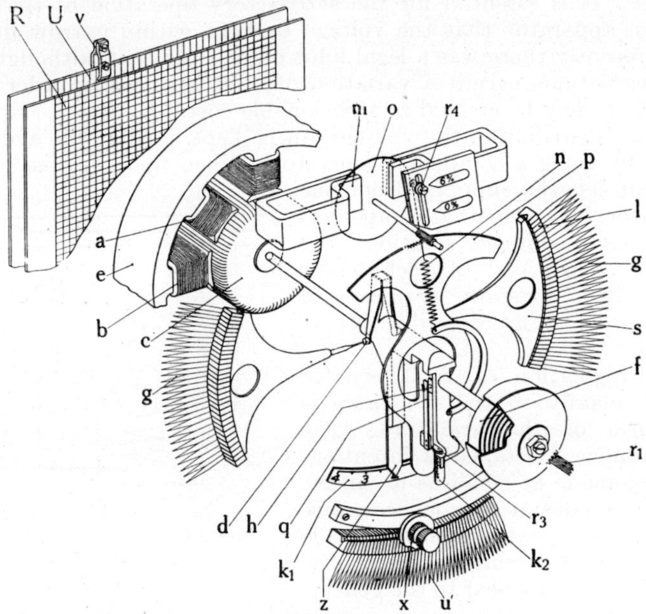

FIG. 131B. BROWN-BOVERI REGULATOR
(*Brown-Boveri*)

carries two windings *a* and *b*. The main winding *a* is connected up to the system to be regulated through the resistances *U* and *u'*, while a resistance *R* is inserted in the circuit of the second winding *b* (see Fig. 131A). By judicious choice of the resistances there is a phase displacement of the currents in the windings, and this results in a torque on the aluminium drum *c*. The main spring *f* and the auxiliary spring *n* provide a mechanical torque which opposes the electrical torque; the screw r_4 can vary the torque due to *n*.

The regulating resistance *g* consists in a pair of resistance elements connected to the contact blocks *l*, which are heavily plated with silver. The contact sectors roll on the inside surfaces of the contact blocks, and make good contact because of the springs *d*. Fig. 131C shows diagrammatically the operation of voltage regulation. Suppose that the voltage is normal at position 1, at which position the mechanical torque has been adjusted to balance the electrical torque. If the alternator voltage rises, the electrical torque increases and the drum rotates to position 3, say. More resistance has been inserted in the exciter circuit, and the field current and alternator voltage drop. Meanwhile the recall spring q has been tightened, and when the voltage has fallen to a lower value, the recall spring is strong enough to force the contact roller back to position 2, which is the equilibrium position. The damping system, m and o, prevents oscillation of the system about the equilibrium position.

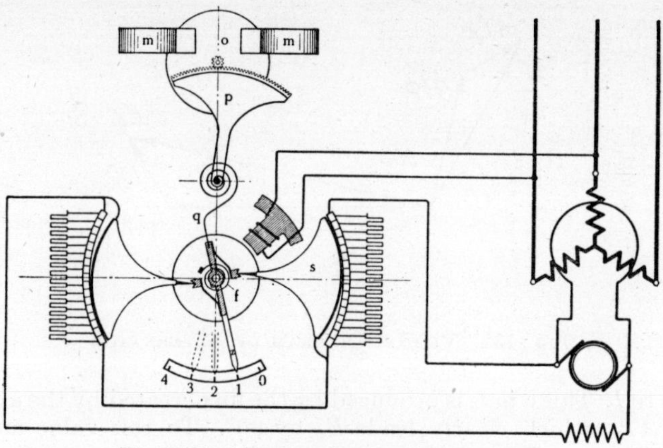

FIG. 131C. BROWN-BOVERI REGULATOR
(*Brown-Boveri*)

The regulator may be compounded for current by means of a current transformer, which is connected across an adjustable part of the resistance u'. In this way the regulator can allow for voltage drop due to the current along a feeder, so that the voltage at the *far end* of the feeder can be kept constant.

Voltage Control by Generator Excitation. In many cases, and in particular in short lines, the voltage at the receiving end is kept within very narrow limits by automatic or hand-operated voltage regulators, which act in the field circuit of the alternator exciter as do the Tirrill and Brown-Boveri types. It is possible to design the windings of the relay, operated by the line voltage and current, to allow for the regulation of a line and of the generator. This method is unsuitable for long lines, as the voltage at the generator terminals will have to vary too much in order that the voltage at the far end of the line may be constant.

The effect of varying the excitation depends upon the system into which the generator is feeding, but in all cases the power output is unaffected as this depends on the fuel supply to the prime mover. In an interconnected system, in which there are two or more alternators, the distribution of load is unchanged by varying the voltages by excitation, but the reactive kVA. can be changed. The sharing of load is, of course, determined by the regulation of the governor of the prime movers.

The excitation required to maintain a given current I and terminal voltage E in an alternator can be found in the following way, shown in Fig. 132. The e.m.f. to be produced by the flux is E_ϕ, where E_ϕ is the vector sum of E and IX, the latter being the voltage drop due to the synchronous reactance of the alternator and is thus at right

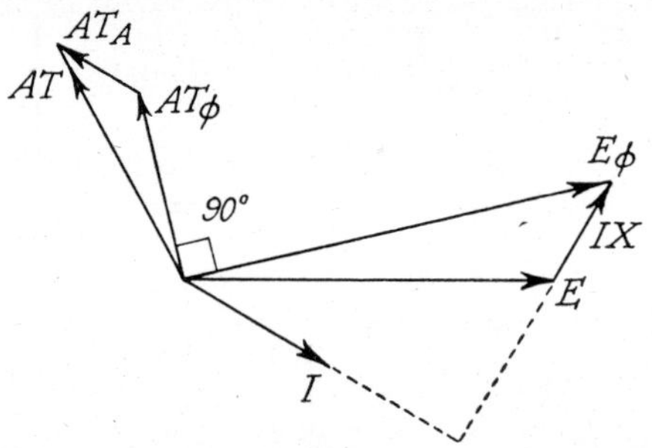

FIG. 132. VECTOR DIAGRAM OF ALTERNATOR

angles to I. This e.m.f. is produced by the flux created by the ampere-turns AT_ϕ, which clearly leads E_ϕ by 90°. To this value must be added the ampere-turns necessary to overcome the armature reaction represented by AT_A, and which is parallel to I when the rotor is cylindrical and nearer in line with AT_ϕ when there are salient poles. In the former case

$$AT_A = XI(I_n/P_n) \times A,$$

where I_n and P_n are the rated current and power and A is the ampere-turns to give normal voltage on open-circuit. The required excitation is given by AT. In the latter case the magnitude and position of AT_A depend on the cross-reluctance factor of the machine; AT_A is smaller than for the cylindrical rotor and is more in line with AT_ϕ.

It is frequently required that the voltage at both ends of a transmission line should be kept constant, sometimes at the same value; this may be desired when loads are taken near the generator end *and* at the far end of the line. It is impossible to achieve this condition by voltage control of the generator, but there are several methods of achieving this requirement which will be described.

Tap-changing Transformers. The voltage can be varied by having a number of tappings on the secondary winding as shown in

Fig. 133, so that the turns ratio can be changed at will. When the movable arm makes contact with stud 1, the secondary voltage is a minimum, and when with stud 5 a maximum. When the load is light the voltage across the primary is not much below the generator

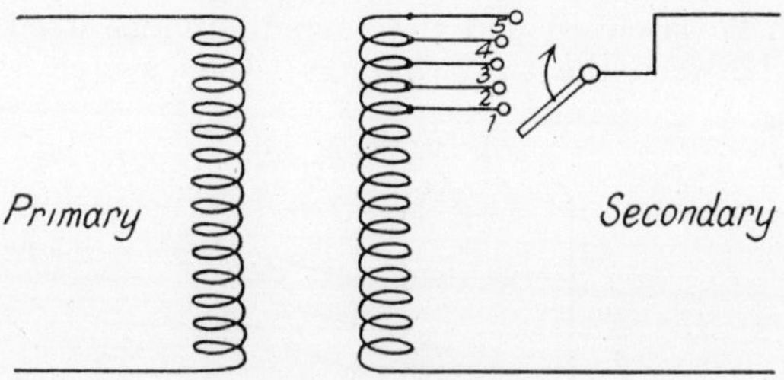

FIG. 133. TRANSFORMER WITH TAPPINGS

voltage and the former position is used. As the load increases, either at this or intermediate load points, the voltage across the primary drops, but the secondary voltage can be kept at the previous value by switching the arm on to a higher stud. This simple type of tap-changing transformer can be varied only when the load is removed. For if the movable arm is narrow and leaves one stud

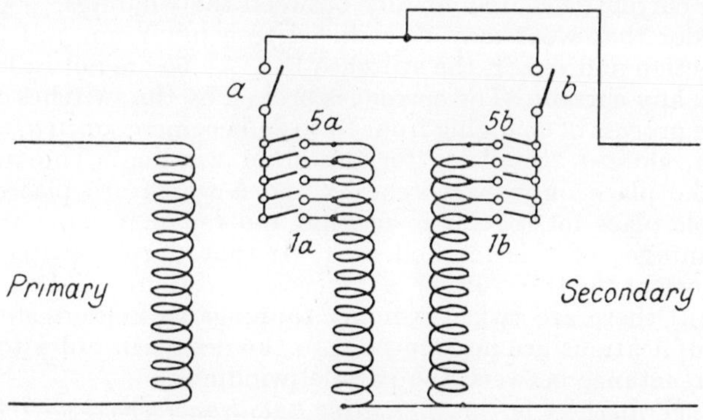

FIG. 134. ON-LOAD TAP-CHANGING TRANSFORMER

before it reaches the next the circuit is broken and arcing results; whereas, if the arm is wide and reaches the next stud before leaving the earlier, the part of the coil between the studs is momentarily shorted and takes a large current, which when broken causes arcing and burning of the contacts.

It is only comparatively recently that on-load tap-changing transformers that work satisfactorily have been made. One method is shown in Fig. 134. The secondary is composed of equal parallel windings, which have similar tappings 1*a* . . . 5*a* and 1*b* . . . 5*b*; we will imagine for the moment that the switches *a* and *b* are closed. When switches 1*a* and 1*b* are closed the secondary voltage is lowest. In order to increase the secondary voltage switch 1*a* is broken and for a short time the *b* winding takes all the current; then switch 2*a* is closed and the current divides itself, somewhat unequally, between the two windings; then switch 1*b* is opened

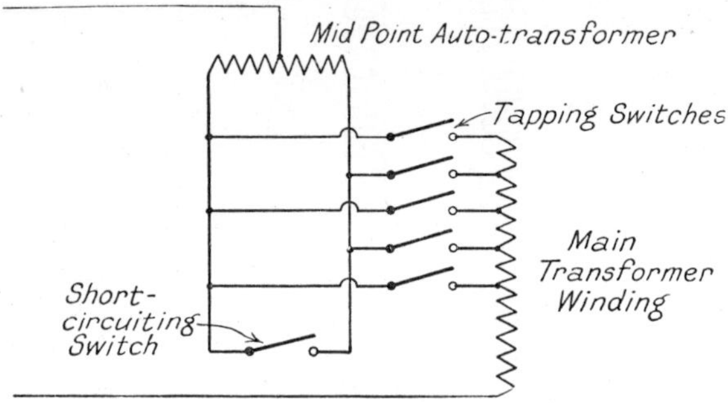

FIG. 135. AUTO-TRANSFORMER TAP-CHANGING

and the *a* winding takes the current; finally switch 2*b* is closed and the current is shared equally between the windings.

In order that wear may be reduced to a minimum and for ease of inspection and repair, the switches 1*a* . . . 5*b* are not called upon to break any current. The current is broken by the switches *a* and *b*; thus the process of changing from 1*a* to 2*a* becomes: open *a*, open 1*a*, close 2*a*, close *a*. Similarly for the other winding. The wear will thus take place only in switches *a* and *b* which are placed in an accessible place for regular inspection and repair if necessary. The disadvantages of this method are: (i) that during switching the impedance of the transformer is increased and there will be a voltage surge, (ii) there are twice as many tappings as voltage steps, and (iii) complications are introduced into the design in order to obtain a high reactance between the parallel windings.

Another method is the *preventive auto-transformer system* shown in Fig. 135. During normal operation the short-circuiting switch is closed; let us suppose that the lowest switch is closed so that the voltage is the minimum. To raise the voltage, the short-circuiting switch is opened, the second switch is closed, the first is opened, and finally the short-circuiting switch is closed. During the time that the first and second switches are closed and the short-circuiting switch is open, a part of the transformer winding is shunted by the auto-transformer; the latter is given sufficient reactance not to cause a large current, but the line current is unaffected by this reactance as it flows in opposite directions in the halves of the auto-transformer. Instead of switches, we may have sliding contacts attached to the ends of the auto-transformer such that one contact makes before the other breaks, and in the normal operating position both contacts touch the same stud. If the auto-transformer is designed to take the full load current continuously, then a working position is possible with two consecutive tapping switches closed. This nearly doubles the number of working positions with a given number of tappings.

Calculation of Transformer Ratios. Let us consider the application of transformer-tapping to the system of Fig. 136, in which it is required to send a given power P to the receiving-end at which the voltage and power factor are E_r and $\cos\phi_r$. The sending-end voltage is fixed at E_s, and the impedance of the line and transformers transferred to line voltage is $Z = R + jX$. Let the transformer ratios be t_1 and t_2 as shown, where t_1 and t_2 are

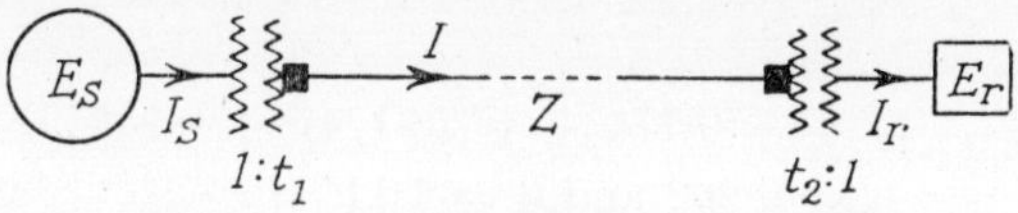

FIG. 136

expressed as fractions of the normal ratios: thus if the normal ratio of the sending-end transformer is 6·6 kV. : 33 kV. and the ratio now is 6·6 : 36 we should have $t_1 = 36/33 = 1{\cdot}09$. We take the receiving-end voltage E_r as the basic vector. Then the receiving-end current is

$$I_r = (P/E_r) - j(P/E_r)\tan\phi_r,$$

where ϕ_r is positive for a lagging power factor. The voltage at the line winding of the receiving-end transformer is t_2E_2, where E_2 is the nominal voltage, i.e. the value when the transformer ratio has the nominal value. The line current is

$$I = I_r \times (E_r/E_2) \times (1/t_2) = (P/t_2E_2)(1 - j\tan\phi_r).$$

The voltage at the line winding of the sending-end transformer is t_1E_1, so that

$$\begin{aligned} t_1E_1 &= t_2E_2 + IZ \\ &= t_2E_2 + (P/t_2E_2)(R + X\tan\phi_r) + j(P/t_2E_2)(X - R\tan\phi_r). \end{aligned}$$

Equating magnitudes we get

$$\begin{aligned} t_1^2E_1^2 = {} & [t_2E_2 + (P/t_2E_2)(R + X\tan\phi_r)]^2 \\ & + [(P/t_2E_2)(X - R\tan\phi_r)]^2. \end{aligned}$$

We may choose a value for t_1 or t_2, then this equation will give the other. The minimum range of tap ratio is required when we choose t_1 and t_2 such that $t_1t_2 = 1$. The equation for t_2 then becomes

$$\begin{aligned} & t_2^4E_2^4 + 2t_2^2WE_2^2(R + X\tan\phi_r) \\ & \qquad - E_1^2E_2^2 + W^2\sec^2\phi_r(R^2 + X^2) = 0. \end{aligned}$$

EXAMPLE. Find the transformer tap ratios when the receiving-end voltage is equal to the sending-end voltage, the high voltage line operates at 132 kV. and transmits 20 000 kVA. at 0·8 lagging p.f., and the impedance of line and transformers is $40 + j200$.

The equation for t_2 when we choose $t_1 t_2 = 1$ is

$$t_2^4 + 2t_2^2(P/E_2^2)(R + X \tan \phi_r) - 1 + (P^2/E_2^4) \sec^2 \phi_r (R^2 + X^2) = 0,$$

since $E_1 = E_2$.

We have $R = 40$, $X = 200$, $R + X \tan \phi_r = 40 + 150 = 190$, $R^2 + X^2 = 41\ 600$, $P = 0{\cdot}8 \times 20\ 000$ kW. $= 16\ 000$ kW., $W \sec \phi_r = 20\ 000\ 000$.

The equation for t_2 is thus

$$t_2^4 + 2t_2^2\ (0{\cdot}174) - 1 + 0{\cdot}055 = 0,$$

i.e.

$$t_2^4 + 2t_2^2\ (0{\cdot}174) - 0{\cdot}945 = 0.$$

This gives

$$t_2^2 = -\ 0{\cdot}174 + \sqrt{[(0{\cdot}174)^2 + 0{\cdot}945]} = 0{\cdot}813$$

so that

$$t_2 = \underline{\underline{0{\cdot}902}} \text{ and } t_1 = \underline{\underline{1{\cdot}11}}.$$

It must be remembered that t_1 and t_2 are the factors by which the nominal transformer ratio is to be multiplied. Thus if the line is connected to a 6·6 kV. bus-bar at the sending- and receiving-ends, the actual ratios are—

$$20t_1 = 22{\cdot}2 \quad \text{and} \quad 20t_2 = 18{\cdot}04.$$

Booster Transformers. In some cases it is not convenient to have tap-changing gear in the main transformer. In other cases it may be desired to increase the voltage at a point in the transmission line at a distance from the main transformers, and it is not worth while installing a main transformer with tap-changing if a much cheaper method is available.

Fig. 137 (*a*) shows a suitable voltage regulating equipment consisting of a regulating transformer with tappings and a booster transformer; the on-load tap-changing gear is omitted from this diagram as it would merely confuse matters. The voltage at *PP* is maintained constant by tap-changing in the main transformer, and intermediate loads make it necessary to have some regulating device at *QQ*, in this case a regulating transformer and a boosting transformer. Fig. 137 (*b*) shows an automatic booster arranged to work on the line current. It is clear that greater accuracy is obtained by working on the line current, which varies from zero to full load value, rather than on the line voltage which may vary by only 10 per cent. An error of 1 or 2 per cent in a voltage transformer will cause a regulation error of 10 or 20 per cent (1 or 2 per cent divided by 10 per cent), whilst a similar error in a current transformer causes only an error of 1 or 2 per cent in the regulation. The contacts *R* and *S* are operated by the same solenoid and are mechanically interlocked so that one is closed when the other is opened.

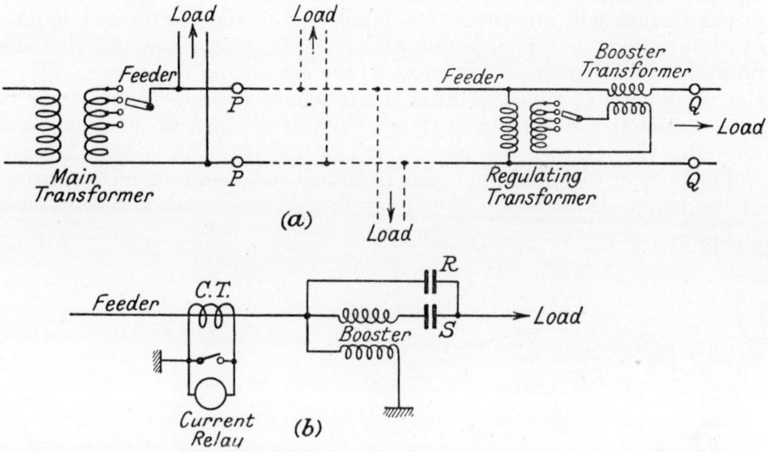

FIG. 137. AUTOMATIC VOLTAGE-REGULATING, BOOSTER TRANSFORMER

The current relay has two sets of contacts, one of which closes when the current reaches the value at which boost is necessary. This contact closes the solenoid circuit, which energizes the solenoid, closing *S* and opening *R* (these contacts are arranged with an auto-

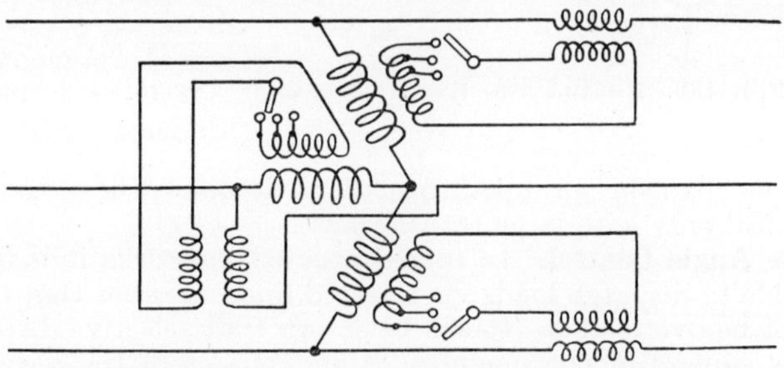

FIG. 138. THREE-PHASE VOLTAGE BOOSTER

transformer device so that there is always a through path for the load current). When the current drops below this value by a certain adjustable amount the other relay contact closes, then *S* is opened and *R* closed.

It is clear that the kVA. rating of this equipment is the product of the line current and the injected voltage, and is thus only 10 or

20 per cent of that of a main regulating transformer. Fig. 138 shows a three-phase regulating system.

The advantages of the booster system are that the regulating equipment is independent of the main transformer so that a failure in the former will not throw the latter out of service for any length of time; and that it is much the cheaper method when there is no main transformer at the point where regulation is desired. When the regulation is required at a point where a main transformer is to be placed, the system is dearer, requires more floor space, and increases the losses.

The cost of tap-changing gear is almost independent of the rating of the transformer to which it is applied, as the cost is mainly that

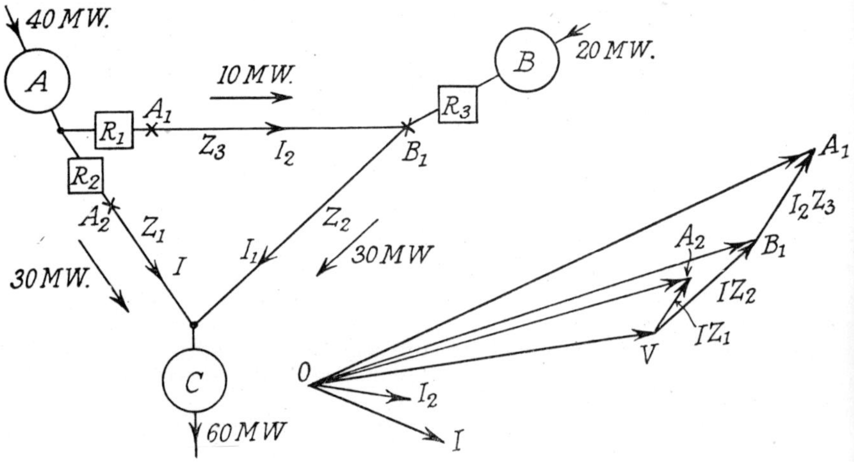

FIG. 139. REGULATING REQUIREMENTS IN INTERCONNECTED SYSTEM

of the mechanism for operating the switches. The gear is thus economical only with large transformers.

Phase Angle Control. In interconnected networks it is essential to be able to dispatch loads via selected lines, in order that the lines may not be overloaded. It will be shown that selective dispatching requires control of the magnitudes and phases of the various line voltages. Thus in the system of Fig. 139, suppose that *C* requires a load of 60 MW., of which *A* and *B* can supply 40 and 20 respectively; suppose, moreover, that the line from *A* to *C* cannot transmit more than 30 MW. It is then necessary to send 10 MW. from *A* to *B* and 30 from *B* to *C*. It may be required that the voltages at *A*, *B*, and *C* should be the same, *V*. It will be shown that in order that this may be so, it will be necessary to have regulation equipment, R_1, R_2, and R_3, of which R_1 or R_2 can control phase.

The currents along A_2C and B_1C must have equal in-phase components, since they carry 30 MW. each; it is most economical that the currents have the same phase. Let these currents be I, and let the current along A_1B_1 be I_2. If the impedances of the interconnectors A_2C, A_1B_1, and B_1C are Z_1, Z_2, and Z_3, the voltages

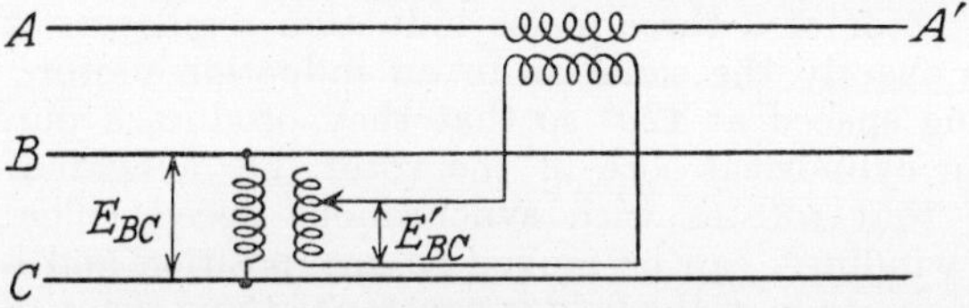

FIG. 140. PHASE-SHIFT TRANSFORMER IN THREE-PHASE SYSTEM

at A_1, A_2, and B_1 are shown in the vector diagram of Fig. 139 as OA_1, OA_2, and OB_1. R_3 must raise the voltage at B to OB_1; the phase will correct itself by synchronizing current. R_1 and R_2 must raise the voltage at A, which is V, to the appropriate magnitudes OA_1 and OA_2; and there must be a phase control in one of them to create the phase difference given by the angle $\phi = \angle A_1OA_2$. If this is done the phase of the generator at A will then correct itself by synchronizing current.

There must be some means available whereby the phase of the voltage can be varied as well as the magnitude.

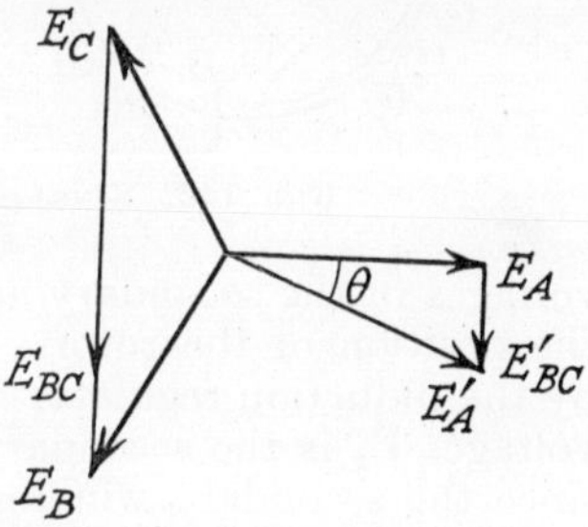

FIG. 141. VECTOR DIAGRAM OF FIG. 140

Phase-shifting Transformer. In a three-phase system, phase-shift can be introduced by a transformer as shown in Fig. 140. The vector diagram of the voltages is given in Fig. 141. The voltage between lines B and C is E_{BC}, which is at right angles to E_A; E'_{BC} is a positive or negative multiple of E_{BC}, and the final line voltage is $E_A' = E_A + E'_{BC}$, making an angle $\theta = \text{arc tan } (E'_{BC}/E_A)$ with the voltage E_A. The magnitude of θ is varied by tappings on the secondary of the transformer between lines B and C.

Induction Regulator. The single-phase induction regulator cannot be used for phase-shifting, but merely for boosting or lowering the voltage. It consists of a transformer with coupling between the primary and secondary varying from $-M$ to $+M$, where M is the maximum mutual inductance. The output voltage lies between $V - v$ and $V + v$, where V is the input voltage and v is the maximum voltage that can be induced in the secondary. Fig. 142 shows a simple sketch of the induction regulator as used in a single-phase circuit. The mutual between the windings is varied by rotating the secondary, which carries the line current and has a maximum voltage of about 10 per cent of the full voltage.

The three-phase induction regulator is similar to the induction motor in construction, except that all the six terminals of the three secondary windings must be brought out. Fig. 143 shows the stator and rotor of a three-phase induction regulator. The stator winding is exactly the same as in an induction motor, the three phases being spaced at 120° so that they produce a pure rotating field in the cylindrical core of the rotor, i.e. a field of constant magnitude that rotates with synchronous speed. The rotor, or secondary windings, can be moved to any position and held there. Since the magnitude of the field is constant at any point the induced

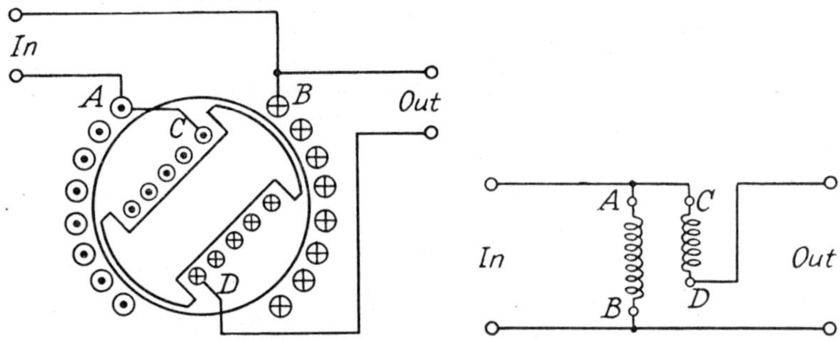

Fig. 142. Single-phase Induction Regulator

voltages in the secondary are constant, but their phases depend on the position of the rotor. The character of the voltage produced by the induction regulator is shown in Fig. 144; V_P is the primary voltage, V_R is the secondary voltage, and V is the output voltage, since the secondary windings are in series with the lines as shown in Fig. 146. The end of the voltage vector V can lie on any point on the circle, whose centre is the end of V_P and radius is the magnitude of the induced secondary voltage. The voltages for the three phases vary together as the secondaries move together, the phase-shifts being positive for a rotation in one direction and negative for the other direction. In Fig. 145, *OA*, *OB*, and *OC* show the phase voltages at the input of the induction regulator, *AA'*, *BB'*, and *CC'* the secondary voltages, and *OA'*, *OB'*, and *OC'* the output phase voltages. The output line voltages are *B'C'*, *C'A'*, and *A'B'*, whilst the input line voltages are *BC*, *CA*, and *AB*. As far as we have considered, the effect of rotating the secondaries is symmetrical about the position in which V_R lies along V_P. Actually the effect of magnetizing current is to make the action unsymmetrical, but we shall not take this into account, although in practice it is not negligible.

(*a*)

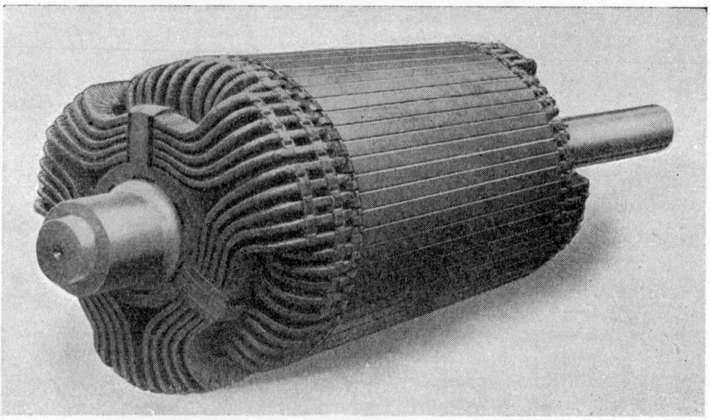

(*b*)

Fig. 143. Three-phase Induction Regulator

(*a*) Stator. (*b*) Rotor.

(*English Electric Co.*)

The three-phase induction regulator just described gives a phase-shift in addition to a voltage boost. In dead-end feeders the phase-shift is unimportant, but in a ring or interconnected system it is important. If it is required to obtain a voltage boost without

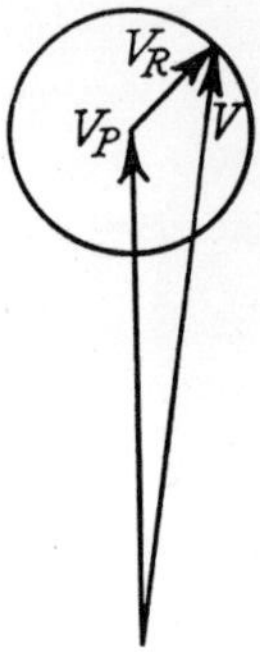

FIG. 144
VECTOR DIAGRAM OF THREE-PHASE INDUCTION REGULATOR

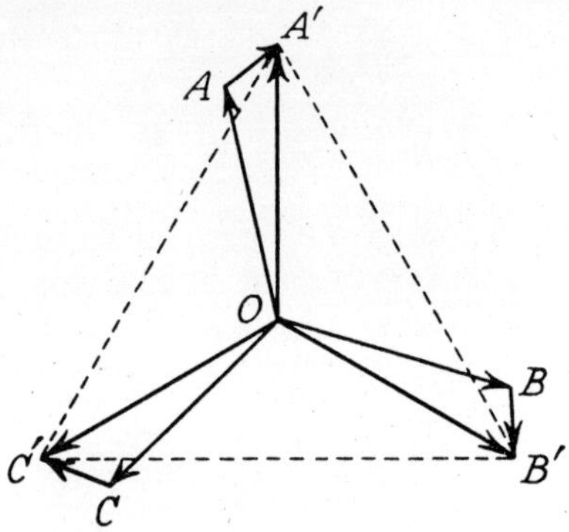

FIG. 145
VECTOR DIAGRAM OF THREE-PHASE INDUCTION REGULATOR

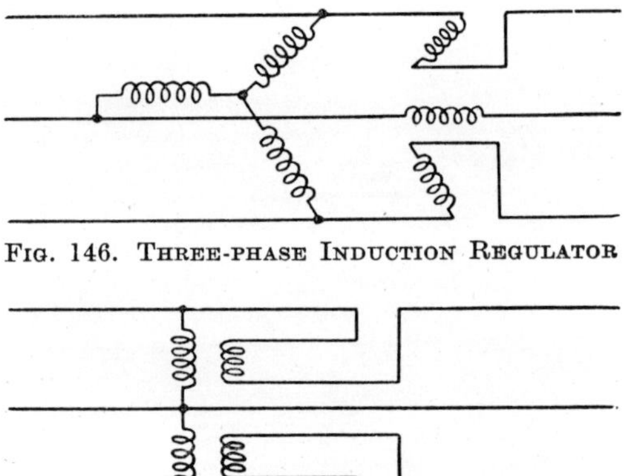

FIG. 146. THREE-PHASE INDUCTION REGULATOR

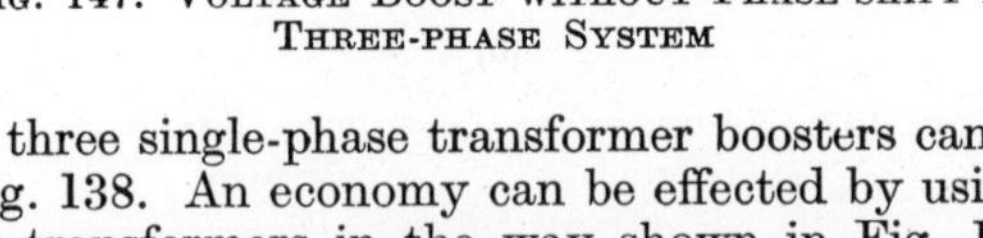

FIG. 147. VOLTAGE BOOST WITHOUT PHASE-SHIFT IN THREE-PHASE SYSTEM

phase-shift, three single-phase transformer boosters can be used, as shown in Fig. 138. An economy can be effected by using only two single-phase transformers in the way shown in Fig. 147; in this method the voltages are unbalanced, and this may be a disadvantage.

Variable voltage boost without phase-shift can be obtained by using a *twin three-phase induction regulator*. In one regulator the voltage has a phase-shift in one direction, in the other in the opposite direction. Fig. 148 shows the primary voltage V_P and the secondary voltages V_{R1} and V_{R2}; the output voltage is $V_P + V_R$, and V_R can take any value between $+ 2v$ and $- 2v$, where v is the magnitude of the secondary voltage in each regulator.

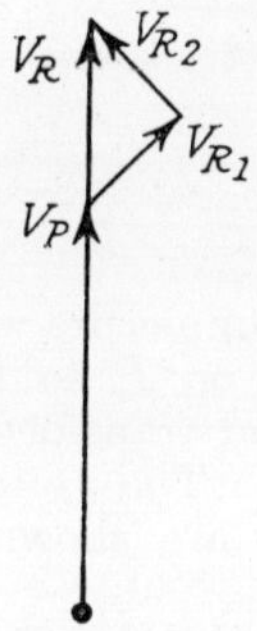

FIG. 148
VECTOR DIAGRAM OF TWIN, THREE-PHASE, INDUCTION REGULATOR

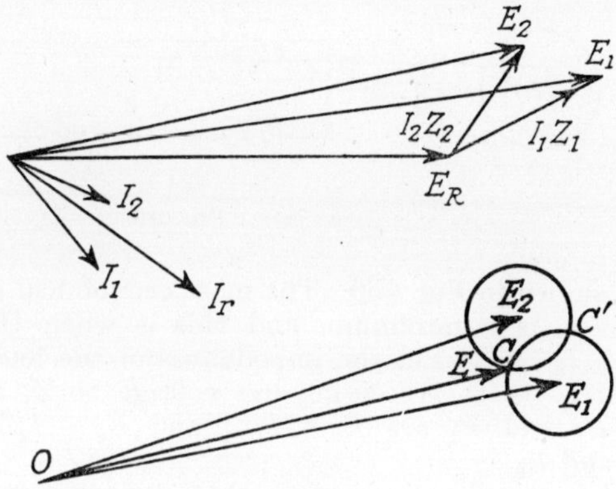

FIG. 150

Methods of using tap-changing transformers and induction regulators are given in the following example, in which two parallel feeders connect two stations.

EXAMPLE. Two stations are interconnected by two parallel three-phase feeders, one working at 33 000 V. and the other at 66 000 V. through the medium of transformers and induction regulators. Explain the requirements for the correct subdivision of the total load between the two feeders and illustrate your explanation by vector diagrams. (*Lond. Univ.*, 1931.)

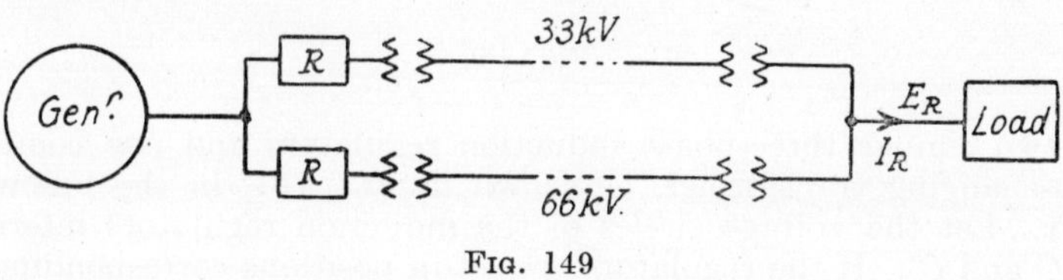

FIG. 149

Fig. 149 represents the system. R denotes any regulating equipment, either tap-changing or an induction regulator. The load voltage E_r and current I_r are fixed by the load conditions at the receiving station. It is simplest to replace the system by the equivalent system reduced to the receiving station voltage. Let the currents I_1 and I_2 along the feeders be postulated by considerations of the current carrying capacities of the feeders: given the magnitudes of I_r, I_1, and I_2, the phase positions of the latter two are fixed in either of two positions by a parallelogram of currents, and

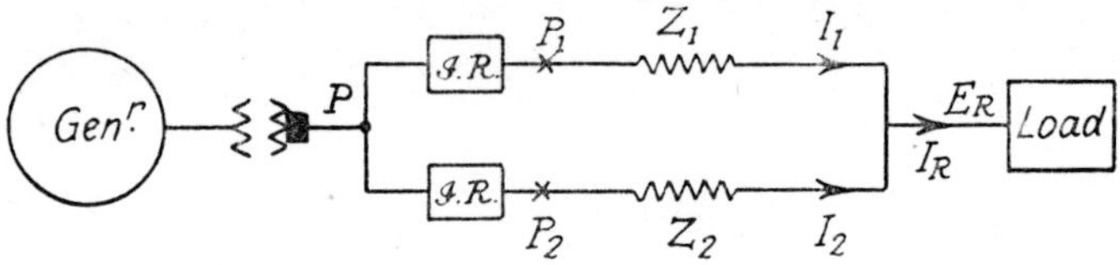

FIG. 151

are shown in Fig. 150. The most economical condition occurs when $I_1 + I_2$ is a minimum, and this is when they lie on I_r so that $I_1 + I_2 = I_r$. Let the impedances of the feeders and transformers referred to the receiving-end voltage be Z_1 and Z_2. The voltages at P_1 and P_2 are $E_r + I_1Z_1$ and $E_r + I_2Z_2$, and are shown as E_1 and E_2.

It is possible to produce these voltages at P_1 and P_2 by means

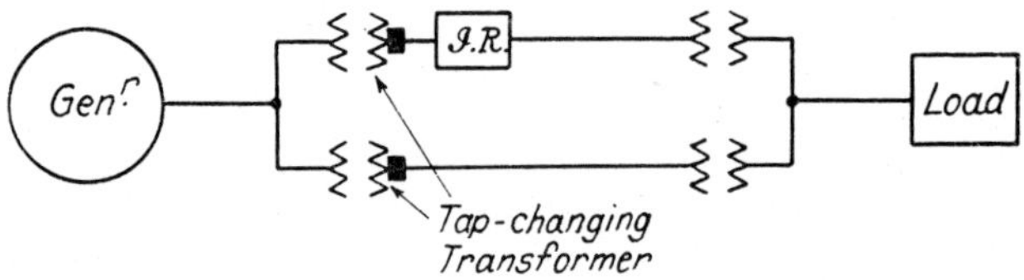

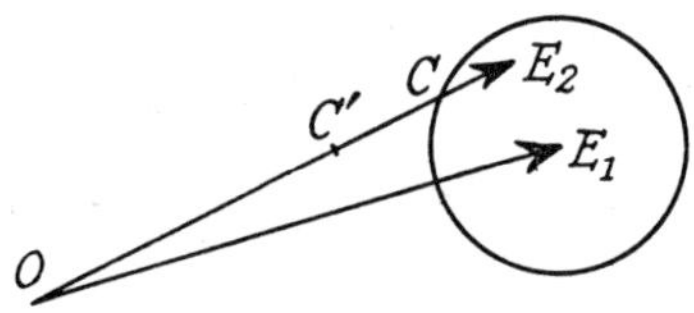

FIG. 152

of two similar three-phase induction regulators and one common tap-changing transformer, as shown in Fig. 151, in the following way. Let the voltage circles of the induction regulators intersect at C and C'. If the regulators are set in positions corresponding to one of these points, the voltage at P is given by OC or OC' and may be achieved by the use of the tap-changing transformer.

Another way is to use two tap-changing transformers and one three-phase induction regulator, as shown in Fig. 152. Let the voltage circle of the induction regulator intersect E_2 at C. Let the generating voltage be represented by OC'. The first tap-changing transformer must increase the voltage from OC' to OC, and the second from OC' to E_2. If the generator voltage can be increased to OC by increasing the excitation, there is no need for a transformer in series with the induction regulator, and then the second transformer needs a tap-ratio of $OC : E_2$; we require then only an

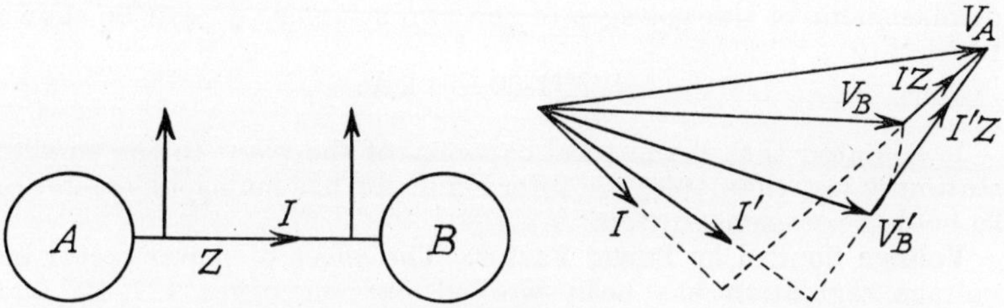

FIG. 153. PARALLEL OPERATION OF GENERATORS CONNECTED BY A LINE

induction regulator in one feeder and a tap-changing transformer in the other. It is, of course, understood that the steam has been regulated to provide the required load.

If the regulation is not very large and the load is not sensitive to voltage, it will be sufficient to keep a fixed generator voltage. Then all other voltages and the load will be reduced in the proportion OC' to OC. The *division of the load* will remain as above merely by the use of one induction regulator and one tap-changing transformer.

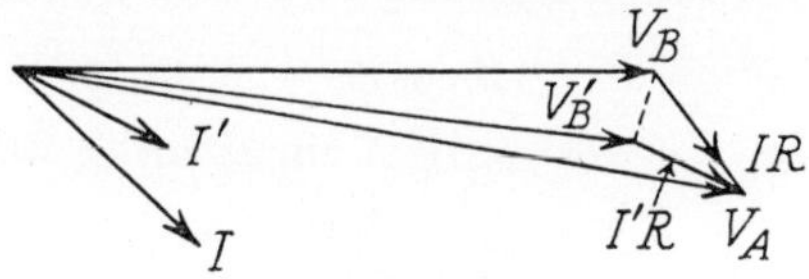

FIG. 154
PARALLEL OPERATION OF GENERATORS CONNECTED BY A LINE

Parallel Operation of Generators. Suppose that the two generating stations A and B are supplying loads and that current I flows from A to B. The voltages V_A and V_B at the stations are given in the vector diagram of Fig. 153. Suppose now that additional load is thrown on to the bus-bars of station B. The generator voltage at B will fall back in phase to V_B', so that the voltage difference becomes $V_A - V_B'$ and the current increases to I', which is larger than I and more in phase with V_A. Thus station A sends more power to B and the steam governor adjusts itself to meet the condition. The reverse happens when load is taken off the bus-bars of B or load is thrown on to the bus-bars at A. In the vector diagram shown in Fig. 153 we have represented Z, the impedance of the interconnector, as a pure reactance. When the interconnector is a pure resistance, the vector diagram is as shown in Fig. 154. It is seen that when load is thrown on to B the current I diminishes so that less power flows from A to B, and B is therefore called upon to supply more load. This is progressive and leads to hunting and instability. The presence of reactance in the interconnector is thus essential for stability.

The synchronous capacity of an interconnector is defined as the change of kilowatts transmitted per radian change of angular displacement of the voltages of the two systems. It can be shown to be

$$(E^2X/1\,000Z^2)\text{ kW.}$$

It is stated that if the total capacity of the plant in the smaller station is less than this, the interconnector has sufficient reactance to hold the stations in step.

Voltage Control by Power Factor. The effect of power factor on voltage regulation has been worked out on pages 137–42, and a line chart for a particular example is shown in Fig. 106. Thus with a given receiving-end voltage of 10 kV. and load of 1 000 kW., the sending-end voltage is 16·8 kV. at a power factor of 0·5 lagging, 14·6 at 0·7 lagging, 13·3 at 0·9 lagging, 10·8 at 0·9 leading, 9·7 at 0·7 leading, and 8·6 at 0·5 leading. By control of the power factor we can therefore vary the voltage of the line over a very wide range of values.

For example, we can make the voltages at the sending- and receiving-ends equal by a proper choice of the power factor in the following way.

Putting $E_s = E_r$ we see that equation (69) gives

$$R\cos\phi_r + X\sin\phi_r = -IZ^2/2E_r, \quad . \quad . \quad . \quad (69)$$

or $$R\cos^2\phi_r + X\sin\phi_r\cos\phi_r = -P(R^2 + X^2)/2E_r^2$$

or $$R\cos 2\phi_r + X\sin 2\phi_r = -(PZ^2/E_r^2) - R.$$

Substituting for R and X in terms of the line impedance angle we get

$$\begin{aligned}\cos\psi\cos 2\phi_r + \sin\psi\sin 2\phi_r &= -(PZ/E_r^2 + R/Z)\\ &= \cos(\psi - 2\phi_r),\end{aligned}$$

so that

$$\phi_r = \tfrac{1}{2}\psi + \tfrac{1}{2}\cos^{-1}[\cos\psi + PZ/E_r^2] - (\pi/2) \quad . \quad (92)$$

For example when $P = 0$ this reduces to

$$\begin{aligned}\phi_r &= \tfrac{1}{2}\psi + \tfrac{1}{2}\psi - \pi/2\\ &= \psi - \pi/2.\end{aligned}$$

Example. Find the power factor for equal sending- and receiving-end voltages in the above system when the load is 1 000 kW.

$R = 16$, $X = 30$, $Z = 34$, $\psi = 90° - 28° 3' = 61° 57'$, $\cos\psi = 0{\cdot}47$.

$$PZ/E_r^2 = 0{\cdot}34.$$

$$\therefore \quad \phi_r = 30° 58' + \tfrac{1}{2}\cos^{-1} 0{\cdot}81 - \pi/2$$
$$= 30° 58' + 17° 57' - 90°$$
$$= -41° 1',$$

and the power factor is 0·75 leading.

From Fig. 106 it is seen that at equal sending- and receiving-end voltages of 10 kV. and load of 1 000 kW. the reactive kVA. is 860 kVAR. leading, giving

$$\tan\phi_r = -0{\cdot}86,$$

or

$$\phi_r = -41°,$$

in agreement with what we have found.

In this case the in-phase current is 100 amperes and the wattless current 86 amperes, so that the total current is 133 amperes.

It may be noted that at unity power factor the sending-end voltage is 12·0 kV. and the current is 100 amperes. It is then a question of economics whether it is worth while raising the current from 100 amperes to 133 amperes in order to achieve zero regulation. If the phase angle is adjusted to the necessary negative value to give zero regulation, no other regulating equipment is required. But the current-carrying capacity of the alternator and line or cable must be greater in the ratio of 133 : 100 than for the case when the power factor is unity; in the latter case, however, regulating equipment, in the form of tap-changing or booster transformers, is required. In practice it is found worth while to operate as near unity power factor as possible, as then the plant can be used to the maximum advantage and the line losses for a given transmitted load are a minimum. Thus in the case just discussed, the losses at unity power factor are only nine-sixteenths of those at the condition of equal sending- and receiving-end voltages.

The advantages of advancing the phase to as near unity power factor as possible are thus, (1) the load output of a given plant is increased and with it the earning capacity, (2) line losses are reduced to a minimum, (3) voltage regulation is considerably reduced, and (4) there is a beneficial effect upon the stability of the system.

Average values of power factor for different kinds of loads are the following—

Lighting	0·95
Lighting and power, mainly the former	0·8–0·85
Lighting and power, mainly the latter	0·75
Power	0·65–0·70
Single-phase power	0·5

In all cases the power factor is lagging, so that the effect on voltage regulation is adverse. The methods of advancing the phase employ static condensers, over-excited synchronous motors running light, and phase advancers of the Kapp vibrator type.

Static Condensers. A static condenser takes a current which leads the voltage by an angle only a little less than 90°. They have small losses, require almost no maintenance, and are very convenient in the small sizes. In order to improve the power factor of a load they can be placed in series with the line, as shown in Fig. 155, or in shunt, as shown in Fig. 156.

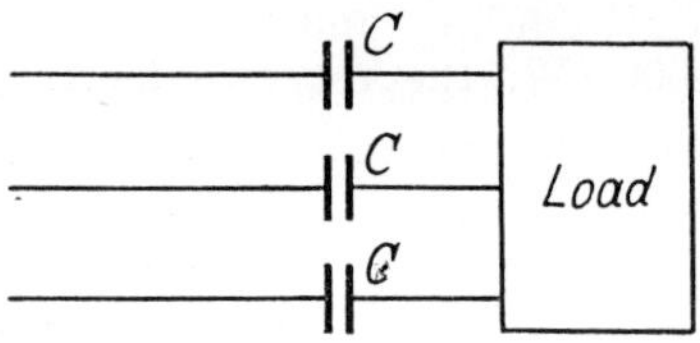

FIG. 155
POWER FACTOR IMPROVEMENT BY SERIES CAPACITANCES

The effect of series condensers on the power factor of a load can be found in the following way. Let the load impedance per phase be

$$Z_0 = R_0 + jX_0,$$

so that the angle of lag is

$$\phi = \tan^{-1}(X_0/R_0).$$

When the series condensers are used the load impedance per phase is

$$Z_0' = Z_0 + 1/j\omega C = R_0 + j(X_0 - 1/\omega C)$$

so that the phase angle is

$$\phi' = \tan^{-1}[(X_0 - 1/\omega C)/R_0].$$

If C is chosen to be equal to $1/\omega X_0$, $\phi' = 0$ and we have unity power factor.

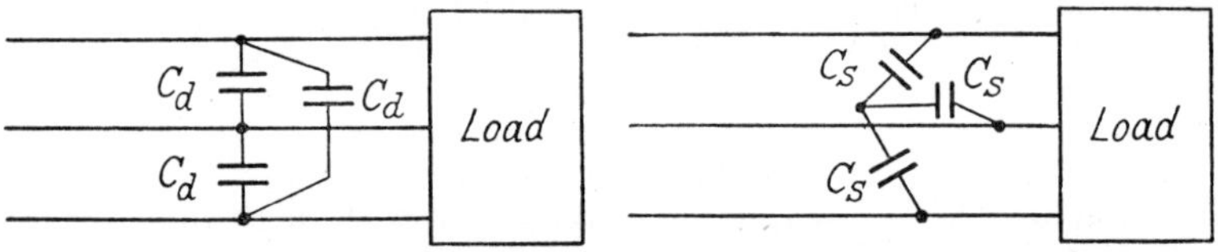

FIG. 156. POWER FACTOR IMPROVEMENT BY SHUNT CAPACITANCES

EXAMPLE. Find the series condensers required to raise the power factor of a load of 100 kW. from 0·8 lagging to unity, the supply being three-phase at 10 kV. Find also the rating of the condensers.

The phase voltage is $10\,000/\sqrt{3}$, and the load per phase is $\frac{1}{3} \times 100$ kW. The in-phase current is thus $10/\sqrt{3}$ and the wattless current is $7{\cdot}5/\sqrt{3}$. The load impedance per phase is

$$Z_0 = \frac{10\,000/\sqrt{3}}{(10 - j7{\cdot}5)/\sqrt{3}} = 640 + j480.$$

Therefore $$C = \frac{1}{2\pi \times 50 \times 480} = 6{\cdot}63\ \mu\text{F}.$$

The voltage across each condenser is $I/\omega C$, so that the rating per condenser is $I^2/\omega C$ and the total rating is

$$3I^2/\omega C = 3I^2X_0 = (10^2 + 7{\cdot}5)480$$
$$= 75 \text{ kVA.}$$

The static condenser is the most economical solution for ratings below 600 kVA.

Series condensers can be used to improve the regulation of a line by neutralizing the reactance. The value of condenser is then $C = 1/\omega^2 L$, where L is the inductance per phase of the line; the value is usually very large, but may be reduced to a reasonable value by a transformer as shown in Fig. 157.

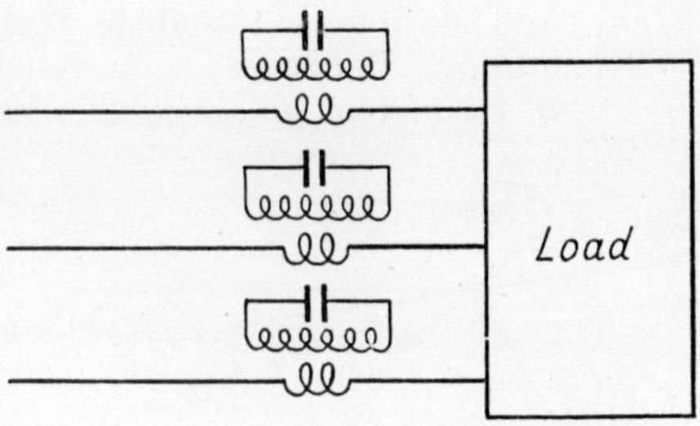

FIG. 157. CAPACITANCE BOOSTER

The value of the shunt condensers for the improvement of the power factor can be found in the following way. Let I_q be the wattless current per phase which is to be neutralized. The value of capacitance in the *star* bank is given by

$$C_s = I_q/\omega V,$$

where V is the phase voltage. The capacitance in the *delta* bank is

$$C_d = \tfrac{1}{3}I_q/\omega V.$$

The rating in both cases is $3I_qV$. The values in terms of the kVAR. and the line voltage V_1 are

$$C_s = \frac{\text{kVAR.}}{\omega V_1^2} \text{ and } C_d = \frac{\text{kVAR.}}{3\omega V_1^2},$$

since kVAR. $= 3I_qV$ and $V_1 = \sqrt{3}V$.

EXAMPLE. Find the shunt condensers required to raise the power factor to unity in the preceding problem.

The kVAR. $= \frac{3}{4} \times 100\ 000 = 75\ 000$, so that

$$C_s = \frac{75\ 000}{2\pi \times 50 \times 10\ 000^2} = 2{\cdot}39\ \mu\text{F}.$$

and $$C_d = 0{\cdot}797\ \mu\text{F}.$$

The values need not be accurate within 5 per cent. The rating is

$$3I_qV = 3 \times (7{\cdot}5/\sqrt{3}) \times (10\ 000/\sqrt{3})$$
$$= 75 \text{ kVA.},$$

which is the same as for the series condensers and is the value of the kVAR. to be neutralized, as it must be.

Phase Advancers. Large induction motors have low power factors on partial load, rising from about 0·6 at a quarter load to 0·88 at full load. They are often provided with a Kapp vibrator, which advances the phase so that the power factor is nearly unity at all loads. For a description of the Kapp vibrator see *Electrical Technology*.*

Other types of phase advancers used in practice are the Leblanc exciter and the Scherbius and the Walker types of phase advancer.

Synchronous Condenser. A synchronous condenser, or, as it is better termed a *synchronous phase modifier*, is a synchronous

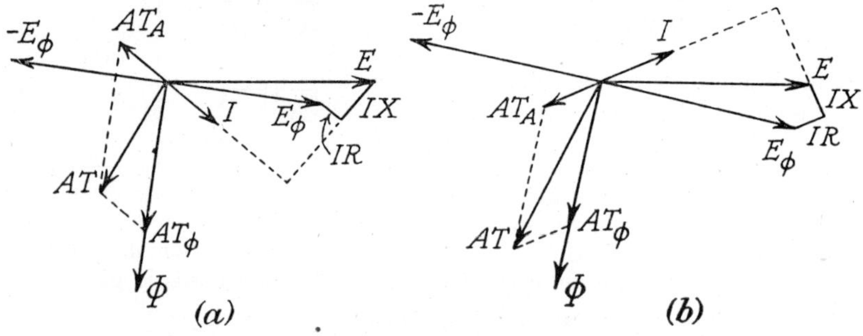

FIG. 158. VECTOR DIAGRAM OF SYNCHRONOUS PHASE MODIFIER

motor operating with no load. It is the property of such a motor that it takes lagging kVA. when the field current is below a certain value, and leading kVA. when the field current is above this value. This can be seen in the following way.

Fig. 158 (*a*) and (*b*) show the vector diagrams for a synchronous motor taking lagging and leading current, respectively. E is the applied voltage, and E_ϕ is this voltage less the drop in the armature resistance R and leakage reactance X. The back e.m.f. $-E_\phi$ is produced by the flux Φ which leads $-E_\phi$ by 90°, since the back e.m.f. is $-d\Phi/dt = -j\omega\Phi$ so that $\Phi = (j/\omega)$ times the back e.m.f. The ampere-turns AT_ϕ are required to produce the flux Φ. Armature reaction produces the ampere-turns AT_A in phase opposition to I, and the field excitation must provide the ampere turns AT which is the vector sum of AT_ϕ and AT_A. (It should be noted that whilst e.m.f.'s and currents are time vectors the ampere-turns are space vectors.) With a given excitation, and the consequent AT, the current adjusts itself to the requisite phase so that the component of I in phase with E_ϕ is a constant. It is clear from the vector

* *Electrical Technology*, by Prof. H. Cotton (Pitman), Sixth Edition, pp. 518–21.

diagrams that the excitation is less for lagging current than for leading current.

Fig. 159 shows the V-curves of a synchronous motor under different loads. For the purposes of power factor correction and voltage regulation the machine is run without load, and the losses due to windage and friction are 4 to 2 per cent for machines of 2 000 kVA. to 20 000 kVA. and over, respectively. The machines are designed to give full-load output at leading power factor, but only about 50 per cent at lagging power factor. Larger lagging loads require a weak field excitation and may cause instability. A further important factor is the effect of the variation of voltage on the kVA. of the machine. This effect is best given in the form shown in Fig. 160, in which the kVA. is given against voltage for various values of the field current.

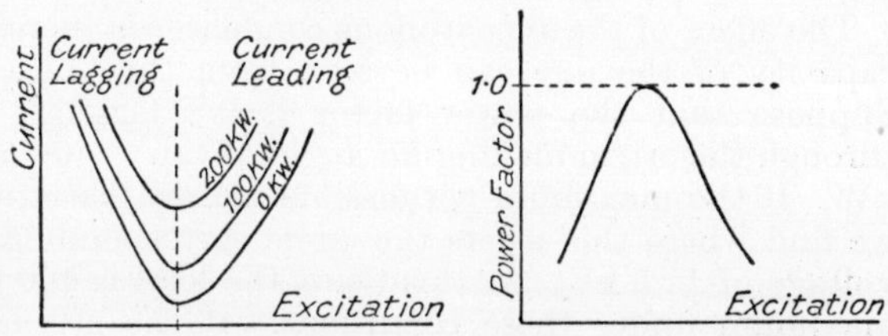

FIG. 159. V-CURVES OF SYNCHRONOUS MOTOR

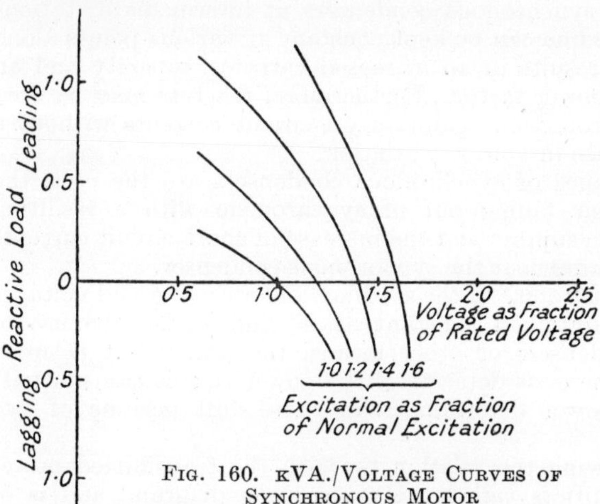

FIG. 160. kVA./VOLTAGE CURVES OF SYNCHRONOUS MOTOR

(*Electrical Power Transmission and Interconnection*)
(Dannatt and Dalgleish)

The effect of the synchronous condenser on regulation is seen from Fig. 105. The receiving-end voltage is $E_r = OA$. The in-phase current causes a voltage drop AC and the wattless current a drop CE, so that the sending-end voltage is $E_s = OE$. By hand or automatic control of the excitation of the synchronous condenser the wattless component of the current can be reduced, made zero, or even made negative. In the last case the sending-end voltage is OG; this can be made equal to E_r, the condition for this case being given in equation (92), so that the leading kVAR. required is $P(-\tan\phi_r + \tan\phi_s)$ where $\cos\phi_r$ is the power factor of the load. The voltage variation of a line limits the amount of power that can be transmitted, and the use of synchronous condensers increases the carrying capacity of the line considerably by maintaining the same voltage at both ends. The limit is then set, not by voltage variation, but by the line losses.

The effect of the synchronous condenser in increasing the carrying capacity of the line can be seen from the line chart of Fig. 106. Suppose that the power factor is 0·8 lagging; we draw a line through the origin making an angle of $\tan^{-1}$ (0·75) with the axis of kW. If the maximum permissible voltage variation is 15 per cent, we find where this meets the circle corresponding to a sending-end voltage of 11·5 kV. At this point the load is 400 kW., which is the maximum under these conditions. If the synchronous condenser neutralizes the wattless current the load is 800 kW., the current is 80 amperes and the line losses are

$$I^2R = 80^2 \times 16 = 103 \text{ kW.},$$

which is about 13 per cent. The load that can be carried with a 10 per cent energy loss is approximately $800/\sqrt{1{\cdot}3} = 700$ kW., so that the carrying capacity of the line is nearly doubled.

By the use of synchronous condensers at intermediate stations the voltage of the line can be kept constant at various points along its length. This results in an increased carrying capacity and an improvement of power factor. Furthermore, reactors may be used on the line for protection against short-circuit currents without a consequent increase in voltage variation.

The disadvantages of synchronous condensers are the cost, the possibility of their falling out of synchronism with a resulting interruption of the supply, and the increase of short-circuit currents when the fault occurs near the synchronous condenser.

Power-angle Diagrams. If the sending- and receiving-end voltages of a line are maintained constant at E_s and E_r by the use of synchronous condensers or other means, the phase-shift between the voltages at the ends determines the power that is transmitted; conversely the power determines the phase-shift (as one of two values).

The curve showing the relation between the transmitted power and the phase-shift is called the *power-angle* diagram, and is of great help in studying the problem of stability.

As a simple example let us find the power-angle diagram of a short line which can be represented by a series impedance; the power-angle diagram for an alternator is of exactly the same kind, and is found by replacing the series impedance by the synchronous impedance of the alternator. The diagram can be found graphically by the use of Fig. 103. Let $OB = E_r$ be fixed; then E_s will lie on a circle of radius E_s and centre O. If we choose a value of phase-shift θ, so that angle $BOC = \theta$, BC represents the voltage drop in the line. We then construct the triangle BCF, where the angle at F is a right angle and the angle at B is $\tan^{-1}(X/R)$. The current is BF divided by R and is along BF, whilst the power component of the current is the projection of the current on OB. Multiplying this by E_r, we get the transmitted power P. This is done for various values of θ, and the curve P versus θ is drawn.

The power-angle diagram can be calculated in the following way. The angle that BF makes with OB is ϕ_r where $\cos\phi_r$ is the receiving-end power factor. Resolving E_s along OB we get

$$\begin{aligned} E_s \cos\theta - E_r &= IR\cos\phi_r + IX\sin\phi_r \\ &= I_wR + I_qX, \end{aligned} \qquad (93)$$

where I_w and I_q are the power and wattless components of the current. Resolving E_s at right angles to OB we get

$$E_s \sin\theta = IX\cos\phi_r - IR\sin\phi_r = I_wX - I_qR \qquad (94)$$

Eliminating I_q by multiplying (93) by R and (94) by X and adding, we get

$$E_s(R\cos\theta + X\sin\theta) - E_rR = I_w(R^2 + X^2).$$

The power is $P = E_rI_w$, so that

$$\left.\begin{aligned} P &= (E_r/Z^2)\,[E_s(R\cos\theta + X\sin\theta) - E_rR] \\ &= (E_r/Z^2)\,[E_sZ\cos(\theta - \psi) - E_rR]. \end{aligned}\right\} \qquad (95)$$

This is the equation of the power-angle diagram, which is seen to be a sine curve, since all are constant except P and θ. When $\theta = 0$, the power is

$$P = (E_r/Z^2)\,(E_s - E_r)R. \qquad (96)$$

P vanishes when

$$E_sZ\cos(\theta - \psi) = E_rR,$$

i.e. $\cos(\theta - \psi) = (E_r/E_s)\,(R/Z) = (E_r/E_s)\cos\psi$,

so that $\theta = \psi + \cos^{-1}[(E_r/E_s)\cos\psi]$

or $\theta = \psi - \cos^{-1}[(E_r/E_s)\cos\psi]$. (97)

For example, the power-angle diagram for the problem on page 143 for $E_r = 10$ kV. and $E_s = 11{\cdot}5$ kV. meets the axis of $P = 0$ at

$$\begin{aligned} \theta &= \tan^{-1}\frac{30}{16} \pm \cos^{-1}\left(\frac{10}{11{\cdot}5} \times \frac{16}{34}\right) \\ &= 61^\circ\,57' \pm 65^\circ\,54' \\ &= -\,3^\circ\,57' \text{ or } 127^\circ\,51'. \end{aligned}$$

The power at $\theta = 0$ is

$$P = (10\,000/34{\cdot}0^2)\,(1500)16 = 207 \text{ kW}.$$

This is seen to be the power corresponding to the point of intersection of the line representing E_r and the circle representing a sending-end voltage of 11·5 kV.; the corresponding kVAR. is about 410 kVAR. lagging. The maximum power that can be transmitted is found by differentiating P with respect to θ in equation (95) and equating to zero. We get

$$-(E_r/Z^2)E_sZ \sin(\theta - \psi) = 0,$$

i.e.

$$\theta = \psi,$$

so that the maximum power is

$$P_{max} = (E_r/Z^2)(E_sZ - E_rR)$$
$$= (E_r^2/Z^2)[Z(E_s/E_r) - R] \quad . \quad . \quad . \quad (98)$$

which is the value found on page 146.

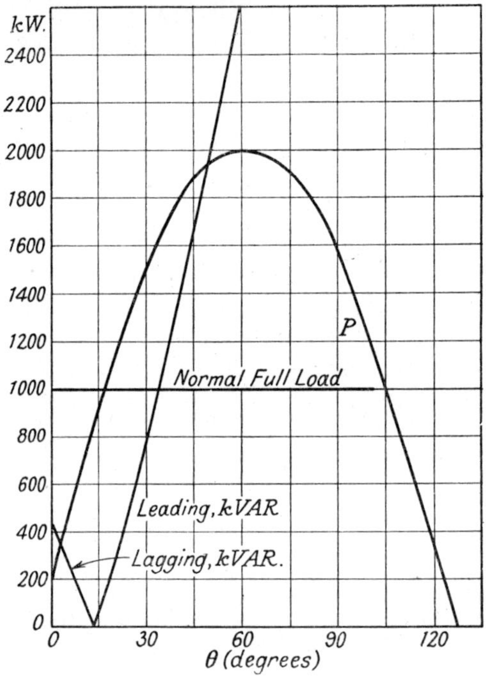

FIG. 161. POWER-ANGLE DIAGRAM OF SHORT LINE

In the case quoted the maximum power is 2 000 kW., which is the value obtained from Fig. 106 at the top point of the circle corresponding to $E_s = 11{\cdot}5$ kV. At this point the reactive power is 2 600 kVAR. leading; it is obvious from Fig. 106 that there is 2 600 kVAR. leading at the position of maximum power for any values of sending- and receiving-end voltages.

Fig. 161 shows the power-angle diagram for the case just discussed, and also the kVAR. curve.

From equation (96) it is seen that if the voltages at the ends are equal, no power can be transmitted without a change of phase. In this case P is zero at $\theta = 0$ and $\theta = 2\psi$; in the problem

$$2\psi = 123° 54'.$$

Maximum power is transmitted still at $\theta = \psi$. The power-angle diagram is thus underneath that shown in Fig. 161, the maximum in each curve occurring at the same value of θ.

If the receiving-end voltage can be varied, the maximum power is found by differentiating equation (98) with respect to E_r. We get

$$(E_s/Z)(1 - 2E_rR/E_sZ) = 0$$
$$E_r = E_s(Z/2R)$$

and the power $P_{max} = E_s^2/4R$.

This is clearly what happens when the load reactance neutralizes

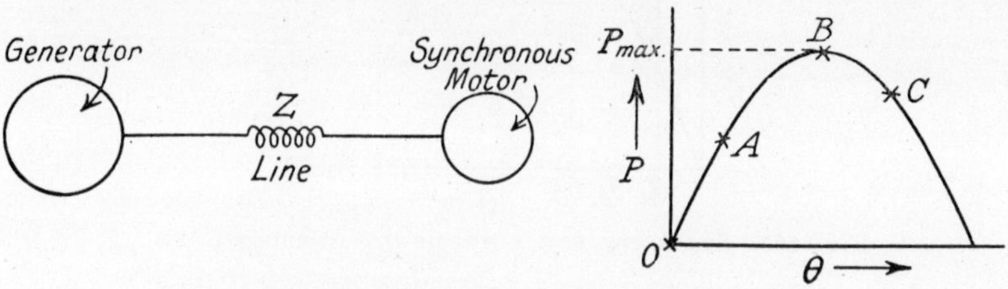

FIG. 162. STABILITY OF A SYNCHRONOUS SYSTEM

the line reactance and the load resistance is equal to the line resistance, as is well known in telephony.

If E_s, E_r, and R are fixed, but the line reactance is variable (by altering the spacing of overhead lines, say), we have $Z = \sqrt{(R^2 + X^2)}$ in equation (98). Differentiating P_{max} in this equation by X and equating to zero we get

$$X = R\sqrt{[4(E_r/E_s)^2 - 1]}$$

and $$P = E_s^2/4R.$$

If the line impedance is considered to be a pure reactance, the power-angle curve has the equation

$$P = (E_rE_s/X)\sin\theta \qquad . \qquad . \qquad . \qquad . \qquad (95a)$$

P is zero at $\theta = 0$ and $\theta = 180°$, and the maximum power is (E_rE_s/X) and occurs at $\theta = 90°$.

The power-angle diagram of a general network can be derived from the general equations (74).

$$E_s = AE_r + BI_r \qquad . \qquad . \qquad . \qquad . \qquad . \qquad (74)$$

If we take E_r as the basic vector, and put

$$A = A_1 + jA_2,\ B = B_1 + jB_2,\ I_r = I_w + jI_q,$$

we get real part of $E_s = E_s \cos\theta = A_1E_r + B_1I_w - B_2I_q$
and imaginary part of $E_s = E_s \sin\theta = A_2E_r + B_2I_w + B_1I_q$.

Eliminating I_q we get

$$B_1(E_s \cos\theta - A_1E_r) + B_2(E_s \sin\theta - A_2E_r) = I_w(B_1^2 + B_2^2).$$

The power is $P = E_rI_w$ so that

$$P = [E_r/(B_1^2 + B_2^2)]\,[E_s(B_1 \cos\theta + B_2 \sin\theta) - E_r(A_1B_1 + A_2B_2)],$$

so that the curve is still a sine curve.

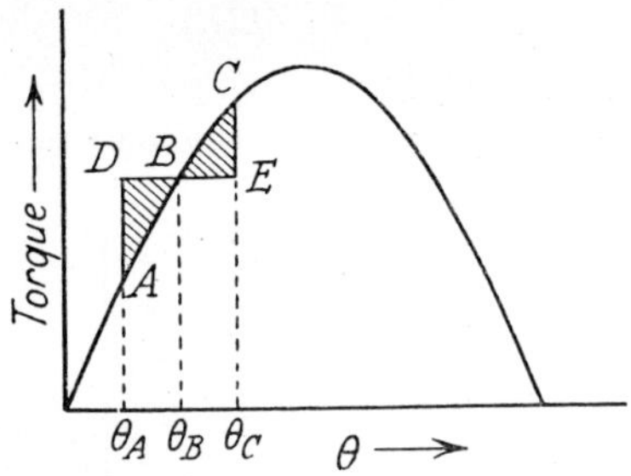

FIG. 163. STABILITY FOR A SUDDENLY APPLIED LOAD

In the nominal-Π method

$$A = 1 + \tfrac{1}{2}YZ = (1 - \tfrac{1}{2}\omega CX) + j(\tfrac{1}{2}\omega CR)$$

and $$B = Z = R + jX,$$

so that $$P = (E_r/Z^2)\,[E_s(R \cos\theta + X \sin\theta) - E_rR],$$

which is the same as equation (95).

Stability. Let us consider a system consisting of a generator, transmission line, and a load which is a synchronous motor, and suppose that Fig. 162 represents the power-angle diagram. As a synchronous motor revolves at a constant speed, the load is proportional to the torque. If we neglect the losses and windage of the motor, there is no current when the machine is running without load. Suppose that the motor is run up to speed, synchronized, and then put on to the line. No current flows and the voltages at the ends of the line are equal and in phase. The position on the power-angle diagram is the point O. Suppose that load is applied very slowly to the motor, that regulating equipment keeps the voltages constant, and that the generator supplies without delay any demand of load. The motor slows down temporarily and falls back in phase until θ reaches the value at which the transmitted load is that which is required to overcome the torque. The motor then runs at synchronous speed with this phase lag, and the point of operation on the power-angle diagram is A. The load may be increased slowly in this manner until the point B is reached. If the load is increased beyond the value P_{max}, the motor slows down still more, so that the point C is reached. This is not, however, a position of equilibrium, for at C the power supplied is less than P_{max} whilst the load is greater. The motor slows down still more, draws less power, and slows down still more rapidly and tends to come to a standstill. The wattless current increases to a large value, limited by the impedance of the line and the stationary motor, and the circuit-breakers open. Thus P_{max} is the maximum load that can be supplied by the generator through the line to a synchronous load which is applied very gradually, and it is called the limit of *static stability*.

In practice, the generator cannot supply any demand instantly, and the load is not applied gradually. Under such conditions the load that can be supplied is very much less, and we have a limit of *transient stability* which is determined by the inertia of rotating parts, governor operation, voltage regulators, and the performance of the electric plant under transient conditions. Before considering the effects of these four phenomena, let us consider an ideal case in which only the inertia of the rotor of the synchronous motor is important, the other conditions being corrected instantly.

Suppose that the power-angle diagram is calibrated in terms of torque T and is shown in Fig. 163, and suppose that the torque is T_A and the phase angle θ_A. Let the load torque be suddenly increased to T_B, which requires a phase angle θ_B for the supply of the necessary power. The rotor cannot fall back instantaneously to a position corresponding to θ_B, and so the power is less than that required for the torque. The rotor therefore decelerates at a finite rate, falls back in phase, and the power torque rises. This goes on until the rotor falls back to the phase angle θ_B, when the power torque becomes equal to the load torque and the deceleration ceases. At this moment, however, the rotor is not revolving at synchronous speed, as it has been decelerating since it left the position of θ_A. The amount by which its speed has been diminished can be found, if we ignore damping and reaction effects, by considering the area ABD. At any position between the positions A and B, the decelerating torque is given by $T_B - T$, i.e. the distance from BD to the power-angle curve. The kinetic energy lost by the rotor is thus given by the area ABD, and the lost angular velocity can be expressed in terms of this energy. The rotor runs thus at this smaller speed, falls back further in phase, and experiences an increased power torque which accelerates it. Its speed increases until it reaches the synchronous speed, when it ceases to fall back in phase and begins to advance. The point at which this happens is C. The energy that has been supplied by the power torque to accelerate the rotor between the positions B and C is given by the area BCE. If we ignore friction and electromagnetic braking, the areas of ABD and BCE are equal, since the speed at A and at C is the synchronous speed. The rotor has thus overshot the position B and has swung from A to C. It will now swing from C to A, and then forwards and backwards until the oscillation is damped out. This is *hunting* or *phase-swinging*. In the case shown in Fig. 163 the load is stable.

Suppose that the case is as shown in Fig. 164. The rotor decelerates from A to B. Then it can accelerate from B to B', but if it passes B' it will decelerate to a standstill. If then there is a point C between B and B' such that area BCE = area ABD, the rotor will overshoot to position C and return. If, however, the area between BB' and the curve is less than ABD, the rotor will not have reached synchronous speed by the time it reaches B'; it will fall back more in phase to a point F and then it will decelerate progressively towards standstill.

The maximum load depends clearly upon the magnitude of the sudden increase which is permitted. It is given by P_B, where the area above BB' equals the area ABD, and AD is the permitted sudden increase.

Let us now consider the effects of governor operation, voltage regulators, and behaviour of the electrical plant. Suppose the load is steady and is being supplied by the generator. When a further shaft load is applied to the motor, the rotor slows down and the e.m.f. falls back in phase as well as decreasing slightly in magnitude. The increase of phase angle results in a heavier line current and transmitted load. Before the governor of the prime mover operates to supply the extra load, the extra load is drawn from the kinetic energy of the generator which also slows down. Thus both the generator and motor voltages decrease, and a larger phase angle is required to supply the load for the extra torque. The falls in voltage in the generator and in the motor are opposed by the transient currents in the fields and amortisseur windings, by Lenz's law. The automatic voltage regulators begin to operate to raise the voltage when the motor reaches the position B in Fig. 164, the rotor begins to accelerate and its slip velocity decreases. If the slip velocity reaches zero, i.e. the speed reaches synchronous speed before the power falls below the shaft load (point B' in Fig. 164), the system is stable. The rotor oscillates about the steady speed with diminishing amplitude due to damping.

The stability of the system is increased in several ways. The prime mover governor is designed to enable the prime movers to follow closely any power demands. The excitation system is designed to give close voltage regulation under transient conditions; the voltage regulators are made quick-acting with a small time constant exciter, so that they take effect before the flux in a synchronous machine has time to fall; relays and regulators have a small time of operation. The lines and terminal apparatus are made stiffer; the maximum static power limit is obtained when $X = \sqrt{3}R$. In cables X may be less than this optimum amount, and series reactors will increase the stability. In overhead lines X is usually greater than the optimum value, and may be reduced by increasing the size of the conductors, or by using series con-

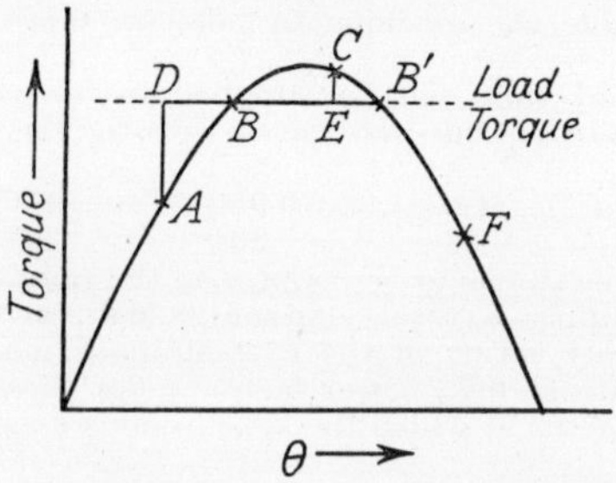

FIG. 164. INSTABILITY FOR A SUDDENLY APPLIED LOAD

densers. The insertion of synchronous condensers at various points of a line keeps the voltage at these points constant, and tends to increase the power limit to the limit of each section.

EXAMPLES VII

1. Describe the procedure to be taken when starting up and paralleling an alternator with others already running. Explain the electrical reactions holding the new machine in step after paralleling. How are the load and power factor on the incoming machine adjusted to the correct values? (*Faraday House*, 1935.)

2. Give a diagram of the essential circuits, and explain the action of an automatic voltage regulator for use with a turbo-alternator. Why is good inherent voltage regulation not called for with modern a.c. generators? (*Lond. Univ.*, 1933.)

3. A three-phase induction regulator is to be used for controlling the voltage on a 3 300 V. feeder. Draw a diagram of connections of the regulator and draw to scale the vector diagram of its currents and voltages when the line voltage on the supply side is 3 300 V., that on the feeder side is 3 650 V. and the load on the feeder is 100 kW. at 0·8 lagging p.f. The voltage induced in each secondary winding with 3 300 V. on the primary is 300 V. and the magnetizing current is 2 A. Resistance and reactance drops may be neglected.

From the diagram find the current in the primary winding and the total current on the supply side of the regulator.

Under what circumstances would the phase difference between the voltages on the supply and feeder sides of the regulator be a disadvantage? (*Lond. Univ.*, 1933.)

4. State concisely the more important methods in use for improving the power factor at the receiving-end of a transmission line.

A three-phase, 50-cycle, 3 kV. motor develops 600 h.p., the power factor being 0·75 lagging, and the efficiency 0·93. A bank of condensers is connected in delta across the supply terminals, and the power factor raised to 0·95 lagging. Each of the capacitance units is built up of five similar 600 V. condensers. Determine the capacitance of each condenser. (*Lond. Univ.*, 1933.)

5. The exciting current of an alternator running on constant voltage and frequency bus-bars is varied over a wide range whilst the steam supply to the

prime mover remains unchanged. Explain with appropriate diagrams the corresponding variations in the armature current and power factor.

Prove that the maximum load which a synchronous machine can supply to an external circuit before falling out of step depends on the value of the the excitation. (*Lond. Univ.*, 1932.)

6. Explain the process of connecting a synchronous generator to operate in parallel with an existing supply, stating the adjustments which must be made subsequently to enable the machine to take its share of both power and reactive load.

The maximum load that can be supplied by a synchronous generator coupled to constant-voltage bus-bars varies with the (lagging) power factor as follows—

Power factor .	1·0	0·95	0·8	0·0
MVA. . .	36	30	25	20

An asynchronous generator is to be coupled to the same bus-bars, so that the combination may supply a total load of 28 MW. at unity power factor. Determine (i) the least rating of the asynchronous machine, which will be constructed to operate at 0·75 power factor under these conditions, and (ii) the load and power factor at which the synchronous machine would operate. (*Lond. Univ.*, 1954.)

7. A 3-phase transmission line has a resistance of 10 Ω per phase and a reactance of 30 Ω per phase.

Determine, graphically or otherwise, the maximum power which could be delivered if 132 kV. were maintained at each end.

Deduce any formula used or explain the graphical construction. (*Lond. Univ.*, 1953.)

8. Explain what is meant by a "power-angle diagram" for a power line.

Line voltages of 6·6 kV. and 6·0 kV. are maintained respectively at the sending- and receiving-ends of a 3-phase power line having an impedance of $0{\cdot}5 + j1$ ohms per core. Determine, for the receiving-end, (*a*) the power delivered at unity power factor, and (*b*) the maximum steady power limit and the power factor at which it is realized. (*Lond. Univ.*, 1947.)

9. Show with the aid of a vector diagram, how the voltage at the receiving-end of a transmission line can be maintained constant by the use of a synchronous phase-modifier.

A 3-phase overhead transmission line supplies a load of 40 000 kW., 0·707 power factor (lagging), at a voltage of 66 kV. If the receiving-end voltage is maintained constant and a synchronous motor is installed to improve the receiving-end power factor to 0·866 (lagging), while maintaining the line current constant in magnitude, determine the kVA. rating and power factor of the synchronous motor. (*Lond. Univ.*, 1949.)

10. State and explain the conditions (*a*) which must be fulfilled, and (*b*) which are desirable for the successful parallel operation of 3-phase transformers.

Two single-phase transformers have equal turns ratios and supply a total load of 700 kVA. at 6·6 kV. and 0·8 power factor (lagging). One transformer is rated at 300 kVA. with a full-load percentage impedance drop of $(1{\cdot}5 + j3{\cdot}5)$ while the other is rated at 500 kVA. with full-load percentage impedance drop of $(1{\cdot}0 + j4{\cdot}5)$. Determine the output kVA. load and power factor of the 300 kVA. transformer. (*Lond. Univ.*, 1949.)

11. A 3-phase transmission line has a resistance of 10 Ω. per phase and a reactance of 30 Ω. per phase.

Determining the maximum power which could be delivered if 132 kV. were maintained at each end.

Derive a curve showing the relation between the power delivered and the angle between the voltage at the sending- and receiving-ends and explain how this curve could be used to determine the maximum additional load which could suddenly be switched on without loss of stability if the line were already carrying, say, 50 000 kW. (*Lond. Univ.*, 1949.)

CHAPTER VIII

SHORT CIRCUITS: SYMMETRICAL COMPONENTS

Introduction. When a fault occurs on a network such that a large current flows in one or more of the phases, it is said that a short circuit has occurred. The fault may be a short between one phase and earth, between two or more phases and earth, between two phases only, or across all three phases; a short between one phase and earth will cause a short circuit only if the neutral point is earthed. It is necessary to know the maximum short-circuit currents that can occur at the various points of a system in order that circuit-breakers may be selected that are adequate to withstand the currents and operate successfully to cut out the faulty section, and also in order that the protective relays may be selected for correct operation. Moreover, it is necessary to be able to calculate, approximately at least, the size of the protective reactors which must be inserted in the system to limit the short-circuit currents to a value which is not beyond that capable of being withstood by the circuit-breakers.

The short-circuit currents in an alternating current system are determined mainly by the reactance of the alternators, transformers, and lines up to the point of the fault in the case of phase-to-phase faults. When the fault is between a phase and earth, the resistance of the earth path may play an appreciable part in limiting the currents.

Percentage Reactance and Short-circuit Currents. The method of specifying the impedance or reactance of a piece of electrical plant, as described in Appendix III, pages 471–2, is very convenient for the calculation of short-circuit currents. The percentage reactance is given by

$$(\% X) = (IX/E) \times 100, \qquad \qquad (99)$$

where X is the reactance, E the rated voltage, and I the full-load current. If the piece of apparatus is the only impedance in the circuit, the short-circuit current is given by E/X, which by equation (99) is

$$I_{sh} = E/X = I \times (100/\% X). \qquad \qquad (100)$$

Thus the short-circuit current for a single piece of apparatus is the full-load current multiplied by (100 divided by the percentage reactance). Thus if the percentage reactance is 10 per cent, the short-circuit current is 10 times the full-load current; if it is 40 per cent, the short-circuit current is 2·5 times the full-load current.

If there are several reactances in series of magnitudes X_1, X_2, and X_3, the short-circuit current is

$$\frac{E}{X_1+X_2+X_3} = \frac{E}{(E/I)(\% X_1) + (E/I)(\% X_2) + (E/I)(\% X_3)} \times 100$$

$$= I \times \frac{100}{(\% X_1) + (\% X_2) + (\% X_3)},$$

by application of equation (99) to the various reactances.

It often happens that the system contains plant of different ratings, and the percentage reactances are given for the respective values. It is then necessary to allow for the different ratings in the following way. Suppose the generator has a rating of 10 000 kVA. and a percentage reactance of 7, and a transformer has a rating of 8 000 kVA. and a percentage reactance of 5. The full-load current of the generator is $(10\ 000\ 000/E) = I_1$, say, so that its reactance is

$$X_1 = (E \times 7)/(I_1 \times 100).$$

Similarly the reactance of the transformer is

$$X_2 = (E \times 5)/(I_2 \times 100)$$

where $I_2 = (8\ 000\ 000/E)$.

The total reactance is

$$X_1 + X_2 = \frac{E \times 7}{I_1 \times 100} + \frac{E \times 5}{I_2 \times 100}$$

$$= \frac{E}{I_1 \times 100}\left[7 + 5 \times \frac{I_1}{I_2}\right] = \frac{E}{I_1 \times 100}\left[7 + 5\frac{EI_1}{EI_2}\right]$$

$$= \frac{E}{I_1 \times 100}\left[7 + 5 \times \frac{10\ 000}{8\ 000}\right],$$

and the total percentage reactance referred to the rating of 10 000 kVA. is

$$(\% X) = \frac{I_1(X_1 + X_2)}{E} \times 100$$

$$= 7 + 5 \times \frac{10\ 000}{8\ 000}$$

$$= 7 + 6{\cdot}25 = 13{\cdot}25.$$

Thus the percentage reactance of the transformer is multiplied by the ratio of the generator rating to the transformer rating in order that it may be expressed with respect to the generator rating.

If there are several pieces of apparatus in the circuit with different ratings, *we choose a basic rating and refer all percentage reactances to this rating by appropriate multipliers.* We can then add the percentages, if the pieces of apparatus are in series, and the short-circuit current is then found by equation (100). The following example illustrates the method.

EXAMPLE. Find the short-circuit current in the single-phase system of Fig. 165, if the fault is a short circuit between lines at the point F which is 10 miles from the transformer T. The reactance per mile is 0·2 Ω. The voltage is 6·6 kV.

We take 2 000 kVA. as the basic rating, so that full-load current is

$$I = 2\,000/6{\cdot}6 = 300 \text{ A.}$$

The percentage reactance of the generator is 8, and of the transformer T, $7 \times (2\,000 \div 1\,200) = 11{\cdot}7$. The line reactance is 2 Ω., so that its percentage reactance is

$$\frac{300 \times 2}{6\,600} \times 100 = 9{\cdot}1.$$

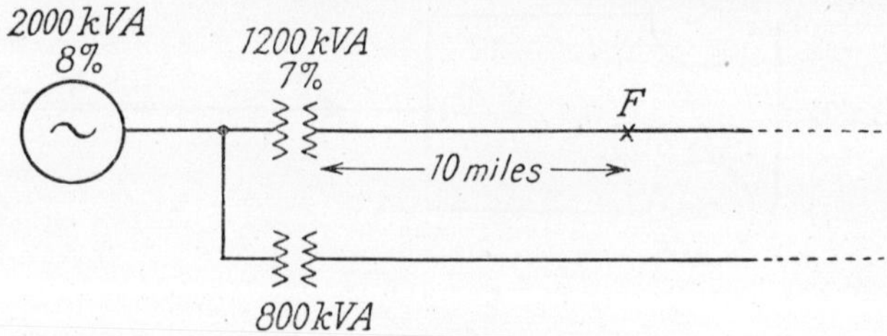

FIG. 165

The total percentage reactance is

$$8 + 11{\cdot}7 + 9{\cdot}1 = 28{\cdot}8,$$

so that the short-circuit current is

$$\begin{aligned} I_{sh} &= I \times (100/\%\, X) = 300 \times (100/28{\cdot}8) \\ &= \underline{\underline{1\,040 \text{ A.}}} \end{aligned}$$

We could use the *direct method*, which consists in the reduction of percentage reactances to actual reactances. Thus the reactance of the generator is

$$\frac{E \times (\%\, X)}{I \times 100} = \frac{6\,600 \times 8}{300 \times 100} = 1{\cdot}76\ \Omega.$$

The reactance of the transformer is similarly

$$\frac{6\,600 \times 7}{300 \times \left(\dfrac{1\,200}{2\,000}\right) \times 100} = 2{\cdot}57\ \Omega.$$

The total reactance is 1·76 + 2·57 + 2·00 = 6·33 ohms, and the short-circuit current is

$$I_{sh} = 6\ 600/6{\cdot}33 = 1\ 040\ \text{A}.$$

If some of the pieces of apparatus are in parallel, their reactances, and hence their percentage reactances, must be compounded by the method used for parallel impedances. Thus a percentage reactance of 10 in parallel with another of 15 gives a resultant value of

$$\frac{1}{\frac{1}{10}+\frac{1}{15}} = \frac{10 \times 15}{10 + 15} = 6.$$

This method can be used for generators in parallel, as shown in the following example. In the calculation of such problems it is

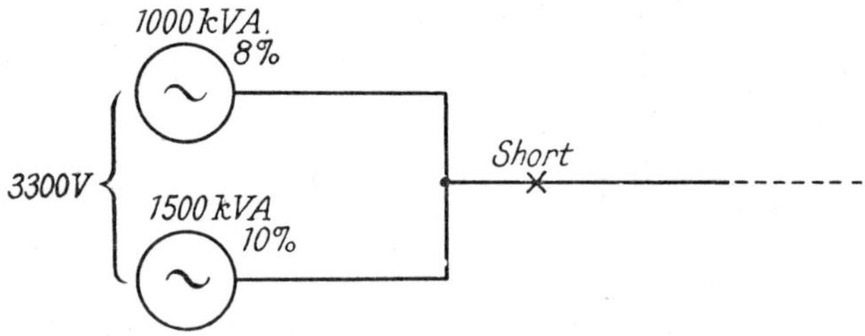

FIG. 166

assumed that the generators have equal voltages which are in phase; the error caused by this assumption should not be serious.

EXAMPLE. Two three-phase generators of ratings 1 000 and 1 500 kVA. and of voltage 3·3 kV. have percentage reactances of 8 and 10, with respect to their ratings. A short across all phases occurs near the common bus-bars. Find the short-circuit current.

The system is shown in Fig. 166.

Let us assume a basic kVA. of 2 500, which is the sum of both ratings. The percentage reactances with respect to this rating are

$$8 \times (2\ 500/1\ 000) = 20$$

and

$$10 \times (2\ 500/1\ 500) = 16{\cdot}7.$$

The resultant percentage reactance is

$$\frac{1}{\frac{1}{20}+\frac{1}{16{\cdot}7}} = \frac{1}{0{\cdot}05 + 0{\cdot}06} = \frac{1}{0{\cdot}11} = 9{\cdot}1.$$

The short-circuit kVA. is therefore

$$2\ 500 \times (100/9{\cdot}1) = 27\ 500.$$

If I_{sh} is the short-circuit current per conductor, the kVA. per phase is

$$I_{sh} \times (3\,300/\sqrt{3}) = \tfrac{1}{3} \times 27\,500\,000,$$

so that $$I_{sh} = \frac{27\,500\,000 \times \sqrt{3}}{3 \times 3\,300} = \underline{\underline{4\,810 \text{ A.}}}$$

The full load current is one-eleventh of this, i.e. 437 A.

Symmetrical Short-circuit Currents. The short-circuit currents are symmetrical, i.e. equal in the different conductors, if the short circuit occurs across both lines in a single-phase system or across the three wires of a three-phase system. The methods developed in the last section are adequate to calculate such short-circuit currents, and the two examples illustrate the methods. If there are several generators

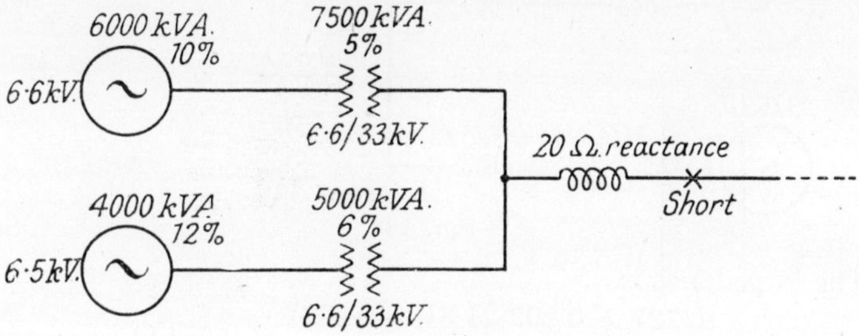

FIG. 167

in the system, it is assumed, as has already been stated, that the e.m.f.'s are equal and in phase with each other. When the system is complicated, the star-delta equivalence of Appendix III is often of great help in effecting a simplification of the network. Sometimes Thévénin's theorem is useful in obtaining the result quickly, especially when generators have unequal e.m.f.'s.

EXAMPLE. Find the short-circuit current in the system of Fig. 167, in which one generator is generating at 6·6 kV. and the other at 6·5 kV. The system is three-phase.

The voltages per phase are 6·6 kV$/\sqrt{3}$ = 3 820 V. and 6·5 kV.$/\sqrt{3}$ = 3 750 V. The full-load current of the larger generator is

$$\frac{6\,000\,000}{3 \times (6\,600/\sqrt{3})} = \frac{6\,000\,000}{\sqrt{3} \times 6\,600} = 525 \text{ A.},$$

so that its reactance per phase is

$$\frac{(\%\,X) \times E}{I \times 100} = \frac{10 \times 3\,820}{525 \times 100} = 0{\cdot}727\ \Omega.$$

In the same way it is found that the reactance of the other generator is 1·31 Ω. It is assumed that the rated voltage in this case also is 6·6 kV., although the generated voltage is 6·5 kV.

Similarly the transformer reactances referred to the low voltage sides are 0·292 and 0·525 Ω. The line reactance transferred to the low voltage side is $20 \div (33/6{\cdot}6)^2 = 0{\cdot}8\ \Omega$. The system is then as shown in Fig. 168. We then apply Thévénin's theorem to the system to the left of *A*. The voltage when the line is disconnected is

$$3\ 820 - \frac{(3\ 820 - 3\ 750)\,(0{\cdot}727 + 0{\cdot}292)}{0{\cdot}727 + 0{\cdot}292 + 1{\cdot}31 + 0{\cdot}525}$$
$$= 3\ 795 \text{ V.}$$

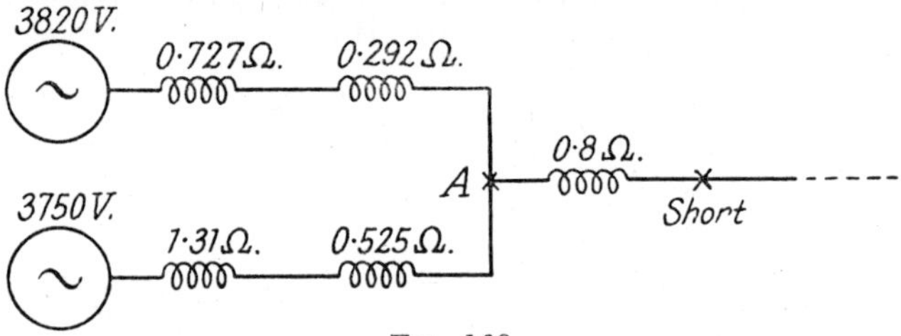

FIG. 168

The impedance is

$$\frac{(0{\cdot}727 + 0{\cdot}292)\,(1{\cdot}31 + 0{\cdot}525)}{0{\cdot}727 + 0{\cdot}292 + 1{\cdot}31 + 0{\cdot}525} = 0{\cdot}654\ \Omega.$$

The short-circuit current is thus

$$\frac{3\ 795}{0{\cdot}654 + 0{\cdot}8} = \underline{\underline{2\ 600 \text{ A.}}}$$

The *actual* short-circuit current is

$$2\ 600 \times (6{\cdot}6/33) = 520 \text{ A.}$$

520 A. is the r.m.s. of the steady short-circuit current, whilst the peak value is $520 \times \sqrt{2} = 735$ A. There may be an increased peak value due to a " doubling effect," which, in circuits of normal value of reactance to resistance, is 1·8 times the steady peak. Thus a maximum peak value of $1{\cdot}8 \times 735 = 1\ 320$ A. may occur.

Short-circuit Current of Alternators. When an alternator is shorted, across all three phases, say, the current rises rapidly to a high value, about 18 times full-load current in turbo-alternators which have cylindrical rotors, and about 12 times in generators with salient poles. The value of the peak current is limited only by the *transient* or *leakage reactance* of the armature. Moreover if the short circuit occurs at an instant at which the voltage is zero there is a doubling effect, and the current wave is offset from the zero. Fig. 169 shows the kind of current wave obtained. If the short circuit persists, the wave becomes symmetrical; then armature reaction reduces the excitation and the current falls to a steady value, which is 4 to 6 times the full-load value. Another way of considering the effect of armature reaction is to consider it as increasing the transient impedance to the synchronous impedance.

The doubling effect may be demonstrated as follows. Let the generator be considered as an e.m.f. $E \sin(\omega t + \theta)$ in series with an impedance (R, L) which is the *transient* or *true impedance*. If a short occurs at $t = 0$, the equation for the short-circuit current is

$$L(di/dt) + Ri = E \sin(\omega t + \theta).$$

The complementary function is given by

$$L(di/dt) + Ri = 0,$$

i.e.

$$i = A\varepsilon^{-(R/L)t} \quad . \quad . \quad . \quad . \quad (101)$$

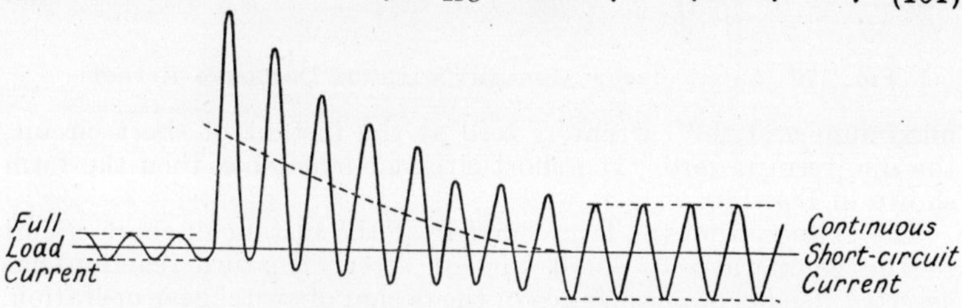

FIG. 169. DOUBLING EFFECT IN SHORT-CIRCUIT CURRENT

The particular integral is

$$i = \frac{E}{\sqrt{[R^2 + (\omega L)^2]}} \sin\left(\omega t + \theta - \tan^{-1}\frac{\omega L}{R}\right), \quad . \quad . \quad (102)$$

which is the steady current under these conditions. The actual current is the sum of the currents given in equations (101) and (102). At $t = 0$ the current is zero. This gives

$$0 = A + \frac{E}{\sqrt{[R^2 + (\omega L)^2]}} \sin\left(\theta - \tan^{-1}\frac{\omega L}{R}\right).$$

The current is thus

$$i = -\frac{E}{\sqrt{[R^2 + (\omega L)^2]}} \sin\left(\theta - \tan^{-1}\frac{\omega L}{R}\right)\varepsilon^{-(R/L)t}$$
$$+ \frac{E}{\sqrt{[R^2 + (\omega L)^2]}} \sin\left(\omega t + \theta - \tan^{-1}\frac{\omega L}{R}\right) \quad . \quad (103)$$

We may consider ωL as much greater than R. Then the first term, which is considered as a d.c. component which decays exponentially, has magnitude

$$-\frac{E}{\omega L} \sin\left(\theta - \frac{\pi}{2}\right)\varepsilon^{-(R/L)t}$$
$$= \frac{E}{\omega L} \cos\theta \,.\, \varepsilon^{-(R/L)t}.$$

If $\theta = 0$, i.e. the voltage is zero and the current is a maximum at $t = 0$, the d.c. component has the initial value of $E/\omega L$. As the alternating part of the current has a magnitude of nearly $E/\omega L$, the d.c. component doubles the current at the instant when the former has its peak value, and reduces it to zero when the former reaches its negative maximum. The current is thus on one side of the zero to begin with. If, however, $\theta = \pi/2$, i.e. the voltage is a

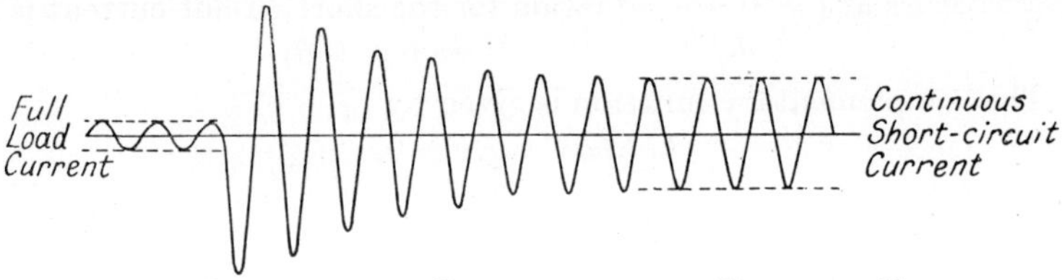

FIG. 170. SHORT-CIRCUIT CURRENT WITHOUT DOUBLING EFFECT

maximum and the current is zero at the instant of short circuit, the d.c. term is zero. The short-circuit current has then the form shown in Fig. 170.

The change from the large current at the instant of short circuit to the comparatively small current after armature reaction has asserted itself is of importance in the design of switchgear operation. The behaviour of an alternator is most easily expressed in terms of *decrement factors*, which are found by extensive tests in the

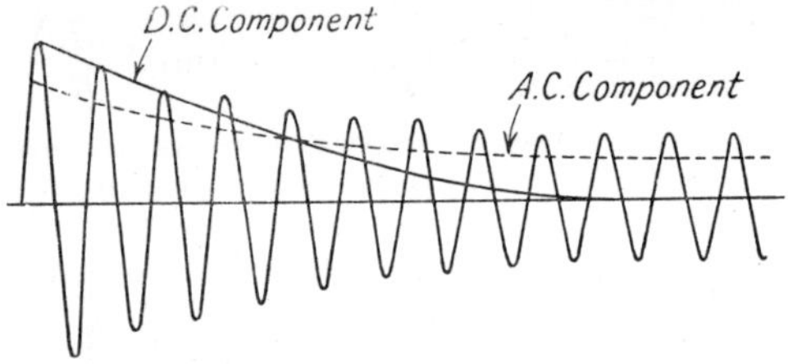

FIG. 171. D.C. AND A.C. COMPONENTS OF SHORT-CIRCUIT CURRENT

following way. The generator excitation is adjusted to the value for full load at 0·8 power factor lagging, and external reactance is put in series with the alternator to bring the value up to some definite amount, 5, 10, 20 . . . per cent. This value includes the transient reactance. A short circuit is applied and an oscillogram of the current taken. In order that the test results should be as severe as possible, it is arranged that the short circuit should take place at an instant when the full doubling effect is incurred, viz. at an instant of zero voltage. The oscillogram is analysed so that the r.m.s. current is found as a function of time. To do this the current wave of Fig. 169 is resolved into the d.c. and a.c. components shown in Fig. 171. The r.m.s. of the a.c. component is shown by the dotted line, and has a value $I_{A.C.}$ at time t; the d.c. component, which decays exponentially, has a value $I_{D.C.}$ at the same instant, and the total current has an r.m.s. value of $\sqrt{[I_{D.C.}^2 + I_{A.C.}^2]}$. Curves are then drawn giving the r.m.s. of the current (as a multiple of full-load current) against time for different values of the total percentage reactance. Fig. 172 shows a set of such curves for a short circuit across all three phases.

When looking up the decrement factor, the transient reactance of the alternator is added to the external reactance to give the appropriate percentage reactance.

EXAMPLE. A 20 000 kVA. generator, whose decrement curves are shown in Fig. 172, has 15% reactance and feeds a line through a step-up transformer of 6% reactance. Find the breaking capacity of the circuit-breakers, which operate in 0·25 sec. and are on the high voltage side of the transformer.

The total reactance is 21%, and from Fig. 172 it is seen that the decrement

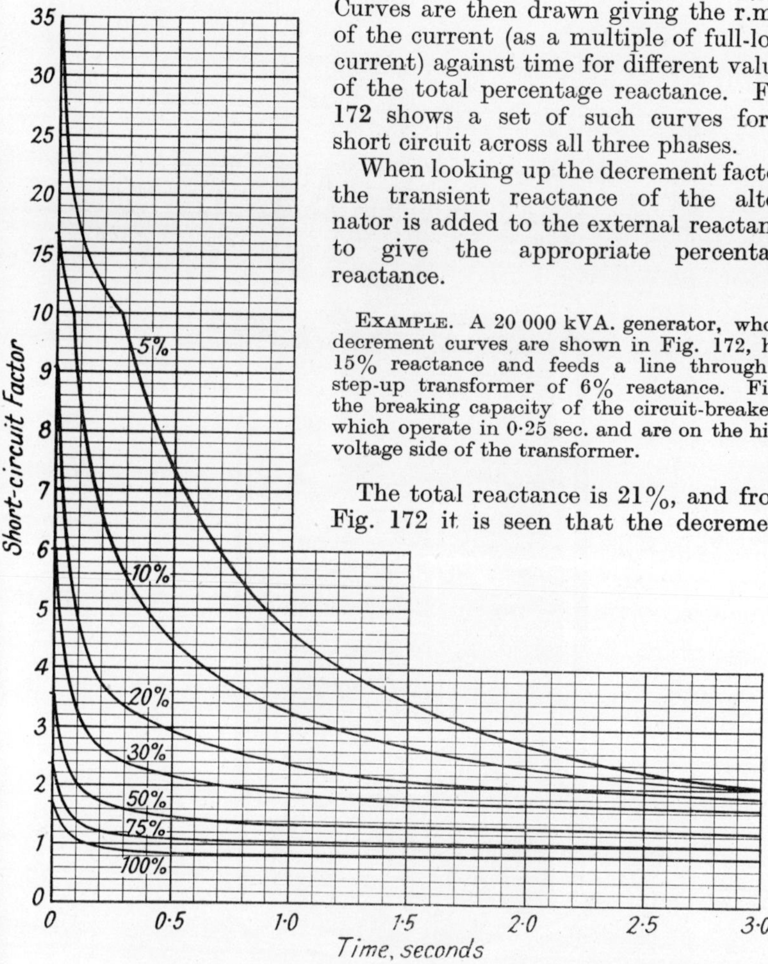

FIG. 172. DECREMENT CURVES FOR ALTERNATOR

factor at 0·25 sec. is 3·4. The current to be interrupted is thus 3·4 times the full-load current. If we assume that the recovery voltage in the breaker is equal to the normal voltage (the matter will be investigated in detail in Chapter IX), the kVA. to be broken is

$$3{\cdot}4 \times 20\,000 = \underline{\underline{68\,000 \text{ kVA.}}}$$

FIG. 173. CURRENT LIMITING REACTOR
(*English Electric Co.*)

Current-limiting Reactors: Sectionalization of Networks. It is clear that the short-circuit currents are decreased by an increase of the percentage reactance in the system. In large interconnected systems the total rating of the generators is very high, and unless precautions are taken, the current fed into a fault will be enormous. The short-circuit current at a fault can be considerably reduced by the judicious placing of protective reactors in the system. It is possible to arrange the reactors so that they do not cause a large voltage drop during normal operation, but prevent a large short-circuit current being fed by most of the generators into the fault. The methods of placing reactors in a system will be considered later.

Reactors are moreover of considerable importance in limiting the currents so that the various circuit-breakers are not called upon to

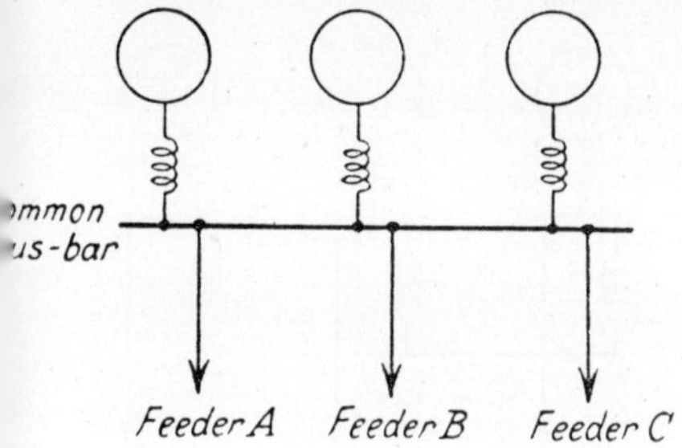

FIG. 174. GENERATOR REACTORS

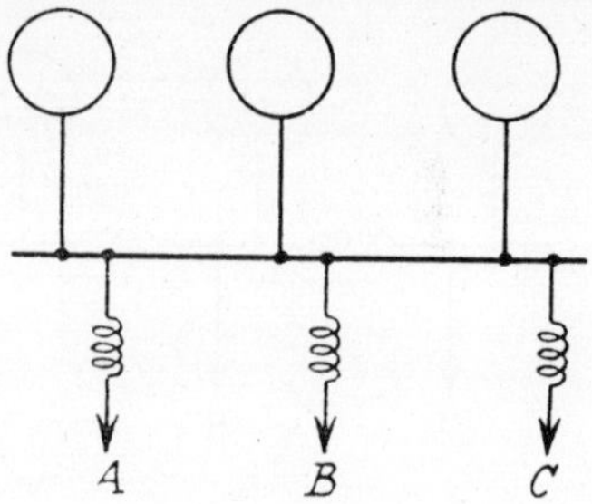

FIG. 175. FEEDER REACTORS

break currents above their rated value. If extensions are made in a system, it is essential that the additional kVA. be virtually segregated from the existing circuit-breakers when a short-circuit occurs. This is done by means of current-limiting reactors.

Fig. 173 shows a reactor. The turns, which are of copper bar or strip, experience large attractive forces under the influence of the short-circuit currents, and they are placed in concrete separators to prevent their being buckled.

Methods of Locating Reactors. Reactors may be inserted in series with each generator, as shown in Fig. 174. The main disadvantage of this method is that if a short occurs on one feeder, the voltage at the common bus-bar drops to a low value and the synchronous machines attached to the other feeders may fall out of step. The whole system is interrupted, and the synchronous machines must be re-synchronized when the faulty feeder is cut out. Moreover in modern alternators the transient reactance is sufficiently large to protect the machine itself against short-circuit currents, and separate reactors are used only with old alternators.

The main disadvantage of the last method is avoided by putting reactances in series with each feeder, as shown in Fig. 175. When a short-circuit occurs on feeder A, the main voltage drop is in its reactor and the bus-bar voltage does not drop unduly. The remaining load and plant are therefore able to continue running. It is true that when a short circuit occurs across the bus-bars, the reactors do not protect the generators. This is, however, of no importance, as bus-bar short circuits seldom occur and the generators are protected by their internal reactances.

A disadvantage from which both the previous methods suffer is that the reactors take the full-load currents under normal operation, so that there is a constant loss and a voltage drop. The voltage drop is eliminated in a new type of reactor in which part of the windings are shunted by a carbon tetrachloride fuse. Under normal conditions the windings are such that they neutralize each other's

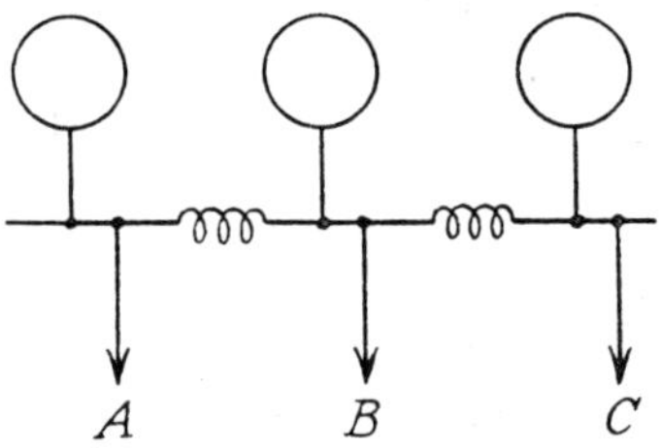

FIG. 176. BUS-BAR REACTORS, RING SYSTEM

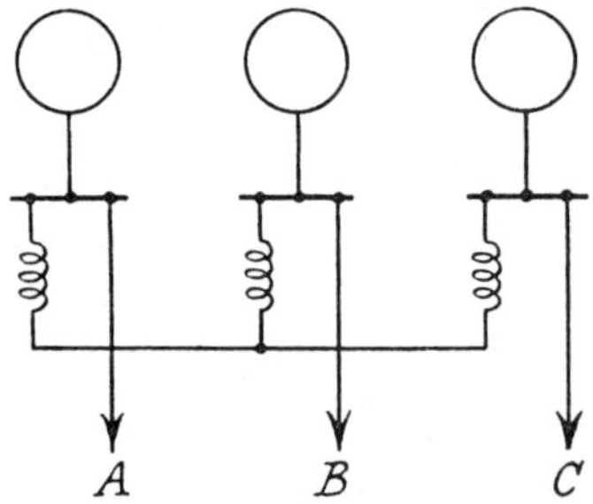

FIG. 177. BUS-BAR REACTORS, TIE-BAR SYSTEM

magnetic field and the reactor has a very small reactance; but when a short circuit occurs and the fuse blows, a large reactance is inserted into the circuit. The constant loss, however, is not eliminated.

The constant loss in reactors can be avoided by inserting the reactors in the bus-bars in the ways shown in Figs. 176 and 177. The former is the *ring* system, and the latter is the *tie-bar* system. In the ring system each feeder is normally fed by one generator, only a small amount of power flowing across the reactances. The reactors can therefore be made with a fairly high ohmic resistance and there is not much voltage drop across it. When a short circuit occurs in one feeder, the current is fed mainly by one generator, the other generators having to feed through the reactances. The tie-bar system acts in the same way, but has the following advantage. If the number of sections in the tie-bar system is increased, the current that flows into the fault will not exceed a certain value which is fixed by the size of the individual reactors. If the switchgear is designed to operate successfully on this limiting value of current, the system can be extended to any number of sections without modification of the switchgear.

EXAMPLE. Find the ratio of the percentage reactance of the reactors to that of the generators in a tie-bar system, if the short-circuit current is not to exceed three times the current of a single section.

Let the percentage reactance of a generator be G and of a reactor X, and suppose there are n sections. When there is a short circuit on a feeder, the remaining reactors and generators are in parallel, so that their percentage reactance is $(G + X)/(n - 1)$. This reactance is in series with the reactor of the faulty feeder, giving a reactance

$$X + (G + X)/(n - 1) = (G + nX)/(n - 1).$$

This reactance is in parallel with the reactance of the generator which is connected to the faulty feeder, so that the total reactance is

$$\frac{G \times \dfrac{G + nX}{(n - 1)}}{G + \dfrac{G + nX}{(n - 1)}} = G\frac{G + nX}{nG + nX}.$$

The short-circuit current is thus

$$I \times \frac{100}{G} \times \frac{nG + nX}{G + nX}.$$

where I is the normal full-load current.

When $n = 1$, the current is

$$I \times (100/G).$$

The last factor gives the effect of the remaining sections, and increases from unity when $n = 1$ to $(G + X)/X$ when n is infinitely large. Thus if the current is not to exceed three times the short-circuit current due to a single section

$$(G + X)/X = 3$$

i.e.

$$X = \tfrac{1}{2}G.$$

If it is certain that the number of sections will not exceed a known number n we have

$$(nG + nX)/(G + nX) = 3$$

i.e.

$$X = [(n - 3)/2n]G.$$

Thus if n will not exceed 6, X need not be greater than $\tfrac{1}{4}G$.

Choice of Interconnection to Limit Currents. The cost of reactors is large and their installation is avoided if possible. It is sometimes practicable to make use of the reactance of feeders and transformers so that reactors are unnecessary.

For instance, suppose that two parallel feeders are fed by four transformers, as shown in Fig. 178, and suppose that a short circuit occurs at B. If the parallel feeders are not connected at their ends, the reactance from A to B is

$$\frac{X(3X + 2F)}{X + 3X + 2F} = \frac{X(3X + 2F)}{4X + 2F}.$$

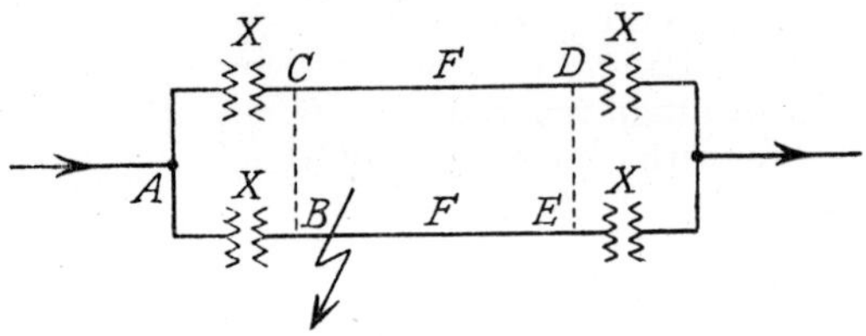

FIG. 178

If the feeders are connected at their ends (between B and C, D and E), the reactance from A to B is $\frac{1}{2}X$. The latter reactance is considerably less than the former; thus if $F = X$, the former is $\frac{5}{6}X$ and the latter only $\frac{1}{2}X$.

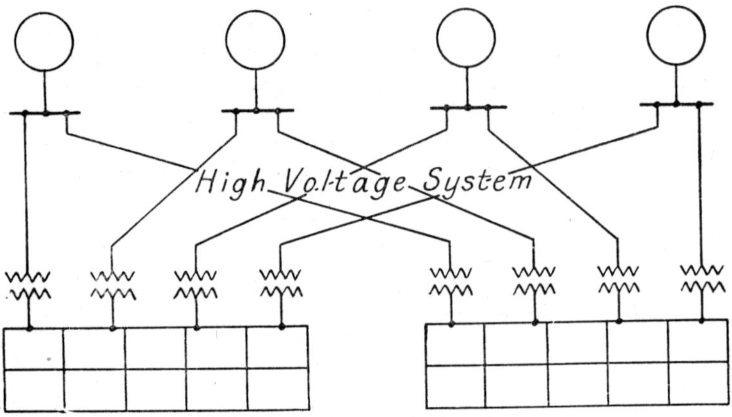

FIG. 179. SYNCHRONIZATION AT THE LOAD

As a general rule it is advisable to keep the parallel connections as few as possible.

An interesting and important application of this rule is shown by the method of interconnection of Fig. 179. The generators are unconnected in the high-tension system, but connected only at the low voltage system. This system is said to be *synchronized at the load.*

Protection of Switchgear by Reactors. A generating station may be extended by the addition of alternators or by a supply from the Grid. It is uneconomical to scrap the existing switchgear, which was adequate for the former output but is not of sufficient rating to meet the extensions. In such a case a protective reactor may be placed between the old system and the extensions to limit the short-circuit currents to a permissible value. An example will show how the requisite reactance is calculated.

EXAMPLE. A small generating station has two alternators of 3 000 and 4 500 kVA., and percentage reactances 7 and 8. The circuit-breakers are rated at 150 000 kVA. It is intended to extend the system by a supply from the Grid via a transformer of 7 500 kVA. rating and 7·5% reactance. Find the reactance necessary to protect the switchgear.

Let us take as the basic rating 7 500 kVA. The reactances of A and B are then

$$(7\,500/3\,000) \times 7 = 17{\cdot}5\%$$

and

$$(7\,500/4\,500) \times 8 = 13{\cdot}3\%,$$

so that their combined reactance is

$$\frac{17{\cdot}5 \times 13{\cdot}3}{17{\cdot}5 + 13{\cdot}3} = 7{\cdot}55\%$$

The short-circuit kVA. with respect to these alone is

$$7\,500 \times (100/7{\cdot}55) = 99\,300 \text{ kVA.}$$

If no protective reactor is present the short-circuit rating due to the Grid supply is

$$(7\,500/7{\cdot}5) \times 100 = 100\,000 \text{ kVA.}$$

so that the total is 199 300 kVA. In order to keep the kVA. down the rated value of 150 000 kVA. suppose that a reactor of percentage reactance X is interposed as shown in Fig. 180. The short-circuit kVA. of the Grid supply is then $7\,500/(7{\cdot}5 + X)$, and this must not exceed the difference between the ratings of the circuit-breakers and the generators A and B.

Therefore

$$\frac{7\,500}{7{\cdot}5 + X} \times 100 = 150\,000 - 99\,300$$

$$= 50\,700,$$

giving

$$7{\cdot}5 + X = \frac{7\,500 \times 100}{50\,700}$$

$$= 14{\cdot}8,$$

so that

$$X = 7{\cdot}3.$$

If the voltage is 3 300 V., the full-load current per phase corresponding to 7 500 kVA. is

$$\frac{7\ 500\ 000}{3\ 300 \times \sqrt{3}} = 1\ 310 \text{ A.},$$

so that the actual reactance of the reactor per phase is

$$\frac{7{\cdot}3 \times (3\ 300/\sqrt{3})}{1\ 310 \times 100} = \underline{\underline{0{\cdot}106\ \Omega}}.$$

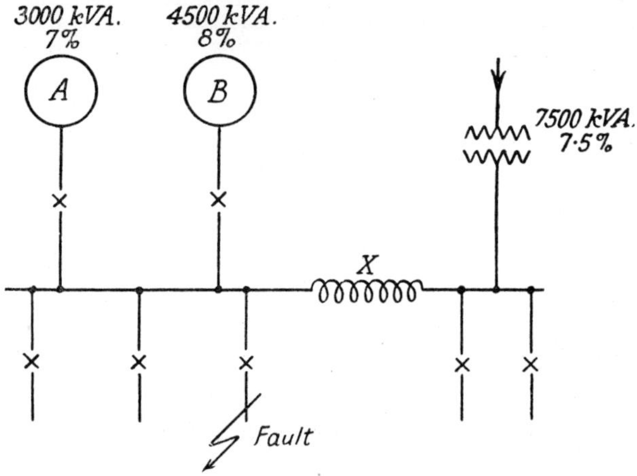

FIG. 180. REACTOR PROTECTING SWITCHGEAR

corresponding to an inductance of

$$\frac{0{\cdot}106}{2\pi \times 50} = 0{\cdot}000337 \text{ H.}$$

$$= 0{\cdot}337 \text{ mH.}$$

Unsymmetrical Short Circuits: Symmetrical Components. The methods of calculating short-circuit currents developed in the previous sections apply only to the cases when the fault occurs across all phases and the currents in the phases are equal. When the fault occurs across two of the three phases or between one or two phases and earth, the currents are unequal and the methods are inadequate. The currents can be found by Kirchhoff's laws, but this method is usually laborious. A method has been devised which uses *symmetrical components* of the currents and voltages.

It will first be shown how three unbalanced vectors can be expressed as the sum of three systems of balanced vectors. Three *balanced vectors* are vectors which are equal in magnitude and differ in phase by 120°, so that they can be represented by

$$E_1,\ E_1' = E_1\varepsilon^{j(4\pi/3)},\ E_1'' = E_1\varepsilon^{j(2\pi/3)},$$

if E_1 is taken as a reference vector. Fig. 181 shows the vectors. It should be remembered that multiplication of a vector by the factor $\varepsilon^{j\theta}$ results in a vector of unchanged magnitude but with a phase angle increased by θ; for if the vector has magnitude r and angle α it is represented by

$$r\underline{|\alpha} = r\varepsilon^{j\alpha},$$

and multiplication by $\varepsilon^{j\theta}$ results in a vector

$$r\varepsilon^{j\alpha} \times \varepsilon^{j\theta} = r\varepsilon^{j(\alpha+\theta)} = r\,\underline{|\alpha + \theta}.$$

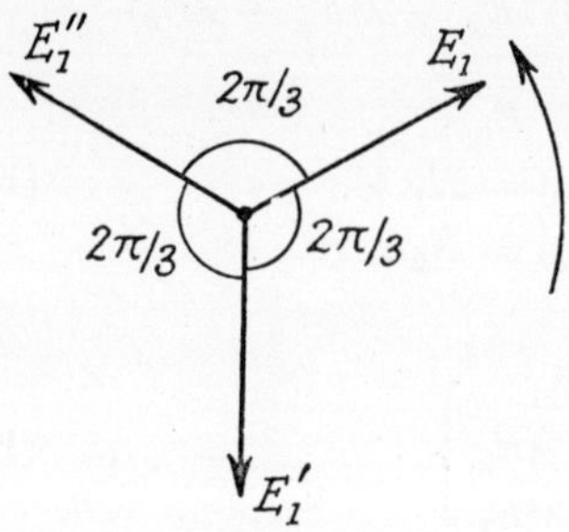

Fig. 181
POSITIVE PHASE SEQUENCE
E_1 leads E_1' leads E_1''

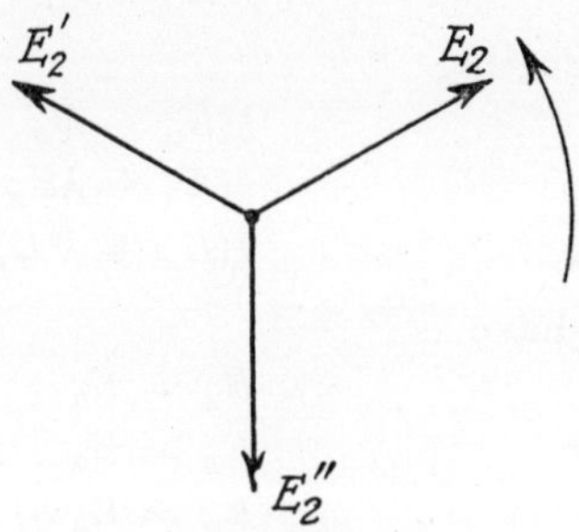

Fig. 182
NEGATIVE PHASE SEQUENCE
E_2 lags E_2' lags E_2''

As is pointed out in Appendix III, the values of the alternating quantities are the horizontal components of these vectors, which are assumed to rotate in the positive (anti-clockwise) direction with an angular velocity $\omega = 2\pi \times$ frequency. *Thus E_1, E_1', and E_1'' correspond to alternating quantities which reach their maxima in the order given.* The system just given is called a system of *positive phase sequence.* It is usual to replace $\varepsilon^{j(2\pi/3)}$ by λ, so that

$$\left.\begin{aligned} \lambda &= \varepsilon^{j(2\pi/3)} = \cos(2\pi/3) + j\sin(2\pi/3) = -0{\cdot}5 + j0{\cdot}866 \\ \text{and}\quad \lambda^2 &= \varepsilon^{j(4\pi/3)} = \cos(4\pi/3) + j\sin(4\pi/3) = -0{\cdot}5 - j0{\cdot}866. \\ \text{It follows that}\quad & 1 + \lambda + \lambda^2 = 0. \end{aligned}\right\} \quad (104)$$

A positive phase-sequence system of vectors is thus represented by

$$E_1,\ \lambda^2E_1,\ \lambda E_1. \quad . \quad . \quad . \quad . \quad (105)$$

It would be symmetrical but unconventional to write these as

$$\lambda^3E_1,\ \lambda^2E_1,\ \lambda E_1.$$

A balanced system with *negative phase sequence* is shown in Fig. 182. The order of maxima is E_2'', E_2', E_2, so that the vectors are

$$E_2, \ E_2' = \lambda E_2, \ E_2'' = \lambda^2 E_2. \quad . \quad . \quad . \quad (106)$$

Suppose there is a system of unbalanced vectors E_A, E_B, and E_C. By virtue of the relations $(1 + \lambda + \lambda^2) = 0$ and $\lambda^3 = 1$ we may write

$$E_A = \tfrac{1}{3}(E_A + E_B + E_C) + \tfrac{1}{3}(E_A + \lambda E_B + \lambda^2 E_C) + \tfrac{1}{3}(E_A + \lambda^2 E_B + \lambda E_C),$$

$$E_B = \tfrac{1}{3}(E_A + E_B + E_C) + (\lambda^2/3)(E_A + \lambda E_B + \lambda^2 E_C) + (\lambda/3)(E_A + \lambda^2 E_B + \lambda E_C),$$

and $$E_C = \tfrac{1}{3}(E_A + E_B + E_C) + (\lambda/3)(E_A + \lambda E_B + \lambda^2 E_C) + (\lambda^2/3)(E_A + \lambda^2 E_B + \lambda E_C).$$

Putting

$$\left.\begin{aligned} \tfrac{1}{3}(E_A + E_B + E_C) &= E_0, \\ \tfrac{1}{3}(E_A + \lambda E_B + \lambda^2 E_C) &= E_1, \\ \tfrac{1}{3}(E_A + \lambda^2 E_B + \lambda E_C) &= E_2, \end{aligned}\right\} \quad . \quad . \quad . \quad (107)$$

and

we have

$$\left.\begin{aligned} E_A &= E_0 + E_1 + E_2, \\ E_B &= E_0 + \lambda^2 E_1 + \lambda E_2, \\ E_C &= E_0 + \lambda E_1 + \lambda^2 E_2. \end{aligned}\right\} \quad . \quad . \quad . \quad (108)$$

It is seen from equations (105) and (106) that the terms in E_1 represent a positive phase-sequence system of balanced vectors, whilst the terms in E_2 represent a negative phase-sequence system of balanced vectors. The equal terms E_0 are said to represent a *zero phase-sequence* system of vectors, which are equal. Thus any system of three unbalanced vectors can be resolved into three systems of balanced vectors, a zero phase-sequence system E_0 (1, 1, 1), a positive phase-sequence system E_1 $(1, \lambda^2, \lambda)$, and a negative phase-sequence system E_2 $(1, \lambda, \lambda^2)$. The values of E_0, E_1, and E_2 are found from the unbalanced vectors E_A, E_B, E_C by means of equations (107) by multiplication of complex numbers or by a graphical method. The former method is described in Appendix II. The graphical method follows from the facts that, since $\lambda = \varepsilon^{j(2\pi/3)}$, multiplication by λ turns a vector in the positive direction through an angle of $2\pi/3$ radians and multiplication by λ^2 turns a vector through $4\pi/3$ radians.

Fig. 183 (*a*) shows an unbalanced system of vectors. Fig. 183 (*b*) shows how these are added to give $E_A + E_B + E_C$, one-third of which is E_0. Fig. 183 (*c*) shows how we obtain $E_A + \lambda E_B + \lambda^2 E_C$, one-third of which is E_1; and Fig. 183 (*d*) shows how we obtain $E_A + \lambda^2 E_B + \lambda E_C$, one-third of which is E_2.

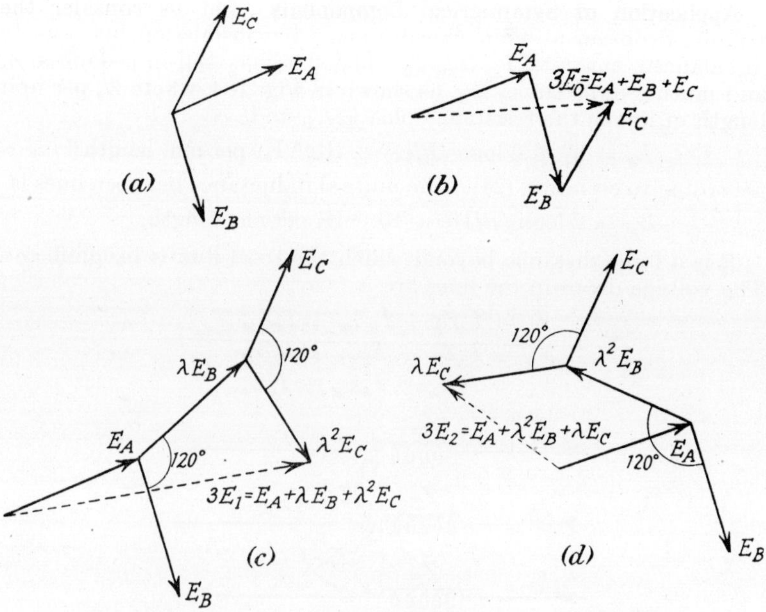

FIG. 183. GRAPHICAL METHOD OF FINDING SYMMETRICAL COMPONENTS

EXAMPLE. Resolve the system $E_A = 1\,500 \angle 30°$, $E_B = 1\,800 \angle -70°$, $E_C = 2\,000 \angle 170°$ into their symmetrical components.

$$
\begin{aligned}
E_0 &= \tfrac{1}{3}(E_A + E_B + E_C) \\
&= \tfrac{1}{3}(1\,500 \angle 30° + 1\,800 \angle -70° + 2\,000 \angle 170°) \\
&= \tfrac{1}{3}(1\,500 \cos 30° + 1\,800 \cos 70° + 2\,000 \cos 170° \\
&\qquad + j1\,500 \sin 30° - j1\,800 \sin 70° + j2\,000 \sin 170°) \\
&= \tfrac{1}{3}(1\,300 + 615 - 1\,970 + j750 - j1\,690 + j347) \\
&= -18 - j198 = 198 \angle 264° 48'. \\
E_1 &= \tfrac{1}{3}(E_A + \lambda E_B + \lambda^2 E_C) \\
&= \tfrac{1}{3}(1\,500 \angle 30° + 1\,800 \angle -70° + 120° + 2\,000 \angle 170° + 240°) \\
&= 1\,250 + j1\,220 = 1\,740 \angle 44° 14'. \\
E_2 &= \tfrac{1}{3}(E_A + \lambda^2 E_B + \lambda E_C) \\
&= \tfrac{1}{3}(1\,500 \angle 30° + 1\,800 \angle -70° + 240° + 2\,000 \angle 170° + 120°) \\
&= -70 - j270 = 280 \angle 252° 30'.
\end{aligned}
$$

It is clear that unbalanced currents can be resolved into symmetrical components in the same way as voltages.

Application of Symmetrical Components. Let us consider the voltage drops in a symmetrically spaced transmission line due to unbalanced currents I_A, I_B, I_C. The line has self-impedances Z_S and mutual impedances Z_M, as shown in Fig. 184, where Z_S per unit length of line is the resistance plus $j\omega L_S$, and

$$L_S = [\tfrac{1}{2} + 2 \operatorname{logh} (R/r)] \times 10^{-9} \text{ H. per cm. length}$$

according to equation (24). The mutual inductance between lines is

$$L_M = 2 \operatorname{logh} (R/D) \times 10^{-9} \text{ H. per cm. length.}$$

R is a large distance beyond which the total flux is insignificant. The voltage drops in the lines are

$$V_A = I_A Z_S + I_B Z_M + I_C Z_M$$
$$V_B = I_A Z_M + I_B Z_S + I_C Z_M$$
$$V_C = I_A Z_M + I_B Z_M + I_C Z_S.$$

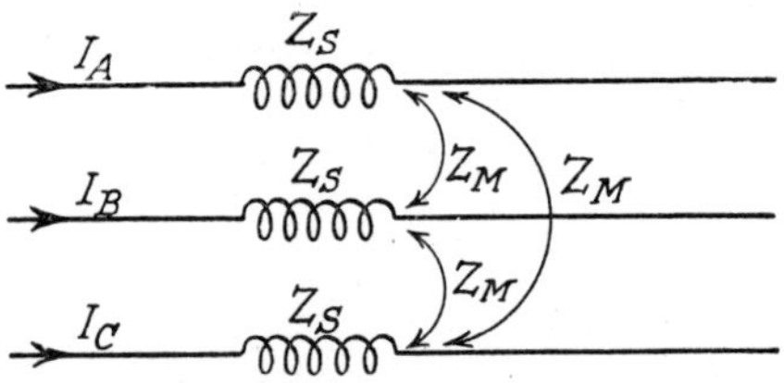

FIG. 184. GENERAL THREE-PHASE LINE

The symmetrical components of the voltages are found in the following way—

$$V_0 = \tfrac{1}{3}(V_A + V_B + V_C) = \tfrac{1}{3}(Z_S + 2Z_M)(I_A + I_B + I_C)$$
$$= I_0(Z_S + 2Z_M),$$
$$V_1 = \tfrac{1}{3}(V_A + \lambda V_B + \lambda^2 V_C) = \tfrac{1}{3}Z_S(I_A + \lambda I_B + \lambda^2 I_C)$$
$$+ \tfrac{1}{3}Z_M(\lambda I_A + \lambda^2 I_A + I_B + \lambda^2 I_B + I_C + \lambda I_C)$$
$$= I_1 Z_S + \tfrac{1}{3}Z_M(- I_A - \lambda I_B - \lambda^2 I_C),$$

using the relation $1 + \lambda + \lambda^2 = 0$. Therefore

$$V_1 = I_1(Z_S - Z_M).$$

Similarly $\quad V_2 = I_2(Z_S - Z_M).$

We have therefore

$$\left.\begin{aligned} V_0 &= Z_0 I_0, \\ V_1 &= Z_1 I_1, \\ \text{and} \quad V_2 &= Z_2 I_2, \end{aligned}\right\} \qquad (109a)$$

$$\left.\begin{aligned} \text{where} \quad Z_0 &= (Z_S + 2Z_M) \\ Z_1 &= (Z_S - Z_M) \\ \text{and} \quad Z_2 &= (Z_S - Z_M). \end{aligned}\right\} \qquad (109b)$$

We have thus the important result that *the phase-sequence voltage drops are due to the separate phase-sequence currents.*

Impedances to the Various Phase-sequence Currents. It is seen from equations (109) that the impedances of a transmission line to the zero, positive, and negative phase-sequence currents are Z_0, Z_1, and Z_2. The last two are equal to $(Z_S - Z_M)$, which is due to an inductance of

$$L_S - L_M = [\tfrac{1}{2} + 2\,\text{logh}\,(D/r)] \times 10^{-9}\ \text{H. per cm. length,}$$

which is the inductance to neutral in a balanced system.

It is thus independent of the distance R, but depends only on the radius of the wires r and their spacing D. The zero phase-sequence impedance depends on R, or more particularly it depends upon the return path for the current I_0 which flows along the three wires in parallel, since the distance R is to be chosen to include the return path. It is difficult to calculate Z_0 and it is best found by experiment. If information is not available we may take Z_0 as twice or three times Z_1. It will be shown later that if the neutral is earthed through an impedance, three times this impedance must be added to Z_0.

In order to calculate unsymmetrical short-circuit currents it is necessary to know the various phase-sequence properties of the generators and transformers in the system.

An alternator generates only a positive phase-sequence system of e.m.f.'s. We have already discussed in detail the impedance (or more simply the reactance) of the alternator to positive phase-sequence currents; there is the initial or transient reactance which increases to the *synchronous reactance* by reason of armature reaction. The initial reactance to negative phase-sequence currents is about 70 per cent of the previous transient reactance, and to zero phase-sequence currents 10 to 25 per cent. Under steady short-circuit conditions, the values are less. The following table gives approximate values in terms of the reactances to positive phase-sequence currents.

	Reactance to Negative Phase-sequence Currents (%)	Reactance to Zero Phase-sequence Currents (%)
Transient . . .	70	10 to 25
Steady Short Circuit		
Salient pole . . .	30	5 to 15
Turbo-alternator . .	15	0 to 5

Decrement factors can be used in the way described above to find the currents at intermediate times.

The impedance of a transformer to negative phase-sequence currents is the same as to the positive system. The impedance to the zero phase-sequence currents is the same as for the other sequences provided there is a through circuit for the earth currents and the compensating currents can flow, otherwise the impedance is infinite. Thus in an unearthed star/unearthed star connection the zero phase currents cannot flow and the zero phase impedance is infinite. In the unearthed star/earthed star connection, if an earth fault occurs on the primary side no zero sequence current can flow and the impedance is infinite; if the earth fault occurs on the secondary side there is a complete path for the zero sequence current, but there is no path for the compensating currents in the primary windings, and the zero phase impedance is again infinite. In the earthed star/delta connection, an earth fault in the

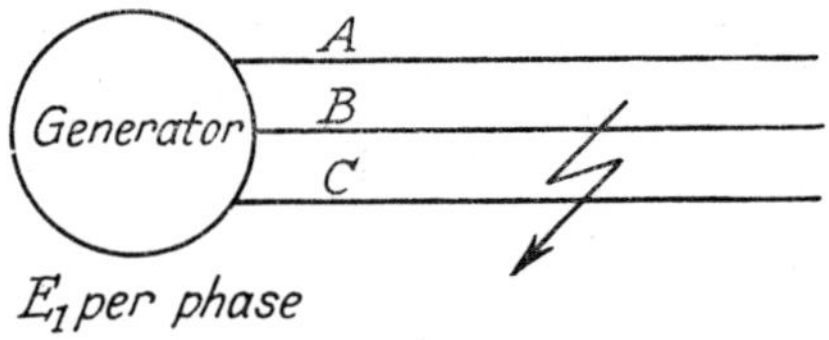

FIG. 185. APPLICATION OF SYMMETRICAL COMPONENTS

primary circuit has a complete path and the compensating currents can flow in the delta, so that the zero sequence impedance is finite; if the fault is in the secondary circuit, the zero phase impedance is infinite.

General Method Using Symmetrical Components. From the conditions of the fault we get three relations between the voltages V_A, V_B, V_C and the currents I_A, I_B, I_C. For instance, if there is an earth fault on line C only, we have $I_A = 0$, $I_B = 0$, and $V_C = 0$. If there is an earth fault on lines B and C, we have $I_A = 0$, $V_B = 0$, and $V_C = 0$. If the earth fault is across all three lines, we have $V_A = 0$, $V_B = 0$, and $V_C = 0$. If there is a short between two lines B and C, we have $I_A = 0$, $I_B + I_C = 0$, and $V_B = V_C$. We can then express the currents and voltages in terms of their symmetrical components. We know that the separate phase-sequence voltage drops are due to the corresponding phase-sequence currents. We thus have two three-phase balanced systems and one single-phase system (the earth return system), and we can apply the simplified method used in a balanced system, in which we replace the system by a single line and impedances to neutral.

As an example let us solve the case when there is a short between two lines, as shown in Fig. 185. The generator produces a positive phase-sequence e.m.f. only of E_1 per phase; it has impedances Z_{G0}, Z_{G1}, and Z_{G2} to the sequence currents. The line has impedances

Z_{L0}, Z_{L1}, and Z_{L2}. Let V_A, V_B, V_C be the voltages and I_A, I_B, I_C the currents near the fault. We have

$$I_A = 0,\ I_B + I_C = 0, \text{ and } V_B = V_C.$$

The phase-sequence currents are given by equations (106) as

$$\left.\begin{aligned} I_0 &= \tfrac{1}{3}(I_A + I_B + I_C) = 0, \\ I_1 &= \tfrac{1}{3}(I_A + \lambda I_B + \lambda^2 I_C) = \tfrac{1}{3}(\lambda - \lambda^2)I_B, \\ I_2 &= \tfrac{1}{3}(I_A + \lambda^2 I_B + \lambda I_C) = \tfrac{1}{3}(\lambda^2 - \lambda)I_B. \end{aligned}\right\} \quad (110)$$

The phase-sequence voltages are similarly given by

$$\left.\begin{aligned} V_0 &= \tfrac{1}{3}(V_A + V_B + V_C), \\ V_1 &= \tfrac{1}{3}(V_A + \lambda V_B + \lambda^2 V_C) = \tfrac{1}{3}(V_A + \lambda V_B + \lambda^2 V_B), \\ V_2 &= \tfrac{1}{3}(V_A + \lambda^2 V_B + \lambda V_C) = \tfrac{1}{3}(V_A + \lambda^2 V_B + \lambda V_B). \end{aligned}\right\} \quad (111)$$

The phase-sequence e.m.f.'s are E_1, 0, 0, so that applying equations (108) we get

$$\begin{aligned} V_0 &= 0 - (Z_{G0} + Z_{L0})I_0 = 0, \text{ since } I_0 = 0. \\ V_1 &= E_1 - (Z_{G1} + Z_{L1})I_1, \\ V_2 &= 0 - (Z_{G2} + Z_{L2})I_2. \end{aligned}$$

From equations (111) we see that $V_1 = V_2$ so that

$$E_1 - (Z_{G1} + Z_{L1})I_1 = -(Z_{G2} + Z_{L2})I_2.$$

Substituting for I_1 and I_2 from equations (110) we get

$$\begin{aligned} E_1 &= (Z_{G1} + Z_{L1})I_1 - (Z_{G2} + Z_{L2})I_2 \\ &= (Z_{G1} + Z_{L1} + Z_{G2} + Z_{L2})\tfrac{1}{3}(\lambda - \lambda^2)I_B, \\ &= (Z_{G1} + Z_{L1} + Z_{G2} + Z_{L2})\,(j/\sqrt{3})I_B. \end{aligned}$$

so that the magnitude of I_B is

$$\frac{(\sqrt{3})E_1}{Z_{G1} + Z_{L1} + Z_{G2} + Z_{L2}}.$$

Interference with Communication Circuits. When a communication circuit runs parallel with a high voltage overhead line, high voltages may be induced in the former resulting in acoustic shock and noise. The induced voltages are due to electrostatic and electromagnetic induction, and are reduced considerably by *transposing* the power lines as shown in Fig. 186. See example 9 on page 236. The effect of transposition is to balance the capacitances of the lines, so that the electrostatically induced voltages balance out in the length of a complete set of transpositions; such a length is called a *barrel.* Transposition results also in a diminution of the electromagnetically induced e.m.f. on the wires, since the fluxes due to the positive and negative phase-sequence currents will add up to zero along the barrel. The flux due to the zero phase-sequence or longitudinal current is unaffected by transpositions of the power system, since it flows along the three wires in parallel. In order to reduce the e.m.f. in the telephone circuit due to the zero phase-sequence current, which is called the *longitudinal current*, the telephone line is transposed, as shown in Fig. 186. By a proper co-ordination of transpositions of the power and communication lines, the induced voltages can be reduced to very small proportions under normal working conditions.

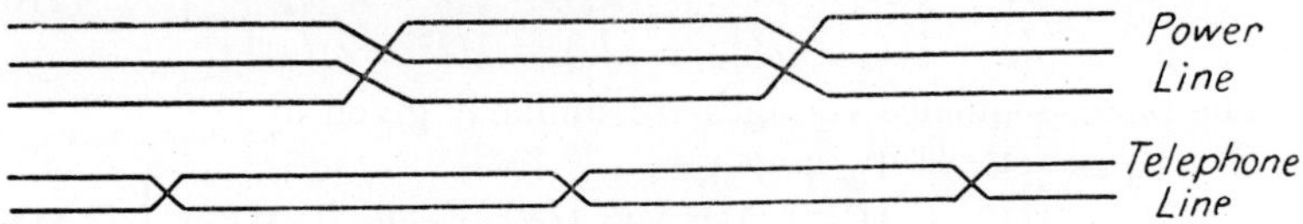

FIG. 186. CO-ORDINATED TRANSPOSITION OF POWER AND COMMUNICATION CIRCUITS

At the ends of a barrel the induced voltage is small and we have *silent points*. At points inside the barrel there may be a high voltage on the telephone line. If it is desired to tap the communication circuit at such a point, it is advisable to insert an isolating trans-

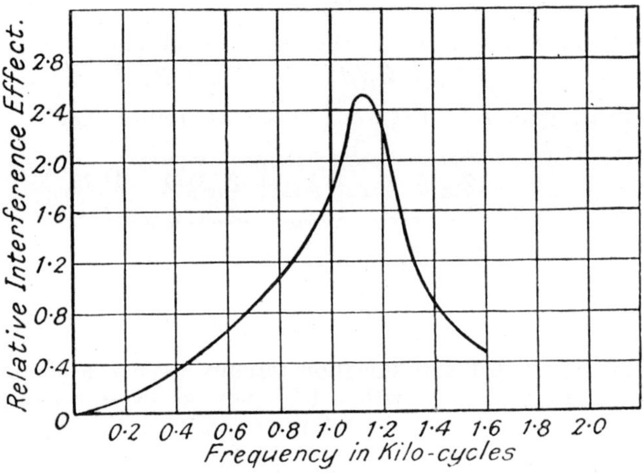

FIG. 187. TELEPHONE INTERFERENCE FACTOR CURVE
(*Gill*)

former, the insulation between the line and telephone windings being adequate to withstand the voltage; the telephone winding is earthed at the mid-point or at one end, so that a high voltage cannot reach the telephone.

The interference effect of an induced voltage or current depends greatly upon the frequency. The relative interference effect of different frequencies is shown in Fig. 187, which shows the *T.I.F.* curve, i.e. the telephone interference factor curve.

It is usual to express the interference currents in terms of a current at 800 cycles per sec. which produces the same degree of disturbance according to the curve of Fig. 187. Thus suppose that we have induced currents of 20 μA. at 250 cycles and 10 μA. at 350 cycles; the disturbances caused by them are the same as caused by 800 cycle currents of $20 \times 0{\cdot}25 = 5$ μA. and $10 \times 0{\cdot}3 = 3$ μA., respectively.

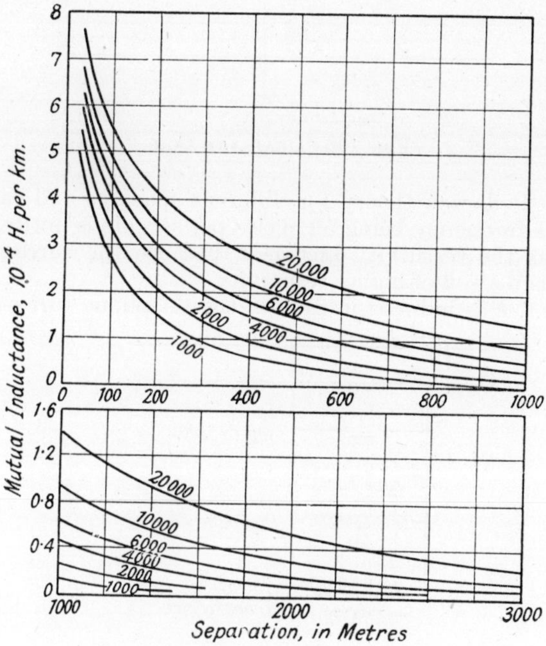

FIG. 188. MUTUAL INDUCTANCE BETWEEN A LINE AND EARTH RETURN AND ANOTHER LINE (CARSON-POLLACZEK FORMULA). FREQUENCY: 50 CYCLES

Curves are for different resistivities of soil (in Ω. per cm. cube).

The total disturbance is considered as due to an 800 cycle current of value $\sqrt{(5^2 + 3^2)} = 5{\cdot}8$ μA.

It is clear that the harmonics in the power system should be kept as low as possible, as they have a high interference factor.

When a short circuit occurs to earth, a large zero phase-sequence or longitudinal current flows along the wires in parallel and through the earth return. In this case the electromagnetic induction is large in magnitude, and depends upon the spacing between the power and telephone lines, the resistivity of the earth, and the frequency of the current. The e.m.f. induced in the telephone line is

$$E = -j\omega MlI,$$

where I is the zero phase-sequence current, $\omega = 2\pi \times$ frequency, l is the length of the parallel, and M the mutual inductance between the power line circuit (with earth return) and the telephone line. Fig. 188 gives the value of M for 1 km. parallel as calculated by

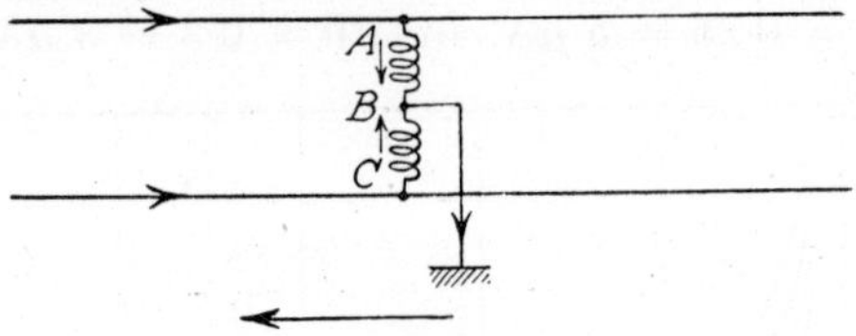

FIG. 189. DRAINAGE COIL

the Carson-Pollaczek theory for different spacings and earth resistivities, the frequency being 50 cycles per sec. It is noticed that M increases as the resistivity increases, because the current spreads out further in a soil of higher resistivity.

The e.m.f. E is induced in each of the telephone wires, so that if

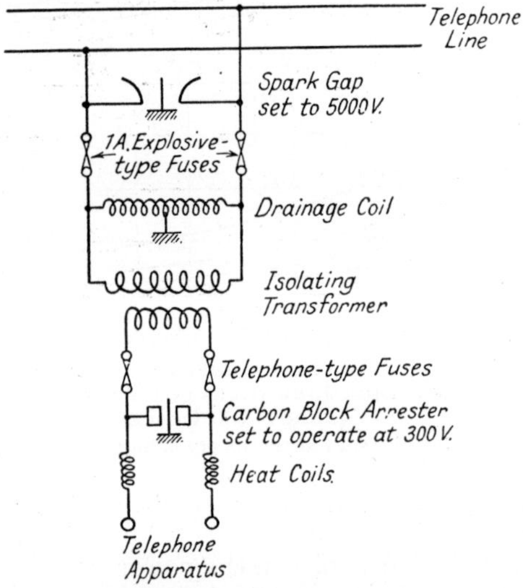

FIG. 190. TELEPHONE CIRCUIT PROTECTIVE SYSTEM

the telephone circuit were perfectly transposed and balanced there would be no voltage *between* the two wires of a circuit. There would, however, be the e.m.f. E between each wire and earth, and the telephone insulators must be capable of withstanding this voltage; for if one wire flashes over to earth, the full e.m.f. is applied to the telephone circuit with a resulting acoustic and electric shock. The potential between the wires of the telephone circuit is kept low by the use of drainage coils, shown in Fig. 189. The windings *AB* and *BC* are series aiding, so that the coil has a high impedance between the terminals *A* and *C* (of the order 6 000 $|\underline{70°}$ at 1 000 cycles), and gives an attenuation of about 2 db. to voice-frequency currents. The mid-point *B* is earthed, so that the impedance of the coil is very low to currents flowing along the lines in parallel and back through earth. In this way the longitudinal potential of the telephone lines is reduced from the high value *E* to a very low value, which depends upon the degree of balance of the windings and their resistance.

Acoustic and electric shock are minimized by the use of isolating transformers, drainage coils, spark gaps, and carbon block arresters. Fig. 190 shows an arrangement for the protection of a telephone circuit, as used in a successful installation which runs parallel with a 132 kV. power line in the Punjab.

EXAMPLES VIII

1. Discuss the phenomena of electrostatic and electro-magnetic induction from power transmission lines to adjacent telephone lines.

State the factors upon which the magnitude of the induction depends in each case and the precautions taken in both the power and communication circuits to reduce it. (*Lond. Univ.*, 1949.)

2. Why is automatic voltage regulation required for modern alternators?

Explain, with a diagram of connections, the operation of an automatic voltage regulator suitable for use with a large turbo-alternator. (*Lond. Univ.*, 1950.)

3. A 3-phase overhead line has the following constants: series impedance per conductor, $20 + j\,50\ \Omega$; shunt admittance of each conductor to neutral, $0 + j\,800$ micromhos. The line supplies, through a star/star transformer with a turn ratio of 4 to 1, a load of 40 MW. at 33 kV., unity power factor. The transformer, at 45 MVA. and 33 kV., has an impedance of $0{\cdot}5 + j\,6$ per cent.

Neglecting iron loss and magnetizing current in the transformer, determine the voltage and current at the supply end of the line, and the phase angle between them. (*Lond. Univ.*, 1954.)

4. Describe and compare the methods of interconnecting the bus-bars of a generating station.

Three generators *A*, *B*, and *C*, each of 12% leakage reactance and of MVA. ratings 25, 50, and 25 respectively, are interconnected electrically by a tie-bar through reactors, each of 10% reactance based upon the MVA. rating of the machine to which it is connected. A 3-phase feeder is supplied from the bus-bars of generator *A* at a line voltage of 11 kV. The feeder has a resistance of $0{\cdot}1\ \Omega$/phase and an inductive reactance of $0{\cdot}3\ \Omega$ phase. Estimate the maximum MVA. which can be fed into a symmetrical short-circuit at the far end of the feeder. (*Lond. Univ.*, 1953.)

5. Explain what is meant by the symmetrical components of a 3-phase 4-wire system.

The p.d.'s to neutral of such a system are $-36 + j\,0$, $0 + j\,48$ and $64 + j\,0$ V. respectively, and the currents in the corresponding line wires *R*, *Y*, and *B* are $-4 + j\,2$, $-1 + j\,5$, and $5 - j\,3$ A. Determine the negative-sequence power and reactive voltamperes. The sequence is *RYB*. (*Lond. Univ.*, 1954.)

6. Explain briefly the advantages to be gained by the insertion of reactances in the bus-bars of a large generating station. A generating station contains four identical three-phase alternators *A*, *B*, *C*, and *D* each of 20 000 kVA., 11 kV. rating and having 20% reactance. They are connected to a bus-bar system which has a bus-bar reactor rated at 20 000 kVA. and having 25% reactance inserted between *B* and *C*. A 66 kV. feeder is taken off from the bus-bar through a 10 000 kVA. transformer having 5% reactance. If a short circuit occurs across all phases at the high voltage terminals of the transformer, calculate the current fed into the fault. (*Nat. Cert.*, 1935.)

7. The bus-bars of a power station are split into two sections *A* and *B*, separated by a 5% reactance (based on 10 000 kVA.). A 30 000 kVA. generator with 10% reactance is connected to section *A* and a 50 000 kVA. generator with 12% reactance is connected to section *B*. Each section supplies a transmission line through a 40 000 kVA. transformer with 6% reactance which steps the voltage up to 132 kV. If a three-phase short-circuit occurs on the high-tension terminals of the transformer connected to section *A*, calculate the maximum initial value of current which may occur at a short circuit.

Describe how you would estimate the current which a circuit-breaker operating after 0·3 sec. would have to interrupt, and explain why this value would be different from the maximum initial value as calculated.

8. A three-phase system of voltages is given by

$$V_A = 1\,000\,\underline{|35^\circ} \qquad V_B = 3\,000\,\underline{|100^\circ} \qquad V_C = 2\,000\,\underline{|270^\circ}$$

Resolve these voltages into their symmetrical components, namely, a balanced positive sequence component, a balanced negative sequence component and a zero sequence component.

Explain how the method of symmetrical components can be used for the calculation of short-circuit currents under unbalanced fault conditions. (*Lond. Univ.*, 1934.)

9. Two 3-phase, 6·6 kV. generators G_1 and G_2 feed into common bus-bars, which are linked by reactors *R* to a second set of bus-bars: the latter are also linked to a grid system of large capacity by an interconnector *I*. From the bus-bars, feeders F_1 and F_2 supply a network *N*, as in Fig. 191. The percentage reactances are as follows—

G_1 and G_2, each 15% at 50 MVA.; *R*, 5% at 10 MVA.; grid system through *I*, 10% at 40 MVA.; *N*, between F_1 and F_2, 10% at 100 MVA. The impedance of each feeder is $0{\cdot}01 + j0{\cdot}02$ ohm per core.

Determine the maximum symmetrical fault current that a circuit-breaker at *X* may have to clear, in the event of either one or the other of the feeders being disconnected. (*Lond. Univ.*, 1949.)

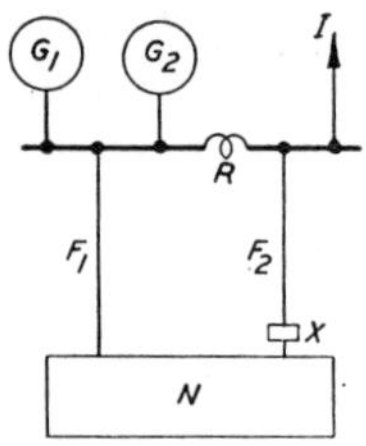

FIG. 191

CHAPTER IX

SWITCHGEAR AND PROTECTION

Introduction. With the continued progress of interconnection and the increasing capacity of generating stations the need for reliable protective devices and switchgear has become of paramount importance. When a short circuit occurs, an enormous power can be fed into the fault with considerable damage and interruption of service. The aims of protective gear are to achieve (1) complete reliability. (2) absolutely certain discrimination, (3) quick operation, and (4) non-interference with future extensions.

Complete reliability is required, as the protective gear is added because it is intended to improve the reliability of the whole system. As protective relays are sensitive and must act within fairly fine limits, they should be simple, robust, and easy to inspect and to maintain; furthermore there should be as few as possible.

Discrimination is important, as it is very inconvenient to cut out healthy sections on account of the necessity for synchronization on re-starting. Moreover the protective gear on healthy sections must not be made to operate as a result of fault currents on faulty feeders. This is specially important, as continuity of supply is expected and demanded by the users of electrical energy.

Quick operation is required in order that no damage may be done to generators, transformers, and cables by the short-circuit currents.

Before the various systems of protection are discussed a brief account of switchgear, circuit-breakers, and protective relays will be given; for it is necessary to know the characteristics of the relays and the circuit-breakers in order that the operation of a protective system may be understood.

Principles of Arc Extinguishing. At the instant when the contacts of a circuit-breaker begin to part there is a large current (several hundreds of thousands of amperes) and a small voltage drop. A small separation of the contacts does not result in an immediate cessation of the current, for as the contacts separate the resistance between them increases and the ohmic loss, I^2R, generates sufficient heat to ionize the air or vaporize and ionize the oil. The ionized air or oil vapour acts as a conductor because of the large number of free electrons present, and the current flows without immediate change, across the arc thus formed. The potential drop between the contacts is just sufficient to maintain the arc and is quite small. Separating the contacts draws the arc out, but it is not practicable to draw the arc out to such a length that the voltage available is insufficient to maintain the arc; in high voltage systems a separation of many yards would be necessary for this purpose.

The conductance (i.e. the reciprocal of the resistance) of the arc is proportional to the number of electrons per cm.^{3} produced by the ionization, the square of the diameter of the arc, and the reciprocal of the length. As we have stated, we cannot do much by increasing the length of the arc to any reasonable value. What can be done is to decrease the density of the free electrons, i.e. reduce the ionization, and decrease the diameter of the arc.

The process of arc extinction in a d.c. circuit is more prolonged than in an a.c. circuit, and will be discussed first. Free electrons are produced in an arc in two main ways: (1) by thermal ionization, i.e. ionization by collision of the atoms due to thermal agitation; (2) by ionization by bombardment of electrons which attain high speeds under the electrostatic field. Free electrons are withdrawn from the arc in two main ways: (3) by re-combination of electrons and positive ions; (4) and by diffusion of electrons out of the arc. Under steady conditions (1) + (2) = (3) + (4). To extinguish the arc it is necessary to decrease (1) and (2), so that

$$(1) + (2) < (3) + (4).$$

The only possible way of doing this is to decrease (1) by cooling the arc. To find the cooling necessary, we imagine that the circuit voltage E acts through an external resistance R across the arc resistance R_a. The rate of heat generated in the arc is

$$E^2R_a/(R + R_a)^2 \text{ watts,}$$

and has a maximum value of $E^2/4R$ when $R_a = R$. When the arc is new R_a is small and the heat generated in the arc is small. If the rate of cooling is high, the arc will get cooler. If the rate of cooling is less than $E^2/4R$, a point will be reached when the rate of heat generated in the arc is equal to the rate of cooling. The arc will then stay at this stable position of equilibrium. If, however, the rate of cooling is greater than $E^2/4R$, cooling is progressive until the arc is extinguished. It is therefore necessary that the rate of cooling be greater than $E^2/4R$ for the extinction of a d.c. arc by cooling.

Arc extinction in a.c. circuits differs from that in d.c. circuits because of the passing of the current and voltage through zero at intervals of $\frac{1}{100}$ sec. It can be shown that if the a.c. arc had to be extinguished by cooling like the d.c. arc, it would be necessary to have a rate of cooling greater than $E^2/2X$, where X is the reactance of the circuit supplying the arc. A circuit-breaker large enough to dissipate heat at such a rate is many times larger than that required in practice. The extinction of an alternating current arc is illustrated in Fig. 192, which represents the main features of many oscillograph records of a.c. arcs. The circuit current is taken as zero to begin with.

A short circuit is applied at the instant when the circuit voltage is zero, so that the doubling effect of the short-circuit current is obtained. The short-circuit current is shown as settling down from the unsymmetrical large value to a somewhat smaller and symmetrical value, until the contacts begin to open 6 half-cycles later. Up to this point the voltage between the contacts is clearly zero, but the circuit voltage is gradually reduced by the effects of armature reaction in the generators. When the contacts open, the arc

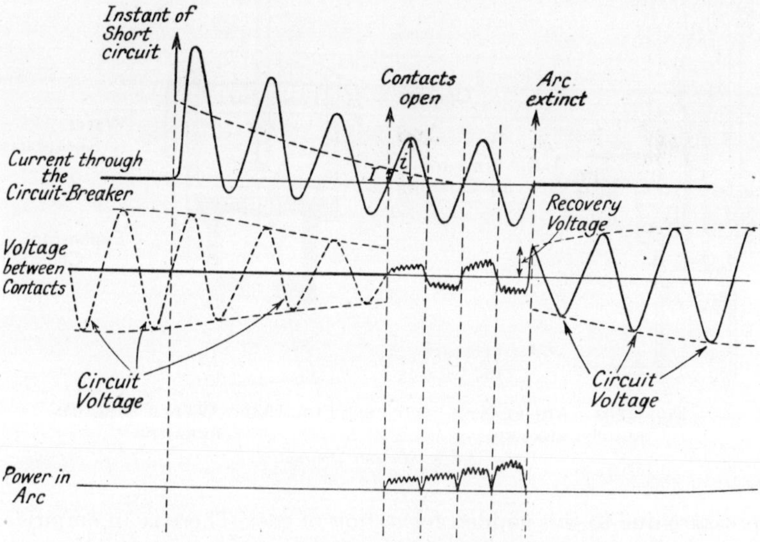

FIG. 192. EXTINCTION OF ALTERNATING CURRENT ARC

is struck, the current is maintained as before, whilst the voltage between the contacts is of the form shown; this voltage is quite small to begin with. When the current reaches its zero position, the arc is automatically extinguished and the voltage between the contacts is zero. The voltage then increases in the opposite direction and soon reaches a value which is sufficient to restrike the arc; during the period of quiescence a certain amount of recombination of ions and cooling takes place, so that the restriking voltage is somewhat greater than the voltage previously required for the maintenance of the arc. This goes on for several half-cycles, at the end of which period sufficient recombination of ions and cooling has taken place for the circuit voltage to be insufficiently large to restrike the arc. The arc is then extinct, the circuit current is zero, and the circuit voltage recovers and increases finally to its normal value. The recovery voltage is somewhat less than the normal circuit voltage, because of armature reaction; its value depends also upon circuit conditions. The energy absorbed by the arc can be measured directly on the oscillograph or it can be calculated from the current and voltage curves; the form of the power curve is shown in Fig. 192. Experiment shows that 1 kW.-sec. of energy liberates about 60 cm.^{3} of gas, and it is necessary to design the size and strength of the circuit-breaker to withstand the explosive

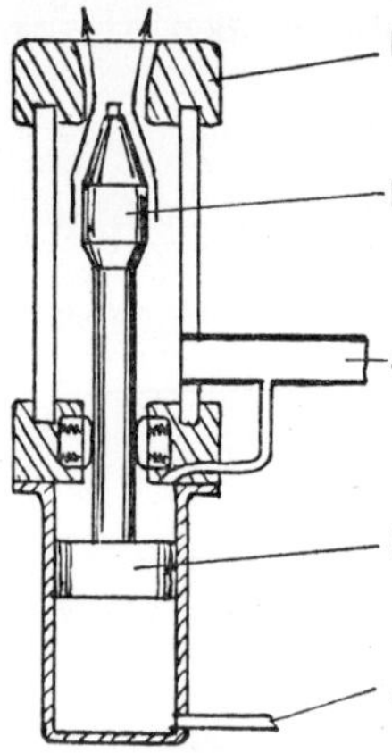

Fig. 193A. Air-blast Circuit-breaker

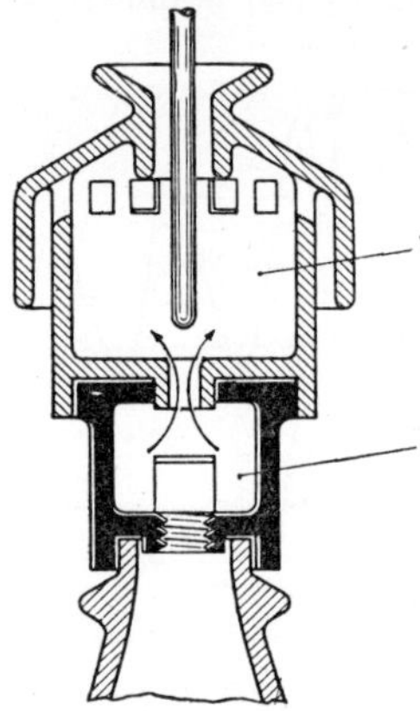

Fig. 193B. Water Circuit-breaker

(*I.E.E. Students' Journal*)

pressure due to this sudden formation of gas. There is an empirical formula for the arc energy, namely,

$$\text{arc energy} = \text{kVA. interrupted} \times 0{\cdot}1 \times \text{arc time in sec.}$$

Thus if the kVA. interrupted is 500 000 and the arc is extinguished in 4 half-cycles, the energy is

$$500\,000 \times 0{\cdot}1 \times 0{\cdot}04 = 2\,000 \text{ kW.-sec.},$$

and the gas liberated is 120 litres or about 4 ft.^{3} In practice the arc may last for as long as 0·1 sec. or more.

Any method of decreasing the diameter of the arc or reducing the density of the ions will result in an increased restriking voltage and a quicker extinction of the arc.

There are various ways of achieving a reduced diameter and density of ions.

Fig. 193A shows a sketch of the *air-blast circuit-breaker*. The breaker is closed by applying pressure at the lower opening, and opened by applying pressure at the upper opening. When the contacts separate, the cold air rushes round the movable contact and blows out the arc. Fig. 193B shows the principle of the water circuit-breaker. The contacts are in water, which is turned into steam by the arc and rushes past the opening to blow out the arc. Both these forms of circuit-breaker are much smaller than the oil circuit-breaker, and operate in one or two half-cycles.

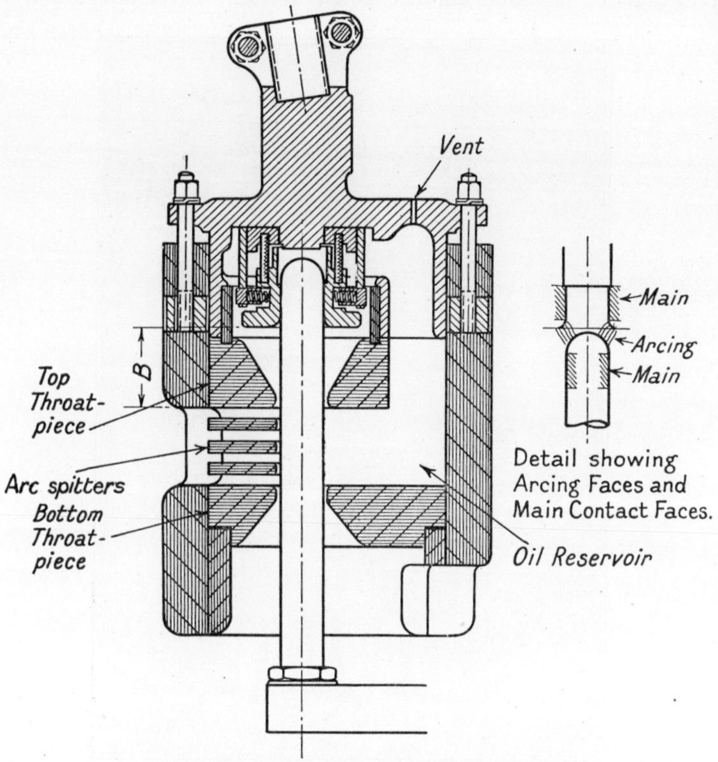

FIG. 194. CROSS-JET EXPLOSION POT
(*Outdoor High Voltage Switchgear* (Todd and Thompson))

Fig. 194 shows a section of a cross-jet explosion pot. Below the female contact there is a shroud, consisting of a stack of insulating plates with channels at right angles to the arc. When the contacts part the arc breaks up the oil surrounding it into gas at high pressure, and the gas expels a cross-jet of cold oil from the side pocket through the channels across the arc as soon as the movable contact passes an opening. If the short-circuit current is heavy, the pressure of the gas is great and the cross-jet is sufficiently powerful to extinguish the arc at the first or second channel. In order to break relatively low currents it is necessary to have more channels. Fig. 194 shows a 66 kV. breaker which has four channels. Fig. 195 shows a three-phase circuit-breaker operating at 33 kV., fitted with cross-jet contacts.

A recent development is the single-break switch, which has the advantage of a much smaller construction. It is not much less

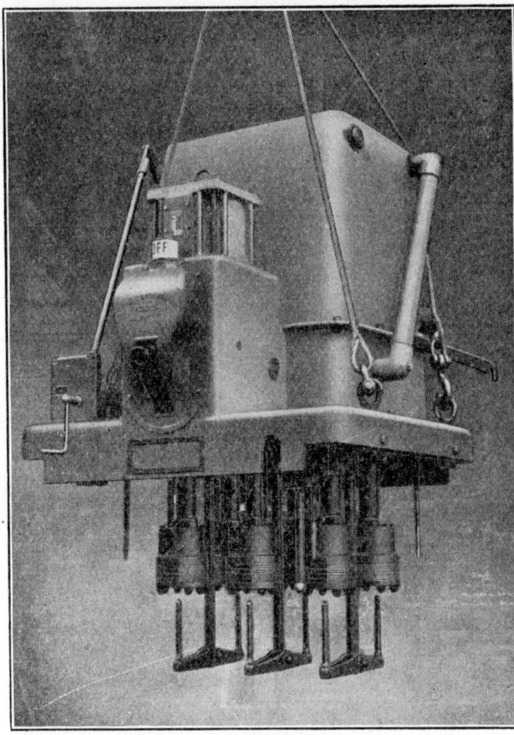

FIG. 195. THREE-PHASE 33 KV. CIRCUIT-BREAKER WITH CROSS-JET CONTACTS
(*Metropolitan-Vickers*)

effective than the double-break switch for the following reason. Suppose that the fixed contacts A and C have capacitances of 10 $\mu\mu$F. each to the moving contact B, which has a capacitance of 40 $\mu\mu$F. to earth (See Fig. 196). If an earth fault occurs on C, the resulting capacitances are 10 $\mu\mu$F. between A and B and 50 $\mu\mu$F. between B and C. The voltage between A and C is then taken as to five-sixths by the gap AB and one-sixth by BC, so that the arc AB does nearly all the work, and a single-break switch would be nearly as effective as the double-break. Fig. 197 shows a single-break switch, with cross-jet pot and gang-operated oil-immersed isolators on the bus-bar and cable sides.

In the *deion circuit-breaker* (Fig. 198), a stack of iron plates is magnetised by the arc current, and the resulting magnetic field forces the arc along the cold surface of a grid of insulating stacks, and this results in a rapid cooling and deionization. The arc is in two parts, each of which flows by a stack of iron plates $\frac{1}{16}$ in. thick, separated from each other by $\frac{1}{16}$ in. and protected by an asbestos arc-resisting material.

Rating of Circuit-breakers. The current is taken as the r.m.s. value at the instant that the contacts open. It is seen in Fig. 192 that the current has a d.c. component which decays rapidly: let the value of the d.c. component be I at the instant of contact separation. Let the a.c. component have peak value i. Then the r.m.s. value is

$$\sqrt{(I^2 + \tfrac{1}{2}i^2)},$$

and this is the value taken as the current broken. Continental practice ignores the d.c. component. The voltage is taken as the recovery voltage, which is somewhat less than the normal circuit voltage; American practice is to take the normal voltage. The product of the current and the voltage to neutral gives the kVA. rating per phase. The American rating is thus higher than the rating given in this country. Thus the ratings of a certain breaker according to Continental, British, and American practices are 385 000, 456 000, and 526 000 kVA. respectively.

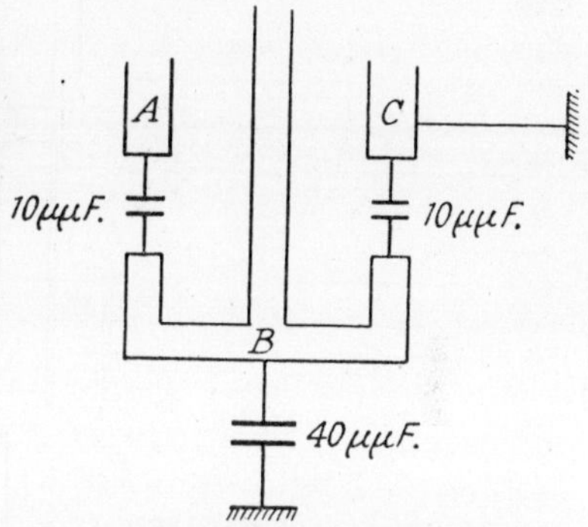

FIG.. 196 DOUBLE-BREAK SWITCH

The *conditions of severity* for circuit-breaker operation are the power factor of the load, the recovery voltage, and the rate of rise of recovery voltage.

If the power factor of the circuit to be interrupted is unity the voltage is zero when the current is zero, so that when the arc is temporarily extinguished there is no voltage immediately available to restrike it. If the power factor is zero, the voltage is a maximum at this instant and the arc is more easily restruck. It is found that the duration of the arc is approximately proportional to sin ϕ when the power factor is low. British practice requires that the breaker be capable of interrupting the rated current at 0·1 power factor, which corresponds to short-circuit currents under the worst conditions.

The recovery voltage depends upon the circuit conditions, and there is a desire to consider it as a condition of severity. Thus if there is a three-phase short to earth, the recovery voltage on the first breaker to open is equal to the phase voltage (less the drop due to armature reaction, etc.) if the neutral point of the generator is earthed, but $\sqrt{3}$ times this if the neutral is not earthed. The

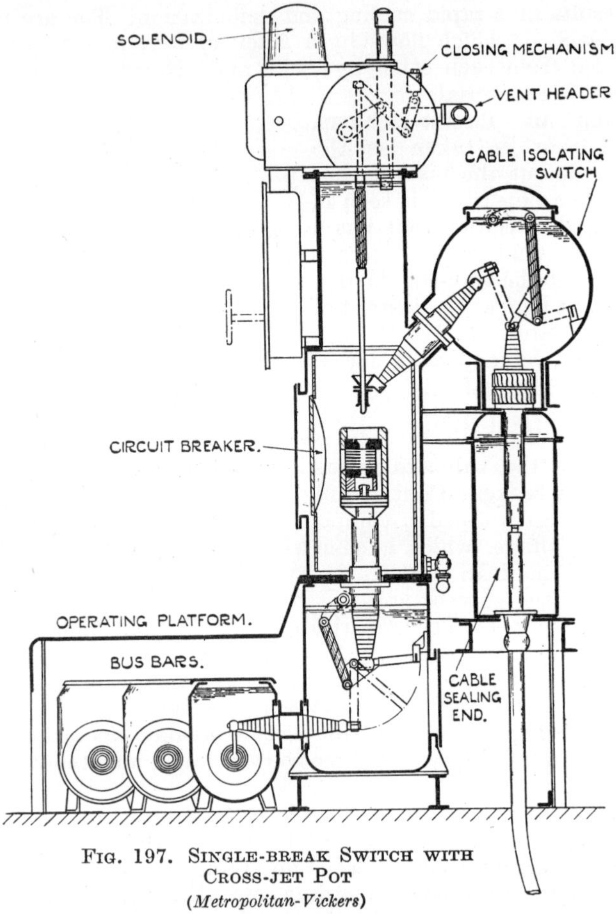

FIG. 197. SINGLE-BREAK SWITCH WITH CROSS-JET POT
(Metropolitan-Vickers)

rate of rise of recovery voltage has an effect on the extinguishing of the arc. When the current falls to zero the ions in the arc recombine and the dielectric strength rises at a rate between 10 and 200 V. per cm. per μsec. At this instant the voltage is zero and is rising. If the rate of rise of voltage is less than the rate of rise of dielectric strength of the gas the arc will not be restruck; if greater, the arc will be restruck. The rate of rise of recovery voltage is affected by the high-frequency oscillations of the electrical system. If the capacitance of the system is high, i.e. the line is long, the frequency of the oscillations is comparatively low, about 300 cycles per sec., and the rate of rise of recovery voltage is smaller than if the line is short, the capacitance low, and the frequency high (say 3 000 cycles per sec.). It is thus easier for the arc to be extinguished if there is a large capacitance in the system. Circuit-breakers are usually tested by direct connection to a generator, so that the

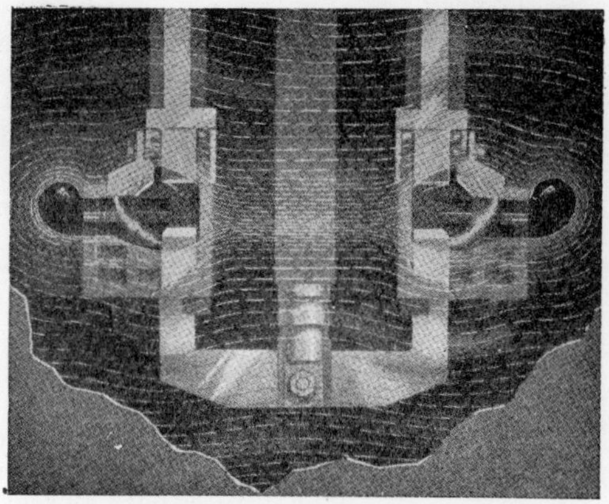

FIG. 198. DEION CIRCUIT-BREAKER
(*English Electric*)

capacitance is low and the conditions are as severe as will be met in practice.

Air Circuit-breakers. These are much used for d.c. circuits and for low voltage alternating currents. Fig. 199 shows a heavy-current air circuit-breaker. It consists of an upper and a lower contact block fixed to the switchboard panel, and a moving brush contact which bridges under pressure the stationary contacts. In addition to the main contacts, auxiliary copper contacts are provided to protect the main contacts, and also springs carrying carbon contacts on which the final break is made.

To obtain proper contact pressure the link mechanism is designed to toggle, and the links are held in position by a hardened latch, which is tripped by the operation of the various automatic release devices. Various automatic attachments may be provided for overload trip with or without time lag, reverse current trip, under voltage trip, shunt trip, or tripping interlock to ensure simultaneous tripping of two or more breakers. The time delay device usually consists of a dashpot and plunger, which is retarded by oil or by air escapement.

Fuses. The low-voltage fuse or fusible cut-out consists of a short length of conductor, which can carry a certain current indefinitely but will fuse if the current is greater. The time to fuse or blow depends upon the magnitude of the excess current. Small current fuses (up to 100 amperes) are usually the only protective devices included in domestic installations. Fuses for large currents (up to 600 amperes) are made for low-voltage supply systems; one very successful fuse for large currents consists of parallel strips of silver; another consists of a copper strip with a waist which is cut and soldered together. These types of fuse are enclosed in cartridges of porcelain or synthetic heat-resisting substance such as bakelite.

FIG. 199. HEAVY-CURRENT, AIR CIRCUIT-BREAKER

In high-voltage rural distribution schemes high-voltage fuses are used as alternatives to the oil circuit-breaker. Recent research has produced fuses with a breaking capacity of 900 000 kVA. at 132 kV. and 500 000 kVA. at 11 kV.

The tetrachloride fuse consists of a strong glass tube, sealed at both ends with brass caps, and filled with carbon tetrachloride. A flexible copper conductor is held by a strong spiral spring (usually of phosphor bronze); one end of the spring is fixed to an end cap, and the other end is held by a high-resistance wire which is in parallel with the copper conductor. When the current exceeds a certain value, the high-resistance wire fuses, the spring pulls back the copper conductor, and the arc is quenched by the carbon tetrachloride vapour. The amount of metal vaporized is very small, and the explosive pressure is not unduly great. *Quenchol* is used sometimes instead of carbon tetrachloride.

There is a cartridge-type fuse consisting of a bakelized canvas tube filled with powdered quartz. The conducting element consists of silver and high-resistance wires in parallel: the vaporized silver forms silver silicate which is non-conducting.

Both the above types of fuse will clear a heavy fault in one or two cycles, and thus they compare favourably with a relay-operated circuit-breaker which takes 0·2 to 0·5 sec. The disadvantage of fuses is their complete lack of discrimination, as they will cut out any section which carries the excess current.

Relays. There are many kinds of relays in service in protective systems, and there are many designs of relay for the same purpose. The name attached to a relay indicates its function, e.g. *overload* or *over-current* relay, *reverse power* relay with graded time-lag, etc.

Fig. 200 shows an *instantaneous relay* of the attracted armature type. The a.c. or d.c. through the winding attracts the armature and makes the contacts C and D. At the same time the catch E is released from the movable blade F which springs into the stationary jaws G, making another contact in parallel with CD. The blade F must be reset by hand, and serves as an indication of the operation of the relay.

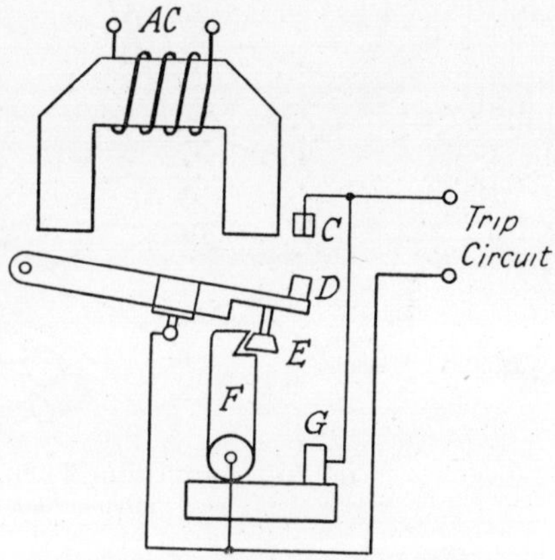

FIG. 200. INSTANTANEOUS RELAY, ATTRACTED ARMATURE TYPE (*B. T.-H.*)

The *solenoid and plunger* type of relay can be used for instantaneous action; a definite time-lag can be obtained by using an oil dashpot, an air escapement chamber, or clockwork mechanism. It is advisable that the dashpot or escapement chamber be widened at one end, so that there should be a free movement over the last part of the stroke: this makes for a good contact.

As the time of blowing of a fuse varies inversely as the overload, the placing of a fuse in shunt with an instantaneous or definite time relay produces a relay system with inverse time-lag. In the former case the time tends to zero as the current becomes very great, in the latter to a definite time.

Figs. 201 and 202 show an *induction* type *overcurrent relay*. The current enters a tapped primary, the turns of which can be chosen by a plug 7, which thus gives a current setting. A closed secondary is wound on the upper and lower magnets. The fluxes due to the primary and secondary windings are separated in phase and space and produce a torque, as in the shaded-pole induction disc motor. The disc experiences a torque, which depends on the current, and

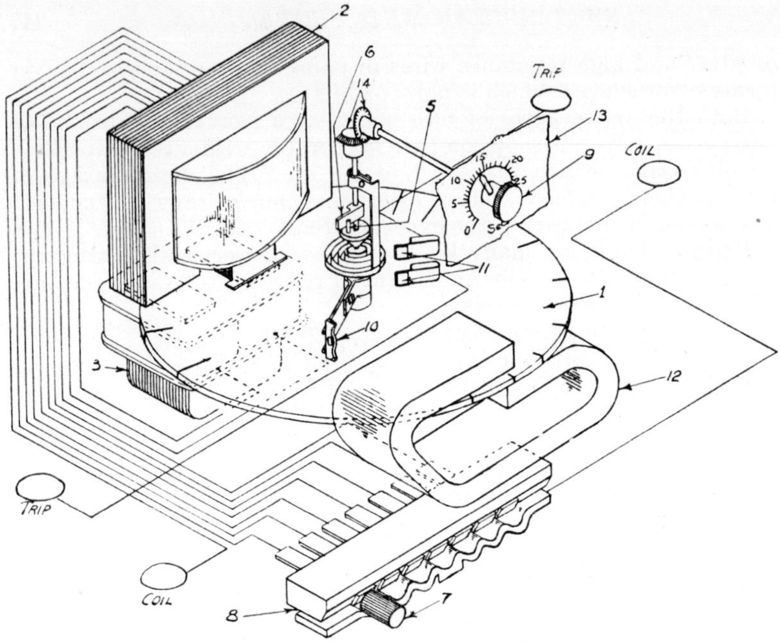

FIG. 201. INDUCTION TYPE OVERCURRENT RELAY
(Metropolitan-Vickers)

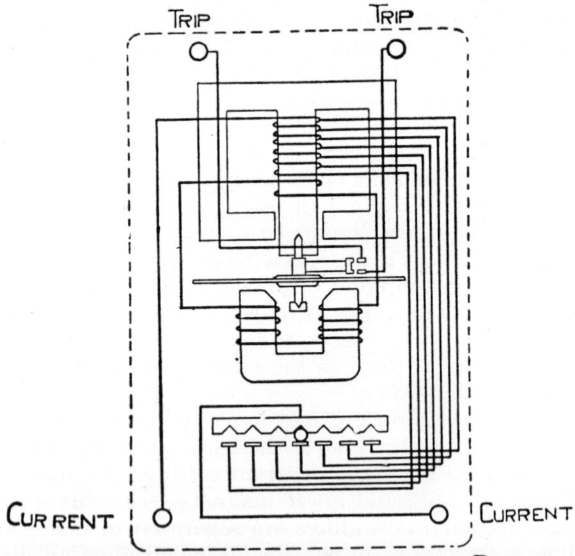

FIG. 202. INDUCTION TYPE OVERCURRENT RELAY
(Metropolitan-Vickers)

will move against a restraining spring provided the current is large enough. The time of travel is adjustable by means of the stop 6, which adjusts the distance of travel to the contacts 11. The torque is kept constant at all positions of the disc by means of graded slits. Since the torque increases with the current, the relay has an inverse time characteristic.

A *reverse power* relay is obtained by having a winding on the

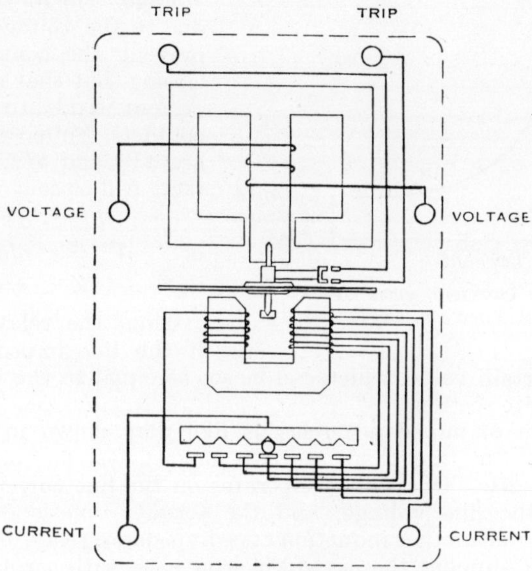

Fig. 203. Reverse Power Relay
(*Metropolitan-Vickers*)

middle limb of the upper magnet operated by the voltage, and a separate winding on the two limbs of the lower magnet operated by the current: the current winding has tappings as before. (See Fig. 203.) When the power flows in the normal direction the fluxes in the windings tend to turn the disc in a direction away from the contacts 11. When the power flows in the reverse direction the torque is in the opposite direction and the contacts 11 are closed. The relay can be made very sensitive by having a very light control spring, so that a very small reversal of power will cause the relay to operate.

A *directional overload* relay can be made of two induction disc type relays, one of which is a simple overload relay as shown in Fig. 202 and the other a reverse power relay as shown in Fig. 203. The two relays are fitted in one case. Their contacts are connected in series, so that the trip-circuit is not energized unless both operate,

Induction type relays have an inverse with definite-minimum time characteristic as shown in Fig. 204. The definite-minimum time is due to the self-braking effect of the fluxes which produce the driving torque.

Distance or *impedance relays* are of several kinds; but a common characteristic of all is the production of a force or torque in one direction by the line current and in the other direction by the voltage. The force or torque due to the voltage tends to prevent the contacts from closing, but that due to the current tends to close the contacts. Suppose the forces are aV^2 and bI^2; the contacts will close if

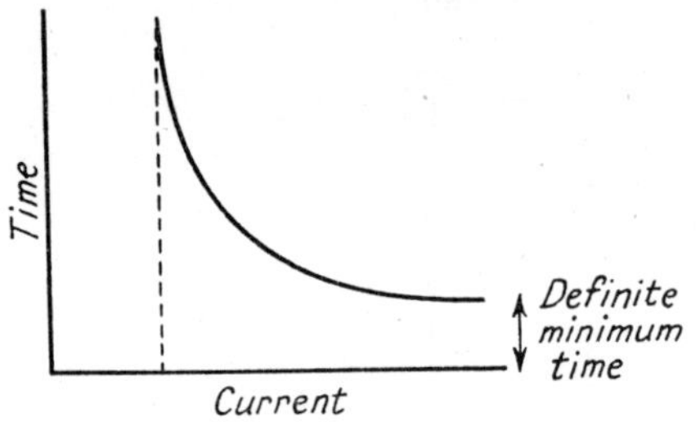

FIG. 204. INVERSE WITH DEFINITE-MINIMUM TIME CHARACTERISTIC

$$bI^2 > aV^2,$$

or $$V^2/I^2 < b/a,$$

i.e. $$Z < \sqrt{(b/a)}.$$

Thus the relay operates if the line impedance falls below a certain value, which can be set as equal to the impedance of a complete section.

One form of impedance relay is like that shown in Figs. 201 and 202, except that there are two magnet systems, one on each side of the disc. One system operates on the line current and the other on the line voltage, and the torques oppose each other. Fig. 205 shows another induction type impedance relay. A potential transformer supplies the cruciform iron path with a rotating flux, which causes drum D to rotate in one direction. A current transformer supplies the iron path X, which has a shaded pole, with a flux which causes D_1 to rotate in the opposite direction. Fig. 206 shows diagrammatically the solenoid-and-plunger type, whose action is obvious.

Impedance relays operate very quickly if the impedance falls below a certain value. *Impedance-time* relays are designed so that the time of operation is small for a certain low impedance, but increases linearly as the impedance increases. In one type of impedance-time relay, the current drives a disc round by induction and a spring is wound up. This spring tends to close the contacts of the trip-circuit but is opposed by an armature attached to the spindle and attracted by a coil carrying a current produced by the line voltage. Until the spring exerts a force as large as the attraction on the armature the spindle does not turn; when the spring is sufficiently wound up, the armature leaves the voltage coil and the trip circuit is made at once.

In many systems of protection it is required that a relay shall operate when the currents at two points of the system are unequal. For example, if the currents at the two ends of an alternator winding are unequal there is either a fault to earth or between phases.

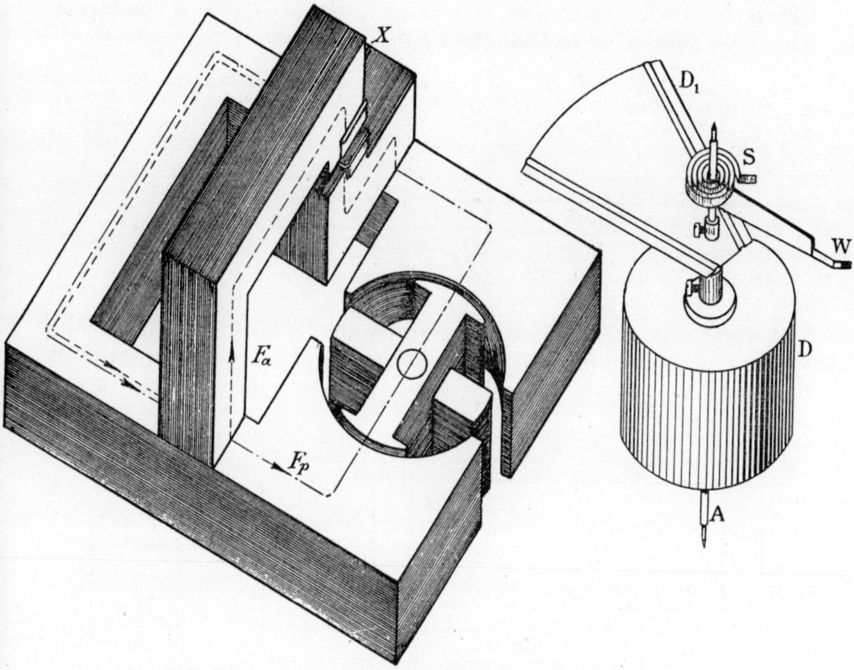

FIG. 205. INDUCTION TYPE IMPEDANCE RELAY
(*I.E.E. Journal*)

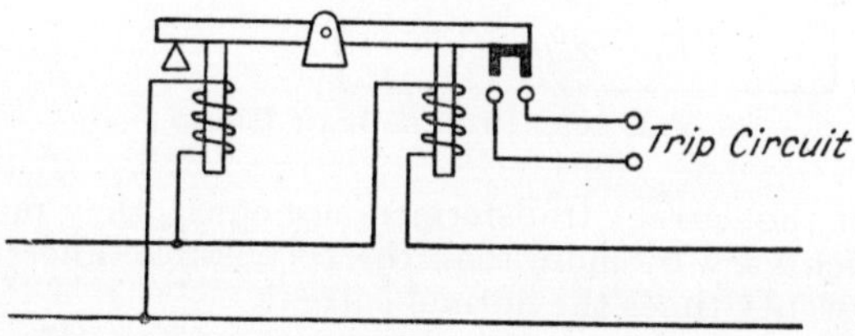

FIG. 206. SOLENOID-AND-PLUNGER TYPE IMPEDANCE RELAY

Fig. 207 shows a method whereby the relay R can be made to operate when I_1 is not equal to I_2. The current transformers, $C.T.$, are identical and are connected so that their secondary e.m.f.'s are in the directions shown. Their secondary currents i_1 and i_2 therefore flow in opposite directions through R, so that there is no current when $I_1 = I_2$ and $i_1 = i_2$. This method is called the *circulating current* method.

Fig. 208 shows an alternative method, known as the *opposed voltage* method. The relay is in series with the two secondaries, which are connected so that their e.m.f.'s oppose.

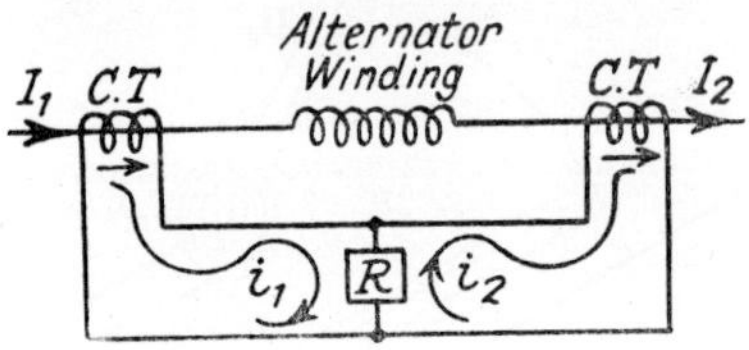

FIG. 207. CIRCULATING CURRENT METHOD OF TESTING BALANCE

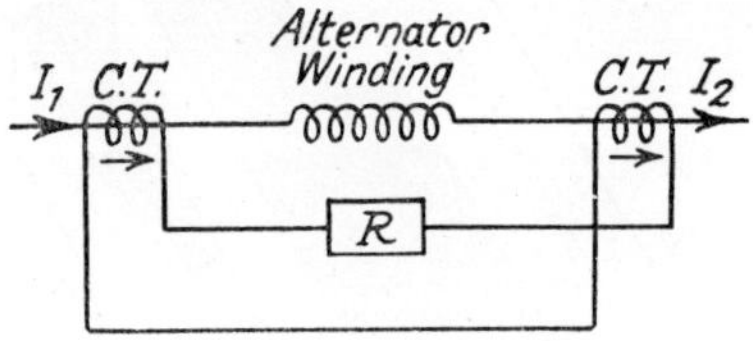

FIG. 208. OPPOSED VOLTAGE METHOD OF TESTING BALANCE

A very important disadvantage in simple balance systems is due to the inequalities of current transformers. Thus an external fault may lead to a large current; then, although the currents in the

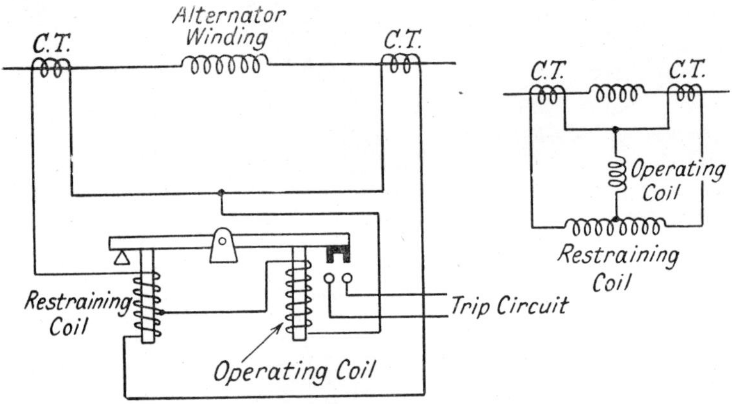

FIG. 209. BIASED BEAM RELAY

primaries of the current transformers are equal, they may produce e.m.f.'s which vary by more than the permissible difference if they are ten or twenty times the normal current.

This disadvantage is overcome by the use of *bias*. Fig. 209 shows a *biased beam* relay. The system is still the circulating current system of Fig. 207, but there is in addition a restraining coil which carries both circulating currents. Thus if the currents are large, there is a comparatively large restraining force which cannot be overcome by an error in the current transformers. It can be seen that the relay operates when the ratio I_1/I_2 differs from unity by more than a certain amount, which is adjustable by varying the number of turns of the restraining coil.

A similar result can be achieved by biasing the relay mechanically by moving the fulcrum nearer to one side, or magnetically by having two operating coils each of a different number of turns.

Protection of Alternators and Transformers. It is not considered advisable to have overload protection for alternators, which are now designed to withstand their complete short-circuit current without danger. They are disconnected by hand.

If the steam supply is cut off, there is no occasion to disconnect the alternator, as it will run as a motor and take a small current.

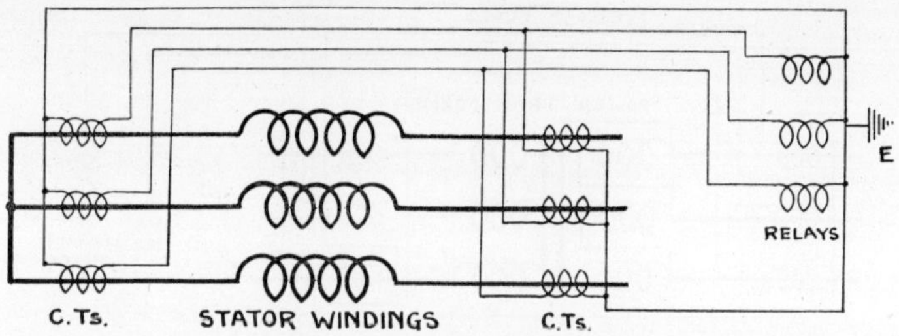

Fig. 210. Merz-Price System for Alternator
(*Automatic Protective Gear* (Henderson))

It will keep in synchronism, and operate when the steam supply is restored.

The result of a failure of the field of an alternator is very uncertain and depends upon the load and the condition of the machine. The machine can be switched out by means of an under-current relay in the field circuit. The occurrence of a field failure is, however, rare, and automatic protection is not often provided.

The most dangerous fault in an alternator is a failure of the insulation, for then the machine will feed current into its fault and destroy itself.

If an earth fault occurs on the stator windings whose star point is earthed, power will flow through the fault to earth. If this power is greater than the load supplied, the machine will receive power from other generators and reverse power protection will suffice. The reverse power relay can contain two wattmeter elements, as used in measuring three-phase power, or three single-phase wattmeter elements. As the reverse power relay will not act unless the power sent into the fault is greater than the output of the machine, the method is not satisfactory and is obsolescent.

Self-balance protection is the most useful method for alternators and transformers. Fig. 210 shows the Merz-Price system, in which three pairs of current transformers are connected to relays. When the currents at both ends of each winding are equal, equal e.m.f.'s are produced in the secondaries of the current transformers and no current passes through the relays. The Merz-Price circulating current system just described is clearly effective for earth faults and faults between phases. If another generator feeds current into the fault, it is seen that the currents in a pair of current transformers are opposing and the total current will flow through the relay, and thus operation is aided. The Merz-Price system is, however, unable to operate when there is a fault between turns on the same phase, as the currents at the ends of the phase are still equal.

The alternator can be protected for leakage by means of the leakage relay shown in Fig. 211.

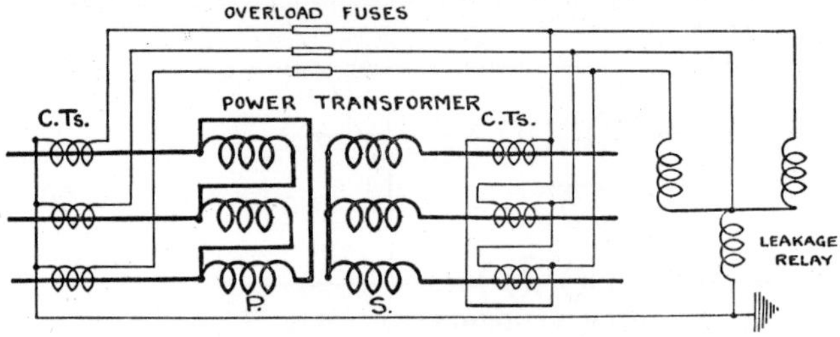

FIG. 211. MERZ-PRICE SYSTEM FOR TRANSFORMERS
(*Automatic Protective Gear* (Henderson))

The Merz-Price system is available for the protection of transformers. Fig. 211 shows the system applied to the protection of a star/delta connected transformer, in which overload and leakage protection is added. The latter is performed by the leakage relay. The former is achieved by the use of overload fuses in the pilot wires between the three pairs of current transformers. When the overload blows a fuse, one of the pair of current transformers is disconnected from the relay, which receives the current from the remaining transformer. The balance is thus upset and the relay operates.

It must be remembered that the currents in the windings of a transformer are not equal, and the current transformers must have the proper numbers of turns, so that their secondaries have equal currents during normal operation. Thus in a star/star connection with a ratio of 11 kV./66 kV., the current transformers must have a turns ratio of 66/11.

Different power-transformer connections demand different connections of the protective transformers. Thus for star/star power transformers, the protective transformers must be connected in delta on both sides, since a fault on one phase of the secondary appears in two phases of the primary. Fig. 212 (*a*) shows the case where the protective transformers are connected in star/star formation for a star/star transformer with earthed neutral in the secondary.

Suppose that an earth fault occurs outside the transformer on the secondary side as shown, and let the fault current be I. We may assume that the transformer has unity ratio for the purpose of explaining the action. Then the currents in the primary windings

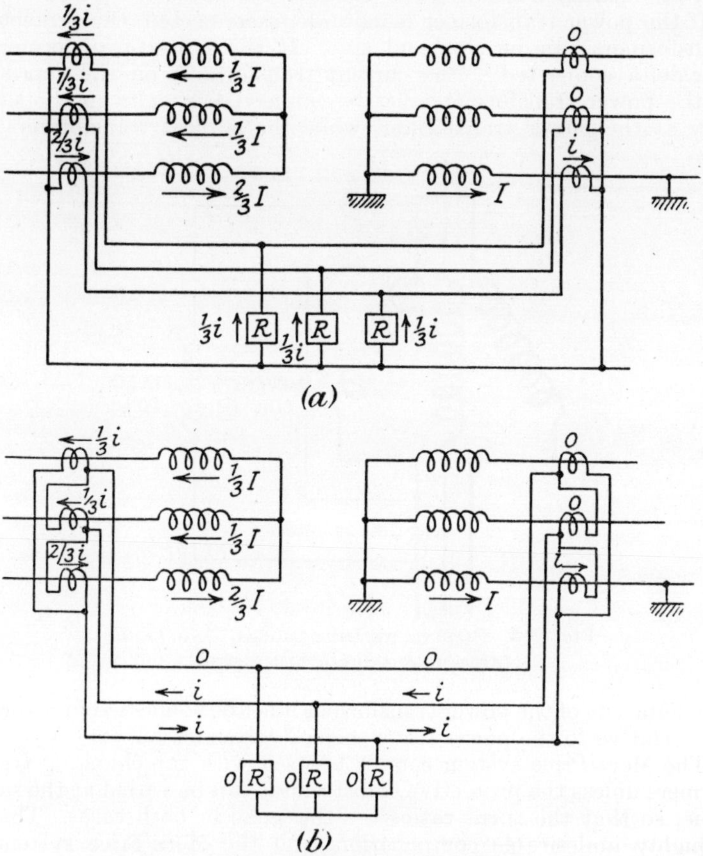

FIG. 212. MERZ-PRICE SYSTEM

(a) In star/star connection. (b) In delta/delta connection.

are $\frac{2}{3}I$, $\frac{1}{3}I$, and $\frac{1}{3}I$, the directions being those in the figure: the m.m.f.'s in the three limbs of the transformer are due to $\frac{1}{3}I$ in a winding and they oppose each other, whilst the primary current has no zero phase sequence current, since it is isolated. The currents in the secondaries of the current transformers are $\frac{1}{3}i$, $\frac{1}{3}i$, $\frac{2}{3}i$, and i as shown, so that each relay has a current $\frac{1}{3}i$ and they operate although the transformer is sound. Fig. 212 (b) shows the case where the protective transformers are connected in delta/delta. With the same power transformer and earth fault as in Fig. 212 (*a*), the distribution of currents in the protective system is as shown in Fig. 212 (*b*). It is seen that no current passes through the relays, and the healthy transformer is not switched out.

If the power transformer is delta/delta connected, the protective transformers can be star and star. If the power transformer is star/delta connected, " the current transformers on the star side of the power transformer must be connected in delta in the same way as the power transformer, while the current transformers on

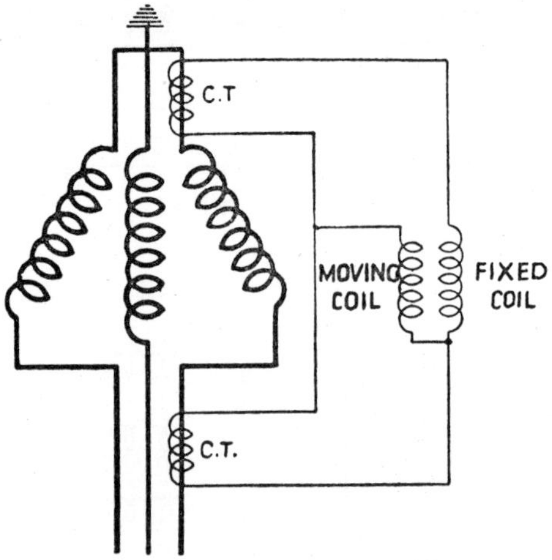

FIG. 213. BIASING BY INTENTIONAL UNBALANCE
(*Automatic Protective Gear* (Henderson))

the delta side of the power transformer must be connected in reversed star relative to the power transformer." (Henderson, *loc. cit.*)

The Merz-Price system cannot be used with tap-changing transformers unless the protective transformers can be varied at the same time, so that the turns ratios are the same in both cases. This is a highly undesirable complication, and the Merz-Price system is replaced by the simpler self-balance method of protection illustrated in Fig. 210; both sets of transformer windings are treated in the same way as the alternator windings.

We have already stated that the self-balance method cannot indicate the presence of short-circuited turns, as the currents at the end of each winding are the same. The Merz-Price system is nominally able to indicate such a fault, since the presence of short-circuited turns causes a change in the turns ratio between the primary and secondary. In practice it is usually too insensitive to do so.

Biased protective systems are being increasingly used, to offset the inaccuracies of current transformers with very high currents. The method is shown in Fig. 209. A biasing method using an intentional unbalance is shown in Fig. 213. The current transformer at the star point is arranged to take 7·0 amperes in the secondary, while the current transformer at the bus-bar takes 7·5 amperes. The relay, which is a reverse-power relay, has under normal conditions a current of 0·5 amperes which keeps the relay in the open position. When a fault occurs, the current near the star-point increases whilst that at the bus-bar end decreases and then reverses; both effects tend to reverse the current in the relay, and if the fault current is greater than the set unbalance of 0·5 ampere, the relay operates.

The Buchholz relay is used for the protection of transformers, and depends upon the formation of gas from the oil. The relay consists of two floats which carry contacts. The upper float indicates the presence of gas which is formed comparatively slowly, and it rings an alarm. The lower float is affected only by a sudden rush of gas formed by a heavy fault current, and when it drops, its contacts complete the trip circuit.

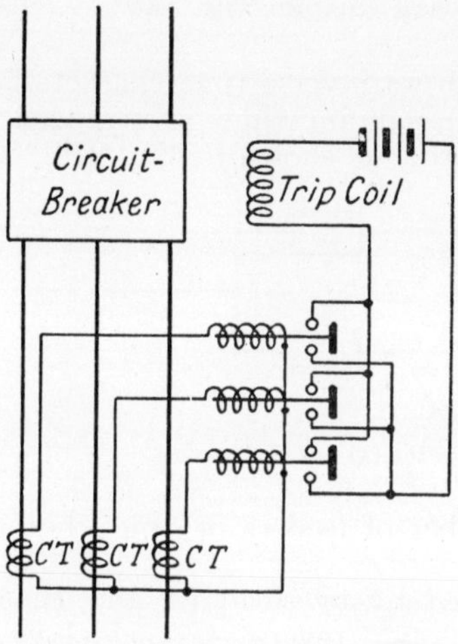

FIG. 214. OVERLOAD PROTECTION

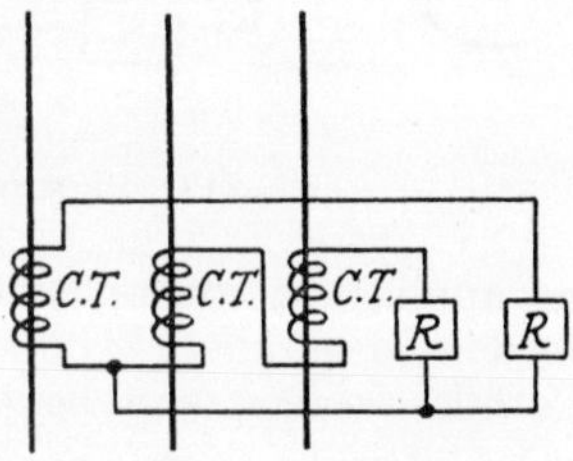

FIG. 215. Z-CONNECTION

Protection of Lines and Cables. The methods used for alternators and transformers are suitable, with modifications, for bus-bars and lines: the modifications required are due to the length of the lines, capacitance currents, and other reasons. Moreover, the need for discrimination in supply networks has called forward many protective schemes which have no application to the comparatively simple cases of alternators and transformers.

OVERLOAD PROTECTION. Fig. 214 shows a simple system of overload protection. Fig. 215 shows the well-known Z-connection, which requires only two relays for the protection of a three-phase circuit.

It is, of course, highly undesirable that all the overload relays in a system should operate as soon as a fault occurs, for then the whole system is shut down.

In a *radial system* the smallest possible part of the system is

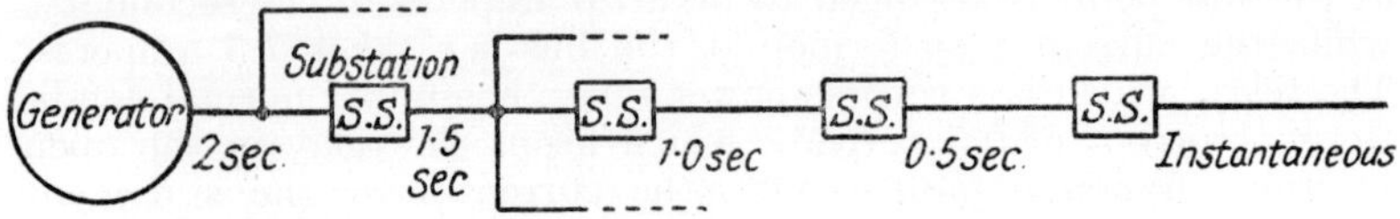

FIG. 216. RADIAL SYSTEM WITH GRADED TIME-LAG

switched out by having relays with an inverse time characteristic with a definite minimum. The minimum time is arranged to decrease from the generator outwards, as shown in Fig. 216. The

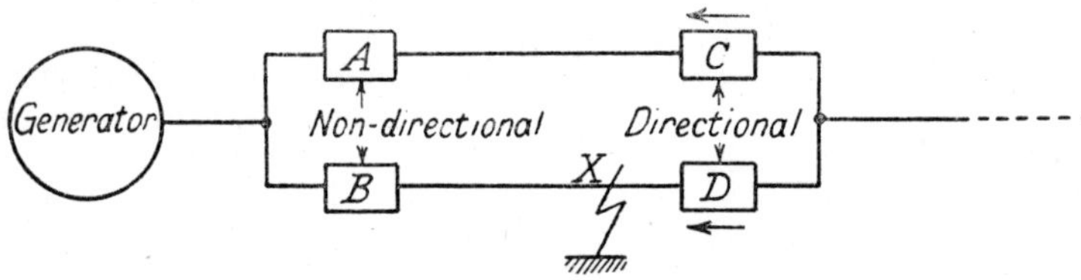

FIG. 217. PROTECTION OF PARALLEL FEEDERS

same applies, of course, to a number of feeders in series, but not to feeders in parallel.

Parallel feeders may be protected by overload (A and B) and reverse power relays (C and D), as shown in Fig. 217. If a fault occurs at X, power flows round $ACDX$ and relay D operates. Then B will operate and the faulty feeder is cut out. This method is not satisfactory unless the fault is heavy and reverses the power at D. It is advisable to apply differential protection to both ends of the parallel feeders in addition to overload protection.

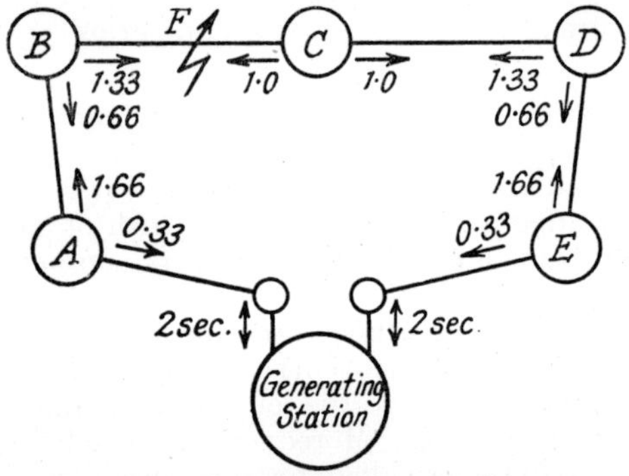

FIG. 218. RING SYSTEM WITH GRADED TIME-LAG

Ring mains can be protected by graded time-lag directional overload relays, as shown in Fig. 218. A relay of this kind is represented by an arrow and a number; the arrow indicates the direction of the power which can make the relay operate, and the number is the definite minimum time of operation in seconds. At the generator station the relays are non-directional, and are indicated by double-headed arrows. The scheme of time-grading is obvious. If a fault occurs at F, the power fed into the fault is fed by two paths, ABF and $EDCF$. The first relay to operate is that between C and F, and then the power can flow only via ABF; then the relay between B and F operates, and the fault has been cleared. The time grading cannot be closer than $\frac{1}{3}$ to $\frac{1}{2}$ sec., and as the longest time that a fault should be fed is 2 sec., the maximum number of sections that can be protected in this way is six.

The *interlock system* has been designed to overcome the disadvantages just described, and it can be used for any number of sections with a very short time-delay. It can be used for radial or ring systems. Fig. 219 shows the method applied to a radial system. If a section is healthy the same current passes at both ends and the overload relays operate. The operation of the relay at the beginning of the section completes the circuit of the trip coil (contact B), but the operation of the relay at the end of the section closes A, and causes the locking relay to operate and thus break the trip circuit. If the fault occurs within the section, the current entering is high, and causes the contact B to close, but the current leaving is small, and the locking relay does not operate.

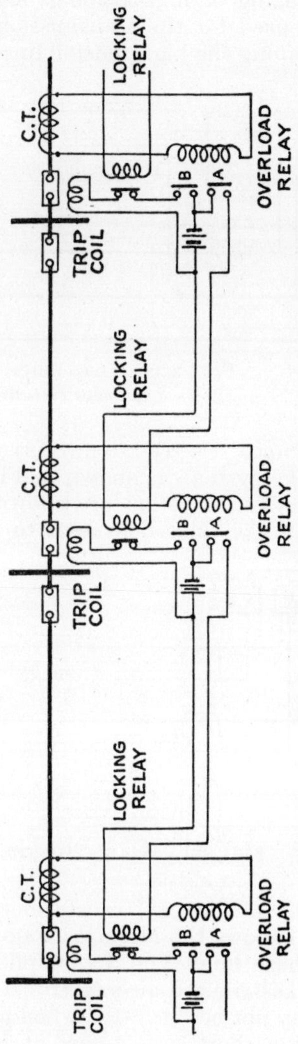

CONTACT A MAKES BEFORE B AND LOCKS IN MORE REMOTE O.C.B.

FIG. 219. INTERLOCK METHOD ON RADIAL SYSTEM
(*Automatic Protective Gear* (Henderson))

In order to apply the method to a ring main or any interconnected system, it is necessary merely to have directional relays to close the circuit for the locking relays.

In the above method it is necessary that the locking relay shall be capable of operating before contact B is closed, and for this reason relay B is made to act in 0·3 to 0·5 sec. Methods have been designed for avoiding this delay by sending tripping impulses instead of stabilizing or non-tripping impulses. Duplex telegraph circuits may be used for the transmission of these impulses. Carrier telegraphy along the high-tension line itself has also been used.

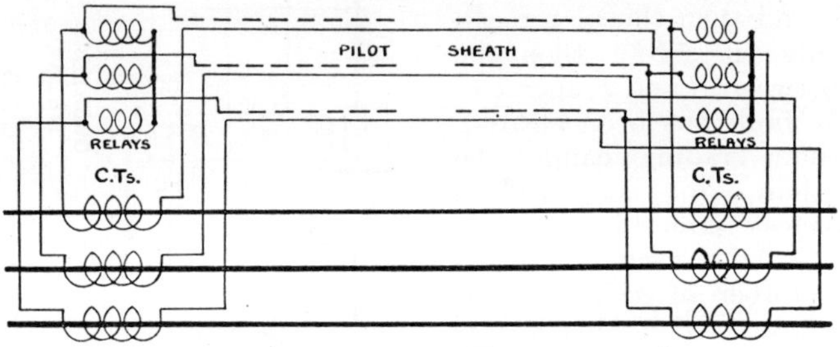

FIG. 220. MERZ-PRICE ON THREE-PHASE LINE
(*Automatic Protective Gear* (Henderson))

BALANCED PROTECTIVE SYSTEMS. The principle of balanced protective systems is shown in Figs. 207 and 208, and examples of the Merz-Price system are shown in Figs. 210 and 211. The use of the McColl biased system to eliminate error due to imperfect

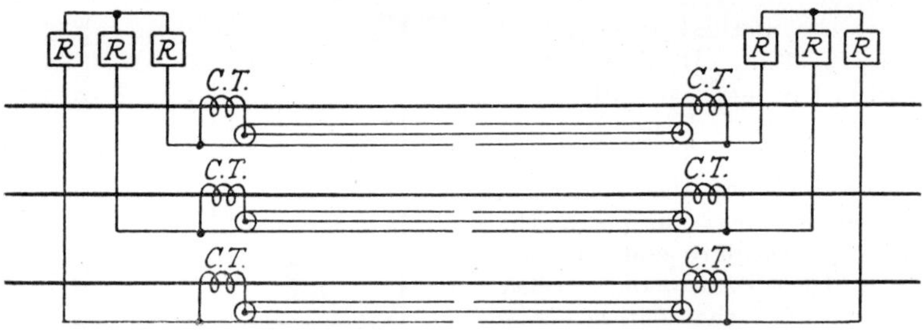

FIG. 221. BEARD-HUNTER SHEATHED PILOT SYSTEM

balancing of current transformers is shown in Figs. 209 and 213. Fig. 220 shows the Merz-Price opposed voltage system applied to a three-phase line. The 3-wire pilot consists of a 3-core, 7/0·029 in. cable. Distributed-air-gap transformers are required in order that they may not be saturated when heavy short-circuit currents flow.

When a short-circuit current passes through a healthy line, the current transformers may have induced in their secondaries equal e.m.f.'s of about 1 000 volts. Although these e.m.f.'s oppose, the capacitance currents that flow may be large enough to cause false operation. The capacitance currents are diverted from the relays by the Beard-Hunter sheathed pilot cable. In this cable each conductor is surrounded by a metallic screen which is divided at the centre of its length, as shown in Fig. 221. The capacitance currents flow in the local circuits formed by the pilot, sheath, and transformer, and do not flow through the relays.

The *translay system* is one of the modern modifications of the Merz-Price opposed-voltage system. Fig. 222 shows a simple form of translay protection for a single-phase feeder. "So long as the feeder is healthy, the line current transformers 10 and 10*a* at opposite ends of the feeder carry equal currents and the coils 11 and 11*a* connected to them induce equal e.m.f.'s in the windings 12 and 12*a* respectively. Coils 12 and 12*a* are connected in opposition by means of the pilot wires, with the operating windings 13 and 13*a* in series with them."

The translay relay is of the induction disc type, the novel feature being of a transformer character.

Fig. 223 shows a modified scheme for the protection of a three-phase feeder, employing type H (single-element) translay relays.

The upper magnetic circuit has three windings, two primaries, and a secondary. The upper and smaller primary is a phase-fault winding and is connected across the red and blue protective current transformers, whilst the mid-point is connected to the yellow. The lower and larger primary acts as a leakage winding, and is connected between the *B* transformer and the star-point of the current transformers. The secondary on the upper magnetic circuit behaves like the opposed-voltage transformer in the Merz-Price system and is connected in opposition to a similar winding, via two pilot wires, at the other end of the feeder. The windings on the lower magnetic circuit are in series with the pilot wires. The moving disc is composed of two sectors. Under normal conditions no current flows in the pilot wires as the opposed voltages are equal. When a fault occurs, the voltages in the windings are unequal and a current flows through the lower elements and the pilot wires. The flux produced in the lower magnetic elements interacts with the leakage flux of the upper magnetic elements to give a forward movement of the disc: the phase relation required for this is obtained as in a watt-hour meter. The capacitance currents lead the voltages and tend to rotate the disc in the opposite direction because of a closed copper ring near the end of the projecting limb of the upper magnetic circuit (see 18 and 18*a* of Fig. 222): thus the main disadvantage of the Merz-Price method has been avoided.

The translay relay can be biased by an unsymmetrical phase adjustment, which gives a backward torque when the flux in the upper element is large.

Split-conductor protection has the advantages of a balanced method of protection without the disadvantage of pilot wires. Fig. 224 shows the basic idea of the Merz-Hunter split-conductor system.

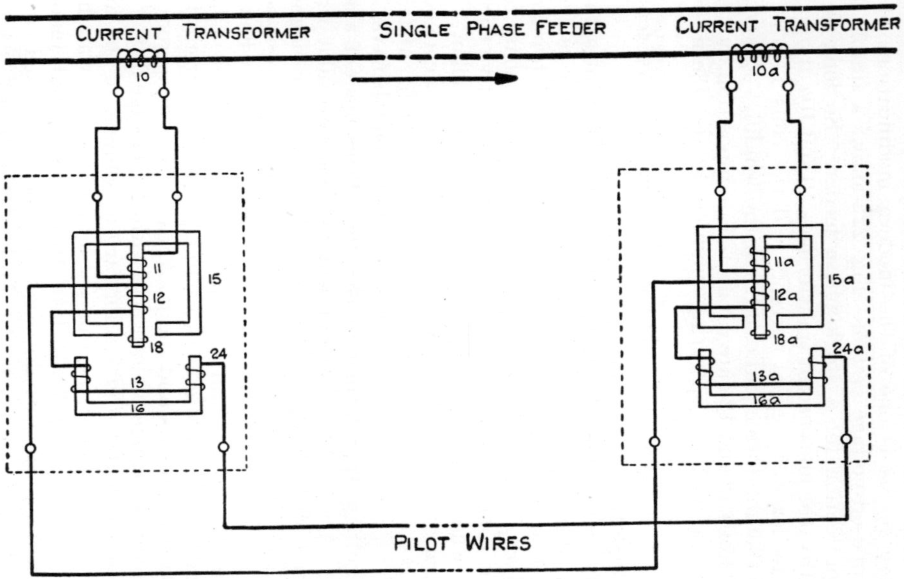

FIG. 222. SIMPLIFIED SCHEME OF CONNECTIONS FOR THE PROTECTION OF A SINGLE-PHASE FEEDER BY MEANS OF THE TRANSLAY SYSTEM

(*Metropolitan-Vickers*)

Each phase of the line is split into two sections which are only lightly insulated from each other. A single-turn current transformer is inserted at each end of the split conductor in the manner shown; the arrangement is identical with that of the well-known clip-on

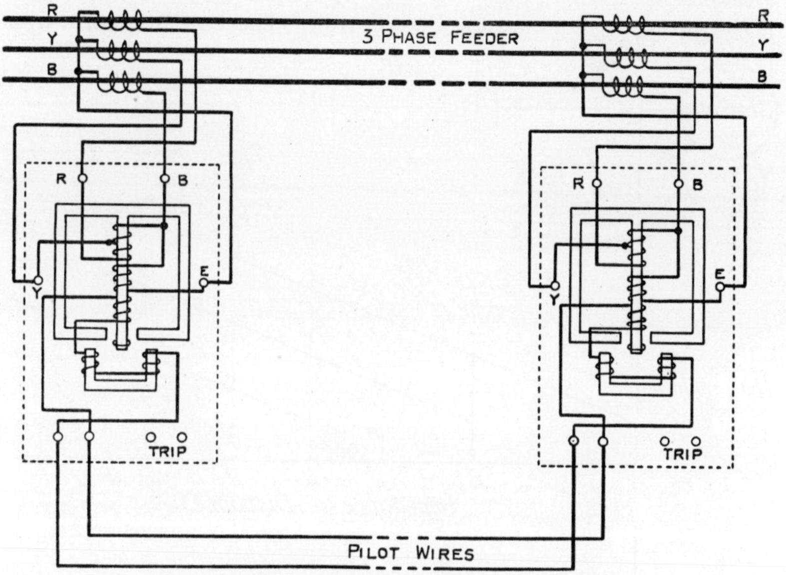

FIG. 223. CONNECTIONS OF TRANSLAY RELAY
(*Metropolitan-Vickers*)

ammeter. When equal currents flow along the two splits, there is zero flux and e.m.f. in the transformers. When a fault occurs one of the splits takes more current than the other, and the relays *R*

FIG. 224. SPLIT-CONDUCTOR PROTECTION

are energized. In the best arrangement, which is shown diagrammatically in Fig. 224, the splits are carried into the circuit-breakers, so that the splits are both opened by the breakers. The reason is as follows. If the fault is at the receiving end of a long line, the receiving-end breakers operate first; then if the breakers merely open the common end of the splits but leave the splits connected to a common end, the sending-end impedances will be nearly the same and the sending-end breakers will not operate. If the splits are both opened, as indicated, the healthy section takes no current and the faulty section takes a current which operates the sending-end breakers.

DISTANCE OR IMPEDANCE PROTECTION. The cost of pilot wires

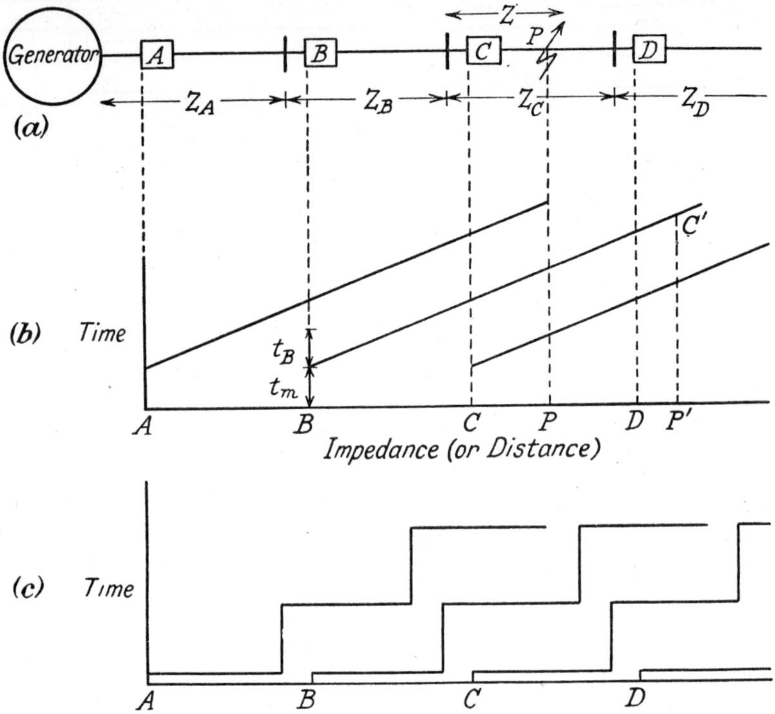

FIG. 225. DISTANCE OR IMPEDANCE PROTECTION

on long overhead systems is prohibitive, and *distance* protection has been designed to give discriminative protection without pilot wires. Fig. 225 (*a*) shows the simplest system consisting of feeders in series, such that the power can flow only from left to right. If a short circuit occurs at P between C and D, the impedances at A, B, and C are $Z_A + Z_B + Z$, $Z_B + Z$, and Z respectively. The relays are set to operate with impedances less than Z_A, Z_B, and Z_C respectively, so that only relay C will operate. Similarly, if the fault occurs between B and C, only relay B operates. The principle of operation and some types of impedance relays have been described. (See Figs. 205 and 206.)

A system with instantaneous impedance relays, set to act on impedances less than or equal to the impedance of a section, would be difficult to adjust; a fault near the junction of two sections is likely to cause the operation of two relays. Furthermore, if a fault of finite resistance occurs near the end of a section, it is possible that the total impedance is greater than that for relay operation. For these reasons it is advantageous to use impedance-time relays, the characteristics of which are shown in Fig. 225 (*b*), for the system of Fig. 225 (*a*).

If a fault occurs on the right-hand side of a junction, *B* say, relay *B* operates in the minimum time t_m and the breaker at *B* operates t_B sec. later. If t_B is made less than the time difference between consecutive relays, only one relay will operate.

Suppose that the fault at *P* has a resistance which causes the total impedance at *C* to be represented by the point *P′* (the fault resistance being *PP′*). Relay *C* operates in time *P′C′*, whereas in the previous system it would not operate at all.

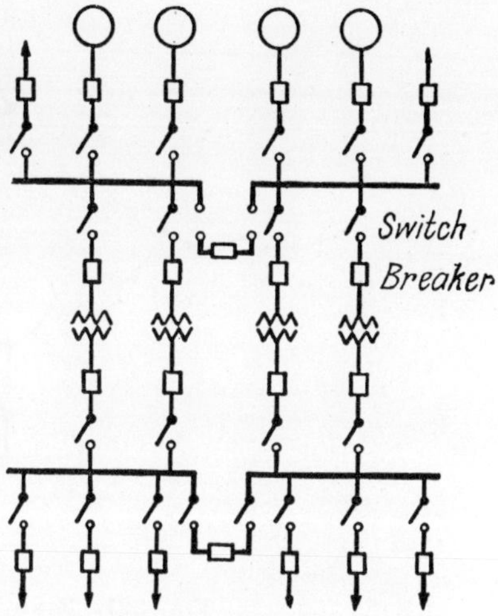

FIG. 226. SINGLE BUS-BAR LAYOUT

An impedance-time relay is a delicate mechanism, and it is considered worth while to replace it by three simple impedance-relays with a definite time of operation. The series combination can be arranged to give a three-step-time characteristic, as shown in Fig. 225 (*c*), which does the same thing as the previous linear characteristic.

The effect of arc resistance or the resistance of the earth return path of a fault can be avoided by the use of a *reactance relay*, which operates when the reactance is less than a certain value and ignores the resistance of the circuit. The reactance relay is made in the same way as the impedance relay, but the flux due to the potential coil is shifted in phase.

Bus-bar Arrangements. There are two types of switches, circuit-breakers and isolating switches. The former must be capable of breaking the maximum short-circuit currents; the latter are operated only under conditions of no current, are much cheaper, and are merely knife switches in air or oil.

In small substations there is a single bus-bar, and the arrangement is as shown in Fig. 226. The bus-bar may be sectionalized by a circuit-breaker and isolating switches, so that a fault on one part does not cause a complete shut-down.

Fig. 227 shows a ring bus-bar system. The circuit-breakers are in series with the ring, and each feeder is supplied by two paths, so that the failure of a section does not cause any interruption of the service.

Fig. 228 shows a single circuit-breaker station which was introduced by the C.E.B. and is working successfully. The protective current transformers are so arranged that, when a line or

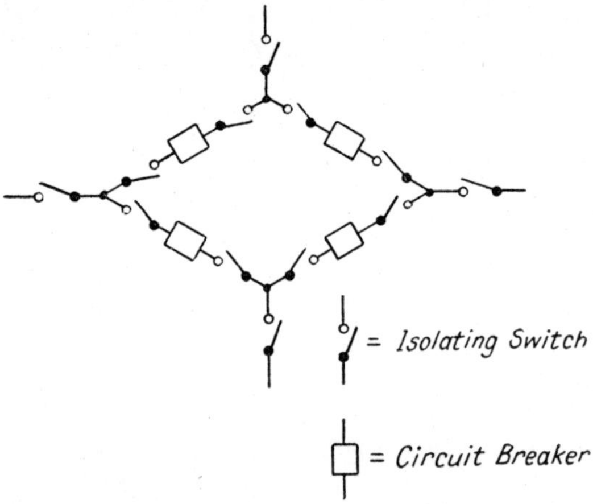

FIG. 227. RING BUS-BAR SYSTEM

transformer fault occurs, both the 132 kV. breaker and the low voltage breaker on the fault side are opened.

On a large system with many circuits it is usual to have duplicate bus-bars. A typical diagram is shown in Fig. 229, in which both the high and low voltage bus-bars are duplicated. In this case it is possible to split the plant into two entirely separate systems, which can be worked at different voltages if so desired. If it were desired to switch a circuit from one bus-bar to another without interruption of service there would have to be two circuit-breakers per circuit. This is too expensive, and the scheme of Fig. 229 is adopted, in which there is only one breaker per circuit and the service must be interrupted for a switch-over. The bus-bar coupling switch is required to effect a quick and easy transfer; the use of a bus-bar coupler and sectionalizing breakers converts the duplicate bus-bars into a ring system, which has great flexibility.

Testing of Transmission Systems. It is necessary to carry out regular and exhaustive tests, in order that the elaborate systems of protection be kept in perfect working order. When a system is

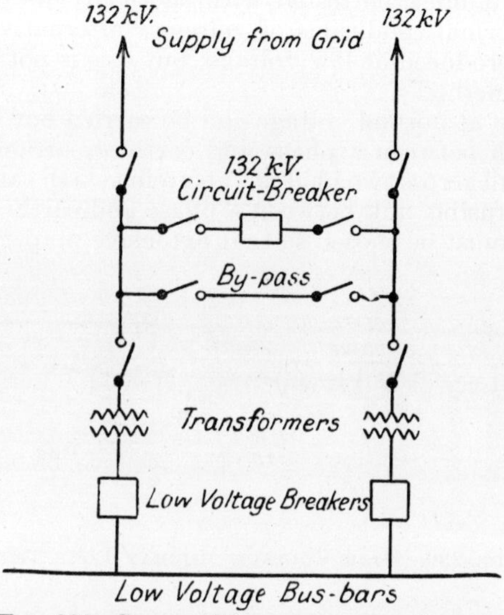

FIG. 228. SINGLE CIRCUIT-BREAKER STATION

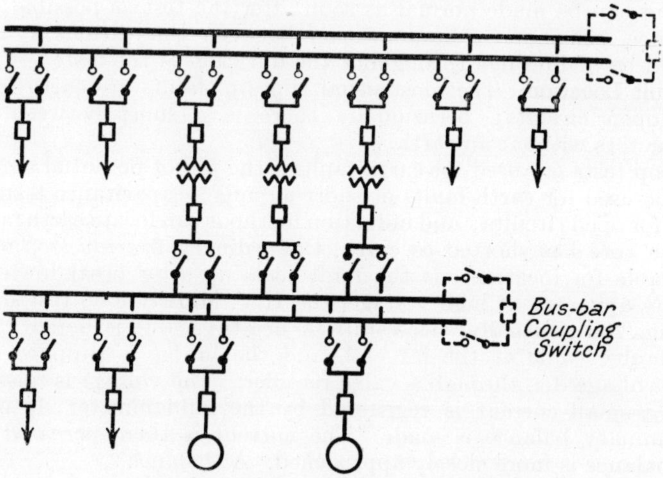

FIG. 229. DUPLICATE BUS-BAR SYSTEM

first installed, it must be tested with short circuits of the various kinds under normal conditions of voltage; alternatively fault currents may be produced at low voltage, but this is not quite so good as the former method.

Staged faults at normal voltage can be carried out by putting an isolating switch between a phase and earth, or arranging a copper conductor to fall across two phases: an arcing earth can be simulated by inserting a fusible link between a phase and earth. The time for a staged fault must be chosen so that extensive preparations against

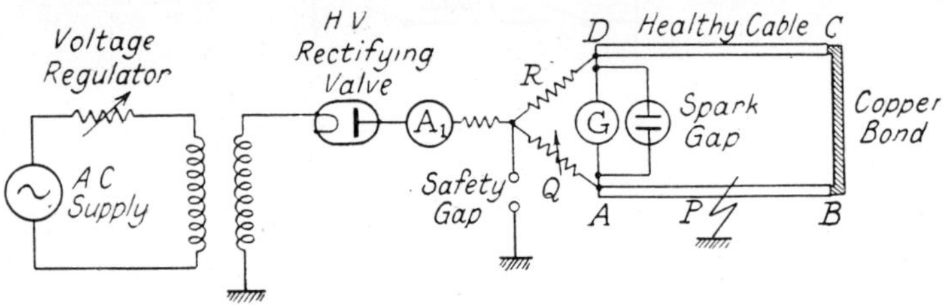

FIG. 230. HIGH VOLTAGE MURRAY LOOP TEST

disturbances can be made; important consumers must be provided with an alternative circuit of supply or they must be advised of the interruption. Considerable organization is required so that the system can be made normal as soon after the test as possible; for instance, telephone communication between the various stations should be maintained throughout the duration of the test.

Fault Location. The most usual types of faults are earth faults and open circuits; occasionally there is a short between two conductors without an earth.

Loop tests are used for earth faults; the fall of potential method can be used for earth faults or short circuits; capacitance tests are used for open circuits; and induction methods can locate earth faults.

If a core has shorted to earth, the ordinary *Murray loop test* is available for location. If the fault seals up after breakdown and is only apparent at high voltage, the H.V. Murray loop test shown in Fig. 230 is usually successful. A healthy cable is bonded with the faulty cable at the far end, and the bridge is supplied with high voltage d.c. through a valve rectifier. The voltage is raised so that a small current is registered by the milliammeter A_1 and a preliminary balance is made. The current is then increased and the balance is more closely approached. At balance

$$\frac{R}{Q} = \frac{\text{Resistance of } (CD + BP)}{\text{Resistance of } AP} = \frac{r_2 CD + r_1 BP}{r_1 AP}$$

$$= \frac{(r_1 + r_2)AB - r_1 AP}{r_1 AP},$$

so that $$AP = \frac{(1 + r_2/r_1)}{(1 + R/Q)} \times AB,$$

where r_1 is the resistance per unit length of AB and r_2 of CD. If equal gauge cables are available, $r_1 = r_2$ and

$$AP = \frac{2}{1 + R/Q} \times AB.$$

Although there is very little voltage on the cable when the fault is present, the voltage will rise to a dangerous value if the fault clears. It is therefore necessary to insulate the controls for the full voltage applied and to use a safety gap. Moreover the galvanometer should be protected by a spark-gap.

Fig. 231 shows the Murray loop test as applied to the location of a short between two cores. The distance AP is found as before.

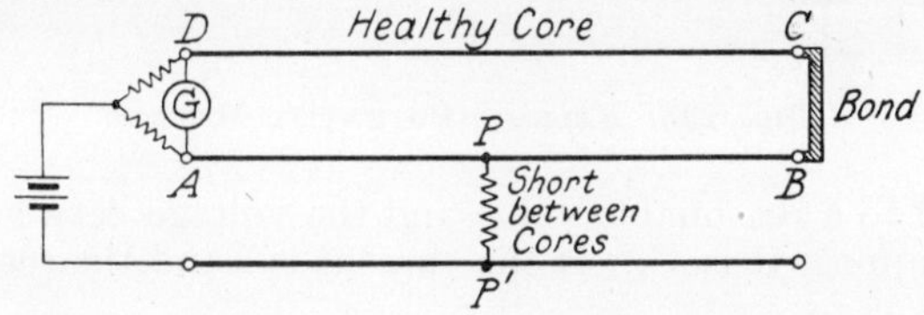

FIG. 231. MURRAY LOOP TEST FOR SHORT BETWEEN CORES

The *Werren (overlap) method* is used to locate a fault involving all conductors, when the conductors are not broken and there is no sound core available. The cores are tested for resistance to earth,

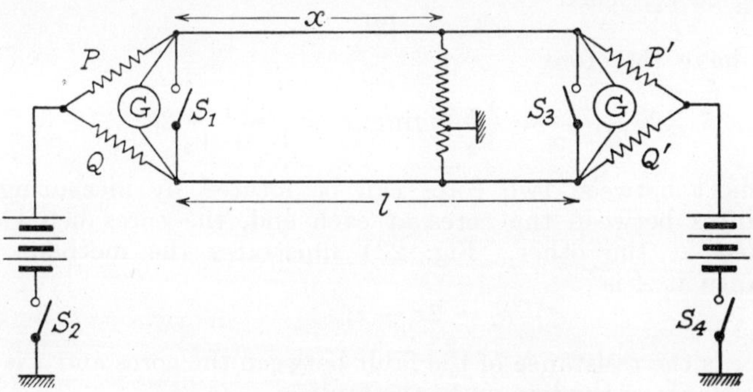

FIG. 232. WERREN OVERLAP METHOD

and the core with the lowest resistance is used in conjunction with that of highest resistance in the way shown in Fig. 232. The loop test is performed from the left end, keys S_2 and S_3 being closed and S_1 and S_4 open; then the test is repeated at the right end, keys

S_1 and S_4 being closed and S_2 and S_3 open. The distance of the fault is given by

$$x = \frac{(P' - Q')}{(P' - Q') + (P - Q)} l.$$

The method fails if the cores have equal resistances to earth, since in that case $P = Q$ and $P' = Q'$.

The fall of potential method is illustrated in Fig. 233. The current

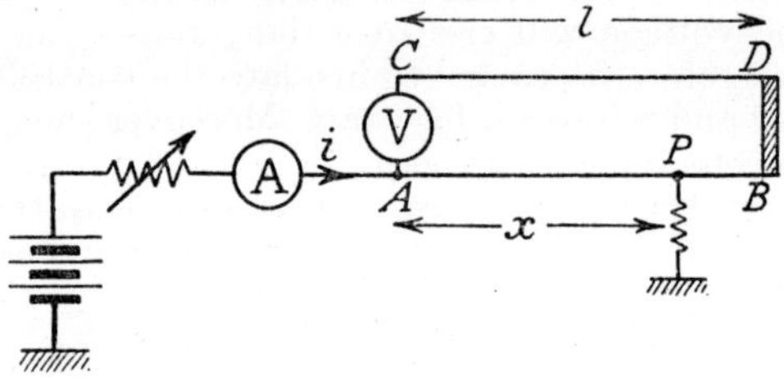

FIG. 233. FALL OF POTENTIAL METHOD

is adjusted to a reasonable value and the voltage between A and C is read; suppose it is V_1. If the current is i and the resistance per unit length is r,

$$V_1 = rxi,$$

since the resistance of the voltmeter is high compared with that of the conductors and all the current flows along AP. Let now the same current be fed in at C instead of A, and let the voltmeter reading be V_2. Then

$$V_2 = r(2l - x)i.$$

We have therefore

$$\frac{x}{2l - x} = \frac{V_1}{V_2} \text{ giving } x = \frac{V_1}{V_1 + V_2} 2l.$$

A short between two cores can be located by measuring the resistances between the cores at each end, the cores being open circuited at the other. Fig. 234 illustrates the method. The resistance at A is

$$r_A = 2x + z,$$

where z is the resistance of the fault between the cores and x is now the *resistance* of the line up to the fault.

Similarly

$$r_B = 2R - 2x + z.$$

Subtraction gives

$$4x - 2R = r_A - r_B$$

or

$$x = \tfrac{1}{2}R + \tfrac{1}{4}(r_A - r_B).$$

Knowing the total resistance and the resistance per unit length we can calculate the distance of the fault.

In the *induction method* of fault locating, current is passed along the faulty cable through an interrupter. A search coil, which is connected to a telephone, is moved along the cable. A buzzing noise will be heard until the search coil passes the fault, when the noise ceases. The presence of a lead sheath or armouring makes this method of detection uncertain. If the position of the fault is known approximately, a very effective method of finding the exact position is to use a telephone diaphragm attached to one end of a long stick, the other end of which has a blunt nail driven into it. The blunt nail is run along the sheath, and when the position of the fault is reached a loud buzzing is heard in the telephone.

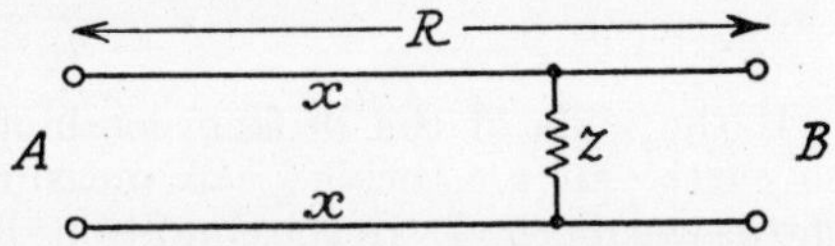

FIG. 234. FAULT BETWEEN CORES BY RESISTANCE MEASUREMENT

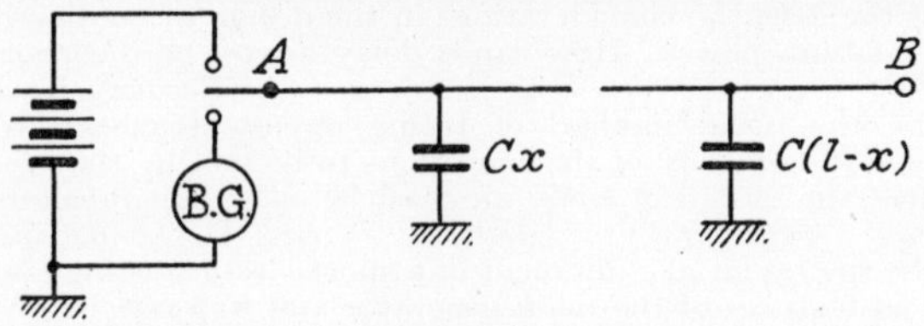

FIG. 235. BALLISTIC METHOD OF FINDING A BREAK

The *capacitance method* of locating a clean break, in which the insulation resistance to earth is high, is indicated in Fig. 235. The cable is charged up to a potential V at end A and then discharged through a ballistic galvanometer $B.G.$ The kick is

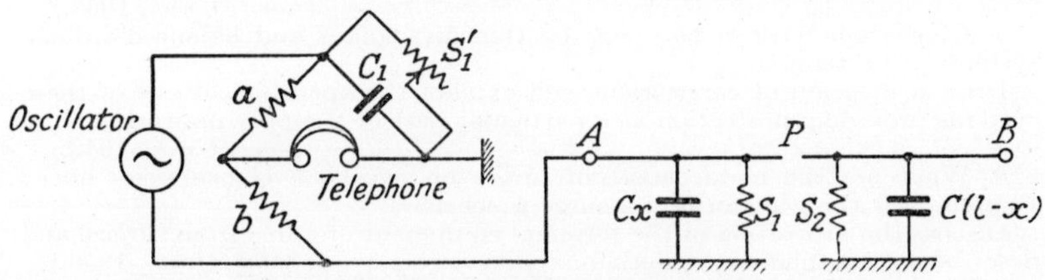

FIG. 236. A.C. BRIDGE METHOD OF FINDING A BREAK

$$d_1 = kCxV,$$

where C is the capacitance per unit length, $AP = x$, and k is a constant of the galvanometer.

The procedure is repeated at end B when the kick is

$$d_2 = kC(l - x)V.$$

By division we have

$$d_1/d_2 = x/(l - x),$$

i.e.

$$x = \frac{d_1}{d_1 + d_2} l.$$

If the ends of the broken conductor have resistances S_1 and S_2 to earth, an a.c. bridge can measure the capacitances. Fig. 236 shows the method. In the condition shown a balance is attained when

$$Cx = (a/b)C_1 \text{ and } S_1 = (b/a)S_1'.$$

$C(l - x)$ can be found from a similar measurement at end B and x can be calculated as before.

EXAMPLES IX

1. Discuss the essential considerations in the design of an air-break circuit-breaker for medium power. How can a delay action be incorporated in the switch itself? (*Faraday House*, 1935.)

2. Describe the usual method of rating circuit-breakers. What circuit features decide the severity of the conditions to be met by the breaker? How do these affect the design of *either* air-blast *or* oil circuit-breakers? (*Lond. Univ.*, 1954.)

3. Describe the essential equipment of a short-circuit testing station. What are the special features of the main generator and its exciter? What records are usually taken in testing a large circuit-breaker? (*Lond. Univ.*, 1954.)

4. Describe the process of arc rupture in (i) an air-blast a.c. circuit-breaker, (ii) an oil-break a.c. circuit-breaker with arc control. Discuss the advantages and disadvantages of air-blast and oil-break circuit-breakers for large powers and high voltage.

Describe, with a sketch, the arrangement of the contacts and operating mechanism of one form of air-blast circuit-breaker. (*Lond. Univ.*, 1953.)

5. Give a comparison between the translay, solkor and balanced-voltage systems of protection.

Draw a diagram of connections and explain the operation of one of these systems providing protection at a particular part of a supply network. (*Lond. Univ.*, 1948.)

6. What are the main causes of surges on overhead transmission lines? Explain how the waveform of a surge is specified.

Discuss the protection of the terminal equipment of a line from surges, and describe one method of protection. (*Lond. Univ.*, 1950.)

7. Describe the principle and construction of a relay whose operation in tripping the circuit-breaker of a faulty feeder depends upon the impedance of the arc between the circuit-breaker and the fault and show how such a relay can be used to give selective protection to a network.

Show that with a system protected in the above manner the minimum possible clearance time for a fault at the far end of a section of a transmission line consisting of a number of sections separated by substations is equal to twice the time taken for the circuit-breakers to operate plus the time taken to start the relay.

What modification to the above system of protection can be made in order to render the operation independent of the resistance of the fault?

8. Discuss the advantages of "end balance" methods of protection and explain the principal difficulties to be overcome in the design and operation of such a system. Describe one system applied to the protection of an alternator, stating what controls must be operated by the protective gear. (*Nat. Cert.*, 1935.)

9. Explain, with the aid of a diagram of connections, the principle of operation of a current-balance system of protection against earth and interphase faults for a star-connected 3-phase generator.

A 3-phase, 2-pole, 11-kV., 10 000-kVA., star-connected alternator has a reactance per phase of 2·4 Ω. and a resistance per phase of 0·25 Ω. and the neutral is earthed through a resistance of 7 Ω. If the machine has current-balance protection, which operates when the out-of-balance current exceeds 20% of full load, determine the percentage of the alternator windings protected against an earth fault on one phase. (*Lond. Univ.*, 1950.)

10. Discuss the protection of an alternator running in parallel with others against reverse power and describe in detail a suitable relay system with inverse time limit for operating the isolating circuit-breaker. (*Faraday House*, 1935.)

11. Describe, with a diagram of connections, a form of biased relay protection for either a generator or a transformer, stating how the correct bias may be given to the protection.

A three-phase transformer of 220/11 000 line volts ratio is connected star/delta and the protecting transformers on the 220 V. side have a current ratio of 600/5. What must be the current ratio of the current transformers on the 11 000 V. side and how should they be connected? (*Lond. Univ.*, 1931.)

12. Write a short account of protection of cables against overload and leakage, giving details of methods employed. (*Faraday House*, 1935.)

13. Describe the principle and draw a connection diagram of the Merz-Price balanced protective system as applied to the protection of a three-phase transmission line.

What are the possible causes of false operation of the above system and what modifications can be made to it to guard against this? (*Lond. Univ.*, 1933.)

14. Explain briefly the difficulties associated with balanced systems and with distance systems of protection for the overhead lines of a power system.

Describe a form of interlock protection designed to overcome these difficulties and state the circumstances in which it can be satisfactorily applied. (*Lond. Univ.*, 1934.)

15. Describe with sketches the construction of an over-current relay of the induction type and explain the action. Describe how the over-current and time-lag settings are adjusted, and sketch a typical current-time characteristic.

What advantages has such a relay over one of the plunger type with time-lag? (*Lond. Univ.*, 1949.)

16. Sketch one form of over-current relay operating on the induction principle. Explain its action, and how the current and time settings are adjusted. Draw a diagram of connections and sketch a typical characteristic curve. Discuss the advantages of a relay of this type over one of the plunger type for the over-current protection of a.c. circuits. (*Lond. Univ.*, 1947.)

17. Describe a method of locating each of the following types of fault on a two-core cable, both ends being available for testing.

(i) A short circuit between two cores, one core being completely broken through but neither core being earthed.

(ii) A complete break in both cores with no earth at the break.

(iii) An earth fault on one core only, which seals itself as soon as the voltage is removed.

In a test to locate a short circuit between the cores of a two-core pilot cable AB having length of 3 miles, the following readings were taken by means of a Wheatstone bridge—

Resistance between cores measured at end A with end B open-circuited $= 10\ \Omega$.

Resistance between cores measured at end B with end A open-circuited $= 8{\cdot}15\ \Omega$.

Resistance per mile of each core $= 1{\cdot}77\ \Omega$.

Determine the position of the fault. (*Lond. Univ.*, 1932.)

18. Give a diagram of connections for, and explain the working of, a system of protection suitable for a 3-phase delta-star distribution transformer. There are three line conductors on each side.

What precautions are advisable in current transformers intended for operating protective gear with the likelihood of heavy through faults, and what is the meaning and purpose of "restraint"? (*Lond. Univ.*, 1948.)

19. Explain the nature and behaviour of the arc in an ordinary circuit-breaker.

Describe briefly methods which are being developed with a view to reducing the duration of the arc in circuit-breakers of both the oil and other types. (*Lond. Univ.*, 1936.)

20. Draw a diagram of connections and explain the principle of operation of a current-balancing system of protecting a 3-phase turbo-alternator against internal faults (earth and inter-phase).

A 3-phase, 20-MVA., 11-kV., star-connected generator is protected by the above system. If the ratio of the current transformers is 1 200/5, the minimum operating current of the relay is 0·75 A. and the neutral-point earthing resistance is 6 Ω., calculate the percentage of each phase of the stator winding which is unprotected against earth faults when the machine is operating at normal voltage. (*Lond. Univ.*, 1947.)

21. A 5 000 kVA., 6 600 V., star-connected alternator has a synchronous reactance of 2 Ω. per phase and 0·5 Ω. resistance. It is protected by a Merz-Price balanced-current system which operates when the out-of-balance current exceeds 30 per cent of full load current. Describe this method of protection and determine what proportion of the alternator winding is unprotected if the star point is earthed through a resistance of 6·5 Ω. (*Lond. Univ.*, 1937.)

22. Explain the terms "symmetrical breaking current," "asymmetrical breaking current," "making current," as applied to oil circuit-breakers, and show how these currents are determined from oscillograms taken during short-circuit tests on a 3-phase circuit-breaker. Explain why at least one of such oscillograms of current in a 3-phase short-circuit test shows asymmetry. What is meant by the rated MVA. breaking capacity of a 3-phase circuit-breaker? (*Lond. Univ.*, 1947.)

23. Describe the "star" or "tie-bar" method of interconnecting bus-bar sections and compare it with other bus-bar arrangements.

A generating station contains three bus-bar sections, to each of which is connected a generating unit of 60 000 kVA. having 15% leakage reactance, the bus-bar reactors having a reactance of 10%. Calculate the maximum kVA. fed into a fault and also the maximum kVA. if the number of bus-bar sections is increased to infinity. (*Lond. Univ.*, 1949.)

24. Explain why it is necessary to provide surge protection in the terminal transformers of an overhead transmission line, and briefly describe two methods whereby such protection is obtained.

An overhead line is connected to a transformer by a short cable. If a 50-kV. surge travels along the line towards the transformer, estimate the voltage of the surge reaching the transformer. The inductance per mile of cable and line are 0·56 mH. and 2·8 mH. respectively while the corresponding capacitances per mile are 0·4 μF. and 0·01 μF. Deduce any formula used other than surge-impedance $= \sqrt{L/C}$. (*Lond. Univ.*, 1949.)

CHAPTER X

VOLTAGE TRANSIENTS AND LINE SURGES

Introduction. There are various ways in which a transmission line may experience voltages greater than the working value, and it is necessary to provide protective apparatus to prevent or minimize the destruction of the plant. Internal causes producing a voltage rise are (1) resonance, (2) switching operations, (3) insulation failure, and (4) arcing earths: a very important external cause is lightning.

Resonance. The effect of resonance is most easily understood by considering the voltage at the end of a lightly loaded cable of short length. The alternator and transformers may be represented by their leakage inductance L, and the cable by a capacitance C. The system is then as shown in Fig. 237, where R represents the resistance of the alternator winding, transformers and cable, and r the resistive load. The total impedance of the circuit is

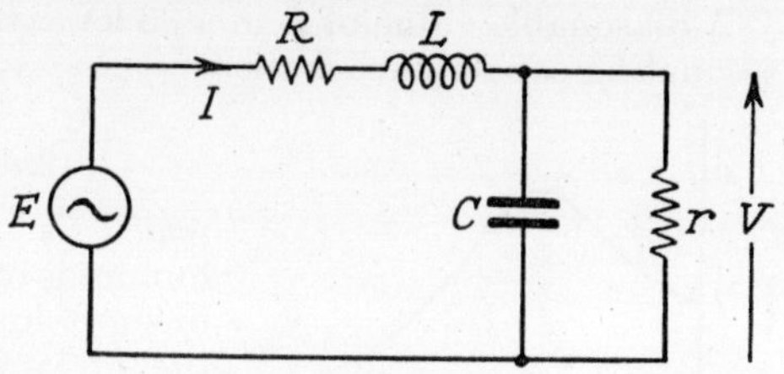

FIG. 237. RESONANCE

$$Z = R + j\omega L + \frac{(1/j\omega C)r}{1/j\omega C + r} = R + j\omega L + \frac{r}{1 + j\omega Cr},$$

the current is

$$I = E/Z,$$

and the voltage on the cable is

$$V = I \times r/(1 + j\omega Cr),$$

since the latter expression represents the impedance of the parallel combination of C and r. Substituting for I in terms of E we get

$$V/E = \left(\frac{r}{1 + j\omega Cr}\right) \div \left(R + j\omega L + \frac{r}{1 + j\omega Cr}\right)$$

$$= \frac{1}{1 + (R + j\omega L)(1/r + j\omega C)}$$

$$= \frac{1}{(1 - \omega^2 LC + R/r) + j\omega(L/r + CR)}$$

The magnitude of (V/E) is

$$|V/E| = [(1 - \omega^2 LC + R/r)^2 + \omega^2(L/r + CR)^2]^{-\frac{1}{2}} \quad . \quad (112)$$

Let us consider the case of an unloaded line first. In this case $r = \infty$, so that

$$|V/E| = [(1 - \omega^2LC)^2 + \omega^2C^2R^2]^{-\frac{1}{2}} \quad . \quad . \quad (112a)$$

If we consider that C can vary, by the insertion of different lengths of cable, $|V/E|$ varies in the manner shown in Fig. 238. The maximum value occurs when

$$C = \frac{1}{\omega^2L + R^2/L} = \frac{1}{\omega^2L(1 + R^2/\omega^2L^2)} \simeq \frac{1}{\omega^2L},$$

$$\text{when} \quad |V/E| = \frac{1}{\omega CR\sqrt{(1 + R^2/\omega^2L^2)}} \simeq \frac{1}{\omega CR} \simeq \frac{\omega L}{R}.$$

A reasonable value of L in a 33 kV. system is 0·05 henry, and the resonating capacitance is then

$$C \simeq \frac{1}{(2\pi \,.\, 50)^2 \times 0{\cdot}05} = 202 \ \mu\text{F.},$$

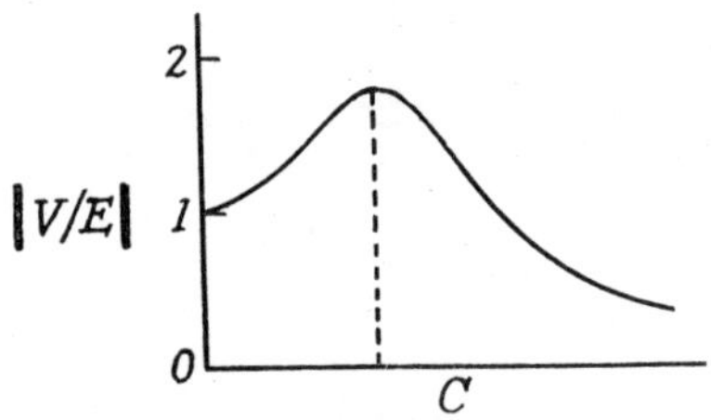

FIG. 238. RESONANCE

which is the capacitance of some hundreds of miles of cable. Resonance in short lines will thus never occur at the fundamental frequency. If we consider the fifth harmonic, which is often present to the extent of 2 or 3 per cent, we see that resonance can occur. The capacitance required is

$$C = \frac{1}{(2\pi \,.\, 250)^2 \times 0{\cdot}05} = 8{\cdot}1 \ \mu\text{F.},$$

which is provided by a cable of length about 28 miles. If we assume a 10 per cent harmonic, the value of V_5 is

$$|V_5| = |E_5| \times 2\pi \,.\, 250L/R = 0{\cdot}10\,|E_1| \times 2\pi \,.\, 250L/R,$$

where E_1 is the fundamental, and E_5 the fifth harmonic. If we take $R = 5$, we find that

$$|V_5| = 1{\cdot}57\,|E_1|,$$

so that the fundamental voltage of $E_1 = 33$ kV. has a fifth harmonic of magnitude 52 kV. (r.m.s.). The peak value between phases may then be $\sqrt{2} \times 85$ kV. in place of the normal value of $\sqrt{2} \times 33$ kV.

The effect of a load is seen by comparing equations (112a) and (112). It is seen that the term (R/r) is an additive constant in the first term on the right-hand side of the equations and alters the condition for the neutralization of reactance, whilst the term (L/r) causes a considerable damping of the resonance. Let us take $r = 200$ ohms, which corresponds to a load of 5 000 kW. Then with the values of L, C, and E_5 taken above, we find that

$$1 - \omega^2LC + R/r \simeq 5/200 = 0{\cdot}025$$

$$\text{and} \quad \omega(L/r + CR) = \omega CR(1 + L/CRr) = 7{\cdot}2\omega CR = 0{\cdot}46.$$

The first term is thus negligible compared with the second, so that we may take

$$|V_5/E_5| \simeq \frac{1}{\omega(L/r + CR)} = \frac{1}{7{\cdot}2\omega CR} \simeq \frac{\omega L}{7{\cdot}2R},$$

so that V_5 is reduced by the factor 7·2 and has a magnitude of $52 \div 7{\cdot}2 = 7{\cdot}2$ kV. The resonance voltage has been therefore effectively damped by the load.

Switching. A switching operation produces a sudden change in the circuit conditions, and is accompanied by a *transient state* which leads from the earlier to the later steady (a.c.) states. The behaviour

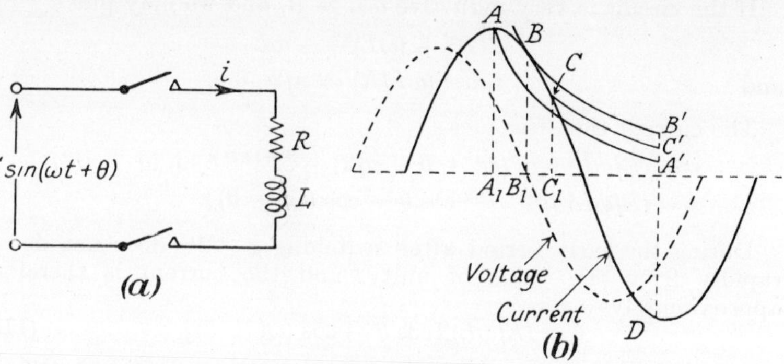

FIG. 239. SWITCHING-IN AN INDUCTIVE RESISTANCE

of the system can be explained with exactness only by means of *travelling waves*, which will be explained later; but in short systems the behaviour is sufficiently well explained if we consider the circuit to be composed of lumped resistances, inductances, and capacitances. The method used is that given on pages 214–16, where we showed that a current of twice the normal peak value can be obtained when an alternator is short-circuited.

Transients in Circuits with Lumped Constants. There are two interesting cases which we will solve, the switching-in of an inductive load and the switching-in of an open-circuited line.

Fig. 239 (*a*) represents the switching-in of a load of inductance L and resistance R. The equation for the circuit is

$$L(di/dt) + Ri = E \sin(\omega t + \theta),$$

of which the solution is (see page 215)

$$i = A\varepsilon^{-(R/L)t} + \frac{E}{\sqrt{[R^2 + (\omega L)^2]}} \sin\left(\omega t + \theta - \tan^{-1}\frac{\omega L}{R}\right)$$

The constant A is determined by the fact that $i = 0$ at the time $t = 0$, so that we find that

$$i = -\frac{E}{\sqrt{[R^2 + (\omega L)^2]}} \sin\left(\theta - \tan^{-1}\frac{\omega L}{R}\right)\varepsilon^{-(R/L)t}$$

$$+ \frac{E}{\sqrt{[R^2 + (\omega L)^2]}} \sin\left(\omega t + \theta - \tan^{-1}\frac{\omega L}{R}\right) \quad . \quad (103)$$

The first term represents the *transient current* which decays exponentially. It has an initial value equal and opposite to that of the a.c. component at the time of switching (so that the initial current is zero).

If the circuit is very inductive $\omega L \gg R$, and we may put

$$\sqrt{[R^2 + (\omega L)^2]} \simeq \omega L$$

and

$$\tan^{-1}(\omega L/R) = \pi/2.$$

The current then becomes

$$i = (E/\omega L)\,[\sin(\omega t + \theta - \pi/2) - \varepsilon^{-(R/L)t}\sin(\theta - \pi/2)]$$

$$= (E/\omega L)\,[\varepsilon^{-(R/L)t}\cos\theta - \cos(\omega t + \theta)].$$

During the early period after switching $\varepsilon^{-(R/L)t}$ does not decay rapidly from the value of unity, and the current is therefore approximately

$$i = (E/\omega L)\,[\cos\theta - \cos(\omega t + \theta)], \quad (113)$$

and varies between the values of $(E/\omega L)\,[\cos\theta - 1]$ and $(E/\omega L)$ $[\cos\theta + 1]$. The peak value is thus

$$(E/\omega L)\,(1 + |\cos\theta|),$$

i.e. $(1 + |\cos\theta|)$ times the normal peak value. The maximum peak is thus obtained when $\theta = 0$ and is twice the normal peak. This condition occurs when the circuit is closed at zero voltage and the current is

$$i = (E/\omega L)\,[1 - \cos\omega t], \quad . \quad . \quad . \quad (114)$$

which varies between zero (at $t = 0$) and $(2E/\omega L)$ (at $t = \pi/\omega$).

It can be shown that, whatever the power factor of the circuit may be, the maximum "doubling" effect is obtained when the circuit is closed at zero voltage. Fig. 239 (*b*) shows the normal sinusoidal current. If the circuit is switched in at A the transient has initial amplitude AA_1, if at B the amplitude BB_1, and if at C the amplitude CC_1. The transients corresponding to these switching points are represented by the curves AA', BB', CC' and must be subtracted from the sine wave. The total current at any instant is thus the vertical distance between the sine wave and the appropriate transient curve. It is clear that if the circuit is switched in at position B the current is greater than if switched at any other position, since the transient curves have the same time factor $\varepsilon^{-Rt/L}$ and have the same decay rate. The topmost curve is clearly seen to be that whose slope at the point of contact with the sine wave is equal to the slope of the sine wave. Let us consider this as the time $t = 0$. Equating slopes we get

$$\left[-\frac{R}{L}\sin\left(\theta - \tan^{-1}\frac{\omega L}{R}\right)\varepsilon^{-(R/L)t}\right]_{t=0}$$

$$= \left[\omega\cos\left(\omega t + \theta - \tan^{-1}\frac{\omega L}{R}\right)\right]_{t=0}$$

i.e. $\quad -(\omega L/R) = \tan[\theta - \tan^{-1}(\omega L/R)],$

which gives $\theta = 0$ or π. If $\theta = 0$ or π the voltage is zero at $t = 0$, i.e. the maximum doubling occurs if the circuit is closed at the instant of zero voltage.

Suppose the load has a power factor of 0·8 lagging,

$$\omega L/R = 0{\cdot}6/0{\cdot}8 = 0{\cdot}75.$$

If the circuit is closed at zero voltage the current is

$$i = \frac{E}{\sqrt{[R^2 + (\omega L)^2]}}\left[\sin\left(\omega t - \tan^{-1}\frac{\omega L}{R}\right) + \sin\left(\tan^{-1}\frac{\omega L}{R}\right)\varepsilon^{-(R/L)t}\right]$$

$$= \frac{E}{\sqrt{[R^2 + (\omega L)^2]}}[\sin(\omega t - 36^\circ\,52') + 0{\cdot}6\varepsilon^{-1{\cdot}33\omega t}].$$

For this case the voltage and currents in Fig. 239 (*b*) must be reversed. The maximum current occurs when $di/dt = 0$, i.e. when

$$\cos(\omega t - 36^\circ\,52') = 0{\cdot}6 \times 1{\cdot}33\varepsilon^{-1{\cdot}33\omega t} = 0{\cdot}8\varepsilon^{-1{\cdot}33\omega t}.$$

Let $\omega t - 36^\circ\,52' = \phi$, so that

$$\omega t = \phi + 36^\circ\,52' = \phi + 0{\cdot}64 \text{ radians}.$$

The equation becomes

$$\varepsilon^{1{\cdot}33\phi}\cos\phi = 0{\cdot}8\varepsilon^{-0{\cdot}853} = 0{\cdot}34.$$

ϕ . . .	1	1·5	1·54
$\varepsilon^{1{\cdot}33\phi}$ . .	3·78	7·39	7·76
$\cos\phi$ . .	0·540	0·0707	0·0308
$\varepsilon^{1{\cdot}33\phi}\cos\phi$.	2·04	0·52	0·24

We may take $\phi = 1{\cdot}53$ radians $= 87° \, 40'$, so that

$$i_{max} = \frac{E}{\sqrt{[R^2 + (\omega L)^2]}}[\sin 87° \, 40' + 0{\cdot}6\varepsilon^{-1.33\omega t}]$$
$$= \frac{E}{\sqrt{[R^2 + (\omega L)^2]}}[1 + 0{\cdot}34\varepsilon^{-1.33 \times 1.533}]$$
$$= \frac{E}{\sqrt{[R^2 + (\omega L)^2]}}[1{\cdot}044].$$

and the peak does not exceed the normal value by more than 4·5 per cent.

Fig. 240 represents the switching-in of an open-circuited line; we assume for simplicity that the e.m.f. is constant and equal to E,

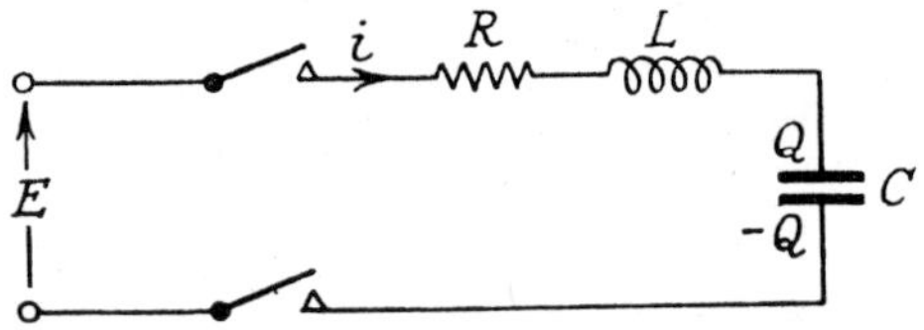

FIG. 240. SWITCHING-IN AN OPEN-CIRCUITED LINE

but the same method is applicable for an a.c. case. The equation for the current is

$$L(di/dt) + Ri + Q/C = E$$

where $i = dQ/dt$.

The voltage at the end of the line is $V = Q/C$. Substituting for i in terms of Q we get

$$L(d^2Q/dt^2) + R(dQ/dt) + Q/C = E,$$

the solution of which is

$$Q = CE + \varepsilon^{-(R/2L)t}(A \cos \alpha t + B \sin \alpha t),$$

where $\alpha = \sqrt{[(1/LC) - (R^2/4L^2)]}$, and A and B are constants which are determined by the initial conditions. At the instant, $t = 0$, of switching-in Q and i are zero. These conditions give

$$A = -CE \text{ and } B = AR/2L\alpha,$$

so that
$$\left.\begin{aligned} V &= Q/C = E - E\varepsilon^{-(R/2L)t}[\cos \alpha t + (R/2L\alpha) \sin \alpha t] \\ &= E - E[1/\alpha\sqrt{(LC)}]\varepsilon^{-(R/2L)t} \cos [\alpha t - \cos^{-1}(\alpha\sqrt{(LC)})], \\ \text{and } i &= dQ/dt = (E/\alpha L)\,\varepsilon^{-(R/2L)t} \sin \alpha t. \end{aligned}\right\} \quad (115)$$

If the resistance is negligible the voltage and current reduce to

$$\left.\begin{aligned} V &= E - E \cos [t/\sqrt{(LC)}] \\ \text{and} \quad i &= [E\sqrt{(C/L)}] \sin [t/\sqrt{(LC)}], \end{aligned}\right\} \quad . \quad . \quad (115a)$$

since $\alpha = 1/\sqrt{(LC)}$ in this case.

The voltage in this case oscillates sinusoidally between 0 and $2E$, whilst the current is a sine wave of peak value $E\sqrt{(C/L)}$. Fig. 241 shows the voltage and current for the case of no resistance (curves A) and for some resistance (curves B).

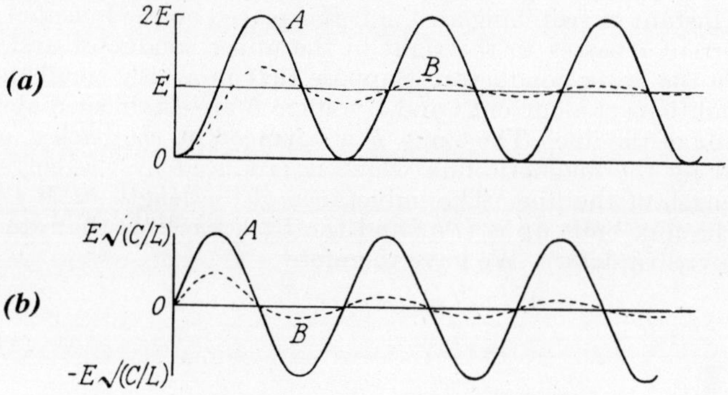

FIG. 241. OPEN-CIRCUITED LINE
(a) Voltage. (b) Current.

Switching Surges. We have found that when an e.m.f. E is switched on to a line, which we replaced by an inductance L and a capacitance C, the voltage oscillates sinusoidally between 0 and $2E$ whilst the current varies similarly between $-E\sqrt{(C/L)}$ and

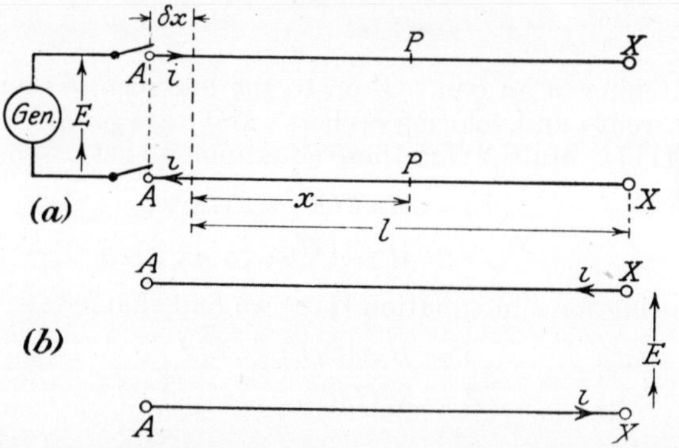

FIG. 242. SWITCHING SURGE ON OPEN-CIRCUITED LINE

$+E\sqrt{(C/L)}$. It is clear that this does not represent the state of affairs with exactness, for any transfer of energy must travel with a velocity less than that of light, so that the far end of a line is unaffected for the finite time that it takes the energy wave to reach it. It therefore follows that part of the line may be passing current and maintaining a voltage whilst a further part has neither current nor voltage. We will consider the case of the switching-in of an unloaded line from this point of view, and will make the simplifying assumption that resistance and leakage are negligible. Fig. 242 (*a*) shows the arrangement; the line has inductance L and capacitance C per unit length and is open at the far end XX.

At the instant of switching an e.m.f. E is placed on the line at AA, and a current i passes to the right in the upper conductor and to the left in the lower conductor. Suppose that in a very small time δt the conditions of a current i and a voltage E are established along a length δx of the line. The e.m.f. E is balanced by the back e.m.f. generated by the magnetic flux which is produced by the current in this length of the line. The inductance of the length δx is $L\delta x$, so that the flux built up is $iL\delta x$ and the back e.m.f. is the rate of build-up, viz. $iL(\delta x/\delta t)$. We have therefore

$$\begin{aligned} E &= iL(\delta x/\delta t) \\ &= iLv, \qquad \qquad (116) \end{aligned}$$

where v is the velocity of the wave.

The current i carries a charge $i\delta t$ in the time δt, and this charge remains on the line to charge it up to the potential E. Since the capacitance of the length δx of the line is $C\delta x$, its charge is $EC\delta x$. We have therefore

$$i\delta t = EC\delta x,$$

or

$$\begin{aligned} i &= EC(\delta x/\delta t) \\ &= ECv. \qquad \qquad (117) \end{aligned}$$

The switching of an e.m.f. E on to the line results therefore in a wave of current i and velocity v where i and v are given by equations (116) and (117). Multiplying these equations we get

$$Ei = iLvECv = EiLCv^2,$$

so that

$$v = 1/\sqrt{(LC)}. \qquad \qquad (118)$$

Substituting for v in equation (118) we find that

$$\left.\begin{aligned} i &= E\sqrt{(C/L)} = E/Z \\ \text{where} \quad Z &= \sqrt{(L/C)}. \end{aligned}\right\} \qquad \qquad (119)$$

Z is called the *surge impedance* or *natural impedance* of the line; it is a pure resistance for a line without resistance or leakage, and has a value of 400 to 600 ohms for an overhead line and 40 to 60 ohms for a cable. The velocity of the wave on an overhead line is approximately equal to the velocity of light, for

$$\begin{aligned} L &= [1 + 4 \operatorname{logh} (D/r)] \times 10^{-9} \text{ H. per cm.} \\ &\simeq 4 \operatorname{logh} (D/r) \times 10^{-9} \text{ H. per cm.} \end{aligned}$$

and $$C \simeq \frac{1}{4 \text{ logh } (D/r)} \text{ cm. per cm.}$$

$$= \frac{1}{9 \times 10^{11}} \frac{1}{4 \text{ logh } (D/r)} \text{ F. per cm.}$$

so that $$v = \frac{1}{\sqrt{(LC)}} = \sqrt{(10^9 \times 9 \times 10^{11})} \text{ cm. per sec.}$$

$$= 3 \times 10^{10} \text{ cm. per sec.}$$

$$= c,$$

the velocity of light.

The velocity in a cable is $c/\sqrt{\varepsilon}$, where ε is the dielectric constant. v is thus about 186 000 miles per sec. on an overhead line, and $186\,000 \div \sqrt{3{\cdot}6} = 98\,000$ miles per sec. in a cable.

We have shown that a wave of voltage E and current $i = E/Z$, travels towards the right along the line with a velocity v. Such a wave is called a *pure travelling wave*. At any part PP of the line nothing happens until the wave reaches it (at time $t = x/v$), and then the current jumps from zero to i and the voltage from zero to E. This goes on until the wave reaches the open end of the line (XX) at time $t = l/v$. When the wave reaches XX, the current there is i; but this current has no capacitance to charge up, so that it must cease immediately.

The open end of the line has thus a disturbing influence which neutralizes the current completely; this disturbing influence then travels back along the line towards AA, and can therefore be represented by a pure travelling wave moving towards the left and carrying a current $-i$. A travelling wave must possess a voltage and a current whose ratio is Z, the surge impedance of the line. If the current is to the left in the upper conductor and to the right in the lower from the end XX, it is seen from Fig. 242 (*b*) that the voltage is E, i.e. the upper conductor is E volts above the lower. For if an e.m.f. E were switched in at XX the current would be in the direction required and as shown. The disturbing effect of the open end of the line is thus to introduce another pure travelling wave, which moves to the left with velocity v, has a voltage E, and a current i in the opposite direction to that previously flowing. It is convenient to consider a current to the right in the upper conductor as positive, and a current to the left in the upper conductor as negative. The new travelling wave, which moves to the left, has therefore a voltage E and a current $-E/Z$. In general, a wave (E_1, i_1) moving to the right satisfies the relation

$$i_1 = E_1/Z, \qquad (120)$$

while a wave (E_2, i_2) moving to the left satisfies the relation

$$i_2 = -E_2/Z. \qquad (121)$$

The result of the new travelling wave is to establish an extra voltage E at any point of the line that it passes so that a resulting voltage of $2E$ is produced, whilst the current is neutralized. Thus the conditions at the point PP of the line are such that its voltage and current values are $(0, 0)$ from $t = 0$ until $t = x/v$, (E, i) from $t = x/v$ until $t = (2l - x)/v$, and $(2E, 0)$ from $t = (2l - x)/v$ onwards. This goes on until the disturbing wave reaches the generator at AA at time $t = 2l/v$; by this time the line has voltage $2E$ and zero current at every point. When this instant occurs, the voltage at the generator terminals is $2E$. But the generator is supposed to maintain a voltage E at AA, so that another wave is called into play to reduce $2E$ to E. This wave must therefore have potential $-E$, and as it moves to the right it must have a current $-E/Z = -i$ by equation (120). As this wave travels from AA to XX it reduces the voltage to E and produces a current $-i$. Thus the voltage drops from $2E$ to E at the point PP at time $t = (2l + x)/v$ and the current jumps from zero to $-i$. When this third wave reaches XX it establishes a current $-i$ there, which must be neutralized by a fourth wave travelling to the left with current $+i$, and voltage $-iZ = -E$ by equation (121). As this fourth wave travels from XX to AA, the current vanishes at any point it passes, and the voltage becomes $E - E = 0$ at every point. The line is thus completely discharged and has no current, and a complete cycle of travelling waves has been finished. If the line were completely without resistance and leakage, this cycle would be repeated indefinitely. The current at AA, the current and voltage at the mid-point of the line, and the voltage at XX are shown in Fig. 243.

It is interesting and instructive to compare the exact description of the switching phenomenon with the approximate description derived by considering the line as composed of a lumped inductance and capacitance. In both descriptions the potential at any point varies between 0 and $2E$; but in the exact description the time-variation of the potential depends greatly upon the point considered (see Fig. 243, last two curves) and changes in jumps, whilst in the approximate method the time-variation is sinusoidal. The current varies between $+i$ and $-i$ in both cases, where $i = E/Z$ and $Z = \sqrt{(L/C)}$; but again the time-variations are radically different. There is one further difference, viz. the periodicity of the two descriptions. In the approximate method the frequency is $1/2\pi\sqrt{(LC)}$; whilst in the exact method a complete cycle is of duration $4l/v$, so that the frequency is

$$v/4l = 1/4l\sqrt{(LC)} = 1/4\sqrt{(L_0C_0)},$$

where L_0 and C_0 are the total inductance and capacitance. The difference is therefore in replacing the 2π by 4.

Before entering on a somewhat more general description of

travelling waves, it is worth while considering the energy properties of the simple waves we have described.

Energy Considerations. A wave of voltage E and current i carries a power of Ei. A simple travelling wave therefore transmits a power Ei with a velocity v. As this wave travels it establishes a magnetic field with energy $\frac{1}{2}Li^2$ per cm. length of the line and an electrostatic field with energy $\frac{1}{2}CE^2$ per cm. length. From equations

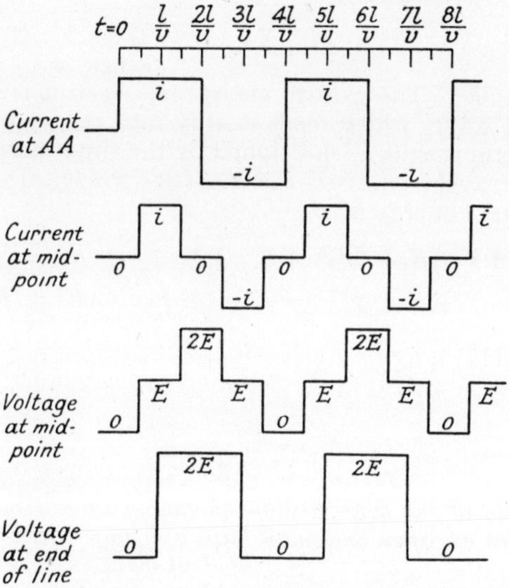

FIG 243. CURRENT AND VOLTAGE IN SWITCHING SURGE

(116) and (117) it is seen that the magnetic and electrostatic energies delivered by a simple wave are equal, for

$$\begin{aligned}\tfrac{1}{2}Li^2 &= \tfrac{1}{2}(iLv)\,(i/v) = \tfrac{1}{2}(Ei/v)\\ &= \tfrac{1}{2}(E/v)\,(ECv) = \tfrac{1}{2}CE^2.\end{aligned}$$

Each of these is equal to $\frac{1}{2}Ei/v$, which is half the total energy delivered by the wave in the time it passes along the part of the line. The energy of the wave is thus half absorbed as magnetic and half as electrostatic energy.

When a pure travelling wave of voltage E and current i moves to the right and meets an open-circuited line, we said that the disturbing effect of the open end is to bring into action a reflected wave of voltage E and current $-i$ (travelling to the left). It will be seen that this is consistent with the conservation of energy, and is in fact demanded by this principle. For suppose that the disturbance engenders a wave with a current $-i$, the latter being required in order to neutralize the current at the open end of the line. Suppose that the voltage attached to this wave is E'. When the wave has travelled a distance XY (Fig. 244), the voltage over XY is $E + E'$ whilst the current is zero. The energy associated with this part of the line is now

$$\tfrac{1}{2}C \,.\, XY \,.\, (E + E')^2,$$

whereas previously it was

$$\tfrac{1}{2}C \,.\, XY \,.\, E^2 + \tfrac{1}{2}L \,.\, XY \,.\, i^2 = C \,.\, XY \,.\, E^2,$$

since $\tfrac{1}{2}Li^2 = \tfrac{1}{2}CE^2$. The gain in energy has been derived from the first (incident) wave, which feeds energy into the section XY at a rate Ei; the gain is thus Ei multiplied by the time that the reflected wave takes to travel from X to Y, viz. $Ei \times (XY/v)$. If the principle of conservation of energy is to hold, then

$$\tfrac{1}{2}C \,.\, XY \,.\, (E + E')^2 = C \,.\, XY \,.\, E^2 + Ei(XY/v),$$

or

$$\tfrac{1}{2}(E + E')^2 = E^2 + Ei/Cv = E^2 + E^2$$

(by equation (117)),

so that $$(E + E')^2 = 4E^2,$$

i.e. $$E' = E.$$

The principle of the conservation of energy thus demands that the reflected wave at an open end shall have a voltage equal to that of the incident wave; the current is equal and opposite to that of the incident wave since no current can leave the open end.

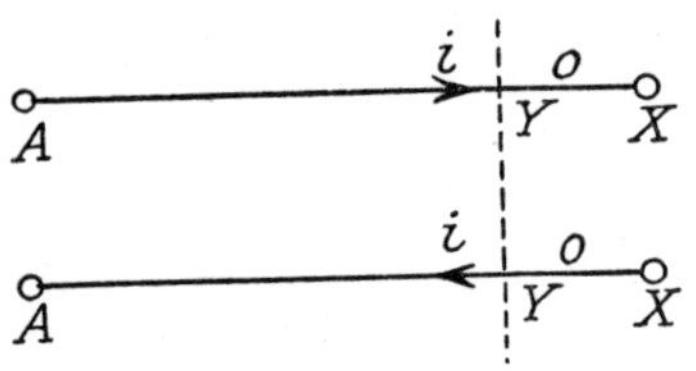

FIG. 244. ENERGY CONSIDERATIONS IN SURGES

Sudden Interruption of a Circuit. We have described in full the surge that takes place when a generator is suddenly switched on to a line that is open at the far end. The phenomenon that takes place when the far end is terminated by a finite impedance will be considered in the section on the reflection and transmission of travelling waves. The method employed above serves to describe the events that occur when a current in a circuit is suddenly interrupted, by the action of a circuit-breaker, say.

Suppose that a circuit has a current i, which is suddenly interrupted by the breakers S, S (Fig. 245). The disturbance produces two travelling waves moving from S, S to the right and to the left. The wave travelling to the right has a current $-\,i$, and must therefore have a voltage $-E$, where $E = iZ$; line A is therefore $-E$ volts above line B. The wave travelling to the left has a current $-i$, and must therefore have a voltage $+E$, where $E = iZ$; C is therefore $+E$ volts above D. These waves progress in a normal manner until they meet abrupt changes in the line, when they are reflected and transmitted in the ways described later. It should be

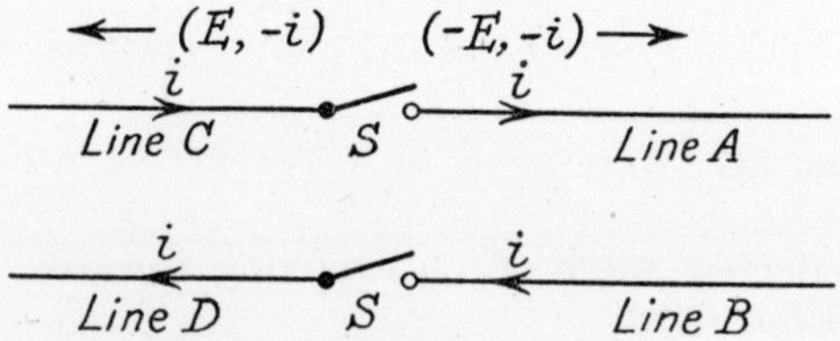

FIG. 245. SUDDEN INTERRUPTION OF A CIRCUIT

noted that if only one break is made, so that B and D are always commoned, the voltage between A and C is $2E$.

The surge voltage E is superposed on the normal voltage in that part of the line which remains connected to the generator.

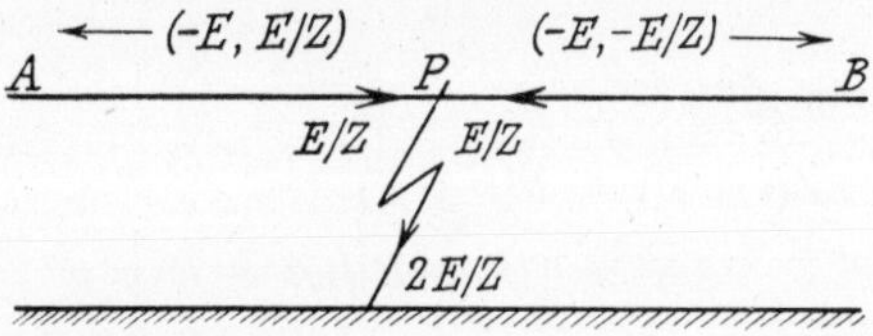

FIG. 246. SURGES DUE TO A FAILURE OF INSULATION

Insulation Failure or Earthing of a Line. Suppose that a line AB, at potential E, is earthed at a point P. The effect of earthing is to introduce a voltage $-E$ at P, and two equal waves of voltage $-E$ travel along PA and PB. The wave travelling to the right has a current of $-E/Z$, and that to the left $+E/Z$. Both these currents pass through P to earth, so that the current to earth is $2E/Z$. Fig. 246 shows the waves and currents in the system.

As these waves travel to the ends of the line they reduce the voltage to zero; and when they reach the open ends, reflected waves are set up which reduce the voltage to $E - E - E$, i.e. $-E$, and the current is neutralized. When the reflected waves reach P, the portions of the line along which they have travelled will be charged to $-E$. The current at P can be reversed by a flashover in the opposite direction, and the result is a periodic flashover with reversals of potential on the line and currents at P until the stored energy is dissipated by damping.

Reflection and Transmission of Travelling Waves. Suppose that a travelling wave (E, i) moves along a line of surge impedance Z and meets a termination of resistance R (Fig. 247). If R is not equal to Z, the end of the line cannot have the voltage E and current i since $E/i = Z$. There is therefore a disturbance which

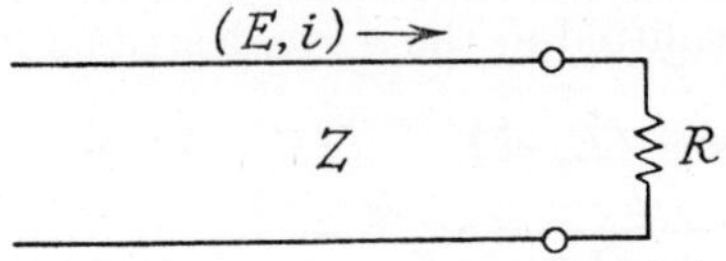

FIG. 247. REFLECTION OF A TRAVELLING WAVE

produces a reflected wave (E', i') moving towards the left. The following relations exist.

$$E = iZ,$$

$$E' = -i'Z.$$

The total voltage at the end is $E + E'$ and the total current is $i + i'$, so that

$$E + E' = R(i + i').$$

These equations give

$$Z(i - i') = R(i + i')$$

so that

$$\left.\begin{aligned} i' &= [(Z - R)/(Z + R)]i \\ \text{and} \quad E' &= -i'Z = [(R - Z)/(Z + R)]E. \end{aligned}\right\} \qquad (122)$$

The total current and voltage are

$$\left.\begin{aligned} i + i' &= [2Z/(Z + R)]i \\ \text{and} \quad E + E' &= [2R/(Z + R)]E. \end{aligned}\right\} \quad . \quad . \quad . \quad (123)$$

If the line is open at the end, $R = \infty$ so that the total current is zero and the total voltage is $2E$, as found before.

If the line is shorted at the end, $R = 0$ so that the current is doubled and the voltage drops to zero.

The case for a finite resistance termination is given by equations (122) and (123). When the termination is not a pure resistance, the result is still given by these equations but they must be evaluated by the operational calculus.

Junction of Two Lines. Fig. 248 shows the case of two lines of surge impedances Z_A and Z_B. A wave (E, i) travels along the left-hand line and meets the junction. So far as a travelling wave is concerned the right-hand line can be considered to have an impedance Z_B, so that the case is the same as that shown in Fig. 247, provided Z is replaced by Z_A and R by Z_B. The reflected wave is thus (E', i') where

$$\left.\begin{aligned} i' &= [(Z_A - Z_B)/(Z_A + Z_B)]i \\ \text{and} \quad E' &= [(Z_B - Z_A)/(Z_A + Z_B)]i. \end{aligned}\right\} \quad . \quad (122a)$$

The transmitted wave must clearly have a voltage equal to the total voltage at the junction and a current equal to the total. Thus the transmitted wave is (E'', i'') where

$$\left.\begin{aligned} i'' &= i + i' = (2Z_A/(Z_A + Z_B)]i \\ \text{and} \quad E'' &= E + E' = (2Z_B/(Z_A + Z_B)]E. \end{aligned}\right\} \quad . \quad (123a)$$

EXAMPLE. Deduce a simple expression for the natural impedance of a transmission line. A transmission line has a capacitance of 0·0125 μF. per mile and an inductance of 1·5 mH. per mile. This overhead line is continued

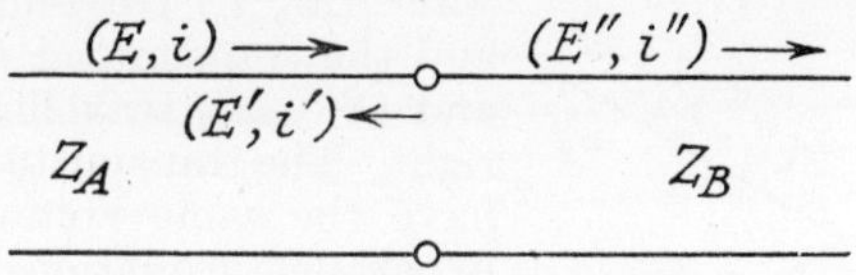

FIG. 248. EFFECT OF A SUDDEN CHANGE IN THE LINE ON TRAVELLING WAVES

by an underground cable with a capacitance of 0·3 μF. per mile and an inductance of 0·25 mH. per mile. Calculate the rise of voltage produced at the junction of the line and cable by a wave with a crest value of 50 kV. travelling along the cable. (*Lond. Univ.*, 1931.)

The natural impedance is $\sqrt{(L/C)}$. The value for the cable is

$$Z_A = \sqrt{\left[\frac{0{\cdot}25 \times 10^{-3}}{0{\cdot}3 \times 10^{-6}}\right]} = \sqrt{833} = 28{\cdot}9\ \Omega.,$$

whilst the value for the overhead line is

$$Z_B = \sqrt{\left[\frac{1{\cdot}5 \times 10^{-3}}{0{\cdot}0125 \times 10^{-6}}\right]} = \sqrt{120\ 000} = 346{\cdot}4\ \Omega.$$

The reflected wave has a crest voltage

$$\begin{aligned} E' &= [(Z_B - Z_A)/(Z_B + Z_A)] \times 50\ \text{kV.} \\ &= (317{\cdot}5/375{\cdot}3) \times 50\ \text{kV.} = \underline{\underline{42{\cdot}3\ \text{kV.}}}, \end{aligned}$$

so that the maximum voltage at the junction is 92·3 kV.

The next example shows the calculation of the reflected and transmitted waves at a point where a line forks.

EXAMPLE. Obtain the law for the behaviour of a voltage surge with vertical wave-front which, after travelling in a transmission line of inductance L and capacitance C per unit length, reaches a fork where the line splits into two sections having line constants L_1C_1 and L_2C_2 respectively. Neglect resistance and attenuation and obtain the distribution of voltage and current immediately after the wave-front has reached the fork.

An overhead transmission line has a surge impedance of 700 Ω., and a voltage wave of 10 000 V. travelling along it. The wave is assumed to be of infinite length and the wave-front is vertical. At a certain point the overhead line terminates and the circuit is continued by two cables in parallel. The surge impedance of one cable is 100 Ω., and that of the other is 200 Ω. Calculate the voltage and current in the overhead line and in the two cables immediately after the travelling wave has reached the fork.

(*Lond. Univ.*, 1927.)

Fig. 249 represents the arrangement schematically. The surge impedances are

$$Z = \sqrt{(L/C)},\ Z_1 = \sqrt{(L_1/C_1)},\ \text{and}\ Z_2 = \sqrt{(L_2/C_2)}.$$

Let the incident wave be (E, i) travelling to the right, the reflected wave (E', i') travelling to the left, and the transmitted waves (E'', i_1'') and (E'', i_2'') travelling towards the right. The transmitted waves clearly have the same voltage as they are in parallel. Equations (120) and (121) give the relations

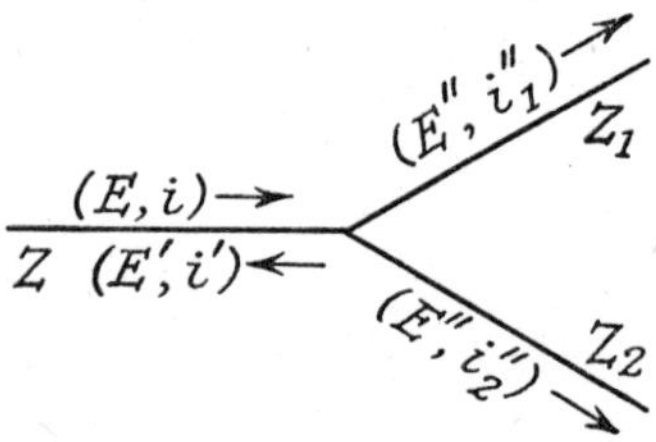

FIG. 249. TRAVELLING WAVES AT JUNCTION OF LINES

$$E = iZ,$$
$$E' = -i'Z,$$
$$E'' = i_1''Z_1,$$
and
$$E'' = i_2''Z_2.$$

The current entering the fork must be equal to the current leaving, so that

$$i + i' = i_1'' + i_2'' \qquad . \quad . \quad . \quad . \quad (124)$$

The voltage at the junction is

$$E + E' = E'' \qquad . \quad . \quad . \quad . \quad . \quad (125)$$

These six equations are sufficient to find E', E'', i, i', i_1'', and i_2'' for an incident wave of given magnitude E. Substituting for the currents in terms of the voltages we see that equation (124) becomes

$$E - E' = E''Z\left(\frac{1}{Z_1} + \frac{1}{Z_2}\right).$$

Adding this to equation (125) we get

$$2E = E''(1 + Z/Z_1 + Z/Z_2),$$

so that the voltage at the fork is

$$E'' = \frac{2E}{1 + Z/Z_1 + Z/Z_2} = 2E\frac{1/Z}{1/Z + 1/Z_1 + 1/Z_2}.$$

The transmitted currents are

$$i_1'' = E''/Z_1 \text{ and } i_2'' = E''/Z_2,$$

whilst the incident current is $i = E/Z$.

The reflected voltage is

$$E' = E'' - E = E\frac{1/Z - 1/Z_1 - 1/Z_2}{1/Z + 1/Z_1 + 1/Z_2}$$

and the current is $i' = -E'/Z$. It is seen that the reflected wave is zero when

$$1/Z = 1/Z_1 + 1/Z_2,$$

i.e. when the parallel combination of the surge impedances of the outgoing lines at the fork is equal to the surge impedance of the line along which the incident wave travels.

In the example $Z = 700$, $Z_1 = 100$, $Z_2 = 200$, and $E = 10\ 000$. We then have

$$i = 10\ 000/700 = \underline{\underline{14{\cdot}3\text{ A.}}},$$

$$E' = 10\ 000\frac{\dfrac{1}{700} - \dfrac{1}{100} - \dfrac{1}{200}}{\dfrac{1}{700} + \dfrac{1}{100} + \dfrac{1}{200}} = \underline{\underline{-\ 8\ 260\text{ V.}}}$$

$$i' = -E'/Z = 8\ 260/700 = \underline{\underline{11{\cdot}8\text{ A.}}},$$

$$E'' = E + E' = 10\ 000 - 8\ 260 = \underline{\underline{1\ 740\text{ V.}}},$$

$$i_1'' = E''/Z_1 = \underline{\underline{17{\cdot}4\text{ A.}}} \text{ and } i_2'' = E''/Z_2 = \underline{\underline{8{\cdot}7\text{ A.}}}$$

The cables thus have the beneficial effect of reducing the surge voltage from 10 kV. to 1·74 kV.

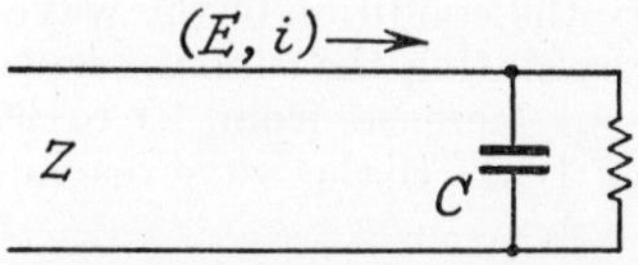

Fig. 250. Effect of Shunt Capacitance on Travelling Wave

Effect of a Capacitance. Suppose that a wave (E, i) meets a termination composed of the parallel combination of a capacitance C and resistance R, as shown in Fig. 250. The problem is the same as that shown in Fig. 247, except that R in equations (123) must be replaced by

$$\frac{(1/pC)R}{1/pC + R} = \frac{R}{1 + pCR},$$

where $p = d/dt$.

The voltage at the termination is thus

$$E_T = E + E' = \frac{2R/(1 + pCR)}{Z + R/(1 + pCR)} E$$
$$= \frac{2R}{Z(1 + pCR) + R.} E.$$

It must be remembered that $p = d/dt$ and E is a voltage which is zero until $t = 0$ and E after $t = 0$. E_T may be found in the following way.

$$E = \frac{Z(1 + pCR) + R}{2R} E_T$$
$$= \tfrac{1}{2}(pCZ + Z/R + 1)E_T$$
$$= \tfrac{1}{2}CZ(dE_T/dt) + \tfrac{1}{2}(Z/R + 1)E_T.$$

This is a linear differential equation for E_T of which the solution is

$$E_T = \frac{2E}{Z/R + 1} + A\varepsilon^{-[(Z + R)/CZR]t},$$

where A is an arbitrary constant and is determined by the fact that E_T can rise at a finite rate from its zero value. This gives

$$A = -\, 2E/(Z/R + 1)$$

so that
$$E_T = \frac{2E}{Z/R + 1} [1 - \varepsilon^{-[(Z + R)/CZR]t}]$$
$$= E_{T0} [1 - \varepsilon^{-[(Z + R)/CZR]t}],$$

where E_{T0} is the voltage at the end when there is no capacitance. Fig. 251 shows the graph of E_T. The effect of the capacitance is to cause the voltage at the end to rise to the full value gradually instead of abruptly, i.e. it flattens the wave front. It is usual to specify the condition of the wave-front by stating the time the wave takes to increase from 10 to 90 per cent of its value and multiplying by 1·25. If the wave reaches x of its value in time t

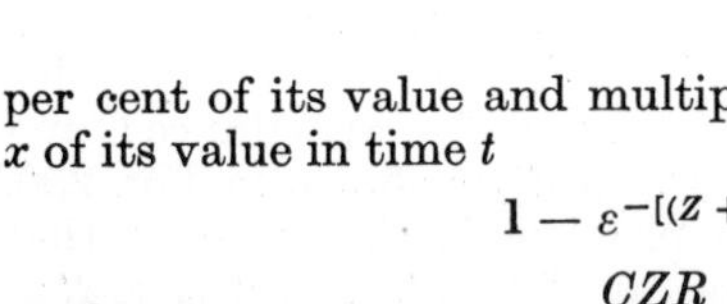

FIG. 251. FLATTENING OF WAVE DUE TO SHUNT CAPACITANCE

$$1 - \varepsilon^{-[(Z + R)/CZR]t} = x,$$

so that
$$t = \frac{CZR}{Z + R} \operatorname{logh}\left(\frac{1}{1 - x}\right).$$

The specifying time in this case is therefore

$$1{\cdot}25 \,.\, [CZR/(Z + R)]\,[\operatorname{logh} 10 - \operatorname{logh} 1{\cdot}11] \text{ sec.}$$
$$= 2{\cdot}75\, CZR/(Z + R) \text{ sec.}$$

In the case of a capacitance at a point of a line which stretches in both directions away from it, $Z = R$ and the time is

$$1{\cdot}37CZ \text{ sec.}$$

Thus a 10 000 $\mu\mu$F. capacitance in a line of surge impedance 500 ohms flattens the wave so that the time of the wave-front becomes

$$1{\cdot}37 \times 10^{-8} \times 500 \text{ sec.} = 6{\cdot}9\ \mu\text{sec.}$$

Flattening the wave-front has a very beneficial effect, as it reduces the stress on the line-end windings of a transformer connected to the line.

Lightning. With the increase of high-voltage overhead lines the problem of lightning is assuming greater importance, and much

FIG. 252. B STROKE

damage is done yearly by lightning. There are two main ways in which lightning affects a line: by a direct stroke, and by electrostatic induction. The way in which thunderclouds get charged up to very high potentials is complicated and not known precisely.

A direct stroke can take place in several ways. In one way the charged cloud induces a charge of opposite sign on tall objects, such as tall masts, church spires, etc. The electric stress at the top points of these objects causes ionization of the air, and eventually a direct stroke takes place between the cloud and the object. Such a stroke is known as the *A stroke*, and is characterized by the comparatively long time taken to produce it and the fact that it strikes the highest point, usually a lightning conductor. Another way results in a much more sudden stroke, which is produced in the manner shown in Fig. 252. Three clouds are involved, and the potential of cloud 3 is decreased by the presence of the charged cloud 2. When cloud 1 flashes over to cloud 2, both these clouds are discharged rapidly; then cloud 3 assumes a much higher potential and flashes to earth very rapidly. This is the *B stroke*, and is characterized by its rapidity and the fact that it ignores tall objects and reaches earth in a random manner. A direct stroke may cause a potential of 10 million volts, and shatter insulators and towers in its vicinity. The most that can be hoped from protective devices is that they will limit the damage and prevent the resulting travelling waves from affecting the plant. Fortunately direct strokes are rare.

The majority of surges in a transmission system are due to lightning, and are caused by electrostatic induction in the manner

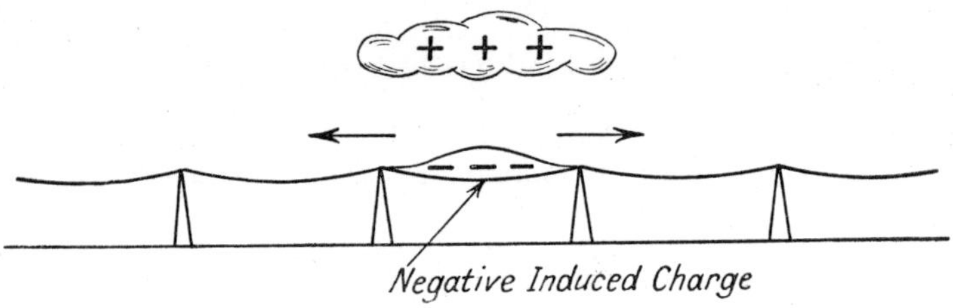

FIG. 253. SURGE DUE TO ELECTROSTATIC INDUCTION

indicated in Fig. 253. A positively charged cloud is above the line and induces a negative charge on the line by electrostatic induction. The induced positive charge leaks slowly to earth via the insulators. When the cloud discharges to earth or to another cloud, the negative charge on the line is isolated as it cannot flow quickly to earth over the insulators. The line thus acquires a high negative potential, which is a maximum at the place nearest the cloud and falls slowly

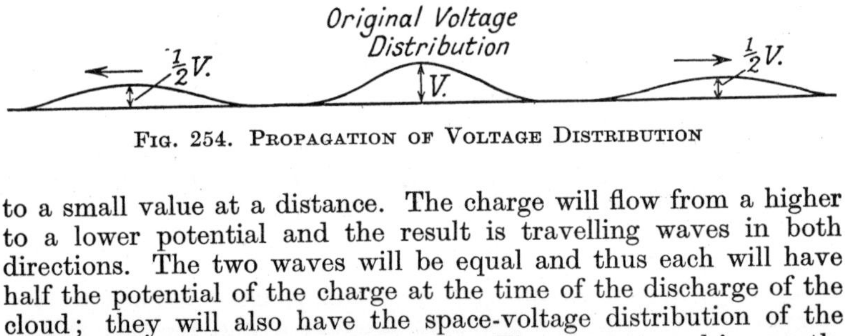

FIG. 254. PROPAGATION OF VOLTAGE DISTRIBUTION

to a small value at a distance. The charge will flow from a higher to a lower potential and the result is travelling waves in both directions. The two waves will be equal and thus each will have half the potential of the charge at the time of the discharge of the cloud; they will also have the space-voltage distribution of the original charge, as shown in Fig. 254. The waves travel in exactly the same way as the waves due to switching, so that the current at any point of the line is the voltage divided by the surge impedance. In a line without resistance or leakage the waves travel without change of shape, but the effect of resistance and leakage is to attenuate the wave and to flatten the wave-front.

The steepness of the wave-front depends upon the space-voltage distribution. If the wave reaches its maximum in 1 000 ft., the time that it takes for the wave to reach the maximum when it passes a point is

$$\frac{1\,000}{186\,000 \times 5\,280} \text{ sec.} = 1{\cdot}02\ \mu\text{sec.}$$

Waves have been recorded with wave-fronts of 1 to 80 μsec. and wave-tails of 3 to 200 μsec. A very steep wave-front may be obtained when a thundercloud is near a building which the line enters. The building screens the line inside from the cloud, so that the induced charge stops abruptly at the building. Extra precautions are therefore necessary where an overhead line enters a building.

Arcing Earths. In the early days of transmission it was the practice to insulate the neutral point of three-phase lines, for then

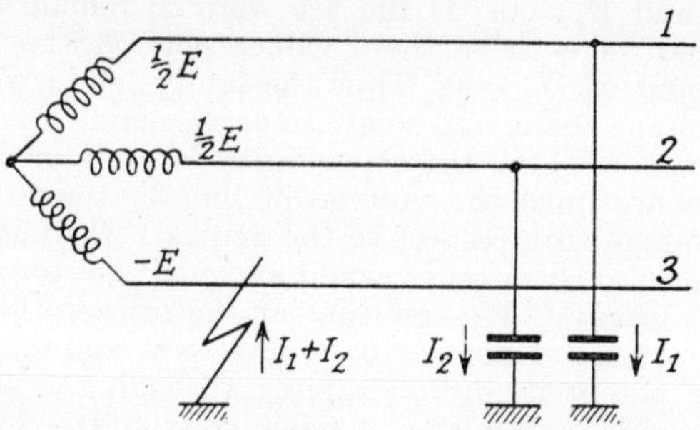

FIG. 255. ARCING GROUND IN THREE-PHASE LINE

an earth on one phase would not put the line out of action; this also eliminated the longitudinal (or zero phase-sequence) current and resulted in a decrease of interference with communication lines. Insulated neutrals gave no trouble with short lines and comparatively low voltages, but it was found that when the lines became long and the voltages high a serious trouble was caused by *arcing earths*, which produced severe voltage oscillations of three to four times the normal voltage. These oscillations were cumulative, and hence very destructive. Arcing earths are eliminated in this country and in America by solid earthing of the neutral, whilst in Germany the neutral is earthed through an inductance (a *Petersen coil*).

There are two accepted theories of arcing earths, in one of which the arc is extinguished at the normal frequency, and in the other at the frequency of oscillation of the line. Let us consider the *normal-frequency arc-extinction theory* for a three-phase line.

Fig. 255 shows a three-phase line. Suppose that line 3 arcs to earth when its voltage to neutral is a maximum $-E$. At this instant lines 1 and 2 have voltage $+\frac{1}{2}E$. Before the arcing earth occurs the capacitances of the lines cause the neutral to be at or near the earth potential, so that the earthing of line 3 causes a sudden voltage of $+E$ to be applied to lines 1 and 2. The ultimate steady state would then be for the lines 1 and 2 to be at potential $\frac{3}{2}E$. But we have shown that when an e.m.f. E is suddenly switched into a circuit of low resistance, the voltage in the circuit oscillates between 0 and $2E$ with a frequency $1/2\pi\sqrt{(LC)}$ (see equations (115*a*) *et seq.*), where L and C are the inductance and capacitance in the circuit. The voltage of lines 1 and 2 will therefore oscillate rapidly between the original value of $\frac{1}{2}E$ and $\frac{1}{2}E + 2E = \frac{5}{2}E$. The high frequency oscillation dies out rapidly. The arc is fed through the capacitances of the lines, as shown in Fig. 255, and will go out when the sum of the capacitance currents passes through zero. The capacitance currents lead the voltages by 90°, so that when their sum $I_1 + I_2$ is zero the line voltages are $E_1 = -\frac{3}{2}E$, $E_2 = -\frac{3}{2}E$, and $E_3 = 0$. If the arc were to remain extinct, the voltages would have to be these values plus E, viz. $E_1 = -\frac{1}{2}E$, $E_2 = -\frac{1}{2}E$, and $E_3 = +E$. Thus the faulty line 3 would have a maximum voltage again, and so arc to earth again. In other words, when line 3 arcs to earth the capacitance currents of lines 1 and 2 maintain the arc until the voltage of line 3 attains its opposite maximum voltage with respect to the neutral; then at the instant when the capacitance currents would allow the arc to go out, line 3 arcs again to ground. We saw that at the instant that the arc is extinct the lines are at potentials $-\frac{3}{2}E$, $-\frac{3}{2}E$, and 0. The charges due to these potentials diffuse rapidly through the system in an oscillatory manner, with the average voltage $\frac{1}{3}(-\frac{3}{2}E - \frac{3}{2}E + 0) = -E$ as the mean position. This is equivalent to an insertion of an e.m.f. of $\frac{1}{2}E$ in lines 1 and 2, so that an added voltage E is applied to these lines. When the arc restrikes, lines 1 and 2 acquire potentials of $-\frac{5}{2}E$ plus this new value $-E$, so that the maximum voltage is $\frac{7}{2}E$. We see therefore that the healthy lines are subjected to a voltage of $3\frac{1}{2}$ times the normal value. As this state can be maintained for a considerable length of time, in a known case 30 min., by the continued arcing, it is very dangerous.

Petersen Coil. We have seen that the capacitance currents I_1 and I_2 maintain the arc even when the voltage of the faulty line 3 is too low to restrike it. In fact these currents have the particularly harmful effect of maintaining the arc until the very moment when the voltage of line 3 is sufficiently high to restrike it. If the neutral is earthed through an inductance L of such a value that the current it passes neutralizes $I_1 + I_2$, the normal frequency follow current through the arc is

$$I_L + I_1 + I_2 = 0.$$

The arc is then extinguished except for the brief moments when the voltage of line 3 passes through its maximum value and can restrike it.

It has been found that the Petersen coil is completely effective in preventing any damage by an arcing earth, and is therefore used extensively on the Continent. The coil is usually provided with tappings, so that its value can be adjusted to suit the capacitances of the system. It is found that effective operation is secured when the inductance is 90 to 110 per cent of the theoretical value for exact neutralization of the capacitance currents.

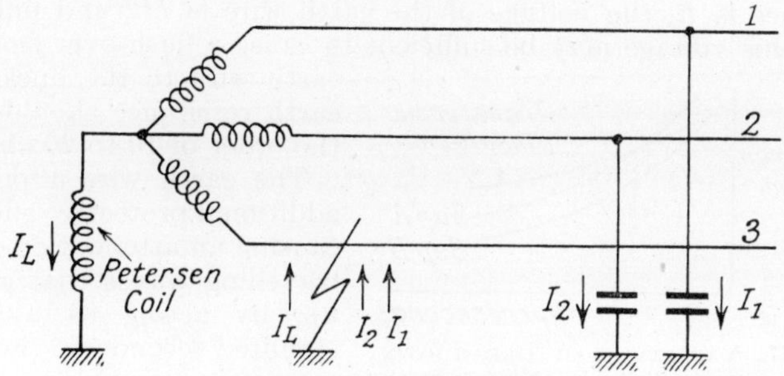

FIG. 256. PETERSEN COIL

Lightning and Over-voltage Protection. The insulation of a transmission system is always designed to withstand voltages of twice the normal value for a reasonable length of time, as switching surges often produce voltages of this magnitude. It is clearly uneconomical to design the system so that the insulation can withstand the very high voltages that may be encountered from extraneous or fault conditions, and recourse is had to protective devices which are adjusted to break down before the insulation, or otherwise prevent a dangerous voltage from damaging the insulation.

Dangerous voltage rises are found to be due to the following: (1) surges due to direct lightning strokes or induced voltages, (2) arcing earths, (3) comparatively low-voltage high-frequency oscillations, (4) static overvoltage. The protective apparatus for these classes are: (1) ground wire and lightning arresters, (2) earthing of neutral solidly or through a Petersen coil, (3) surge absorber or capacitance, (4) water-jet earthing resistance, earthing inductance, or solid earthing of the neutral point.

It is true to say that with the advent of high-voltage overhead lines, such as the Grid, the main cause of damage is lightning. We have seen that most travelling waves due to lightning are caused by electrostatic induction. The latter can be reduced considerably by the use of earth wires running above the transmission line and earthed at every pole or tower. If C_1 is the capacitance of the cloud to the line and C_2 the capacitance of the line to ground, the induced voltage on the line is $C_1/(C_1 + C_2)$ times the cloud voltage. The presence of the earth wire *above* the line causes a considerable increase in C_2 and reduction of the line voltage. The induced voltage could be very much reduced by an array of earth wires above the line, but this is too expensive to install in practice.

The earth wire also provides considerable protection against direct strokes (of the A type), provided the earth resistance of the earth wire is kept low. If the current in the stroke is I and the earth resistance is R, the voltage of the earth wire is IR, and unless R is low this voltage may be sufficient to cause a flash-over from the earth wire to the lines. The earth resistance should be of the order of 10 to 20 ohms.

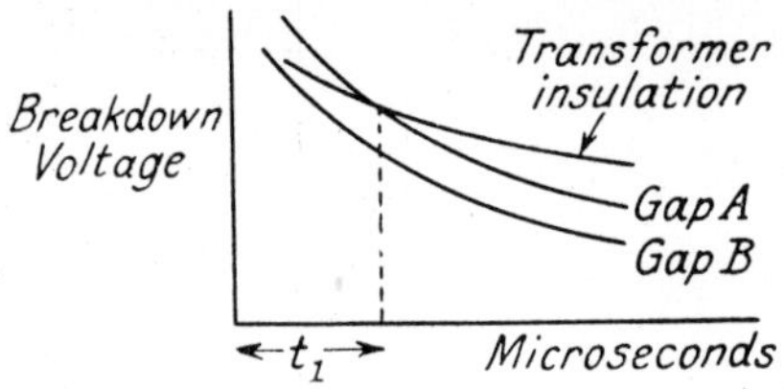

FIG. 257. VARIATIONS OF BREAKDOWN VOLTAGE WITH TIME OF APPLICATION

The earth wire affords an additional protective effect by causing an attenuation of any travelling waves that are set up, by acting as a short-circuited secondary. For this reason its resistance should not be too large. It is usually made of steel, which has a high permeability and thus possesses a resistance which increases with frequency.

Having reduced the magnitude of induced voltages by means of an earth wire, we still find it necessary to install protective apparatus to prevent, or at least minimize, the damage due to the surges that do occur. It is, moreover, essential that the system shall be considered as a whole from the point of view of protection, so that the least essential and most accessible parts protect the more important apparatus; this involves the *co-ordination of system insulation.* The problem is rendered difficult by the fact that the breakdown voltages of the various parts of the system and of the protective apparatus behave differently with time; thus a horn gap which is set to flash-over at 100 kV. at 50 cycles may require 200 kV. in a wave lasting for 20 μsec., or 300 kV. in a wave lasting for 5 μsec. We define the *impulse ratio* of any piece of apparatus as the ratio of the breakdown voltage of a wave of specified duration to the breakdown voltage of a 50-cycle wave; thus the horn gap has an impulse ratio of 2 at 20 μsec., and 3 at 5 μsec. When a method of co-ordinated insulation is considered, the impulse ratio of the various parts must be known or the protection will not be adequate. Fig. 257 illustrates the point. Suppose that the insulation of a transformer to be protected has the breakdown voltage-time characteristic shown. Gap A may be set to break down at a lower voltage than, say, 80 per cent of the breakdown voltage of the insulation at 50 cycles. The gap, nevertheless, does not protect the transformer, as its characteristic rises more rapidly than that of the transformer insulation as the duration of the wave decreases. Then for waves of duration less than t_1 the transformer insulation breaks down before the gap. It is necessary to narrow the gap so that the characteristic is as shown for gap B before the transformer is completely protected. In practice it is not possible to narrow the gap so much that the insulation is protected for waves of the smallest duration, as then the gap would flash over at very low voltages at 50 cycles; a compromise is reached by protecting the insulation for voltages of waves down to a certain minimum time, which is found experimentally to be comparatively harmless.

Sphere Gap. A sphere gap in which the spacing is small compared with the diameter of the spheres has the useful advantage that the

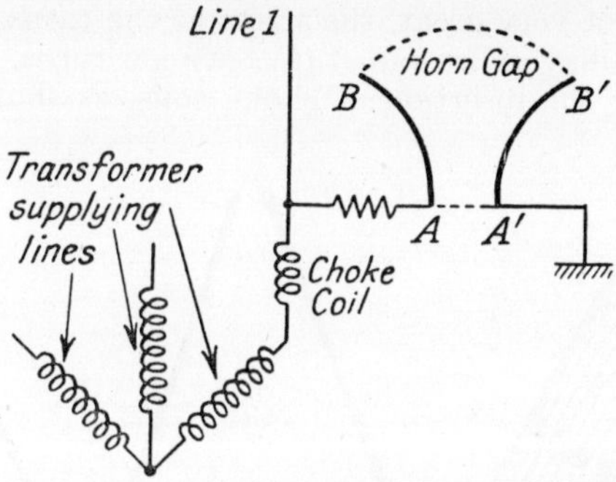

FIG. 258. HORN GAP WITH CHOKE COIL AND RESISTANCE

impulse ratio is unity. If then the apparatus is protected against 50-cycle waves, it is protected against a wave of any duration. Unfortunately, when the sphere gap flashes over, the power current maintains the arc, which requires only a very low voltage to maintain it, and the arc is not self-extinguishing. The circuit-breakers would have to intervene to break the arc current and the service is interrupted. For this reason the sphere gap is not of use.

Horn Gap. Fig. 258 shows a simple sketch of the horn gap. The gap is set so that a flash-over occurs between A and A' at a voltage of 150 to 200 per cent of the normal voltage. The power current creates an arc, which may be considered to be a flexible conductor. A flexible electric circuit moves so as to embrace as many lines of magnetic force as possible, so that the arc is forced up to the position BB'. Another factor tending to blow the arc up to BB' exists when BB' is above AA', for then the arc heats the air and forms a vertical draught. The result is that the arc is forced up to BB', where the gap is wide and the normal voltage is insufficient to maintain it. The arc is thus extinguished, usually in about 3 sec.

The horn gap cannot rupture arc currents much in excess of

10 amperes, and as the arc is a dead short circuit it is necessary to limit the current to a small value. This is done by inserting a resistance, between the line and the horn on the line side, which reduces the current to about 5 amperes. The efficacy of the horn gap is seriously reduced by the resistance. The resistance is a water column, oil-immersed metal wire, carbon rod, or carborundum, and is made as non-inductive as possible.

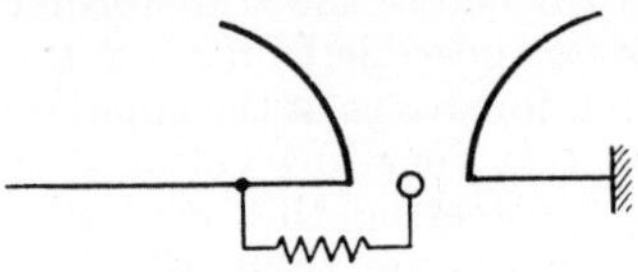

FIG. 259. HORN GAP WITH AUXILIARY ELECTRODE

It is found that high-frequency waves concentrate at the line-end turns of a transformer, so that although the magnitude of the wave on the line is not very great, the stress at the turns near the line is very high and may cause puncture between turns. This difficulty is overcome by the insertion of choke coils, as shown in Fig. 258.

FIG. 260. BURKE ARRESTER
(Metropolitan-Vickers)

The high-frequency wave is then reflected back to the horn gap, where the doubled voltage causes a flash-over. The choke is without effect on the low-frequency power wave.

For small settings the horn gap is sensitive to corrosion or pitting of the horns, so that it does not maintain its setting. This difficulty is overcome in the arrester shown in Fig. 259. The main gap is set for a voltage well above that to be protected. The auxiliary gap has a platinum electrode, which possesses the character of permanence. When an over-voltage occurs the auxiliary gap flashes over and ionizes the air, and then the main gap flashes over.

Burke Arrester. Fig. 260 shows the Burke arrester. The line current passes through a triangular pancake choke coil, one side of which forms half of the main gap. Severe over-voltages flash across the main and auxiliary gap direct to earth. Less severe voltages flash over the main gap only, and the current is then limited by the resistance.

Multi-gap Arrester. This consists of a number of small gaps in series with a limiting resistance. Another resistance is placed across some of the gaps adjacent to the limiting resistance.

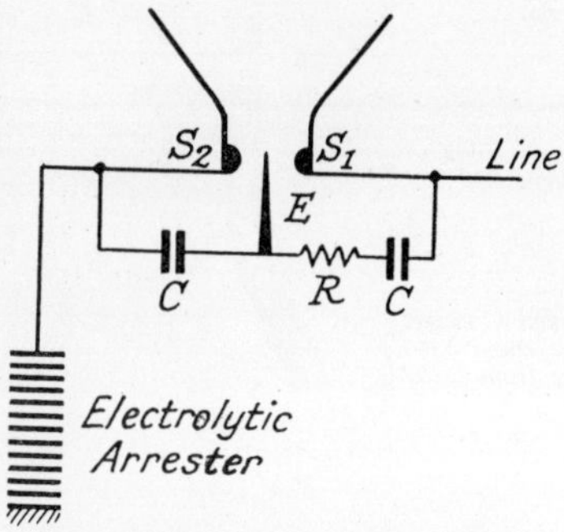

FIG. 261. IMPULSE GAP WITH ELECTROLYTIC ARRESTER

Impulse Protective Gap. It was pointed out that the sphere gap has an impulse ratio of unity, but suffers from the disadvantage that the arc between its electrodes is not self-extinguishing. The horn-gap, however, extinguishes the arc but has a high impulse ratio, 2 or 3. The impulse protective gap is designed to have a low impulse ratio, even less than unity, and to extinguish the arc. Fig. 261 shows a diagram of the impulse gap. S_1 and S_2 are sphere-horn electrodes, and are connected to the line and an electrolytic arrester, respectively. An auxiliary needle electrode E is placed mid-way between S_1 and S_2, and is connected to them via (R, C) and C. At the power frequency the impedance of the capacitances C is very much greater than that of R, so that the potential of E is mid-way between those of S_1 and S_2 and the electrode has no effect on the flash-over between them. At very high frequencies the impedance of C is small, so that E is at the potential of S_2 and the gap is effectively half the previous value. Flash-over takes place between S_1 and E at a voltage less than that required to flash-over between S_1 and S_2. An impulse ratio less than unity can thus be obtained. The electrolytic arrester on the earth side extinguishes the arc.

Electrolytic Arrester. This is the earliest type of arrester with a large discharge capacity. The action depends upon the fact that a thin film of aluminium hydroxide immersed in electrolyte presents a high resistance to a low voltage, but a low resistance to a voltage above a critical value. The critical breakdown voltage is about 400 volts, and voltages higher than this cause a puncture and a free flow of current. The insulating film of hydroxide is formed by applying a direct voltage up to the critical value to aluminium plates immersed in the electrolyte; during the formation of the

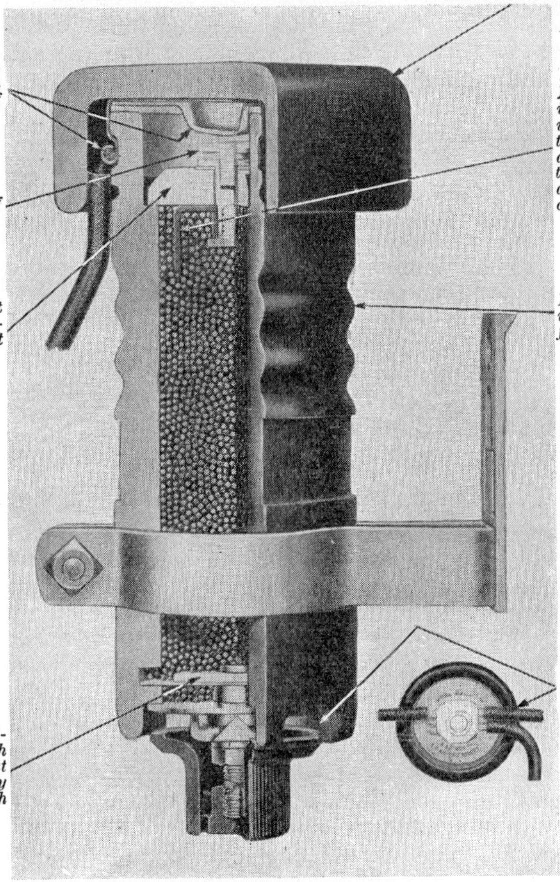

FIG. 262. OXIDE-FILM ARRESTER
(*B. T.-H.*)

film, current passes fairly readily, but when the film is formed the current ceases.

Stacks of films are arranged one above the other and the total critical voltage is equal to the critical voltage of each film multiplied by the number of films.

Daily supervision and reforming of the films is essential, and for this reason the arrester is being replaced by the more robust oxide-film and auto-valve arresters. The electrolytic arrester is used in conjunction with an impulse gap, for the continual leakage and capacitance currents would damage the arrester.

Oxide-film Arrester. Fig. 262 shows the construction of the oxide-film arrester of the pellet type. The lead peroxide pellets are in a column of $2\frac{1}{4}$ in. diameter, the length of the column being 2 in. per kV. of rating. The tube contains a series spark-gap. A single tube system is available for voltages up to 25 kV. when the neutral is solidly earthed, and 18 kV. when the neutral is isolated or earthed through an inductance coil. For higher voltages several units are placed in series.

The pellets have a diameter of approximately $\frac{3}{32}$ in. and are made of lead peroxide with a thin porous coating of litharge.

Auto-valve Arrester. This consists of a number of flat discs of a porous material stacked one above the other and separated by thin mica rings. The material is made of specially prepared clay with a small admixture of powdered conducting substance. The discharge occurs in the capillaries of the material and is thus constrained to be a glow discharge, in which there is a voltage drop of about 350 volts per unit. The narrow gaps between the blocks are of sufficient total width to prevent flash-over due to the normal voltage, so that no current flows in the arrester under normal conditions. This arrester is very effective, robust and cheap, and is being rapidly introduced into modern high voltage systems.

Thyrite Arrester. Thyrite is a dense inorganic compound of a ceramic nature, which has a resistance that decreases rapidly from a high value at low currents to a low value at high currents. The current increases 12·6 times when the voltage is doubled; thus if the current-voltage relation for a given block of thyrite is

$$E = kI^n,$$

then $$2E = k(12{\cdot}6I)^n,$$

so that $$2 = 12{\cdot}6^n,$$

i.e. $$n = \log 2 \div \log 12{\cdot}6 = 0{\cdot}27.$$

Thus the voltage varies approximately as the fourth root of the current. Fig. 263 shows the current-voltage curve of the 11 kV. thyrite arrester of Fig. 264. There are eleven thyrite discs sprayed on both sides to provide a good surface contact; each disc has a diameter of 6 in. and thickness $\frac{3}{4}$ in., and will discharge several thousand amperes without the slightest tendency to flash over the outside edge. When passing 2 000 amperes each disc has a voltage of only 5 kV. At the normal voltage of 11 kV. to earth, the peak voltage is $(11\sqrt{2} \div \sqrt{3})$ kV. = 9 kV. and the current in the arrester is only 3·2 amperes. When one phase is earthed the peak voltage on the other phases is $11 \times \sqrt{2} = 15{\cdot}6$ kV. and the arrester passes

25 amperes. A series gap is provided to prevent current from flowing at the normal voltage. The value of k for the stack is 6 500, or 600 per disc.

When the gap and arrester flash over, a high current flows for the duration of the surge, which is discharged to earth rapidly as is shown by oscillographic records; there appears to be absolutely no time-lag in the thyrite itself. The normal frequency follow-current is very small, 3·2 amperes in a healthy system, and only 25 amperes in a system with an earthed phase. The gap is easily able to clear this small follow-current.

Some modern modifications of the thyrite arrester include a type in which resistance blocks of a ceramic nature are spaced at equal distances from one another. The total gap length is adjusted so

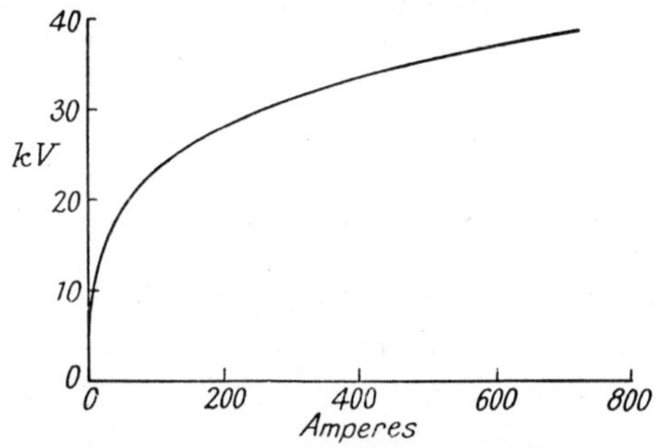

FIG. 263. VOLTAGE-CURRENT CURVE OF THYRITE ARRESTER

that the gaps flash over at twice normal voltage; it is claimed that the distributed gaps behave better than a single gap. Round knobs are provided between the electrodes of the gaps so as to reduce the time-lag. The action of the resistance blocks is similar to that of thyrite.

Condensers. We have shown on pages 290–22 that the effect of a condenser, placed between the line and earth, on a travelling wave is to reduce the steepness of the wave-front. This effect protects the windings of a transformer near the line, since a steep wave-front causes very high stresses in these turns.

The condenser, moreover, protects the transformer against comparatively low-voltage, high-frequency waves. The normal-frequency voltage produces only a very small current in the condenser, so that negligible loss is caused during normal operation.

The latest type of condenser used for protective purposes has a dielectric of acetyl cellulose, the electrodes being silver plating on the strips of the dielectric.

Surge Absorber. A pure condenser of the type described in the previous section cannot dissipate the energy in the wave-front of a travelling wave or in a high-frequency oscillation. It merely reflects the energy away from the apparatus to be protected, and the energy is dissipated in the resistance of the line conductors and the earthing resistances. If a resistance is placed in series with the condenser, the combination can dissipate part of the energy in addition to diverting it from the apparatus. Such a combination is called a *surge absorber*.

Another type of absorber consists of an inductance across which is placed a resistance. This combination is placed in series with the

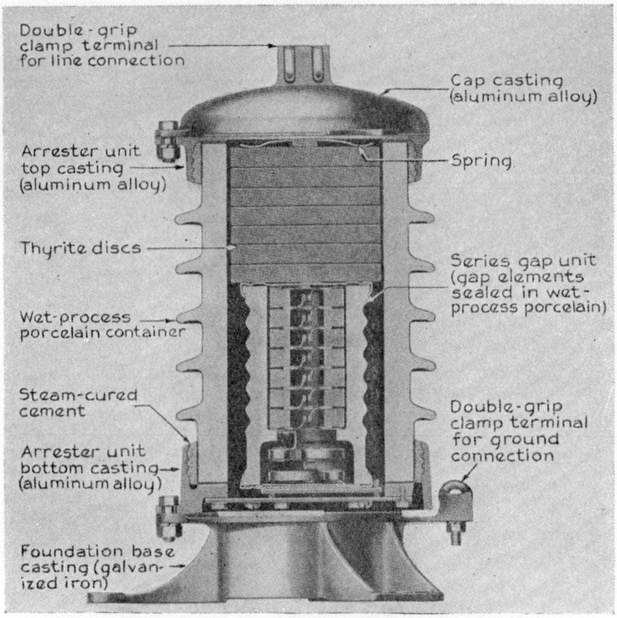

FIG. 264. 11 kV. THYRITE ARRESTER
(International General Electric Co. of New York, Ltd.)

line. Steep wave-fronts or high-frequency waves find the inductance a high impedance path and are forced through the resistance, where they are dissipated. The normal-frequency currents find the inductance a low impedance path and pass through it without much loss.

The *Ferranti surge absorber* consists of an inductance coil, which is coupled magnetically, but not electrically, to a metal shield and/or the steel tank which contains it. The coil is of a cylindrical or pancake form, depending upon the voltage; for voltages above 33 kV. the coil is cylindrical and has inside it a metal shield in which currents are induced. The absorber is enclosed in a cylindrical boiler-plate tank, provided with porcelain-guarded terminals, and is vacuum-impregnated with a light transformer oil. Fig. 265 shows a 66 kV. surge absorber of this kind. The equivalent circuit of this absorber is shown in Fig. 266. There is a filter effect which prevents high frequency currents from passing freely through the absorber;

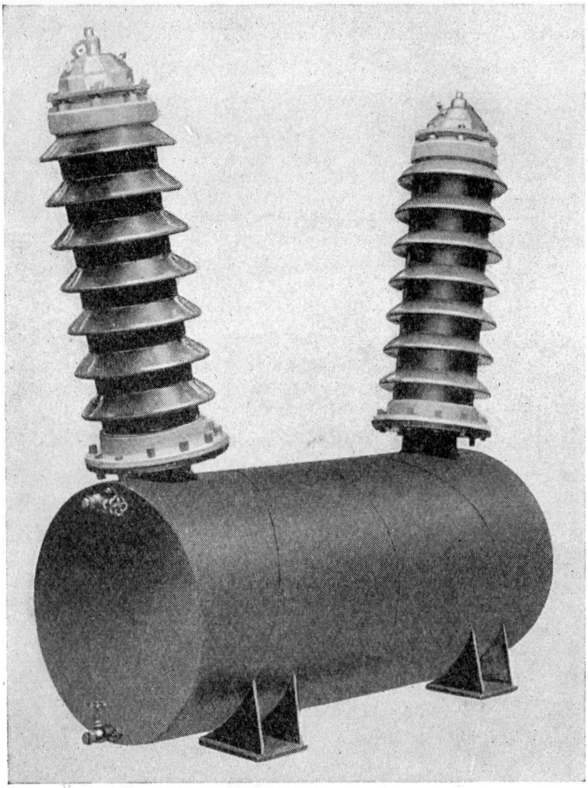

FIG. 265. FERRANTI SURGE ABSORBER
(I.E.E. Students' Journal)

also energy is transferred from the wave by the mutual induction between the coil and the shield and tank into the latter two, where the energy is dissipated as heat.

Recording of Transmission Line Surges. There are three methods of recording transmission line surges, by the high-voltage cathode-ray oscillograph, the klydonograph, and the surge-crest ammeter. These will be described briefly.

HIGH-VOLTAGE CATHODE-RAY OSCILLOGRAPH. This is the only

instrument capable of delineating the voltage-time characteristic of a wave. Fig. 267 shows a high-speed cathode-ray oscillograph manufactured by Metropolitan-Vickers. The tube is continuously evacuated and the pressure in the deflection tube is 10^{-4} mm. of mercury or less. The cathode is cold and at a potential of 50 or 60 kV. above the anode, which is earthed.

The essential process is the following. A supply of electrons is obtained by the ionization of the residual gas in the discharge tube, and these are made to travel with an enormous velocity under the accelerative effect of the applied voltage. The electrons pass through a hole in the anode and proceed in a straight line, until they pass between the time deflection plates. The time deflection plates have applied between them a voltage which varies rapidly and uniformly

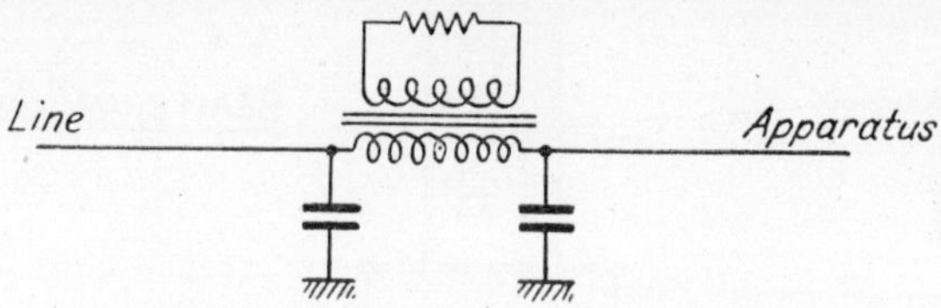

Fig. 266. Equivalent Circuit of Fig. 268

from zero to a maximum value; the electron beam then undergoes a deflection, that is proportional to the time from a given instant. The beam then passes between the voltage deflection plates, between which the wave (or a fraction of it) is applied. The voltage deflection plates produce a deflection at right angles to the time deflection, so that the electron beam, which strikes the photographic plate at the end of the tube, traces out the voltage-time curve of the wave.

In order to photograph waves of only a few microseconds duration the utmost sensitivity is required. This sensitivity is achieved in the following way. The electron beam impinges directly on the sensitive plate, which must therefore be inside the evacuated tube. The velocity of the electrons must be very great, and a high voltage of 50 kV. or more is used to accelerate the electrons. It is quite clear that the electron beam must not impinge on the plate when there is no wave, otherwise the plate would be completely fogged. The beam is diverted from the photographic chamber by beam trap plates and a beam trap tube. When there is no wave, there is a voltage between the beam trap plates which deflects the beam from the straight path that leads through a small hole in a diaphragm at the bottom of the beam trap tube. It is seen that the axis of the discharge tube is inclined at an angle to the axis of the main tube. The reason for this is that although the electron beam is prevented from reaching the photographic plate by means of the beam trap plates during the absence of a surge or wave, there are retrograde

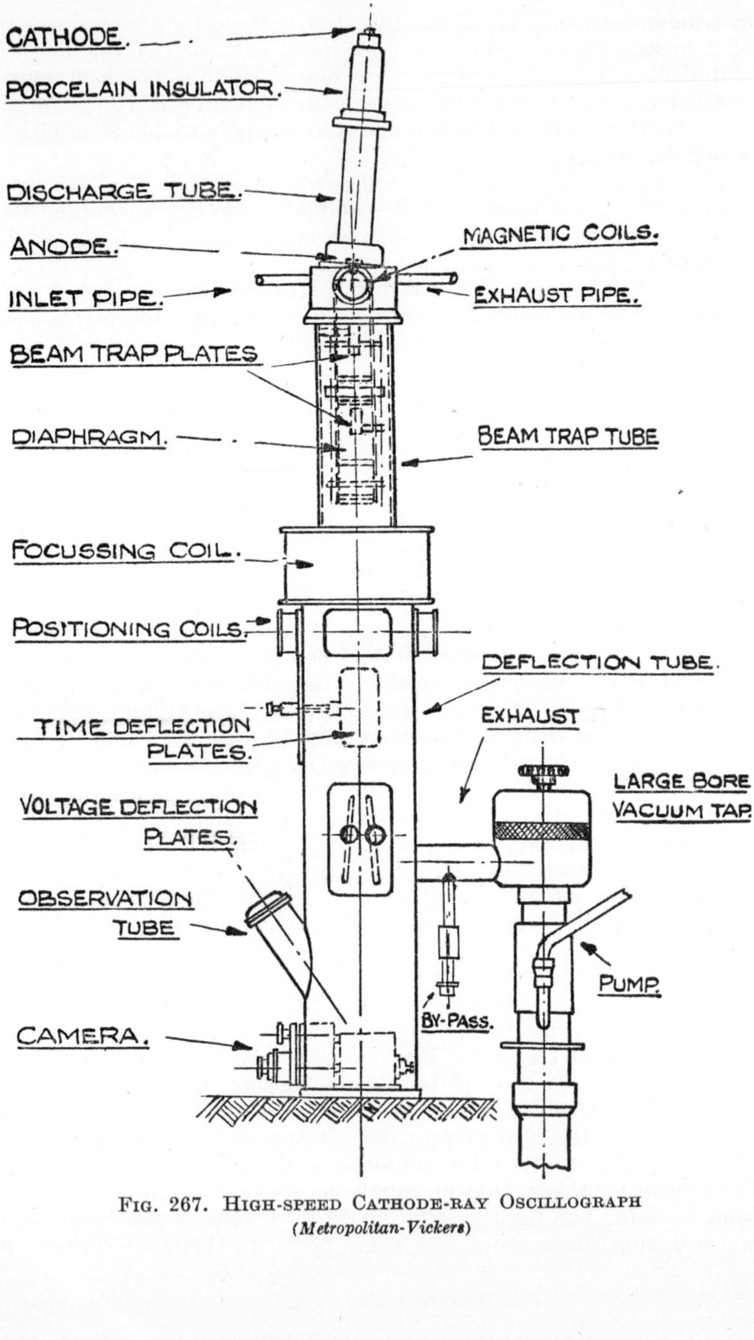

FIG. 267. HIGH-SPEED CATHODE-RAY OSCILLOGRAPH
(Metropolitan-Vickers)

rays consisting of atoms that are not much affected by the beam trap. These rays consist of relatively heavy particles and are thus not easily deflected, so that if they are moving along the axis towards the plate they will do so whether the beam trap is operating or not. Their very property of not being deflected easily is used to get rid of them by inclining the axis of the discharge tube. The

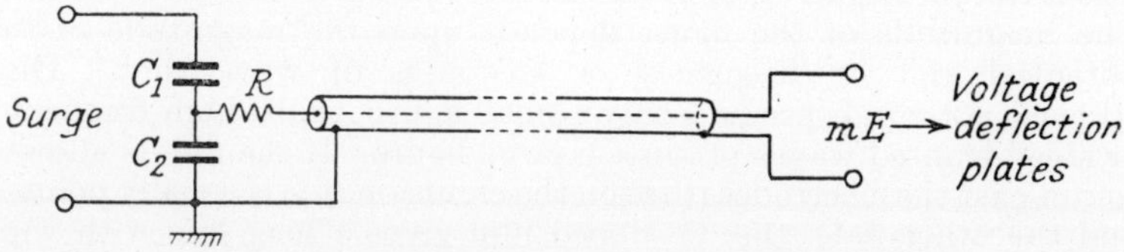

FIG. 268. POTENTIAL DIVIDER AND DELAY CABLE

retrograde rays and the electron beam travel along the axis of the discharge tube towards the anode. Magnetic coils then deflect the electron beam along the main axis, so that the beam can enter the beam trap tube; but the retrograde waves are not deflected from their inclined path and are prevented from entering the main tube.

The electron beam is focused and positioned by magnetic coils.

When a surge arrives it is sent direct to a trigger device which removes the voltage between the beam trap plates, and the electron beam travels to the plate. Meanwhile the surge is put across a potential divider connected to a *delay cable*, which transmits a known fraction of the wave to the voltage deflection plates after a delay of a fraction of a microsecond. The delay cable is a concentric cable, with air or rubber dielectric. Fig. 268 shows the arrangement of the potential divider and the delay cable; R is equal to the surge impedance of the cable. If the capacitance C_1 is ten times the capacitance of the cable, no distortion is introduced and the ratio of step-down is $C_1/(C_1 + C_2)$.

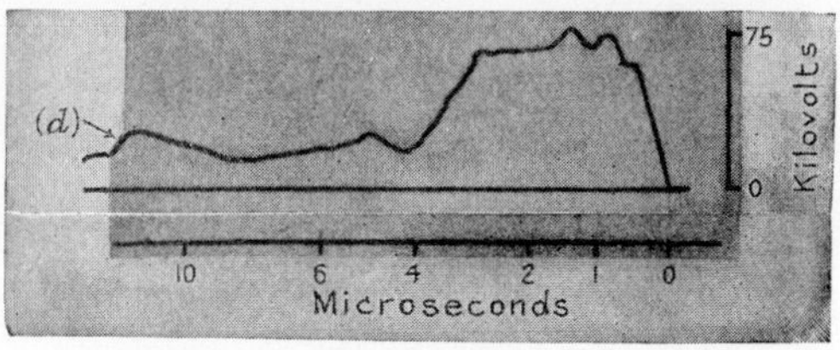

FIG. 269. VOLTAGE ACROSS LINE TURNS OF TRANSFORMER
(*I.E.E. Journal*)

Fig. 269 shows a cathode-ray oscillograph of the voltage appearing across 10 per cent of the line end turns of a transformer winding.* It is seen that in this case the time base is not quite linear.

* Reproduced by kind permission of the Institution of Electrical Engineers from the paper by Miller and Robinson, *Journal of the I.E.E.*

KLYDONOGRAPH. It is found that if a potential difference is applied between the faces of a photographic plate, the emulsion is affected and on developing a figure is obtained. When the emulsion side is at a higher potential than the other side, the figure consists of fine lines radiating from the point of contact; when it is at a lower potential, the figure is a complete and fairly definite circle. The latter, or negative, figure is the more useful as its size is definite. The magnitude of the figure depends upon the magnitude of the potential and its frequency or steepness of wave-front. Thus 50-cycle potentials produce only a small figure, whilst high-frequency or steep-fronted waves produce a large figure. If the film is allowed to run past the electrodes (that on the emulsion side is usually pointed and the other flat), the developed film gives a long line with wide bands. The long narrow line corresponds to the normal operating voltage, and the wide bands to high-frequency discharges or steep-fronted surges. Useful qualitative information has been obtained by the use of the klydonograph, but because of the dependence of the size of the figure on frequency or steepness of wave-front the results are not quantitative.

SURGE CREST AMMETER. The principle of this instrument is the measurement of the residual magnetism in a piece of magnetic material, which has been magnetized by the surge current. From the residual magnetism the peak of the surge current is deduced.

EXAMPLES X

1. Explain what is meant by the surge impedance of a transmission line and derive its value in terms of the line constants. Derive expressions for the values of the transmitted and reflected waves of current and voltage relative to those of the incident waves at a point where the surge impedance changes from Z_1 to Z_2.

A rectangular wave of 200 kV. amplitude travels along a line having a surge impedance of 500 Ω. to a transition point where it is connected to a line of 50 Ω. surge impedance. Determine the values of the transmitted and reflected voltage and current waves. (*Lond. Univ.*, 1954.)

2. Describe and explain the occurrences immediately following the sudden application of a steady voltage to one end of a transmission line open at the far end.

A surge voltage e is travelling along a line of surge impedance Z_A connected at its far end to a line of surge impedance Z_B. Show how to calculate the magnitude of the voltage surges transmitted through and reflected from the junction, explaining all assumptions and approximations.

(*B.Sc. Lond. Univ.*, 1933.)

3. Describe with the aid of sketches one good type of lightning arrester. What auxiliary equipment is used in conjunction with the arrester to safeguard the apparatus in the power stations? (*Nat. Cert.*, 1935.)

4. An overhead line is joined to a three-phase underground cable. What apparatus is necessary to protect the cable against surges? Give a diagram of connections. Knowing the surge impedance of each circuit show how to calculate the proportion of the surge that enters the cable.

(*B.Sc. Lond. Univ.*, 1931.)

5. Enumerate and explain briefly the causes of surges in a transmission line. Describe methods of preventing such surges and of protecting substation apparatus against damage due to them other than by the use of lightning arresters which discharge the surge to earth.

(*Lond. Univ.*, 1932.)

6. Explain the reasons leading to the general practice of earthing the neutral point of a power system and discuss the relative merits of earthing it (*a*) solidly, and (*b*) through an impedance.

An earth electrode consists of a pipe 6 ft. long and 1 in. dia. buried vertically with its upper end at ground level in soil having a uniform resistivity of 10 000 Ω. per cm. per cm.^{2} Estimate the potential difference between the electrode and a point on the ground 5 ft. away from it when 100 A. are flowing through the electrode to earth. (*Lond. Univ.*, 1934.)

7. Explain the principle of the cathode-ray oscillograph and describe briefly the construction of such an instrument suitable for recording transmission line surges.

What means are employed in an instrument used for this purpose to secure good photographic sensitivity and to prevent fogging of the recording plate by the ray before and after the passage of the surge?

(*Lond. Univ.*, 1933.)

8. Two single transmission lines A and B with earth return are connected in series and at the junction a resistance of 2 000 Ω. is connnected between the lines and earth. The surge impedance of line A is 400 Ω. and of B 600 Ω. A rectangular wave having an amplitude of 100 kV. travels along line A to the junction.

Develop expressions for and determine the magnitude of the voltage and current waves reflected from and transmitted beyond the junction. What value of resistance at the junction would make the magnitude of the transmitted wave 100 kV.? (*Lond. Univ.*, 1949.)

9. An underground cable having an inductance of 0·3 mH. per mile and a capacitance of 0·4 μF. per mile is connected in series with an overhead line having an inductance of 2·0 mH. per mile and a capacitance of 0·014 μF. per mile.

Calculate the values of the reflected and transmitted waves of voltage and current at the junction due to a voltage surge of 100 kV. travelling to the junction (*a*) along the cable, and (*b*) along the overhead line.

Explain how the waves would be modified if the cable and line were of considerable length. (*Lond. Univ.*, 1947.)

10. Explain the function and principle of operation of an arc-suppression coil for use on a 3-phase system.

A 33-kV., 3-phase, 50 c/s, overhead line, 50 miles long, has a capacitance to earth for each line of 0·016 μF. per mile.

Determine the inductance and kVA. rating of the arc-suppression coil suitable for this system. (*Lond. Univ.*, 1947.)

11. Describe the construction and explain the operation of a modern type of surge or lightning arrester, and explain at what part of the circuit it would be most satisfactory. (*Lond. Univ.*, 1947.)

12. Describe with the aid of diagrams, the function and operation of the Petersen coil protective device, and derive an expression for the reactance of the coil in terms of the capacitance of the protected line. What are the merits and demerits of the system? (*Lond. Univ.*, 1949.)

CHAPTER XI

ILLUMINATION

Introduction. Illumination was one of the earliest applications of electrical engineering, and is still to-day the most widespread field of use of electrical energy. Illumination by electric lamps is cheaper and more convenient than by any other method; in fact, it is likely that the electrical industry could not have developed at more than a fraction of its speed if it were not for the early profits it made from electric lighting. In return the industry has provided the community with abundant light at a very low cost, and the field of illumination has been extended in ways impossible without the use of electricity.

The lighting of streets and arterial roads with discharge lamps makes motoring at night less fatiguing to the driver and safer to the motorist and pedestrian. The cinema and theatre owe much to electric lighting, and the ordinary house is much cleaner and brighter because of it.

Industry has found that proper illumination reduces the strain of the worker and improves his efficiency, especially in occupations such as type-setting; thus it was found that when the illumination was 2 f.c. the total error in type-setting was 1·4 per cent, and when 25 f.c. only 0·5 per cent. The cost of providing the extra illumination is paid for many times over by the improved output.

The elementary mistake of confusing good illumination with bright lights should not be made; the reader may recollect some shop windows, usually displaying jewellery or glass-ware, in which bright naked lights make viewing a painful task. The problem of illumination is concerned with the disposition and character of the lights as well as with their magnitudes.

Light. Light is an electromagnetic radiation like radio waves, but of a much smaller wavelength. The relation between the wavelength and the frequency is

$$f\lambda = 3 \times 10^{10} \text{ cm. per sec.},$$

where 3×10^{10} is the velocity of the wave in cm. per sec., f is the frequency and λ the wavelength. Thus if f is a million cycles per sec., $\lambda = 3 \times 10^4$ cm. $= 300$ metres; if $\lambda = 6 \times 10^{-5}$ cm., $f = 5 \times 10^{14}$. An electromagnetic wave of wavelength 300 metres is a medium length wireless wave, whilst that of wavelength 6×10^{-5} cm. is light of an orange-yellow colour. Fig. 270 shows how the character of an electromagnetic wave varies with the wavelength or frequency, and Fig. 271 shows how visible light lies in a range of wavelength between $\lambda = 4\,000$ and $\lambda = 7\,500$ Å. The letter Å. stands for the *Ångström unit,* which is 10^{-10} metre or 10^{-8} cm.; sometimes the wavelength is expressed in terms of the *micron* (μ) which is 10^{-6} metre or 10^{-4} cm. Thus the wavelength of yellow light is 5 700 Å. or 0·57 μ.

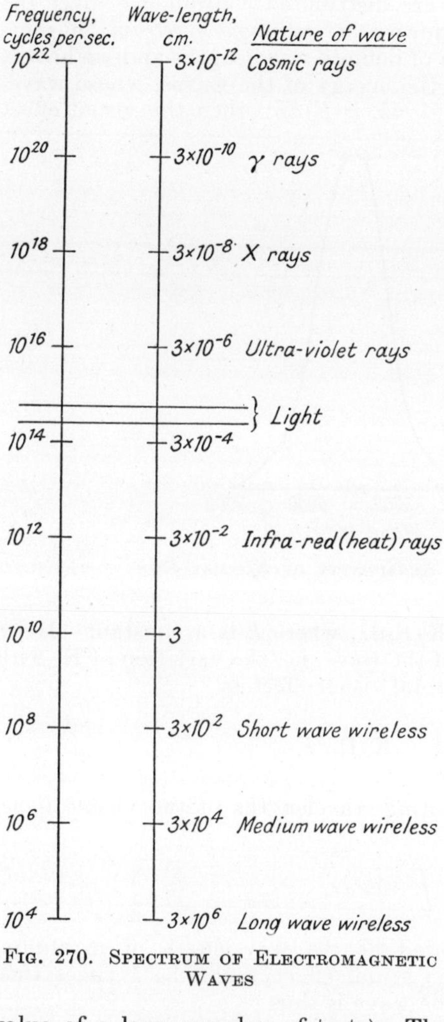

FIG. 270. SPECTRUM OF ELECTROMAGNETIC WAVES

The Spectrum of Light. Visible light can have a wavelength between 4 000 and 7 500 Å., and the colour varies in the way shown in Fig. 271. As one would expect, the sensitivity of the eye to lights of different wavelengths varies, and Fig. 272 shows the relative sensitivity of the "normal" eye (i.e. the average value of a large number of tests). The value of unity is given to the wavelength of greatest sensitivity, which is 5 500 Å. and corresponds to yellow-green light; light of this wavelength is the most efficient, but its colour makes it unsuitable for most purposes. The relative sensitivity at a wavelength λ is written K_λ and is called the *relative luminosity factor.*

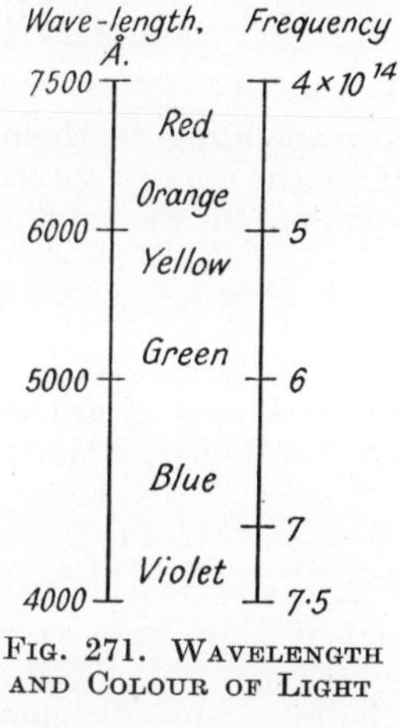

FIG. 271. WAVELENGTH AND COLOUR OF LIGHT

The energy radiated by a hot body or other source is usually composed of waves with wavelengths distributed over a band. In most cases the wavelengths are distributed continuously over the band, and we have what is known as a *continuous spectrum*; in some cases there are several waves of definite wavelengths and we have a *line spectrum*. Suppose that the energy of the waves, whose wavelengths lie between λ and $\lambda + d\lambda$, is $E_\lambda d\lambda$; then the visual effect

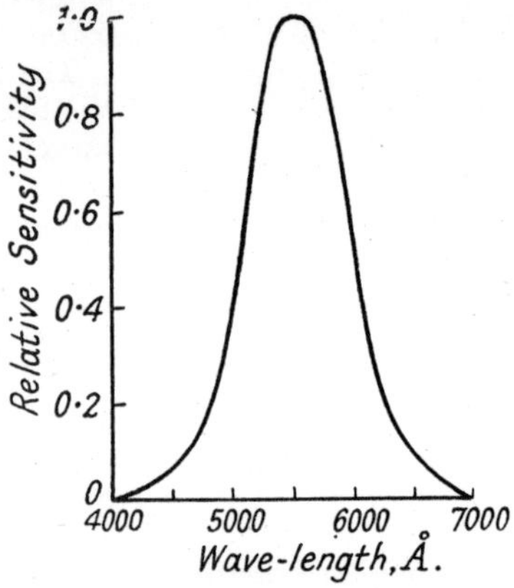

FIG. 272. VISUAL SENSITIVITY OF NORMAL EYE

corresponding to them is $k\mathrm{K}_\lambda \mathrm{E}_\lambda d\lambda$, where k is a constant. If we know the energy spectrum of the wave, i.e. the variation of E_λ with wavelength, we can find the total visual effect as

$$k\int_{\lambda_1}^{\lambda_2} \mathrm{K}_\lambda \mathrm{E}_\lambda d\lambda,$$

where λ_1 and λ_2 are the limiting wavelengths of the visible light. The total energy emitted is

$$\int_0^\infty \mathrm{E}_\lambda d\lambda,$$

and if this were concentrated at the wavelength of maximum sensitivity (i.e. 5 500 Å.) the visual effect would be k times this. The luminous efficiency of the wave is thus

$$[k\int_{\lambda_1}^{\lambda_2} \mathrm{K}_\lambda \mathrm{E}_\lambda d\lambda] \div [k\int_0^\infty \mathrm{E}_\lambda d\lambda]$$

$$= [\int_{\lambda_1}^{\lambda_2} \mathrm{K}_\lambda \mathrm{E}_\lambda d\lambda] \div [\int_0^\infty \mathrm{E}_\lambda d\lambda].$$

This must not be confused with the fraction of the energy in the visible spectrum, which is clearly

$$[\int_{\lambda_1}^{\lambda_2} E_\lambda d\lambda] \div [\int_0^\infty E_\lambda d\lambda]$$

and is called the *radiant efficiency* of the body. Fig. 273 shows the frequency spectra of the waves emitted by a blue sky, a vacuum

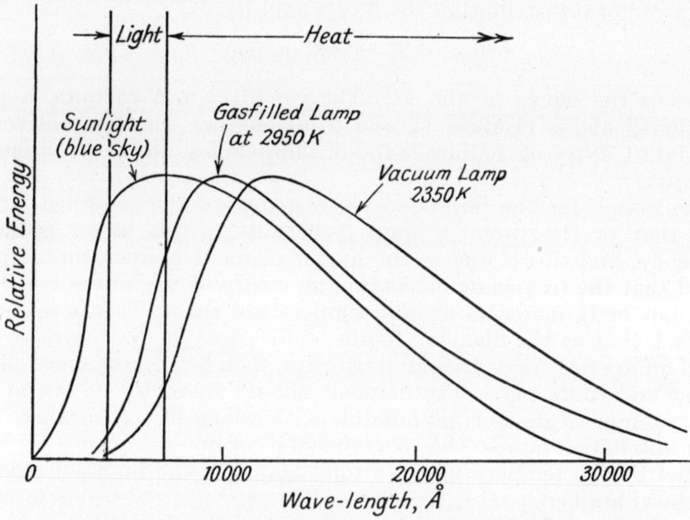

FIG. 273. FREQUENCY SPECTRA OF VARIOUS LIGHT SOURCES

lamp whose filament is at 2 350° K., and a large gas-filled lamp whose filament is at 2 950° K. The radiant efficiencies are about 20, 2, and 5 per cent, respectively.

The unit of light flux is the lumen, which will be defined later. It is found that a power of 1 watt of a wave of wavelength 5 500 Å. produces 670 lumens, so that $k = 670$. The visual effect of 1 watt of sunlight is thus

$$670 \int_{\lambda_1}^{\lambda_2} K_\lambda E_\lambda d\lambda,$$

where E_λ represents the spectrum of sunlight, and the total energy is

$$\int_0^\infty E_\lambda d\lambda = 1.$$

E_λ is substantially constant for sunlight over the visible band, and we put it equal to E_{λ_0}. The visual effect is thus

$$670 E_{\lambda_0} \int_{\lambda_1}^{\lambda_2} K_\lambda d\lambda = 670 E_{\lambda_0} \overline{K}_\lambda (\lambda_2 - \lambda_1),$$

where $\bar{K}_\lambda$ is the mean of K_λ over the visible band and is found to be about 0·37 from Fig. 273. $E_{\lambda_0}(\lambda_2 - \lambda_1)$ is the energy in the visible range and is 20 per cent of the total energy, viz. 0·2 watt. The right flux produced by 1 watt of sunlight is thus

$$670 \times 0{\cdot}2 \times 0{\cdot}37 = 50 \text{ lumens.}$$

If the whole of the energy were concentrated in the visible band with a uniform distribution the flux would be

$$670 \times 0{\cdot}37 = 250 \text{ lumens.}$$

This is the figure to aim at. The gas-filled and vacuum lamps mentioned above produce 12 and 5 lumens per watt respectively. The latest type of sodium vapour lamp gives about 50 lumens per watt.

The reason for the improved performance of the gas-filled lamp over that of the vacuum lamp is because of the larger radiant efficiency, and this is due to the higher filament temperature. It is found that the frequency of maximum energy of the waves emitted by a hot body increases as the temperature rises. This is seen by the fact that as the filament temperature rises, the first rays to be given off are heat rays, then dull red rays, then bright red, and finally yellow and white rays. The filament should therefore be run at as high a temperature as is permissible with a long life of working.

In discharge lamps the phenomenon of incandescence is not limited by the temperature of a solid filament, and high efficiencies are thus obtained.

Standards of Light. *Luminous flux* is defined as "the rate of passage of radiant energy evaluated according to the luminous sensation produced by it." The unit of luminous flux is the *lumen* (l.), and is the luminous flux emitted in a unit solid angle (steradian) by a source of 1 *candle-power* (c.p.).

The British *standard of candle-power* is provided by the Harcourt pentane lamp, for a description of which see *Electrical Measurements and Measuring Instruments*,* by E. W. Golding. This lamp is costly and requires careful maintenance, and sub-standard electric lamps are often used: the sub-standard lamps must be used with the stipulated voltage, and they often have carbon filaments. The lamp does not radiate energy equally in all directions, but this is not important so long as the luminous flux in a small solid angle† in a

* Chapter X, Third Edition.

† Fig. 274 illustrates the meaning of a solid angle. Let a cone of any shape have an apex O, and let a sphere of radius r cut the cone in the area A. The solid angle of the cone is defined as $\omega = A/r^2$, and is clearly independent of r since A is proportional to r^2. The solid angle subtended by a complete sphere is $4\pi r^2/r^2 = 4\pi$ units of solid angle, since the area of the surface of a sphere is $4\pi r^2$: this is called a *complete solid angle*. The unit of solid angle is called the *steradian*.

given direction is constant. If the small solid angle is ω and the candle-power is unity in this direction, the flux in the solid angle is $1 \times \omega = \omega$ lumens.

As a complete solid angle is 4π, the flux emitted by a source which is 1 candle-power in all directions is 4π lumens.

The *luminous intensity* (I) in any given direction is the luminous flux per unit solid angle emitted in that direction. If a flux F is emitted in a solid angle ω, the luminous intensity is $I = F/\omega$. The standard candle-power has a luminous intensity of 1 lumen per steradian in a horizontal direction; and the lumen is the luminous flux per unit solid angle of the standard in the horizontal direction.

The inclusion of the words "evaluated according to the luminous sensation produced by it" in the definition of luminous flux is necessary, since the visual sensation depends upon the wavelength as well as the rate of emission of energy. If we consider the wavelength of optimum sensitivity, we can express the luminous flux (in lumens) in terms of the rate of emission of radiant energy (in watts). It is found that at this wavelength of 5 500 Å.

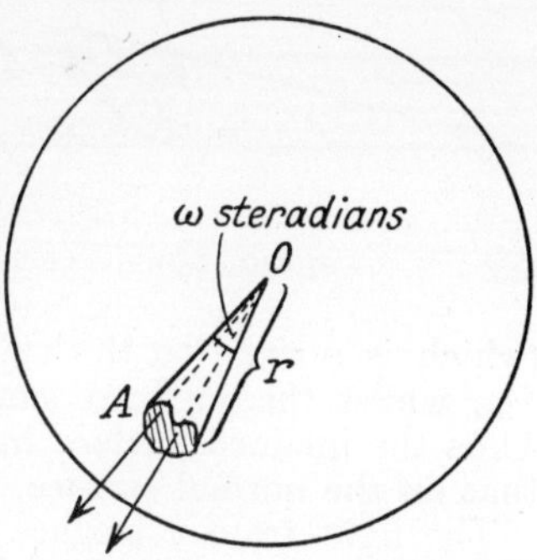

FIG. 274. SOLID ANGLE

$$1 \text{ lumen} = 0{\cdot}0015 \text{ watts}$$

or

$$1 \text{ watt} = 670 \text{ lumens.}$$

This relation is called the *mechanical equivalent of light.*

The *illumination* at a point of a surface is defined as the luminous flux per unit area at the point, the value being expressed as lumens per square metre or lumens per square foot. An illumination of 1 l. per m.^{2} is called 1 *lux,* whilst 1 l. per ft.^{2} is called 1 *foot-candle.* The distinction between illumination and luminous intensity is easily seen from Fig. 274. Suppose that a standard candle-power is situated at O and the narrow cone is along the horizontal direction, so that the luminous flux in this cone is ω lumens. The luminous intensity at any point along the cone is ω lumens per ω steradians, i.e. 1 lumen per steradian. The illumination at any point of the cone, at a distance r from O, is

$$\omega/A \text{ lumens per unit area.}$$

Since the solid angle is ω, $A = \omega r^2$, so that the illumination is

$$\omega/\omega r^2 = 1/r^2 \text{ lumens per unit area.}$$

The illumination is $1/r^2$ f.c. if r is in feet, or $1/r^2$ lux if r is in metres. The illumination thus varies as the inverse square of the distance from the diverging source, but the luminous intensity is constant along a fixed direction.

In general, *if the luminous intensity in a given direction of a diverging source is I lumens per steradian, the illumination at a point distance r from the source is* I/r^2 *f.c.* (r being in feet). This is known as the *inverse-square law*, and is widely used in calculations of illumination.

In the case considered above, the surface was normal to the axis of the light cone. If the surface is not perpendicular to the axis the illumination is less. Suppose that the normal to the surface makes an angle θ with the axis, as shown in Fig. 275. The area A then receives the same light as the area shown dotted, which has an area $A \cos \theta$. If the luminous intensity is I lumens per steradian, the dotted area (which is normal to the rays) has an illumination of $(I/A \cos \theta)$ f.c., whilst the inclined surface has an illumination of (I/A) f.c. Thus the inclined surface has an illumination which is $\cos \theta$ times that on the normal surface. This is the *cosine law*.

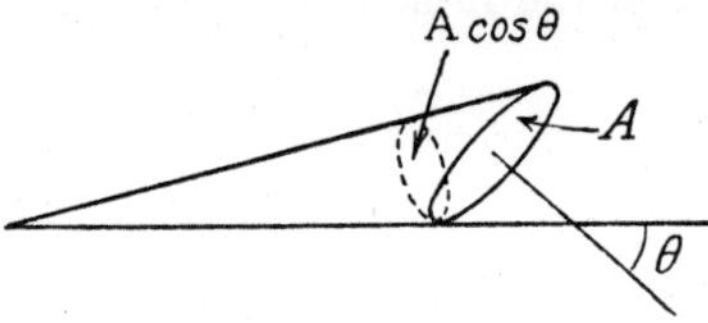

FIG. 275

The light from a surface may be due either to the emission of light or to reflection. The *brightness* of a surface is defined as the luminous intensity per unit projected area of the surface, i.e. the candles per unit area as viewed. A piece of white paper with an illumination of 4 f.c. has a brightness of about 1 c. per ft.^{2} The brightness of a surface depends largely upon the character of the surface if it is itself not an emitter. Thus if light is shining upon a polished surface, the brightness depends upon the angle of viewing;

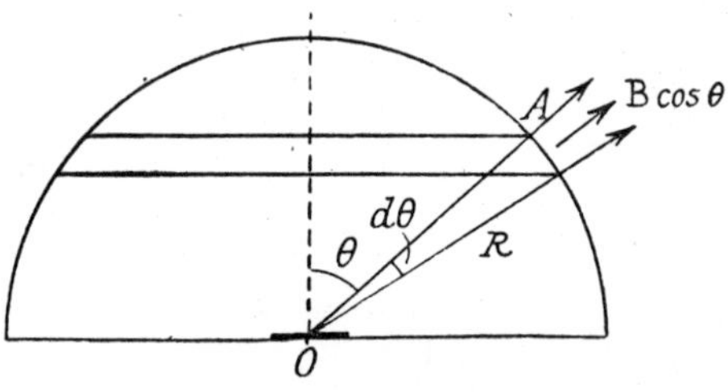

FIG. 276

but if the surface is matt and the diffusion is good, the brightness is approximately independent of the angle of viewing.

Suppose that we have a surface which is perfectly diffusing, so that the brightness is constant, B say, at all angles. Fig. 276 shows the surface, which is considered small and has area a, at O. At a point A on a hemisphere with centre O, the brightness is B. The luminous intensity at A is $\mathrm{B}a \cos \theta$, since B is the brightness and the area has a projected value which is $\cos \theta$ times its actual value a. Consider a zone of the hemisphere between θ and $\theta + d\theta$. The width of this zone is $Rd\theta$ and its length is $2\pi R \sin \theta$ so that its area is $2\pi R^2 \sin \theta \, d\theta$ and its solid angle is $2\pi \sin \theta \, d\theta$. The luminous flux passing through this zone is

$$\mathrm{B}a \cos \theta \times 2\pi \sin \theta \, d\theta.$$

The total luminous flux leaving the surface is then

$$\int_0^{\pi/2} 2\pi \mathrm{B}a \cos \theta \sin \theta \, d\theta = \pi \mathrm{B}a.$$

Thus the brightness of a perfectly diffusing surface is $(1/\pi)$ times the flux leaving it from unit area. Thus if the surface receives 4 l. per ft.^{2} and its coefficient of reflection is 80 per cent, 3·2 l. per ft.^{2} leave it and the brightness is $3{\cdot}2/\pi = 1{\cdot}02$ c. per ft.^{2}

Values of Illumination. The following list gives the value of illumination required for various purposes.

Purpose	Illumination (foot-candles)
Average thoroughfare at night . .	0·02 to 0·5
Main thoroughfare at night . .	0·5 to 2·0
Waiting rooms, bedrooms, etc. .	2 to 4
Living rooms by night . . .	4 to 8
Living rooms by daylight . .	20 to 30
Offices by night	10 to 20
Shop windows at night . . .	20 to 50

It may be mentioned that full sunlight at noon provides 10 000 f.c., a cloudy sky about 500.

The visual effect is approximately proportional to the logarithm of the illumination, so that a rise from 1 to 2 f.c. is as effective as a rise from 4 to 8. This is the Weber-Fechner law.

Polar Curves. In most lamps the luminous intensity is not the same in all directions. Suppose that the lamp is held with its axis vertical and we measure the candle-power (i.e. luminous intensity) at all directions in a horizontal plane through the lamp. The curve of candle-power against direction is plotted as shown in Fig. 277, which shows the horizontal polar curve of a gas-filled lamp with a horizontal circular filament; the drop in candle-power along OA' is due to the break in the ring where the current enters and leaves. The mean of the candle-power in this curve gives the *mean horizontal candle-power*, and is found by taking the mean at the angular positions, 0, 10°, 20° . . . 350°. The luminous intensity (or candle-power) in any given direction is measured by means of a Bunsen, Lummer-Brodhun, or flicker photometer. For detailed descriptions of these instruments and their use see *Electrical Measurements and Measuring Instruments*,* by Golding. Recently photo-cells and electronic cells have been used for the measurement of illumination.

The *mean spherical candle-power* (m.s.c.p.) is the mean of the candle-power in all directions radiating from the lamp, and is the total flux divided by 4π. It would be a very difficult thing to measure the candle-power in all directions and take the mean. The measurement of the total luminous flux emitted by a lamp is therefore measured by the *integrating or Ulbricht sphere* (see Golding

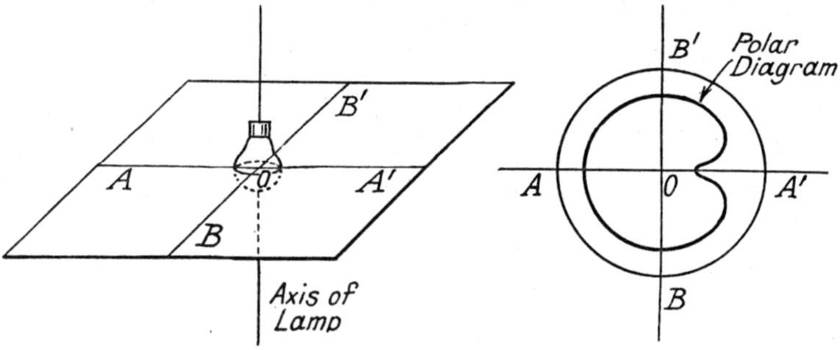

FIG. 277. POLAR DIAGRAM

for details). The principle of the test is the following. The lamp is put inside a sphere, the inside of which is painted white and diffuses the light. A small screen is viewed through a hole in the sphere, and its brightness is compared with that produced by a comparison lamp outside by means of a photometer. The lamp inside the sphere is replaced by a standard lamp and the process is repeated. The brightness of the screen is proportional to the luminous flux produced by the test or standard lamp, so that the ratio of the luminous fluxes is found for the two cases, and thence the flux for the test lamp since the standard lamp has a known flux.

The ratio of the mean spherical candle-power to the mean horizontal candle-power is the *reduction factor* of the source.

Rousseau Diagram and Russell Angles. There is a special case for which the mean spherical candle-power can be found from a polar curve, viz. when there is an axis of symmetry. Suppose that the axis of symmetry is vertical, so that the candle-power is constant in any horizontal direction; the polar diagram in any horizontal plane is then a circle. We have then to find the polar curve in any one vertical plane XOY, as shown in Fig. 278. To do this we can either move our photometer round a circle in this plane, or, more

* Chapter X, Third Edition.

conveniently, we can keep our photometer fixed on the bench and rotate the lamp about the axis OZ which is perpendicular to OX and OY. Suppose that the polar curve is as shown in Fig. 279. Consider a unit sphere with the point O as centre. The zone $PQP'Q'$ between the angles θ and $\theta + d\theta$ has an area

$$2\pi \, . \, RP \, . \, PQ$$
$$= 2\pi \cos\theta \, . \, d\theta,$$

so that the solid angle subtended by this zone at O is $2\pi \cos\theta \, d\theta$. Let the luminous intensity along the direction θ be I; then the

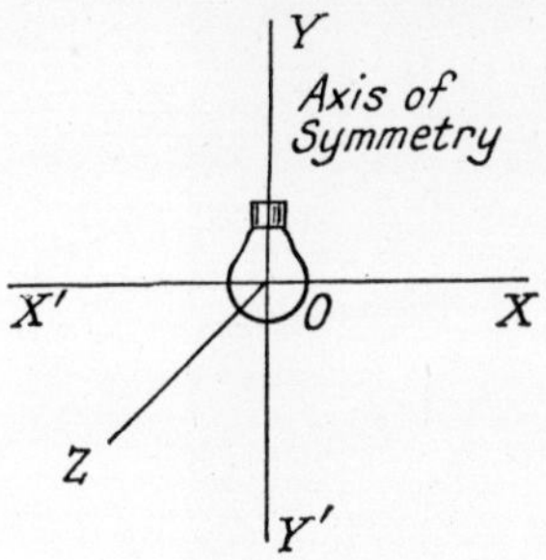

FIG. 278

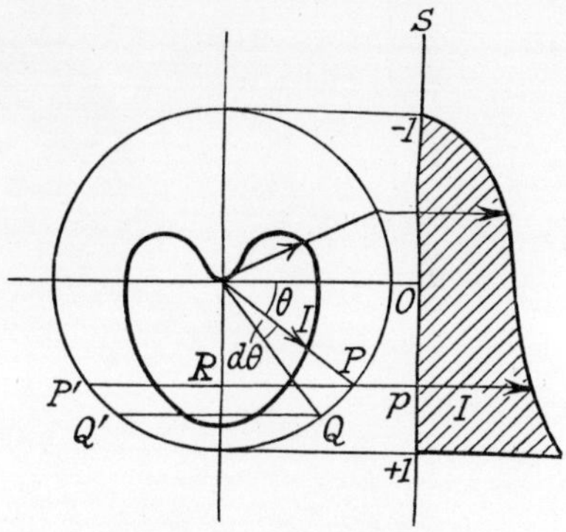

FIG. 279. ROUSSEAU DIAGRAM

luminous flux across the zone is $2\pi\text{I} \cos\theta \, . \, d\theta$, and the total flux emitted by the lamp is

$$\int_{-\pi/2}^{\pi/2} 2\pi\text{I} \cos\theta \, . \, d\theta.$$

This integration is best performed graphically by replacing $\sin\theta$ by s. The integral becomes

$$2\pi \int_{-1}^{+1} \text{I} ds.$$

If we therefore plot I against s, i.e. $\sin\theta$, the integral is the area under this curve. The plot is performed by Rousseau's method shown in Fig. 279. The radial line is drawn to meet the circle, which is the section of the unit sphere, and a perpendicular is drawn to the axis of s. At the point p of the axis of s a line of height I is drawn. The extremity of this line is on the I—s curve, which is found from a number of these constructions. The method is rendered simple by having a card on which the circle, radial lines, and perpendicular lines are printed. The area under the curve can be found by means of a planimeter, but it can be found also by taking the mean of I at a number of close values of s. Thus we can take the values of I at $s = -0{\cdot}95, -0{\cdot}85, -0{\cdot}75, \ldots -0{\cdot}05, +0{\cdot}05, +0{\cdot}15, \ldots 0{\cdot}85, 0{\cdot}95$, and we multiply the mean of these values by $2\pi \times 2 = 4\pi$, since the length of the s-axis is 2. If we take the OY axis as the line from which the angles are measured, the angles corresponding to these values of s are 18·2°, 31·8, 41·4, 49·5, 56·6,

FIG. 280. "OSIRA" 400 W. LAMP
(*G.E.C.*)

63·3, 69·5, 75·5, 81·4, 87·1, 92·9, 98·6, 104·5, 110·5, 116·7, 123·4, 130·5, 138·6, 148·2, and 161·8. These are called the *Russell angles.* The total luminous flux is then 4π times the mean of the luminous intensities at the Russell angles.

Light Fittings and the Polar Curve. The polar curve of most lamps is widespread and has an axis of symmetry; there are certain directions of low luminous intensity due to breaks in the filament or the shading by a cap. The polar distribution is unsuitable for almost any use, and light fittings are used to achieve desirable distributions of the light.

Thus in many industrial cases it is required that the light should give a uniform illumination on a horizontal surface and no light upwards. Fig. 280 shows an "Osira" lamp (400 watts, 17 000 lumens) and an industrial reflector unit. The polar curves of the lamp without a reflector, and with reflectors of different angles, are shown in Fig. 281. Curve C gives uniform illumination over a wide

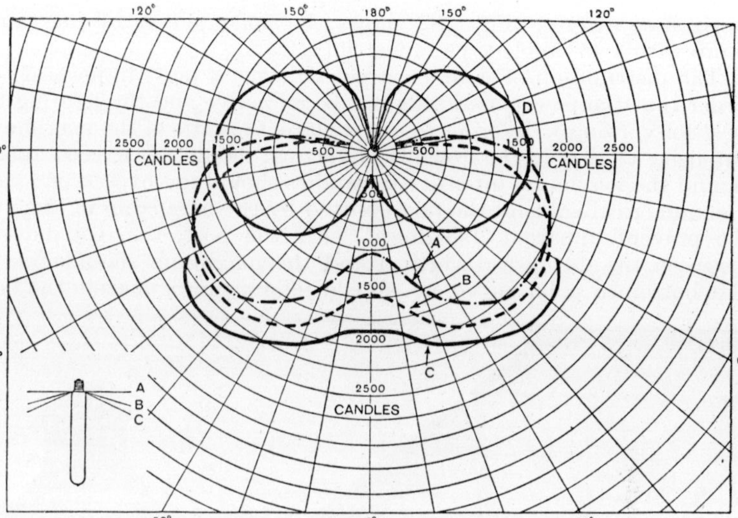

FIG. 281. POLAR DIAGRAMS OF "OSIRA" LAMP WITHOUT AND WITH REFLECTORS
(*G.E.C.*)

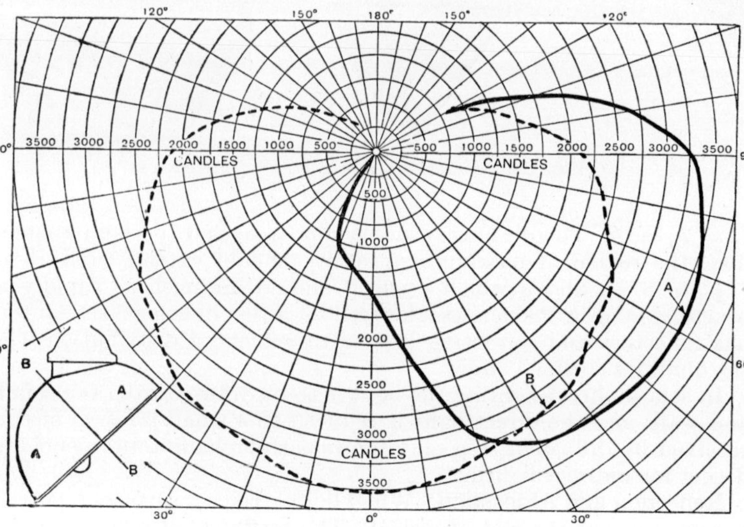

FIG. 282. POLAR DIAGRAM OF "OSIRA" LAMP WITH INDUSTRIAL REFLECTOR
(*G.E.C.*)

horizontal surface. Fig. 282 shows the polar curves of the same lamp with a 45° angle-type industrial reflector.

The distribution of light in street lighting is most important, in order that the power used may not be excessive and the lighting be best for viewing.* In order to avoid glare, the ratio of the maximum intensity to that in the downward vertical should not exceed about 6, and the maximum intensity in candles should not exceed one-half the numerical value of the flux emitted by the bare lamp in lumens. To prevent apparent "shadows" near the post, the distribution between the downward vertical and the maximum should give a flat-bottomed polar curve, i.e. should follow approximately the law

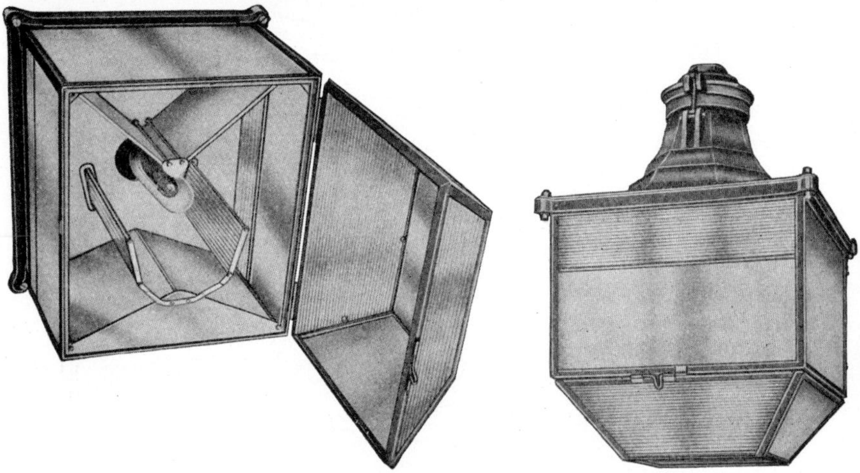

FIG. 283. LANTERN FOR STREET LIGHTING
(*I.E.E. Journal*)

$I_\theta = I_0 \sec \theta$. To give long, wide, bright areas and to avoid waste of flux, the region of maximum intensity should extend from about $77\frac{1}{2}°$ to $87\frac{1}{2}°$, above which angle it should diminish as rapidly as practicable. Fig. 283 shows a lantern for street lighting: the lantern has reflectors and refracting prisms to achieve a distribution with the qualities outlined above.

In semi-indirect lighting the light is to be diffused from the ceiling and walls and the direct light is reduced to a small fraction of that received by diffusion. Fig. 284 shows a Holophane totally-enclosed fitting for indirect lighting.

Reflection and Absorption. When light strikes a surface, a portion is reflected and the rest absorbed. The *coefficient of reflection* is the ratio of the luminous flux leaving the surface to the flux incident

* See "The High-pressure Mercury Vapour Lamp in Public Lighting," by Wilson, Dament, and Waldram, *Journal of the I.E.E.*, 1936.

on the surface. Good white paper reflects 80 per cent, common white paper and white wood about 40 per cent, dark brown paper 15 per cent. A white ceiling reflects 40 to 70 per cent.

When light passes through a transparent material some is absorbed. The ratio of the light absorbed to the light entering is called the *coefficient of absorption*. *Frosting* the surface causes an absorption of 5 to 10 per cent. Modern internal frosting forms no

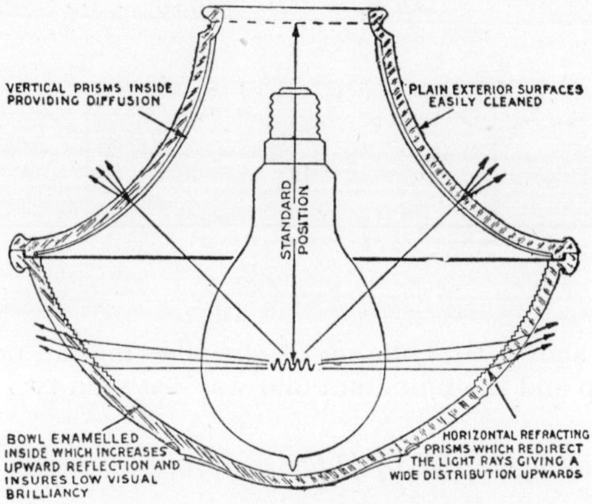

FIG. 284. HOLOPHANE TOTALLY-ENCLOSED FITTING
(*Holophane, Ltd.*)

sharp scratches and the absorption is less than 5 per cent; the outer surface is smooth and easily cleaned.

Lighting Calculations. It is often required to calculate the illumination provided by a given set of lamps and reflectors (or shades). It is necessary to know the polar curve of the sources and fittings, and then the use of the inverse square and Lambert's cosine laws provides a simple solution. The following two examples illustrate the method.

EXAMPLE. Gas-filled lamps with suitable shades are employed to illuminate a corridor. The lamps are spaced 25 ft. apart and are suspended 16 ft. above the centre line of the floor. If each lamp emits 100 c.p. in every direction below the horizontal, determine (*a*) the maximum and (*b*) the minimum value of the illumination on the floor along the centre-line. Make a sketch to scale showing the distribution of the illumination along the floor between a pair of lamps. (*Lond. Univ.*, 1928.)

The polar curve is given as a semi-circle of magnitude 100 c.p. Fig. 285 shows five of the lamps, B, B_1, B_2, B_3, B_4. At a point P at distance x from B_2 the illumination due to B_2 is

$$\frac{100}{B_2P^2}\cos\theta_2 = \frac{100}{16^2}\cos^3\theta_2,$$

since $B_2P = 16/\cos\theta_2$.

The total illumination at P is thus

$$(100/16^2)\,(\cos^3\theta_1 + \cos^3\theta_2 + \cos^3\theta_3 + \cos^3\theta_4 + \ldots),$$

where θ_1, θ_2, θ_3, θ_4 . . . are the angles which B_1P, B_2P, B_3P, B_4P . . . make with the vertical, so that

$$\tan\theta_1 = (25 + x)/16,\ \tan\theta_2 = x/16,\ \tan\theta_3 = (25 - x)/16,$$
$$\tan\theta_4 = (50 - x)/16,\ \ldots$$

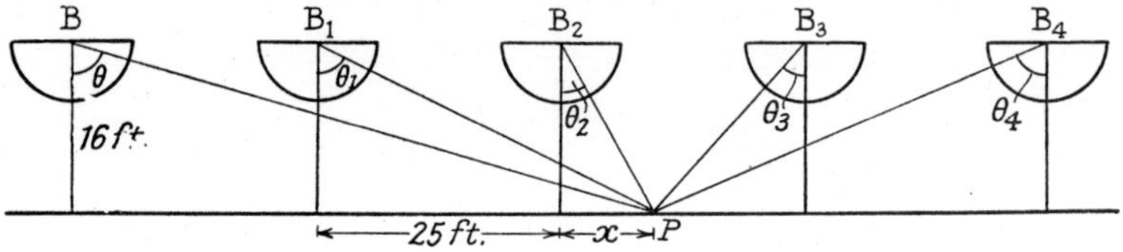

FIG. 285

It can be shown that the maximum illumination occurs directly under a lamp and the minimum mid-way between two lamps.

For

$$(d/dx)\,(\tan\theta_1) = \sec^2\theta_1\, d\theta_1/dx = 1/16,$$

so that

$$d\theta_1/dx = \cos^2\theta/16$$

and similarly

$$d\theta_2/dx = \cos^2\theta/16,\ d\theta_3/dx = -\cos^2\theta_3/16,$$
$$d\theta_4/dx = -\cos^2\theta_4/16.$$

The illumination is a maximum or minimum when

$$\frac{d}{dx}\left[\frac{100}{16^2}(\cos^3\theta_1 + \cos^3\theta_2 + \cos^3\theta_3 + \cos^3\theta_4 + \ldots)\right] = 0$$

$$\text{i.e. when } -\frac{100}{16^2}\times 3\left(\cos^2\theta_1 \sin\theta_1 \frac{d\theta_1}{dx} + \ldots\right) = 0$$

$$= -\frac{100\times 3}{16^2\times 16}(\ldots \cos^4\theta_1\sin\theta_1 + \cos^4\theta_2\sin\theta_2 - \cos^4\theta_3\sin\theta_3$$
$$- \cos^4\theta_4\sin\theta_4 \ldots).$$

When P is mid-way between B_2 and B_3 we have $\theta_2 = \theta_3$, $\theta_1 = \theta_4$, etc., so that the terms cancel out in pairs; there is thus a maximum or a minimum, and it is clearly the latter. When P is directly underneath B_2, $\theta_2 = 0$, $\theta_1 = \theta_3$, etc., so that the θ_2 term is zero and the rest cancel out in pairs; the illumination is clearly a maximum.

The illumination directly under B_2 is found from the general expression by putting $x = 0$, so that $\theta_2 = 0$, $\theta_1 = \theta_3 = \tan^{-1}(25/16)$, $\theta = \theta_4 = \tan^{-1}(50/16)$, etc. We then construct the following table—

$\theta_2 = 0$	$\cos\theta_2 = 1$	$\cos^3\theta_2 = 1$
$\theta_3 = 57° 23'$	$\cos\theta_3 = 0{\cdot}539$	$\cos^3\theta_3 = 0{\cdot}157$
$\theta_4 = 72° 15'$	$\cos\theta_4 = 0{\cdot}305$	$\cos^3\theta_4 = 0{\cdot}028$
$0_5 = 77° 57'$	$\cos\theta_5 = 0{\cdot}209$	$\cos^3\theta_5 = 0{\cdot}009$

It is unnecessary to go further, and we find that the maximum illumination is

$$(100/16^2)(1 + 2 \times 0{\cdot}157 + 2 \times 0{\cdot}028 + 2 \times 0{\cdot}009)$$
$$= (100 \times 1{\cdot}39)/16^2 = \underline{\underline{0{\cdot}54 \text{ f.c.}}}$$

The minimum illumination occurs at the point mid-way between B_2 and B_3 at which point $\theta_2 = \theta_3 = \tan^{-1}(12{\cdot}5/16)$, $\theta_1 = \theta_4 = \tan^{-1}(37{\cdot}5/16)$, etc. We find

$\theta_3 = 38°$	$\cos\theta_3 = 0{\cdot}788$	$\cos^3\theta_3 = 0{\cdot}489$
$\theta_4 = 66° 57'$	$\cos\theta_4 = 0{\cdot}392$	$\cos^3\theta_4 = 0{\cdot}062$
$\theta_5 = 75° 39'$	$\cos\theta_5 = 0{\cdot}248$	$\cos^3\theta_5 = 0{\cdot}015$

so that the minimum illumination is

$$(100/16^2) \times 2 \times (0{\cdot}489 + 0{\cdot}062 + 0{\cdot}015)$$
$$= (100 \times 2 \times 0{\cdot}566)/16^2 = \underline{\underline{0{\cdot}44 \text{ f.c.}}}$$

At a point mid-way between the positions of maximum and minimum illuminations $x = 6{\cdot}25$ ft.

$\theta_2 = \tan^{-1}(6{\cdot}25/16) = 21° 21'$	$\cos\theta_2 = 0{\cdot}931$	$\cos^3\theta_2 = 0{\cdot}807$
$\theta_3 = \tan^{-1}(18{\cdot}75/16) = 49° 33'$	$\cos\theta_3 = 0{\cdot}649$	$\cos^3\theta_3 = 0{\cdot}273$
$\theta_1 = \tan^{-1}(31{\cdot}25/16) = 62° 54'$	$\cos\theta_1 = 0{\cdot}456$	$\cos^3\theta_1 = 0{\cdot}095$
$\theta_4 = \tan^{-1}(43{\cdot}75/16) = 69° 57'$	$\cos\theta_4 = 0{\cdot}343$	$\cos^3\theta_4 = 0{\cdot}040$
$\theta = \tan^{-1}(56{\cdot}25/16) = 74° 9'$	$\cos\theta = 0{\cdot}273$	$\cos^3\theta = 0{\cdot}020$
$\theta_5 = \tan^{-1}(68{\cdot}75/16) = 76° 54'$	$\cos\theta_5 = 0{\cdot}227$	$\cos^3\theta_5 = 0{\cdot}012$

The illumination at this point is

$$(100/16^2) \times 1{\cdot}247 = 0{\cdot}487 \text{ f.c.}$$

Fig. 286 shows the distribution of illumination along the centre line of the corridor.

Example. A space is illuminated by a lamp, which is suspended 10 ft. above the ground, and which gives uniform candle-power in all directions in the lower hemisphere. Find (*a*) the ratio of the ground illumination at a point directly under the lamp to that at a point 20 ft. away, (*b*) the position of a similar lighting unit in order that the minimum illumination on the ground along a line joining the points vertically below the lamps shall not be less than half the maximum value.

Indicate your view of the relative importance in practice of the illumination falling vertically on the ground and that incident on a surface perpendicular to the ground. In case (*b*) above, estimate the relative values of these quantities (i) at the base of a lamp, and (ii) half-way between two lamps.

(*Lond. Univ.*, 1933.)

Let the intensity of luminous flux be I in all directions in the lower hemisphere. At a point directly under the lamp the illumination is $I/10^2$. At a point 20 ft. away the illumination is

$$(I/10^2)\cos^3\theta,$$

where $\tan\theta = 20/10 = 2$. The ratio is thus

$$1/\cos^3\theta = \sec^3\theta = (1+\tan^2\theta)^{\frac{3}{2}} = 5^{\frac{3}{2}} = \underline{\underline{11{\cdot}2}}.$$

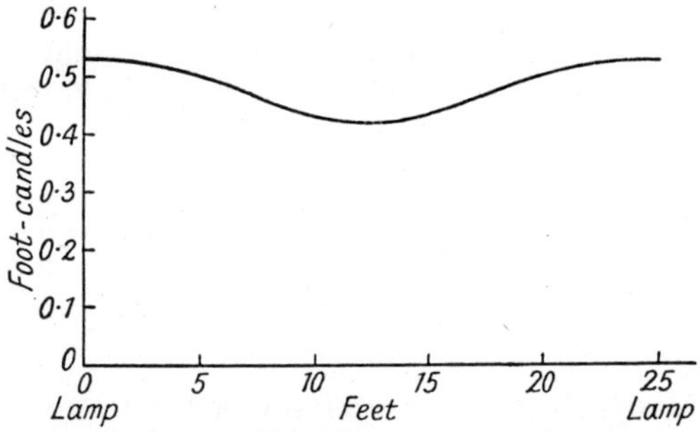

FIG. 286

Let a similar lamp be placed at distance $2x$. The maximum illumination occurs under each lamp and is

$$I/10^2 + (I/10^2)\cos\phi$$

where $\tan\phi = 2x/10$. The minimum illumination is mid-way between the lamps and is

$$2(I/10^2)\cos^3\psi,$$

where $\tan\psi = x/10$. (See Fig. 287.) The minimum illumination is to be one-half of the maximum, so that

$$1 + \cos^3\phi = 4\cos^3\psi$$

or

$$1 + \frac{1}{(1+4x^2/100)^{\frac{3}{2}}} = \frac{4}{(1+x^2/100)^{\frac{3}{2}}}.$$

It is difficult to solve this equation for x by an exact method, but an approximate value of x is easily found. For

$$\frac{4}{(1+x^2/100)^{\frac{3}{2}}} > 1,$$

so that $\qquad 1 + x^2/100 < 4^{\frac{2}{3}} = 2{\cdot}52,$

or $\qquad x^2/100 < 1{\cdot}52,$

i.e. $\qquad x < \sqrt{152} = 12{\cdot}3$ ft.

If we try $x = 10$, we find that the maximum illumination is

$$\frac{I}{10^2}(1 + \cos^3\phi) = \frac{I}{10^2}\left[1 + \frac{1}{5^{\frac{3}{2}}}\right] = \frac{I}{10^2} \times 1{\cdot}0895,$$

whilst twice the minimum illumination is

$$\frac{I}{10^2}\,\frac{4}{(1+1)^{\frac{3}{2}}} = \frac{I}{10^2}\,\frac{4}{2^{\frac{3}{2}}} = \frac{I}{10^2} \times 1{\cdot}414.$$

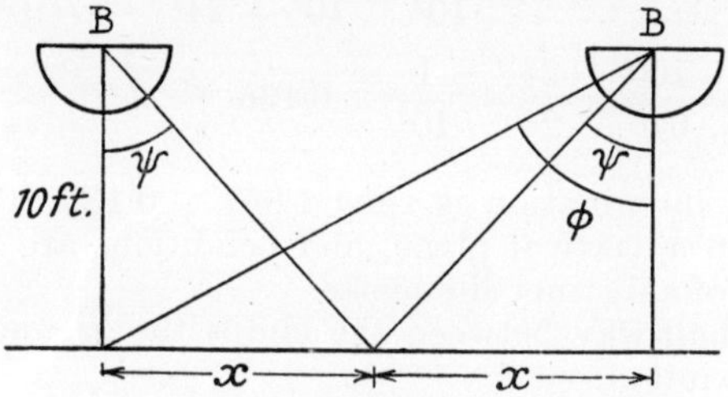

FIG. 287

The ratio

$$\frac{\text{Twice minimum illumination}}{\text{Maximum illumination}} = \frac{1{\cdot}414}{1{\cdot}0895} = 1{\cdot}30.$$

If we try $x = 12$, this ratio is found to be

$$\left(\frac{4}{2{\cdot}44^{\frac{3}{2}}}\right) \div \left(1 + \frac{1}{6{\cdot}76^{\frac{3}{2}}}\right) = \frac{4}{3{\cdot}82 \times 1{\cdot}057} = 0{\cdot}990.$$

x is thus just less than 12 ft., so that the distance between the lamps is <u>24</u> ft.

In modern street lighting the conscious aim is to view people and objects as silhouettes against a light background. It has been found that most objects reflect only a small amount of the light incident on them; thus a fawn-coloured mackintosh reflects only 15 per cent of the light. The method then is to illuminate the road surface as much as possible, even by giving the road a polished surface. If the road were to extend in all directions, the best method would be to leave the people and objects perfectly black, i.e. to have no light in a plane perpendicular to the ground. But roads have a finite width, and it is necessary to illuminate the walls and houses running along the road so that they form a bright background for the silhouettes of the people and objects. The method aims, therefore, at having a bright road surface, small illumination on planes which are perpendicular to the road and in the road, and bright illumination on the walls and houses bounding the road.

In case (*b*) the illumination on the ground at the base of a lamp is

$$\frac{I}{10^2}\left[1+\frac{1}{(1+5{\cdot}76)^{\frac{3}{2}}}\right]=\frac{I}{10^2}\times 1{\cdot}057.$$

The illumination at this point on a surface perpendicular to the ground is

$$(I/10^2)\sin\phi + (I/10^2)\cos^2\phi\,.\,\sin\phi$$

$$=\frac{I}{10^2}\cos^2\phi\sin\phi=\frac{I}{10^2}\times\frac{10^2}{10^2+24^2}\times\frac{24}{\sqrt{(10^2+24^2)}}$$

$$=\frac{I}{10^2}\times\frac{100}{676}\times\frac{24}{26}=\frac{I}{10^2}\times 0{\cdot}136.$$

The ground illumination is thus $1{\cdot}057 \div 0{\cdot}136 = 7{\cdot}8$ times the illumination on a vertical plane, and conditions are very good for the formation of a distinct silhouette.

At a point half-way between the lamps the ground illumination is half the maximum, viz.

$$\tfrac{1}{2}\frac{I}{10^2}\times 1{\cdot}057=\frac{I}{10^2}\times 0{\cdot}528.$$

The illumination on a vertical plane depends on the position of the plane; the plane of brightest illumination is that which is perpendicular to the line joining the lamps, and it receives the light of only one lamp on each of its faces. The illumination is thus

$$(I/10^2)\cos^2\psi\sin\psi$$

$$=\frac{I}{10^2}\times\frac{10}{10^2+12^2}\times\frac{12}{\sqrt{(10^2+12^2)}}=\frac{I}{10^2}\times\frac{10}{244}\times\frac{12}{15{\cdot}6}$$

$$=(I/10^2)\times 0{\cdot}0314.$$

The ratio of ground and vertical plane illuminations is thus 16·8, which is again satisfactory.

Calculation of Light Required. An important problem of frequent occurrence is the calculation of the light required for a desired illumination. For this purpose one must be given the *coefficient of utilization*, which is the ratio of the light incident on the illuminated surface to that emitted by the sources. A usual value is 70 or 80 per cent. The value of the coefficient is considerably diminished by the deposition of dirt on the lamps, shades and reflectors, which should therefore be easy to clean and be cleaned as often as required.

Another problem is the design of reflector (or calculation of the polar curve) to give a desired distribution of illumination. The following example illustrates the methods.

Example. It is desired to illuminate a circular space 30 ft. in diameter at a uniform intensity of 10 f.c. by means of a single lamp suspended 10 ft. above the centre of the space. If the lamp has an output of 19 l. per W. and is used with a reflector having a coefficient of utilization of 70 per cent, find the size of lamp required and draw the polar curve which the lamp and reflector should have. (*Lond. Univ.*, 1934.)

The area to be illuminated is

$$(\pi/4) \times 30^2 = 708 \text{ ft.}^2,$$

so that the total light required is

$$708 \times 10 = 7\,080 \text{ l.}$$

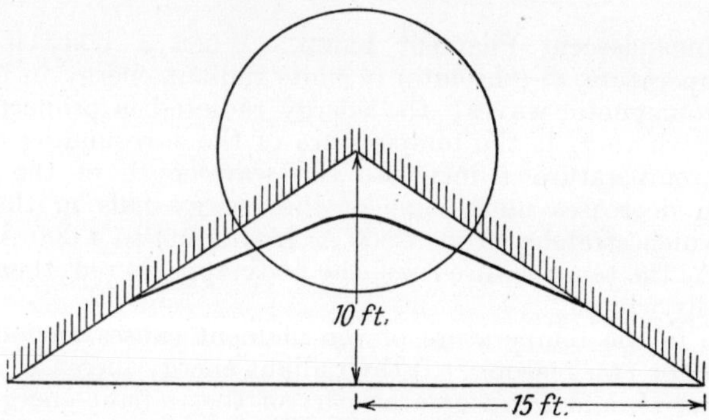

Fig. 288. Polar Curve

The light output of the lamp must therefore be

$$7\,080 \times (100/70) = 10\,100 \text{ l.},$$

which necessitates a power of

$$10\,100/19 = 530 \text{ W.}$$

Let the luminous intensity of the lamp and reflector be I at an angle θ with the vertical. The illumination on the surface is

$$(\text{I}/10^2) \cos^3 \theta = 10 \text{ f.c.},$$

so that

$$\text{I} = 1\,000 \sec^3 \theta.$$

This is the equation of the polar curve within the limiting angle of $\tan^{-1} 1{\cdot}5$; outside this angle I is zero. Fig. 288 shows this polar curve.

Sources of Light. There are three practicable ways of producing light, by incandescence of a heated filament, by an arc, and by a glow discharge. All three methods are used in modern illumination, and modern research on a large scale is continuously devoted to the first and third methods.

The incandescent filament was made in a practical form in 1878 by Swan and Edison simultaneously and independently, and has since then been improved and developed without interruption. Its main field of use in the future is in domestic lighting.

The arc lamp is mainly useful in large outdoor lamps, but a modern descendant in the form of the "Pointolite" lamp is applicable to indoor use.

The discharge lamp was first made in a useful form by Moore in 1905, and suffered neglect for about twenty years, because of the rapidly improving incandescent filament lamp. In the last ten years, however, the discharge lamp has come to the fore, and has already revolutionized the illumination of streets and public lighting generally.

The Incandescent Filament Lamp. When a body is heated to a temperature $\tau°$ (absolute) it emits radiant energy in the form of electromagnetic waves; the energy radiated is proportional to $\tau^4 - \tau_0{}^4$, where τ_0 is the temperature of the surrounding medium. As the temperature τ increases the wavelength of the emitted radiation decreases until some of the energy falls in the visible range, which stretches from 7 500 Å. (red light) to 4 000 Å. (violet light). As the temperature rises, the body appears red, then yellow, and finally white.

A rise in the temperature of the filament causes an increase of the light for two reasons: (i) the radiant energy increases approximately as τ^4, and (ii) a greater part of the radiant energy lies in the visible part of the spectrum. Fig. 273 illustrates the latter effect; the radiant efficiency of a filament at 2 350° absolute is about 2 per cent, and at 2 950° about 5 per cent. If the same filament were concerned, the energy radiated at the higher temperature would be $(2\,950/2\,350)^4 = 2{\cdot}5$ times that at the lower temperature, and the light flux would be $2{\cdot}5 \times (5/2) = 6{\cdot}25$ times as great. For practical lamps it is found that the light emitted is proportional to the twelfth power of the temperature, and this shows the importance of operating at as high a temperature as possible.

The influence of the dimensions of the filament on the heating is found easily if it is assumed that the heat lost by convection is negligible compared with that lost by radiation. Let the filament be of length l, diameter d, and resistivity ρ. The power supplied by a current I is

$$I^2R = I^2(\rho l/\tfrac{1}{4}\pi d^2),$$

and this equals the power radiated, which is

$$\begin{aligned} &ek \times \text{area} \times (\tau^4 - \tau_0{}^4) \\ &= ek \times \pi dl \times (\tau^4 - \tau_0{}^4) \end{aligned}$$

where e is the emissivity of the surface and is unity for a black body and less for a polished body; and k is a constant. Equating these powers we get

$$I = \left[\frac{ek\pi^2}{4\rho}(\tau^4 - \tau_0^4)\right]^{\frac{1}{2}} d^{\frac{3}{2}}.$$

If we consider two filaments of the same material at the same temperature, we see that I is proportional to $d^{\frac{3}{2}}$ or d is proportional to $I^{\frac{2}{3}}$.

Example. Find the ratio of the radii and lengths of the filaments of lamps of the same candle-power, efficiency and material, operating at 200 V. and 240 V.

As the efficiencies are the same, their temperatures must be the same. [This is true of lamps at reasonably high voltages such as 200 V., but it is not true for lamps at very low voltages such as 2 or 4 V.; for in the latter case the temperature varies along the filament from a minimum near the ends to a maximum at the centre, and the temperature distribution is very different in filaments of different lengths.]

Let d_1 and d_2 be the diameters of the filaments, I_1 and I_2 their currents. Then by the law just proved and the fact that the temperatures are equal,

$$d_1/d_2 = (I_1/I_2)^{\frac{2}{3}}.$$

As the watts dissipated are the same in both cases

$$200I_1 = 240I_2,$$

so that

$$d_1/d_2 = (240/200)^{\frac{2}{3}} = \underline{\underline{1{\cdot}13}}.$$

Since the radiated power is the same in both cases

$$d_1l_1 = d_2l_2,$$

so that

$$l_1/l_2 = d_2/d_1 = \underline{\underline{0{\cdot}885}}.$$

The filament for 200 V. is shorter and thicker than that for 240 V.

Carbon Filament Lamp. The method of making this lamp to-day is based on the work of Swan. Swedish filter paper is dissolved in zinc chloride solution, and the viscous solution is squirted slowly through a fine die into a jar of acidified alcohol. Tough cellulose threads are obtained and are wound on formers, which are packed into a crucible filled with finely powdered graphite. The crucibles are baked in a furnace at 1 400° C., and the cellulose threads become tough threads of pure carbon. The threads are mounted on their supports by a graphite paste, and are then heated by a current in coal-gas or benzene. This process, which is known as *flashing*, removes any irregularities in the filament; for a thin part of the filament runs hotter than a thick part and receives a larger deposit of carbon from the coal-gas which is decomposed. The filament is placed in a glass bulb, which is evacuated till the pressure falls below 0·002 mm. of mercury. This is done in order to diminish the losses from the filament due to convection.

Although carbon is the most refractory substance known and can be raised to a temperature of 5 500° abs. in an arc, it commences to evaporate at about 2 000° abs. in a vacuum. This is the limiting temperature, for at higher temperatures the carbon evaporates and blackens the bulb. The efficiency of the latest type of carbon filament lamp, operating at 2 073° abs., is only 3·6 lumens per watt.

Tungsten Filament Lamp. Attention was then paid to the study of metals as filaments, and osmium was found to be the most suitable until Coolidge developed the method of making drawn tungsten wire (1906–9); since then tungsten has replaced all except carbon filaments, which are used only for special purposes.

The process of obtaining tungsten wire suitable for lamp filaments is as follows. Pure tungsten oxide is produced from Sheelite or Wolframite by a number of chemical processes, and is reduced to pure tungsten by heating in hydrogen. The tungsten produced in this way is a grey powder, which is compressed by hydraulic power to form fragile rods of the size of sticks of sealing-wax. A current of 1 500 to 1 600 amperes is passed along the rod, which is kept in an atmosphere of hydrogen, and this causes the rod to become thoroughly coherent. The resulting rod or *slug* of tungsten is *swaged* by a machine, which consists of a rotary hammer that strikes the slug, which is kept red hot, at the rate of 4 000 to 5 000 times per minute. The effect of swaging is to consolidate the particles of tungsten and to produce a ductile, thin wire about 30 ft. long and $\frac{1}{32}$ in. diameter. This wire is drawn through a series of diamond dies until it possesses the required diameter, which may be as small as 0·0006 in.

Tungsten has a melting-point of 3 670° abs., but it can be run at a temperature of 2 420° abs. in a vacuum without undue evaporation. Because of the higher temperature, the vacuum type tungsten-filament lamp produces 9·9 l. per W., as against 3·6 by the carbon filament lamp. (The critical effect of temperature will be appreciated if it is considered that a vacuum tungsten-filament lamp operating at 2 350° abs. gives only 5 l. per W.—see page 339.)

Gas-filled Filament Lamp. Until 1913 filaments were heated in a vacuum in order to avoid the loss of heat by convection. If the bulb is filled with an inert gas, say nitrogen or argon, the tungsten filament can run at 2 950° abs., instead of 2 400° abs., without undue evaporation. The loss of heat by convection will, however, more than neutralize the gain in luminous efficiency due to the higher working temperature. Langmuir pointed out that the effect of convection can be considerably diminished by coiling the filament, in the way shown in Fig. 289; the spiral filament is supported on rods which radiate from a glass knob. A lamp of this kind gives 12 l. per W.

A more recent improvement is the coiled-coil filament. A very fine tungsten wire is wound on a fine iron wire to form a spiral or coil of tungsten. This complete assembly is again wound on a thicker iron wire, and the iron is dissolved out by acid. The result is a coiled-coil of tungsten, of the form indicated in Fig. 290. One

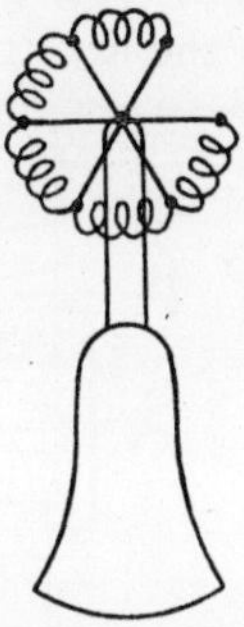

FIG. 289.
COILED FILAMENT

FIG. 290.
COILED-COIL FILAMENT

inch of the coiled-coil will stretch out into a tungsten wire of four feet. The convection loss is greatly reduced and luminous efficiencies of 10 to 30 l. per W. are now obtained.

Tungsten has a resistance which increases with temperature, the resistance when cold being one-fourteenth of that when at 2 900° abs. This means that on switching-on, a current of fourteen times the normal current is taken. The increase of resistance with temperature is not without advantage, as it provides a stabilizing effect on the power consumed and the luminosity with varying voltage; for if the voltage increases, the current and temperature increase, so that the resistance increases and causes less power consumption than if the resistance kept constant. The reverse is the case with the carbon filament lamp, since carbon has a negative temperature coefficient. It is found that

$$\text{light output} \propto E^m$$

and

$$\text{power input} \propto E^n,$$

where $m = 4$ or 5 and $n = 1{\cdot}8$ for a tungsten filament,

and $m = 6$ or 7 and $n = 2{\cdot}1$ for a carbon filament.

The spectral distribution of filament lamps is of the kind shown in Fig. 273, and the colour is yellowish-white.

Arc Lamps. In an arc lamp a current flows between two electrodes which are drawn apart. An arc is struck, which maintains the current, and this is a very efficient source of light. There are various forms of such lamps employing a carbon arc, a flame arc, or a metallic electrode arc.

Carbon Arc Lamp. This is the earliest type and is still used. The electrodes are of hard carbon; the arc consists of carbon vapour and is surrounded by an orange-red zone of burning carbon and pale green flames. The positive carbon forms a crater which is white hot and emits 85 per cent of the light. The temperature is the highest attainable, about 4 200° abs. in the arc, 3 700 in the crater, and 2 700 in the negative carbon. The efficiency is about 9 l. per W., and the polar curve is as shown in Fig. 291.

The voltage required to maintain the arc is

$$E = (39 + 28l) \text{ volts},$$

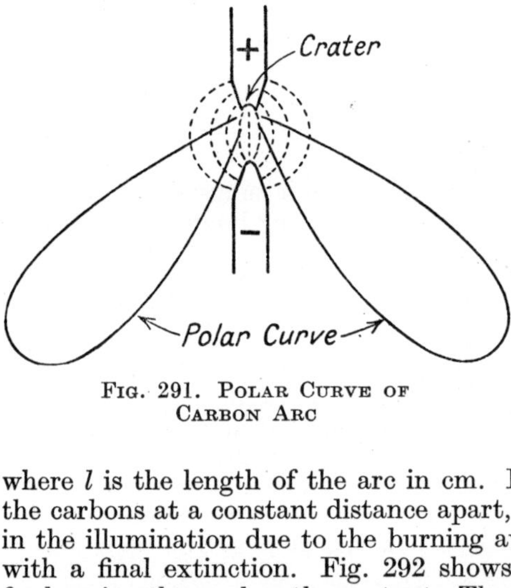

FIG. 291. POLAR CURVE OF CARBON ARC

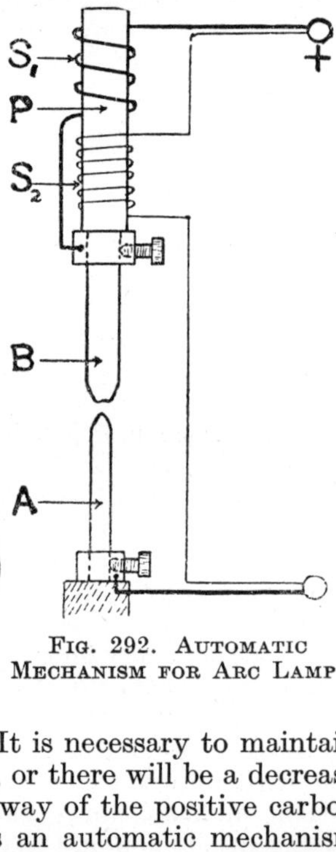

FIG. 292. AUTOMATIC MECHANISM FOR ARC LAMP

where l is the length of the arc in cm. It is necessary to maintain the carbons at a constant distance apart, or there will be a decrease in the illumination due to the burning away of the positive carbon with a final extinction. Fig. 292 shows an automatic mechanism for keeping the arc length constant. The negative carbon A is fixed, and the positive B is connected to an iron plunger P, which is the movable core of the solenoids S_1 and S_2. The former solenoid is in series with the arc and attracts the plunger against its gravitational force, whilst the latter acts in the opposite direction. When the carbons are in contact, a large current flows through S_1 and exerts a force which separates the carbons. A position of equilibrium is reached at which the force due to the current in S_1 equals the sum of the forces due to gravity and the current in S_2. As the carbons burn, their distance apart increases and a greater voltage drop exists between them. There is less voltage for S_1, in which the current decreases. The positive carbon then drops to the equilibrium position.

The carbon arc is unstable, for as the current increases the amount of vaporized carbon increases, and the resistance decreases so much that the voltage drop (current × resistance) decreases. A steadying resistance must therefore be placed in series with the arc.

EXAMPLE. The voltage and current in a carbon arc are related by the equation $E = 60 + 35/I$. Find the minimum value of steadying resistance to make the arc stable with 8 A.

The total voltage across the arc and resistance R is

$$e = E + RI = 60 + (35/I) + RI.$$

$$de/dI = -(35/I^2) + R.$$

If we equate this to zero we find the current which causes the lowest voltage across the combination. This is to occur at $I = 8$ A., and we get

$$R = (35/8^2) = \underline{\underline{0{\cdot}55\ \Omega}}.$$

To keep an arc burning for a long period, it is often the practice to enclose the arc in a closed vessel filled with nitrogen or carbon dioxide. The vessel has asbestos rings to protect it. The light from a carbon arc is rich in ultra-violet light and is used for photographic purposes.

Flame Arcs. These use carbon electrodes with 5 to 15 per cent of various fluorides. The main part of the light comes from the arc, which is spread out by means of a magnet, as shown in Fig. 293. The flame arc is very efficient, but the colour is objected to and the discharge lamp is replacing it for public lighting.

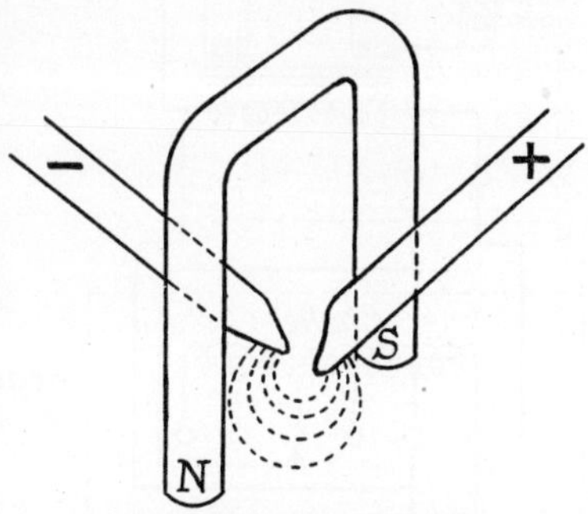

FIG. 293. FLAME ARC

Metallic Electrodes Arc. This is known as the "Magnetite" Arc Lamp and is used in America. The positive electrode is of copper, and the negative of magnetic oxide of iron, titanium and chromium dioxide. The negative burns for 150 to 200 hours, whilst the positive lasts very much longer.

Gas Discharge Electric Lamps. A tungsten filament lamp suffers from two disadvantages: the efficiency of a 100 W. gas-filled lamp working at 2 900° A. is about 12 lumens/watt, and the light is yellowish-white. These disadvantages have been overcome by the design of lamps using gaseous discharges. There are two main classes of discharge lamps: in one class the highest efficiency is sought at the possible expense of poor colour, whilst in the second class colour is more important then efficiency. The former class is used in street lighting and the latter in homes, offices and factories. There are also the gaily coloured lamps or tubes used for advertisement and entertainment.

STREET LIGHTING. The highest efficiency is sought, and the favoured designs have sodium or mercury vapour. The Philora and Sieray lamps containing sodium give a yellow light with an efficiency

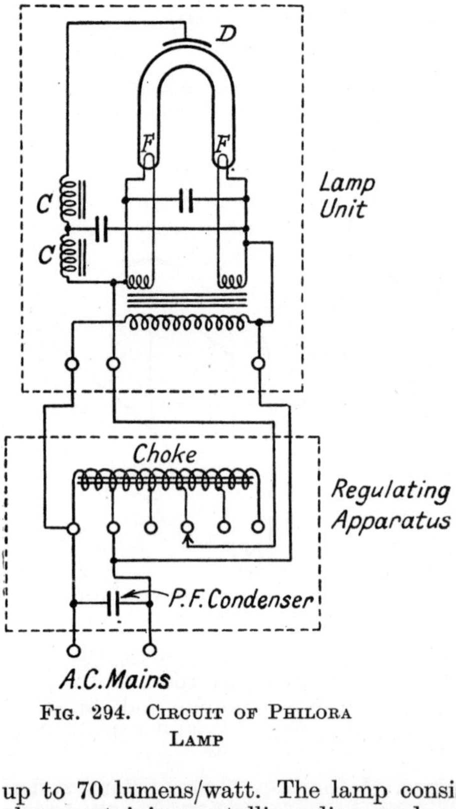

FIG. 294. CIRCUIT OF PHILORA LAMP

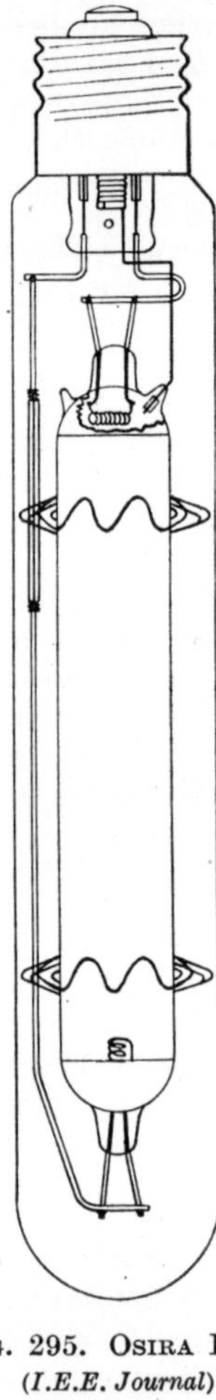

FIG. 295. OSIRA LAMP (*I.E.E. Journal*)

up to 70 lumens/watt. The lamp consists of a U-tube of two-ply glass containing metallic sodium and argon-neon at a low pressure, the whole being enclosed in a vacuum jacket to conserve heat. The inner ply of the U-tube is of a special low silica content glass which resists the action of hot sodium. The discharge is initiated by the argon or neon: the tube warms up, the sodium vaporizes and the colour changes from red to yellow. It takes about 10 minutes to obtain full light output.

Fig. 294 shows the circuit associated with a Philora lamp. Two heated filaments F form the electrodes. A resonant device, chokes C and plate D, give twice normal voltage for starting. A choke is

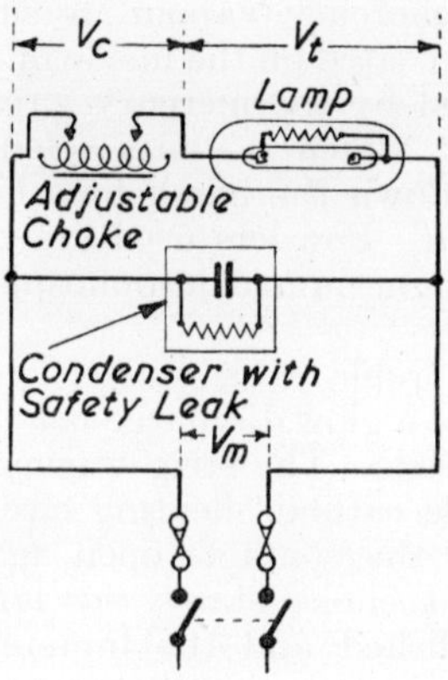

FIG. 296. OSIRA LAMP CIRCUIT
(*I.E.E. Journal*)

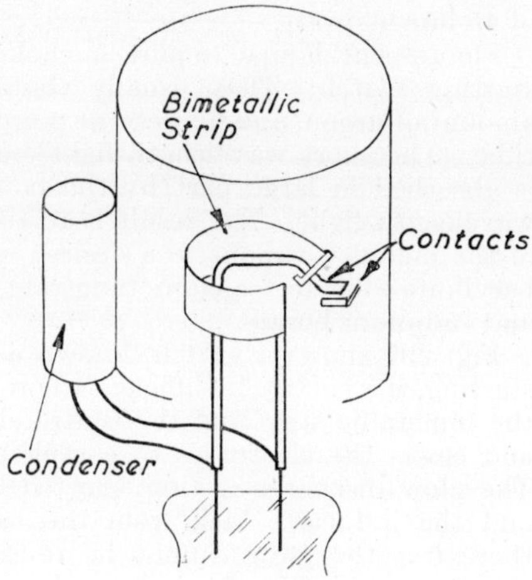

FIG. 297

included to limit the current, and a capacitance is used to increase the power factor to a reasonable value. The efficiency is 40 lumens/watt for a 100 W. lamp.

Fig. 295 shows the Osira lamp, which uses mercury vapour.

The inner tube is the lamp proper, and has at each end an electrode consisting of a stick of rare earths held in a tungsten spiral. The oxides are activated during the construction of the lamp by heating the tungsten spiral. The tube contains argon at a low pressure and a small quantity of mercury. The outer bulb contains oxygen at a low pressure, as this prevents the formation of an absorbing film on the inside of the inner tube. An auxiliary electrode is connected through a high resistance to the lower electrode, whilst it is placed near the upper electrode. When the voltage is applied, full voltage appears between the auxiliary and the upper main electrode, between which a discharge occurs at once. This facilitates starting. The electrodes are heated by the discharge and the mercury vaporizes. As the temperature rises, the luminous column becomes brighter and narrower. The light is bluish-white, which is preferable to that from the sodium light, and the luminous efficiency is 38 l. per W. for a 400 W. lamp. Fig. 296 shows the connections of the complete discharge-lamp unit.

INDOOR LIGHTING. Long, tubular, fluorescent lamps have been introduced for indoor lighting: in this application colour is of the utmost importance, the preferred colour being a rosy-tinted daylight.

As an example, the Philips MCF/U 80 W. lamp is a tube 5 ft. long and 1½ in. diameter. It takes 96 W. at 230 V., and has an efficiency of 45 lumens/watt.

Fluorescent lamps require a choke to limit the current and a starting switch. They usually contain mercury vapour, a small amount of argon, and fluorescent powders coated on the inside of the tube. The short wavelength light emitted by the mercury vapour is absorbed in large part by the powder, which re-emits a longer wavelength light. The result is to tone down the harsh blue light of the mercury vapour to a desired colour. Powders used are zinc beryllium silicate, calcium tungstate, magnesium chlorophosphate and cadmium borate.

Fig. 297 shows a switch, known as a "replaceable starter." On starting, the voltage is sufficient to produce a glow discharge between the bimetallic strip and the centre electrode. The strip warms up and closes the electrodes to complete the cathode heating circuit. The glow discharge is thus shorted out, the contacts open again, and the inductive kick from the ballast choke starts the lamp, thereafter the glow cannot be re-established and the bimetallic strip remains cold. A small condenser is included with the starter inside an aluminium cylinder in order to minimize radio interference.

Summary of Sources of Light

Source	Lumens per Watt	Absolute Temperature	Colour
(Theoretical Maximum)	670	—	Yellow-green, 5 500 Å.
(Theoretical Maximum)	250	—	White (uniform) visible spectrum
Black body radiator	90	6 500	White
16 c.p. carbon filament lamp	2·9	2 000	Yellowish-white
60 W. vacuum lamp	9·2	2 400	Yellowish-white
60 W. gas-filled lamp	11	2 900	Yellowish-white
500 W. gas-filled lamp	15·5	2 900	Yellowish-white
10 kW. gas-filled lamp	31	2 900	Yellowish-white
40 W. coiled-coil lamp	10·5	—	Yellowish-white
100 W. sodium vapour lamp	40-70	—	Yellow
400 W. mercury vapour lamp	40-60	—	Bluish-white
Fluorescent lamps, 80 W.	40	—	Daylight
Neon tubular sign	12	—	Red

EXAMPLES XI

1. Define luminous flux, luminous intensity, and illumination, and state and define the units in which they are measured.

Describe the construction and use of the improved form of Lummer-Brodhun photometer. (*Lond. Univ.*, 1933.)

2. Sketch and describe a good reliable form of photometer. Explain how the

apparatus is set up to determine the polar curve of candle-power in a vertical plane for an incandescent electric lamp. What are the principal difficulties in a measurement of this kind, and how can they be overcome?

(*Lond. Univ.*, 1933.)

3. An open space is illuminated by three lamps each having a uniform distribution of C c.p. in all directions below the horizontal plane through the lamps. The lamps are supported at a height of y ft. above the ground, and arranged at the corners of an equilateral triangle of side x ft. Calculate the illumination on the ground at a point (a) below the centre of the triangle, (b) directly under one of the lamps. (*Lond. Univ.*, 1932.)

4. Compare the spectra of tungsten-filament and mercury-discharge (high and low pressure) lamps, and explain why the luminous efficiency of the discharge lamps is higher than that of the tungsten lamp. Describe developments in tungsten lamps to increase the luminous efficiency without seriously affecting their life. (*Lond. Univ.*, 1950.)

5. A street is lighted by means of a row of lamps suspended in the centre of the street at a height of 20 ft. and at a distance apart of 200 ft. Find the polar distribution of candle-power each lamp should have in order that the illumination at all points along the centre line of the street may be constant at 0·2 f.c. Assume that the light of each lamp is cut off by a shade at an angle of arc tan 5 from the vertical.

How many lumens will each lamp give? (*Lond. Univ.*, 1932.)

6. A lamp of 250 m.s.c.p. is suspended at a height of 16 ft. above the working plane and is provided with a reflector which gives approximately uniform distribution over an area 16 ft. in diameter of the working plane below the lamp. Assuming that the efficiency of the reflector is such that 45% of the total emitted light is directed on to this circular area, calculate the illumination produced.

What would be the illumination at the outer part of the same area if the reflector were removed from the lamp? (*Nat. Cert.*, 1935.)

7. A thoroughfare 50 ft. wide is illuminated by lamps supported 16 ft. above the edge of the road, the lamp standards being placed at alternate sides of the road and spaced 250 ft. apart on each side. The illumination on the ground in the centre of the road opposite a lamp is 0·4 f.c. Calculate the candle-power of each lamp, assuming a uniform distribution in the lower hemisphere. What is the ratio of the maximum to the minimum illumination along the centre line of the road between two lamps? (*Lond. Univ.*, 1931.)

8. It is desired to illuminate a horizontal circular space 25 ft. dia. by lamps suspended at a height of 15 ft. The illumination must not be less than approximately 12 f.c. at any point. Estimate the power consumption of the lamps required if—

(a) a large number of small lamps is used having an efficiency of 10 l. per W., these being spaced uniformly over the area to be illuminated;

(b) a single lamp suspended over the centre of the area is used, the lamp having an efficiency of 20 l. per W. and being fitted with a reflector which directs the light output over the surface to be illuminated and gives a uniform candle-power over this angle.

Assume the coefficient of utilization to be 0·4 in each case.

(*Lond. Univ.*, 1933.)

9. A square of 20 ft. side is illuminated by lamps of 300 c.p., one at each corner and 20 ft. above ground level. Calculate the illumination on the ground at a point half-way along one side of the square.

If the light is emitted uniformly from each of the above lamps, calculate the total flux of light. (*Nat. Cert.*, 1932–3.)

10. Describe the characteristics and operation of a fluorescent mercury discharge lamp, including the starting arrangement.

Draw a circuit diagram for a single 80 W. lamp supplied from 230 V. a.c. mains. (*Lond. Univ.*, 1947.)

11. Describe, with sketches, the construction of a tubular fluorescent lamp, in which the tube is of small diameter and 4 ft. to 5 ft. long. Explain the action of the lamp, and the necessity for auxiliary equipment, and draw a diagram of connections. Compare the spectrum of the lamp with that of a tungsten-filament lamp, and explain why the luminous efficiency of the fluorescent lamp is higher than that of a tungsten-filament lamp.

(*Lond. Univ.*, 1948.)

12. Compare the advantages and disadvantages of tungsten-filament and 5 ft. fluorescent lamps for industrial lighting.

A workshop 65 ft. by 32 ft. is to be lighted by 5 ft. fluorescent lamps each rated at 2 200 lumens. Determine the number of lamps required to give an average illumination of 10 ft.-candles on the working plane, assuming a coefficient of utilization of 0·52. Sketch the connections for one lamp and show how the lamps could be arranged and connected to reduce the stroboscopic effect if the supply is (i) single-phase, (ii) three-phase.

(*Lond. Univ.*, 1949.)

13. Draw a diagram of connections for an 80 W. hot-cathode fluorescent lamp. Explain its starting and operating characteristics and compare them with those of a metal filament lamp.

A room, 80 ft. by 40 ft., is to be lighted with 80 W. fluorescent lamps. The average horizontal illumination over the whole room is to be about 30 ft.-candles. Calculate the number of lamps required if each lamp has a luminous output of 3 000 lumens and the coefficient of utilization is 0·6.

(*Lond. Univ.*, 1950.)

14. Describe, with sketches, the construction of a tubular fluorescent lamp. Draw a diagram of connections and explain the action of the lamp (i) at starting, (ii) when operating normally. Compare the spectrum of such a lamp with that of (i) a tungsten-filament lamp, (ii) a mercury discharge lamp, and explain why the luminous efficiency (lumens per watt) of the fluorescent lamp is so much higher than that of the tungsten-filament lamp.

(*Lond. Univ., B.Sc. Eng.*, 1947.)

15. Show that the maximum luminous intensity of a short uniform linear light source, such as a fluorescent mercury vapour lamp, when observed from a distance which is large compared with the length of the source, is F/π^2 candle power, F being the total flux of the source in lumens.

Indicate on a sketch (without calculation) the approximate distribution of illumination from one such source on a horizontal plane below it when h is about 1·5 times the length of the source. (*Lond. Univ.*, 1949.)

16. A fluorescent lamp 5 ft. long emits 3 000 lumens. It is suspended horizontally, without a reflector, 6 ft. above a level surface. Determine approximately the direct illumination at a point on the surface vertically beneath the centre of the lamp.

Discuss briefly the advantages and disadvantages of this type of lamp.

(*Lond. Univ.*, 1948.)

17. Describe the construction of (i) a low-voltage, (ii) a high-voltage, tubular fluorescent lamp. Compare their luminous efficiencies and life, and discuss their relative advantages and disadvantages for industrial lighting.

Draw a diagram of connections (for each type of lamp) and explain how it is started. Explain also the function of each piece of apparatus in the control gear. (*Lond. Univ.*, 1953.)

18. Explain briefly the construction and purpose of an isocandle diagram for an exterior lighting fitting.

A floodlight has a beam such that the luminous intensity is proportional to $\cos^4 \theta$, where θ is the deviation from the axis of the beam. The maximum luminous intensity is 12 500 candelas (candles). Determine the maximum and minimum illumination provided by the floodlight on a surface 50 ft. away and normal to the beam axis, over a circular area 20 ft. in diameter centred on the axis. Determine also the total light flux reaching this area from the floodlight. (*Lond. Univ.*, 1953.)

CHAPTER XII

ELECTRIC TRACTION

Systems of Electric Traction. Systems of electric traction may be divided into two main groups; in one group the vehicles receive their power from a distributing network fed by a few large generating stations, in the other the vehicles generate or carry their own energy. The first group may be further subdivided into systems operating with d.c., such as tramways, trolley-buses, and railways, and systems operating with a.c., viz. railways. The second group may be subdivided according to the nature of generation or storage; thus there are the diesel-electric trains and ships, the petrol-electric trucks and lorries, and the battery-driven road vehicles.

Advantages and Disadvantages of Electric Traction. One of the most obvious advantages of electric traction, especially of the first group, is the cleanliness it possesses above all other systems. This alone makes it essential for use in the underground and tube railways.

Another well-known advantage is the rapid and smooth acceleration and braking possible with electric traction; an electric locomotive has an acceleration of 1·0 to 2·0 m.p.h. per sec., compared with the 0·4 to 0·5 of a steam locomotive. This is of special importance in suburban traffic, where very frequent trains must be run at early morning, mid-day, and evening, and where frequent stops occur. With a given track and stations, electric traction can carry up to 100 per cent more people than steam traction, because of the higher average speed over short runs with frequent stops.

The size of stations in towns is limited very strictly by financial considerations, and the superior manoeuvrability of the electric locomotive enables twice as many to be used in a station of a given size.

An electric locomotive needs much less time for maintenance and repair than a steam locomotive, so that fewer are required for a given volume of traffic; also the cost of maintenance and repair per locomotive is less by about 50 per cent. It can be used immediately, whereas a steam locomotive takes about two hours to get up steam; this results in a better utilization of drivers' time.

Because of the absence of smoke and sparks, there is a greater safety in driving and an absence of damage to the buildings and apparatus due to the corrosive smoke fumes.

A saving is caused by the absence of coaling and water depots, and also the time of coaling.

The superior braking methods allow less wear on the brake shoes, and in some cases a saving of energy, which is returned to the supply instead of being wasted as heat in the brake shoes.

The main disadvantage is the capital outlay required to convert from steam to electric traction. It is certain that if this difficulty were overcome, the other disadvantages would not prevent a rapid conversion to electric traction.

Another disadvantage is that a failure of the power supply for a few minutes may cause a disorganization of the service for one or two hours. Increased reliability of supply will render failures very infrequent, and improved organization will diminish the time of interruption of service from each failure. It is known that a layer of ice on the conductor rails may prevent the train from collecting the power which is available; this trouble is easily overcome by running a service locomotive up and down the line to prevent the formation of ice.

Steam locomotives use their steam for heating the compartments very cheaply, whereas electric locomotives require to draw power for this purpose at a greater cost.

In many cases telephone and telegraph lines run along the track, and these will experience considerable interference from the power lines. Either the lines must be moved away from the track, or they must be replaced by cables, and a considerable expense—up to 15 per cent of the total cost—may be incurred.

As already stated, the main difficulty is the capital cost to change from steam to electric traction on the railways. By far the major part of the cost is in the overhead equipment and feeders, and this is avoided in the use of diesel-electric traction. In this system the locomotive carries diesel engines which drive a d.c. generator that supplies power to the motor. The diesel engine is run at a constant speed so that its power output is always available, whilst the electric drive makes this power available at all speeds of the locomotive. Diesel-electric locomotives have been made efficient and streamlined, so that very high speeds are available. The main disadvantage in this country is that the oil fuel has to be imported; if the extraction of oil from coal becomes an economic process, it is probable that the main railway lines will be converted to diesel-electric traction.

Petrol-electric traction has been used so far in heavy lorries and buses. The advantage is that the electric conversion produces a very fine and continuous control; thus the lorry can move slowly at an imperceptible speed and yet it can creep up the steepest slope without throttling the engine.

Battery-driven vehicles are being quickly introduced, as they are found very useful as light delivery vans and platform trucks. They are easy to control and very convenient to use. As road vehicles they suffer from the disadvantage of having a limited range and speed.

Electrification Systems. D.C. and a.c. systems are used, the latter being single-phase or three-phase.

For tramcars the supply is about 600 volts d.c., and the rails act as the return circuit. There are regulations relating to the return circuit in order to prevent damage due to leakage currents. The track is designed to have good electrical continuity and conductivity so that the return current does not spread out much. The track is connected to the negative pole of the supply system, and must be such that the potential difference between any two points on it is not greater than 7 volts. When the current is high, it is not practicable to limit the potential difference by having an enormous return rail, but instead use is made of negative boosting in a way that will be described later. When the return circuit is near pipes, the potential of the return must not be more than 4 volts above

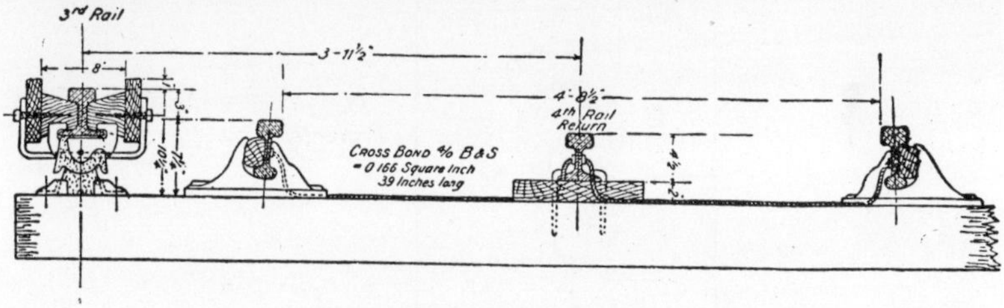

FIG. 298. CROSS-SECTION OF THE PERMANENT WAY

earth or more than 1 volt below earth potential. The supply is either underground in a conduit or overhead by a trolley wire. When a trolley wire is used, the voltage at the generating station must not exceed 650 volts and at the trolley wire 550 volts. The trolley wire must be divided into sections of not more than half a mile, between every two of which there must be emergency switches. When the track is run on private ground 1 500 volts d.c. is favoured. For trolley buses the supply is at 600 volts, both lines being overhead and insulated from ground. As the return circuit is not earthed there is no fear of electrolysis, and negative feeder boosters are not required.

The underground railways in London use 600 volts d.c., and the Southern Region has adhered to this voltage because conditions make it economic and satisfactory. A third and insulated rail provides the positive pole of the supply, whilst the running rails are used as the return. Where leakage currents threaten to cause damage, precautions are taken by having an insulated fourth rail for the return (see Fig. 298). Any new services use 1 500 and 3 000 volts d.c.

The railway line between Shildon and Newport operates at 1 500 volts d.c. The a.c. power is converted at substations by rotary converters and transformers.

Locomotives in Poland, Belgium, and Italy use 3 000 volts d.c., in Montreal 2 400 volts d.c., Paris-Orléans-Midi 1 350 volts d.c.

Single-phase main-line locomotives use 15 kV. at $16\frac{2}{3}$ cycles in Austria, Germany, Sweden, and Switzerland and other countries; in Pennsylvania single-phase of 11 kV. at 25 cycles is used.

Three-phase is used in some mountainous districts, e.g. in Italy. The voltage is 3 600 volts between phases; two overhead conductors are used with the rail as the third phase. The necessity for two collectors is a disadvantage. No transformers are required as the induction motors run at the line voltage. The frequency of the supply is $16\frac{2}{3}$ cycles. Regeneration is automatic, and this is very

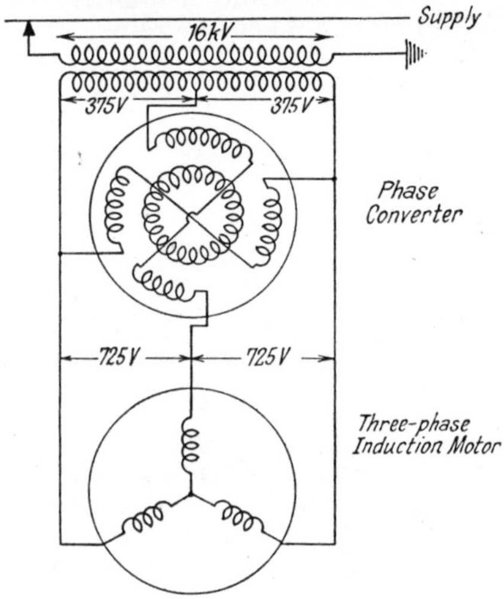

FIG. 299. SPLIT-PHASE TRACTION SYSTEM (KANDO)

useful in mountainous districts. The absence of commutators is a great advantage and lowers the cost of upkeep. As the induction motor is sensitive to speed variations, it is impossible to use the multiple unit method; for the motors running on worn wheels would rotate faster than those on new wheels and would do little work, or even no work if the wheels were worn enough. A locomotive provides all the tractive effort.

Split-phase is used in the Kando system in which the supply is single-phase of 16 kV. at 50 cycles. A phase converter supplies the motor, which is a three-phase induction motor (see Fig. 299).

The trend seems to be to install no more three-phase systems, but either high voltage d.c. or single-phase industrial frequency; the latter is likely to become a serious rival of the older systems.

Mechanical Considerations. In the early days of electric traction the single-phase motor was made in large sizes because of commutator maintenance; and as little experience was available concerning gears, common practice was to use one or two large motors with rods connecting the driving wheels. With the improved performance and commutation of smaller single-phase motors and the availability of good gears, rod drives are being superseded and individual drive is being used. It appears that rod drive secures a better adhesion than individual drive, and is favoured for heavy, low-speed goods locomotives, especially when steep gradients are encountered.

The disposition of the driving and trailing axles in a locomotive is described in the following manner. Numbers are used to indicate

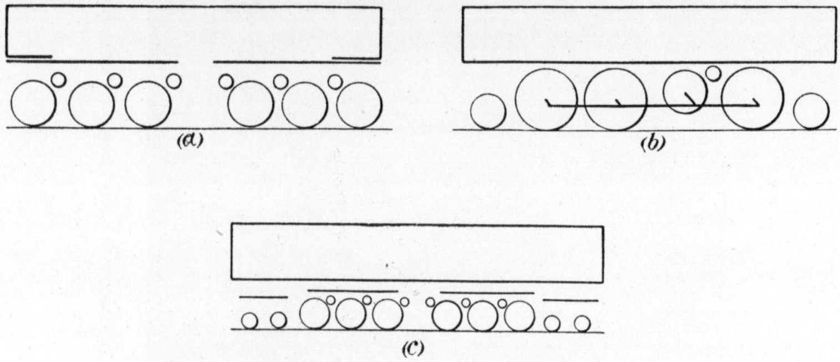

FIG. 300. ARRANGEMENTS OF LOCOMOTIVE DRIVES

(*a*) Six-axle, double-bogie locomotive of the New York Central Railroad. $C_0 + C_0$. Nose suspension, single cushion.
(*b*) Swedish State Railway. $1C1$ Jackshaft and coupling rods. 16 kV., $16\frac{2}{3}$ cycles.
(*c*) Articulated express locomotive of the Pennsylvania Railroad. $2 - C_0 + C_0 - 2$. Safe speed 100 m.p.h.

the trailing axles, letters the driving axles, and a subscript zero is used to indicate individual drive of the driving axles. Fig. 300 shows some examples of the arrangements of locomotives.

Fig. 301 shows a standard, mixed-traffic locomotive of the Swedish State Railways, the drive arrangement being that of Fig. 300 (*b*). The supply is 16 kV., $16\frac{2}{3}$ cycles, and there is one twin-armature motor of 1 770 h.p. continuous rating and 1 975 1-hour rating. The gear ratio is 3·19, with 4·24 on some units used for freight traffic. The continuous rating tractive effort is 16 700 lb., whilst the maximum tractive effort is 37 500 lb., and the maximum safe speed is 75 m.p.h.

Transmission of Drive. There are many ways in which the drive to each axle may be applied, and only a few will be described in general terms.

The simplest way is to use gearless bipolar motors, whose armatures are mounted direct on the axles. Fig. 302 shows a rough sketch

Fig. 301. Mixed-traffic Locomotive (Swedish State Railway)

of such a mounting developed by the G.E. Co. of America for high-speed locomotives. The armature diameter can be kept relatively small for a given output and speed. The field magnets are supported from the locomotive frame and the pole faces are nearly flat, so that

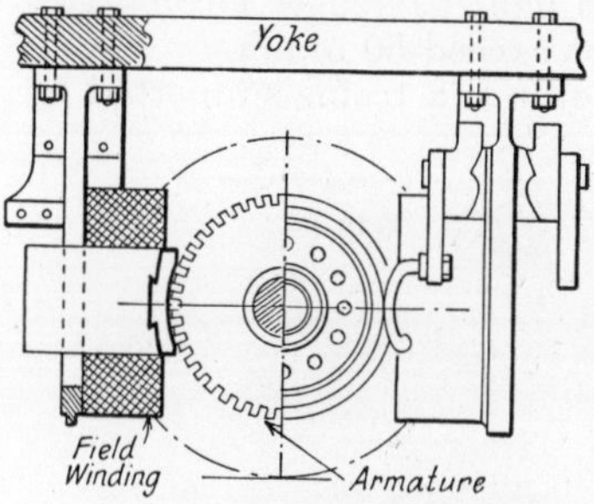

Fig. 302. Axle-mounted Motor

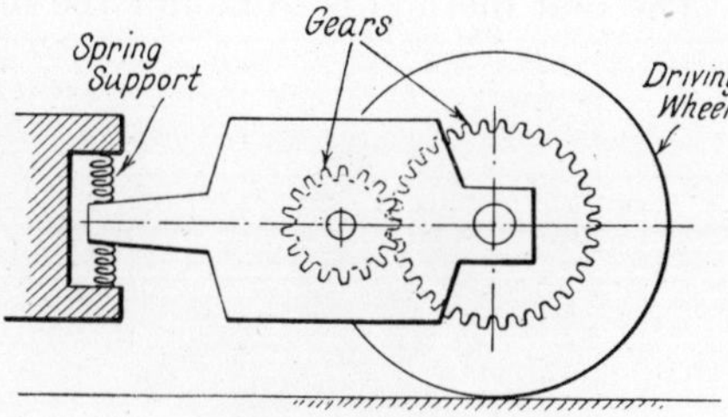

Fig. 303. Nose-suspended Motor

the armature has vertical play without striking the pole faces. Few locomotives have been fitted with such a drive in the past twelve years, and it is unlikely that more such units will be built because of

Fig. 304. Westinghouse Quill Drive

the relatively large unsprung weight, the poor utilization of material, and the low centre of gravity of the locomotive.

The next simple arrangement is the geared, axle-mounted motor, with nose suspension. The use of gearing (3 to 5 : 1) allows a higher speed motor and thus a smaller motor for a given power output. Fig. 303 indicates the method of suspension. For motors above 300 h.p. twin gearing is usually fitted. Because of the relatively large uncushioned load on the driving axle, the method is unsuitable for high speeds and is used for freight and light passenger locomotives for which the maximum speed does not exceed 50 m.p.h.

The best method is to mount the motor in a frame supported by

FIG. 305. BROWN-BOVERI LINKWORK DRIVE

springs from the locomotive. It is then necessary to have a flexible coupling between the armature shaft and the driving axle to take up the play in distance and alignment between these axes. There are several methods of doing this.

Fig. 304 shows the "quill drive" developed by the Westinghouse for geared motors. The photograph gives an arrangement for twin motors. It is seen that the gear wheels are mounted on a hollow shaft, the "quill," which surrounds the driving axle and is connected by six springs at each end to the driving wheels.

It is possible to have a gearless quill drive, in which case the motor armature is directly mounted on the quill.

It is also possible to couple the gear wheels to the driving wheels without the use of a quill, and Fig. 305 shows the Brown-Boveri link drive which achieves this purpose.

Mechanics of Train Movement. Fig. 306 shows a diagram of the essential driving parts of an electric locomotive. The armature of the motor experiences a torque T (in lb. ft.), and it has attached to it a pinion of diameter p. There is a *tractive effort* F′ at the edge of the pinion, where $T = \frac{1}{2}pF'$. This tractive effort is transferred to the driving wheel (diameter D) by means of the gear wheel (diameter d), so that tractive effort on the driving wheel is

$$\left.\begin{aligned} F &= \eta F'(d/D) = \eta T \times (2/p) \times (d/D) \\ &= \eta T \times (2G/D) \end{aligned}\right\} \quad . \quad . \quad (126)$$

where η is the efficiency of the gear and G is the *gear ratio* d/p.

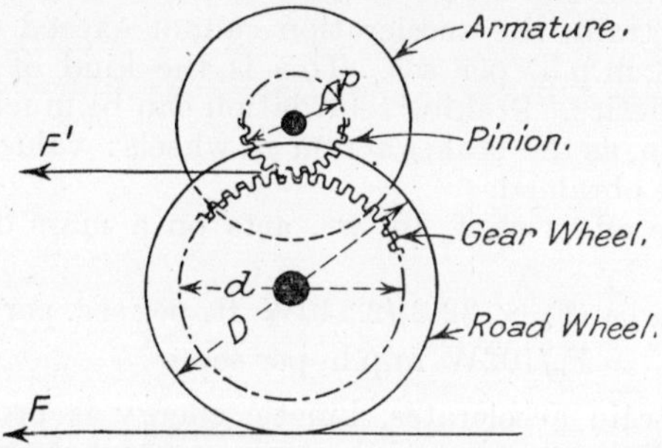

FIG. 306. DRIVING PARTS OF ELECTRIC LOCOMOTIVE

The magnitude of the tractive effort that can be usefully employed depends upon the weight on the driving wheels and the adhesion of the driving wheels to the rails. The *coefficient of adhesion* is defined as

$$\frac{\text{Tractive effort to slip the wheels}}{\text{Adhesive weight}},$$

and the following table gives values for electric tractors on dry rails.

Speed, m.p.h. . .	0	10	20	30	40	60
Coefficient of adhesion.	0·25	0·18	0·14	0·12	0·10	0·09

If the rails are greasy the value may be as low as 0·08. A very important advantage of electric traction is that in a motor coach 100 per cent of the weight is on the driving wheels, in an electric locomotive 70 per cent or more, but in a steam passenger locomotive less than 50 per cent. Moreover, the coefficient of adhesion in electric traction is greater than in steam traction; this is because (i) the torque in electric traction is continuous while in steam traction it is pulsating, and the uneven torque sets up a jolting and skidding, and (ii) in electric traction the driving wheels are distributed along the length of the train, whilst in steam traction they are close together. Thus the maximum possible tractive effort is much greater in electric traction than in steam traction.

The maximum possible acceleration can be found from the coefficient of adhesion. Suppose that the whole of the weight is on the driving wheels and the locomotive is running alone; then the maximum tractive effort is 0·25 times its weight, and the acceleration is 0·25 times g, viz.

$$0{\cdot}25 \times 32{\cdot}2 = 8{\cdot}1 \text{ ft. per sec. per sec.}$$
$$= 5{\cdot}5 \text{ m.p.h. per sec.}$$

If the weight of the motor coaches is only one-third of the total weight of the train, the acceleration cannot exceed one-third this value, viz. 1·8 m.p.h. per sec. This is the kind of value that is obtained in practice. Braking retardation can be much greater than the acceleration, as the brakes act on all wheels: values of 3·2 m.p.h. per sec. can be obtained.

If a tractive effort of F_a lb. wt. acts on a mass of W tons, the acceleration is

$$\alpha = (F_a \times 32{\cdot}2)/2\,240W \text{ ft. per sec. per sec.}$$
$$= F_a/102W \text{ m.p.h. per sec.} \quad . \quad . \quad . \quad (127)$$

When the train accelerates, kinetic energy is produced in two ways, by the linear motion of the train, and by the rotation of the wheels and motors: the former is $\frac{1}{2}Wv^2$, where v is the velocity of the train, and the latter is

$$\Sigma\tfrac{1}{2}I\omega^2 = \Sigma(\tfrac{1}{2}Iv^2/r^2) = \tfrac{1}{2}mv^2,$$

where $m = \Sigma(I/r^2)$, I being the moment of inertia of a rotating part and ω its angular velocity.

The sigma is taken for all rotating parts. This means that the effective value of the mass of the train is $W + m$; in practice m is from 8 to 15 per cent of the dead weight W. Equation (127) then becomes

$$\alpha = F_a/102(W + m) \text{ m.p.h. per sec.} \quad . \quad . \quad (127a)$$

or $$F_a = 102\alpha(W + m) = 102\alpha W_e, \quad . \quad . \quad . \quad (128)$$

where $W_e = W + m$, and is called the *effective* or *accelerating mass of the train*. The tractive effort F_a is that required for acceleration; in practice the total tractive effort supplied by the motors must be equal to this plus the effort to overcome the train resistance, and gravitation if the train is on a slope. The tractive effort to overcome train resistance is

$$F_r = Wr,$$

where r = *specific train resistance*, and is a function of the velocity for a given train. The tractive effort to overcome gravity on a slope of percentage gradient G′ is

$$\begin{aligned} F_g &= \pm\, WG'/100 \text{ tons} \\ &= \pm\, 22{\cdot}4WG' \text{ lb. wt.,} \end{aligned}$$

where the positive sign is used for an up-gradient and the negative for a down-gradient. The total tractive effort is

$$\begin{aligned} F_t &= F_a + F_r + F_g \\ &= (102\alpha W_e + Wr \pm 22{\cdot}4WG') \text{ lb. wt.} \qquad (129) \end{aligned}$$

The power output of the driving axles is

$$P = F_t v \text{ ft. lb. wt. per sec.,}$$

where v is in ft. per sec., so that

$$\begin{aligned} P &= F_t V \times \frac{5\,280 \times 0{\cdot}746}{60 \times 33\,000} \text{ kW.} \\ &= 0{\cdot}00199\, F_t V \text{ kW.,} \qquad (130) \end{aligned}$$

where V is in m.p.h.

EXAMPLE. A motor-coach train weighing 200 tons is accelerated up a gradient of 1 in 200 at a mean acceleration of 1·2 m.p.h.p.s. up to a speed of 30 m.p.h. Find (1) the tractive effort required, and (2) the output at the end of the accelerating period. The train resistance is 10 lb. per ton and the effective weight is 10% more than the dead weight.

In this case $\alpha = 1{\cdot}2$, $W = 200$, and $m = 0{\cdot}1 \times 200 = 20$, so that $W_e = 220$, $r = 10$, and $G' = \frac{1}{2}$ (1 in 200).

By equation (129) the required tractive effort is

$$\begin{aligned} F_t &= 102 \times 1{\cdot}2 \times 220 + 200 \times 10 + 22{\cdot}4 \times 200 \times \tfrac{1}{2} \\ &= 27\,000 + 2\,000 + 2\,240 \\ &= \underline{\underline{31\,240 \text{ lb. wt.}}} \end{aligned}$$

At the end of the accelerating period V = 30, so that the power is

$$\begin{aligned} P &= 0{\cdot}00199 \times 31\,240 \times 30 \\ &= \underline{\underline{1\,870 \text{ kW.}}} \end{aligned}$$

Speed-time Curves. If a curve is plotted with time (in seconds or minutes) as the abscissa and the speed (in miles per hour) as the ordinate, the complete information of the motion of the train is represented. The acceleration at any instant or speed is found by drawing a tangent at the corresponding point on the curve and calculating the slope of this tangent; the acceleration is given usually in miles per hour per second (1 m.p.h.p.s. is equal to 1·47 ft. per sec. per sec.). The distance covered in a given time is represented by the area between the curve, the time axis, and the ordinates through the instants between which the time is taken. Fig. 307 shows the speed-time curves for city and main-line services.

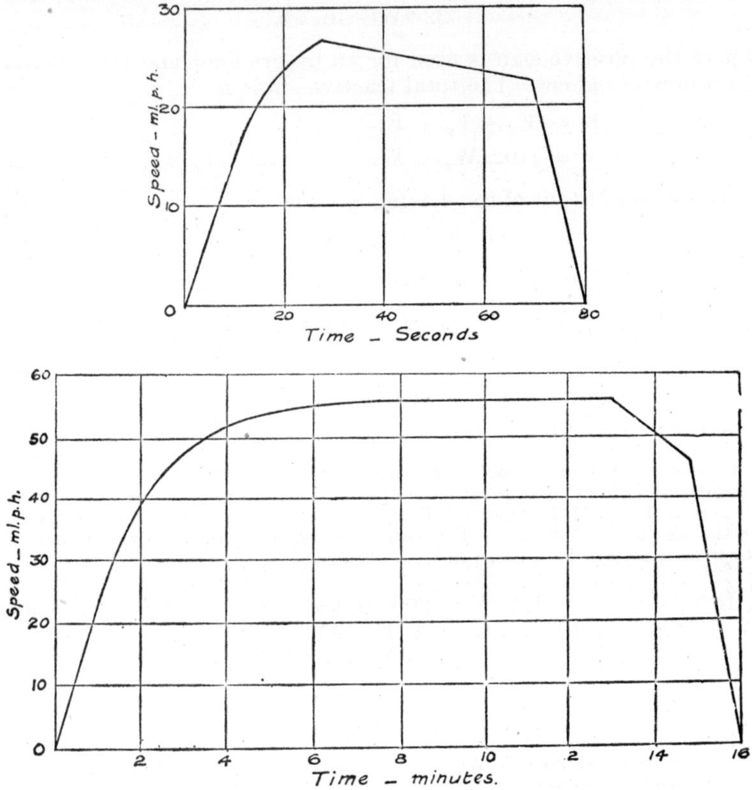

FIG. 307. SPEED-TIME CURVES: CITY SERVICE, MAIN-LINE SERVICE
(*Electric Traction* (Dover))

The initial acceleration in the city service is seen to be 10 m.p.h. per 8·3 sec. = 1·21 m.p.h.p.s. The total distance between stops in the main-line service is represented by the area of 37·5 squares, each of which corresponds to a distance of

$$10 \text{ m.p.h.} \times 2 \text{ min.} = \frac{10 \times 2}{60} \text{ miles} = \tfrac{1}{3} \text{ mile},$$

so that the total distance is $37{\cdot}5 \div 3 = 12{\cdot}5$ miles.

There are usually four periods in the run, viz. *acceleration*; constant speed or *free running*; *coasting*, when the power is shut off and the train slows down gradually because of resistances to motion; and *braking*. In the speed-time curve shown for a main-line service these periods are 6, 7, $1\frac{3}{4}$ and $1\frac{1}{4}$ min. respectively. In short runs, such as the city and suburban services, the free running period may not exist. The acceleration period consists of two parts. In the first part the motor tractive effort is kept constant by means of resistance notching, or more recently by the metadyne; this occurs until all the resistances are switched out and a speed V_1 is reached (see Fig. 308). The tractive effort available for acceleration, and climbing if necessary, is F, where F is the difference of the motor tractive effort and the train resistance. The acceleration in this period is nearly constant. In the second part of the acceleration period the motor tractive effort is the maximum that the motor can give at the speed and falls rapidly with the speed. The train resistance, however, increases, slowly to begin with and then rapidly, until at a certain speed V_2 the motor tractive effort is equal to the train resistance. At any speed between V_1 and V_2 the tractive effort available for acceleration (and climbing) is f, which decreases from F at V_1 to zero at V_2. The acceleration between these speeds therefore decreases from the maximum value (about 1 or 2 m.p.h.p.s.) to zero. V_2 is the *maximum possible speed*, and requires a motor tractive effort F_1 to maintain it. If the power is shut off, the train resistance slows the train; at a speed V the decelerating force is due to the train resistance at that speed. If the curves of motor tractive effort and train resistance versus speed are known, the foregoing method enables the acceleration and deceleration of the train at any speed to be found. It will be shown later how the speed-time and distance-time curves of the train can be calculated from the acceleration—or deceleration—speed curves.

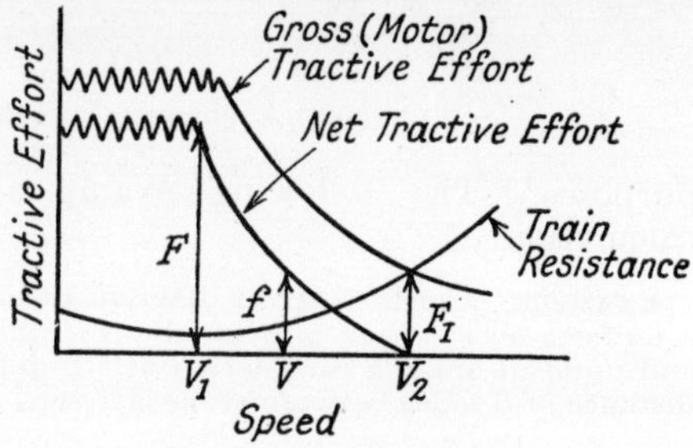

FIG. 308. TRACTIVE EFFORT VERSUS SPEED

There are three speeds of importance: the *crest speed*, which is the maximum speed attained on the run; the *average speed*, which is the mean speed from start to stop; and the *schedule speed*, which is the mean speed when the stop period is included. Thus in the speed-time curve shown for a main-line service the crest speed is 56 m.p.h., and the average speed is

$$(12{\cdot}5 \times 60)/16 = 46{\cdot}8 \text{ m.p.h.}$$

If the stops are 2 min., the schedule speed is

$$(12{\cdot}5 \times 60)/(16 + 2) = 41{\cdot}6 \text{ m.p.h.}$$

SIMPLIFIED SPEED-TIME CURVES. The speed-time curve of a city service can be replaced by a quadrilateral (Fig. 309 (*a*)) or a trapezoid (Fig. 309 (*b*)), whilst that of a main-line service is best and most easily replaced by a trapezoid (Fig. 309 (*c*)). It is much easier to calculate the performance of the train from the simplified speed-time curves, and the results are accurate enough for most practical

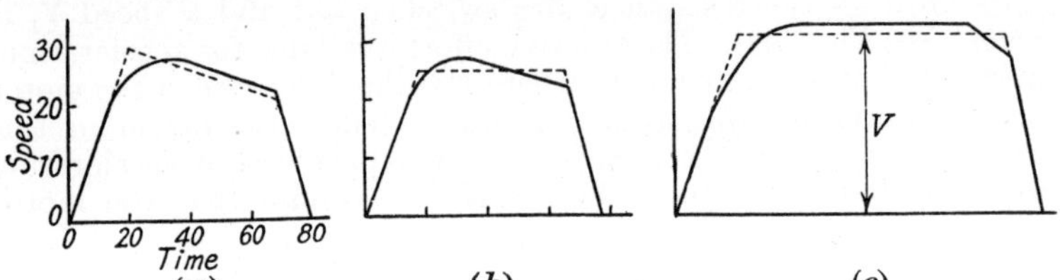

(*a*) (*b*) (*c*)

FIG. 309. APPROXIMATE SPEED-TIME CURVES

purposes. The following examples illustrate the method of calculation.

EXAMPLE. The time-speed diagram of an electric train is represented by a uniform acceleration of *a* m.p.h.p.s., constant coasting speed of V m.p.h., and uniform braking retardation of *b* m.p.h.p.s. If the time taken to run a distance of S miles between stops is T sec., show that

$$V = \frac{1}{2k}[T - \sqrt{(T^2 - 14\,400Sk)}], \text{ where } k = (a + b)/2ab.$$

(*Lond. Univ.*, 1931.)

In this case it is assumed that the coasting speed is constant—it would be better to call this free running—so that the speed-time curve is as shown in Fig. 309 (*c*). The acceleration is *a* and the final speed V, so that the duration of acceleration is V/*a* and the distance travelled in this time is

$$\tfrac{1}{2}a(V/a)^2 = \tfrac{1}{2}(V^2/a).$$

Similarly the duration of braking is V/*b* and the distance travelled in this time is $V^2/2b$. The time of free running is thus

$$T - V/a - V/b,$$

and the distance travelled in this time

$$V(T - V/a - V/b).$$

The total distance is thus

$$\begin{aligned} S' &= V^2/2a + V^2/2b + V(T - V/a - V/b) \\ &= VT - (V^2/2a + V^2/2b). \end{aligned}$$

This is a quadratic equation for V, viz.

$$V^2(1/2a + 1/2b) - VT + S' = 0.$$

In this equation a and b are in m.p.h.p.s., V in m.p.h., T in sec., so that S′ is in

$$\text{miles per hour} \times \text{seconds} = \frac{1}{3\,600}\ \text{miles}.$$

If we want the distance in miles, S say, we have

$$S'/3\,600 = S \text{ or } S' = 3\,600\,S.$$

The equation for V becomes

$$\left.\begin{aligned} &kV^2 - VT + 3\,600S = 0, \\ \text{where}\quad &k = 1/2a + 1/2b = (a + b)/2ab \end{aligned}\right\} \quad . \quad . \quad . \quad (131)$$

The solution is

$$\begin{aligned} V &= T/2k \pm (1/2k)\sqrt{[T^2 - 4 \times 3\,600kS]} \\ &= (1/2k)\,[T \pm \sqrt{(T^2 - 14\,400Sk)}]. \end{aligned}$$

To determine the correct sign we note that the time of free running is

$$T - V/a - V/b = T - 2kV = \mp \sqrt{(T^2 - 14\,400Sk)}.$$

It is thus necessary to take the lower sign, and we have

$$V = (1/2k)\,[T - \sqrt{(T^2 - 14\,400Sk)}]$$

and the time of free running is $\sqrt{(T^2 - 14\,400Sk)}$.

Effect on Schedule Speed of Acceleration, Braking and Distance. Equation (131) gives a general relation, for the trapezoidal speed-time curve, between the maximum speed, acceleration, braking retardation, distance, and time of running. Its main use is for finding the maximum speed necessary *or* the acceleration required for a desired schedule speed on a given line. The following example shows how this is done.

Example. An electric train operating on a suburban service has a maximum running speed of 38 m.p.h. The average distance between stops is 2 200 yd. and the schedule speed including a station stop of 20 sec. is 25 m.p.h. Calculate the necessary acceleration, allowing a maximum retardation of 2·5 m.p.h.p.s. (*B.Sc., Lond. Univ.*, 1928.)

As the distance S is 2 200 yd. = 1·25 miles and the schedule speed is 25 m.p.h., the time of travel plus the stop of 20 sec. is 1·25/25 = 0·05 hr., i.e. 3 min. or 180 sec. The time of travel, T, is thus 180 − 20 = 160 sec. The maximum speed V is 38 m.p.h. Equation (131) can be written as

$$k = (VT - 3\,600S)/V^2 = 1/2a + 1/2b,$$

so that

$$\begin{aligned} 1/a &= 2(VT - 3\,600S)/V^2 - 1/b \\ &= \frac{2(38 \times 160 - 3\,600 \times 1{\cdot}25)}{38^2} - \frac{1}{2{\cdot}5} \\ &= 1{\cdot}76 \end{aligned}$$

and the required acceleration is

$$a = 1/1{\cdot}76 = \underline{\underline{0{\cdot}57}}\ \text{m.p.h.p.s.}$$

CALCULATION OF SPEED-TIME CURVE. We have seen that the tractive effort available for acceleration is the total tractive effort less that required to overcome train resistance and gravity. We can rewrite equation (129) as

$$F_a = F_t - F_r - F_g$$

or $$102\alpha W_e = F_t - Wr \mp 22{\cdot}4WG',$$

so that the acceleration is

$$\alpha = (F_t - Wr \mp 22{\cdot}4WG')/102W_e. \qquad (129a)$$

The total tractive effort and the train resistance are given, in the form of curves, as functions of the speed, as shown in Fig. 308. We can then calculate the acceleration a at any speed v. As

$$\alpha = dv/dt,\ dt = dv/\alpha$$

and thus $$t = \int \frac{1}{\alpha} dv. \qquad (132)$$

Equation (132) expresses time as a function of the speed, i.e. it gives the time to attain a certain speed under the varying acceleration. By making t the abscissa and v the ordinate we obtain the speed-time curve, from which the complete performance is easily found in the way shown above.

During coasting and braking both α and dv are negative, but the method is just the same.

If α is a simple function of v the integration may be possible in known functions; otherwise a graphical method must be used. The method described is applicable to rotating machinery as well as to traction, in which case α is the angular acceleration, v is the angular velocity, and we replace mass by the moment of inertia.

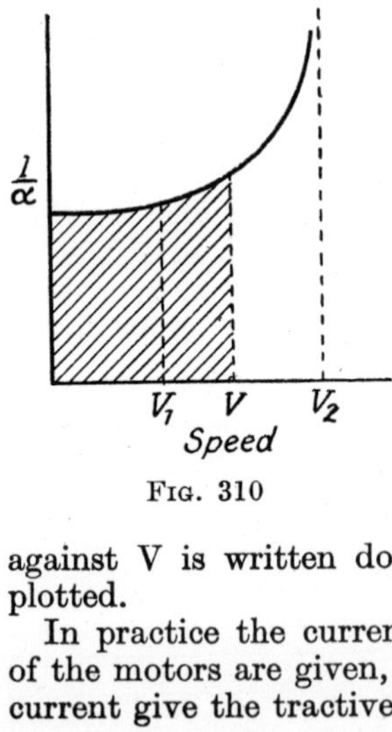

FIG. 310

The graphical method of obtaining the speed-time curve is the following. We plot $1/\alpha$ against v; Fig. 310 shows this curve for the traction system represented by the curves of Fig. 308. $1/\alpha$ is approximately constant up to the speed V_1, and increases to ∞ at speed V_2. The time t to reach a speed V is given by the shaded area shown in Fig. 310. This is done for several values of V, and a table of t against V is written down from which the speed-time curve is plotted.

In practice the current-speed and current-tractive effort curves of the motors are given, and these with the condition of maximum current give the tractive effort-speed curve of the motors.

Example. A train has a total weight of 116 tons and is equipped with four motors each of 275 h.p. The characteristics of the motor and the train resistance are given by the following table—

Current (A.)	100	200	300	400	500
Speed (m.p.h.)	51	31·4	26·4	23·9	22·1
Tractive effort (lb. wt.)	390	1 600	2 960	4 330	5 690
Train resistance (lb. wt per ton)	11	10	9	9	10

The ratio of the effective weight to the dead weight of the train is 1·1 to 1. The mean accelerating current is 425 A. per motor, and the braking retardation is 2 m.p.h.p.s. A run of 0·86 mile is to be made in 115 sec., there being an average up grade of 0·119 per cent. Calculate the r.m.s. current per motor for the run.

Fig. 311 shows the current-speed and current-tractive effort curves per motor. The current is not allowed to exceed 425 A. by

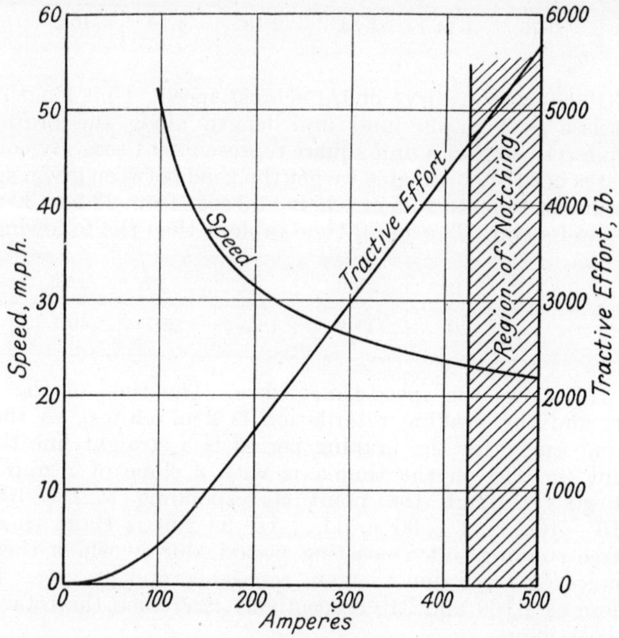

Fig. 311

notching, so that the maximum tractive effort per motor is 4 650 lb. wt. and this acts until the speed reaches 23·5 m.p.h. The tractive effort required to overcome gravity is 22·4 × 116 × 0·119 = 310 lb. wt. We now construct a table of F_t, F_r, F_g and F_a against speed, and from F_a we calculate the acceleration a from equation (128), viz.

$$a = \frac{F_a}{102W_e} = \frac{F_a}{102 \times 1{\cdot}1 \times 116} = \frac{F_a}{13\,000},$$

since W_e is 1·1 × the dead mass.

Speed .	< 23·5	26	28	30	35	40	45
F_t .	18 600	12 800	9 760	7 600	4 800	3 200	2 080
F_r .	1 160	1 050	1 050	1 160	1 350	1 580	1 860
F_g .				310			
F_a .	17 130	11 440	8 400	6 130	3 140	1 310	– 90
a .	1·32	0·88	0·65	0·47	0·24	0·10	– 0·007
$1/a$.	0·76	1·14	1·54	2·13	4·16	10·0	

Fig. 312 shows the curve of $1/a$ against speed. Unit length along the abscissa is 1 m.p.h. and unit length along the ordinate is 1/10 m.p.h.p.s., so that a unit square represents 0·1 sec. By counting the squares between ordinates we get the time between given speeds; thus the time from 0 to 23·5 m.p.h. is 17·3 sec., from 23·5 to 30 m.p.h. 8·4 sec., and so on. The speed-time table is then the following.

Speed . .	0	23·5	25	30	35	40
Time . . .	0	17·9	18·5	25·7	40·7	86·2

Fig. 313 shows the speed-time curve. The time of the run is 115 sec. and the braking retardation is 2 m.p.h.p.s., so that the speed-time curve for the braking period is a straight line through the point $t = 115$ on the time-axis with a slope of 2 m.p.h.p.s.; thus it goes through the point corresponding to $t = 100$ and $v = (115 - 100) \times 2 = 30$ m.p.h. As the run is short there will be no free running but a coasting period, during which there will be a retarding force due to train resistance and gravity. Let us take these as 1 160 and 310 respectively, and then the retardation during coasting is

$$1\,470/13\,000 = 0{\cdot}113 \text{ m.p.h.p.s.}$$

The coasting period is represented by a line of this slope. The position of this line is such that the area under the composite curve, consisting of acceleration, coasting and braking curves, corresponds to the distance travelled, viz. 0·86 mile.

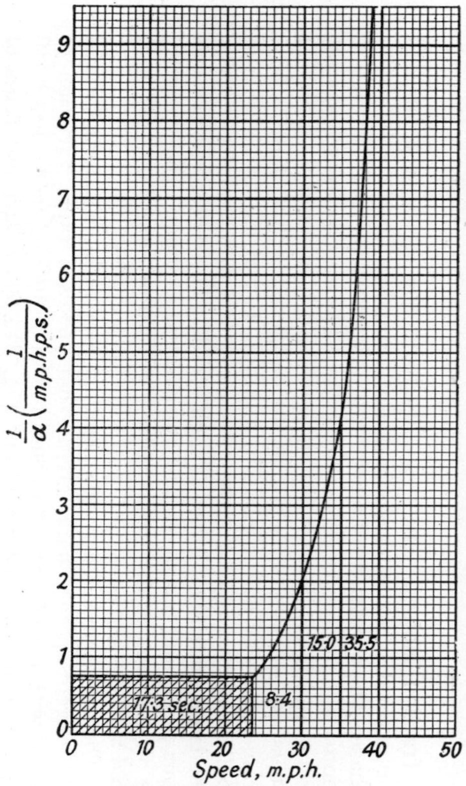

FIG. 312. GRAPHICAL METHOD OF FINDING SPEED-TIME CURVE

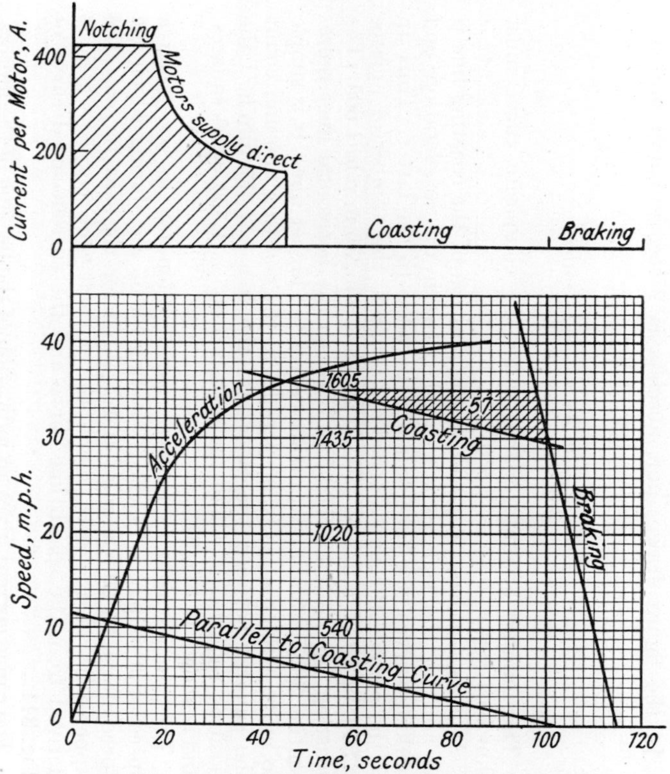

FIG. 313. SPEED-TIME AND CURRENT CURVES CORRESPONDING TO FIG. 312

In the figure drawn, unit abscissa is 2 sec. and unit ordinate 1 m.p.h., so that unit square represents

$$1 \times \frac{2}{3\,600} \text{ miles} = \frac{1}{1\,800} \text{ miles.}$$

The number of unit squares under the composite curve must thus be $1\,800 \times 0{\cdot}86 = 1\,550$.

A quick and easy way to find the position of the coasting line is to add up the squares contained between the acceleration and braking curves, the time axis and parallels to the latter. Thus up to the parallel through a speed of 10 m.p.h. the number of squares is 540, up to 20 m.p.h. 1 020, 30 m.p.h. 1 435, 35 m.p.h. 1 605. The coasting line must lie therefore between the horizontal lines representing 30 and 35 m.p.h., rather nearer the higher line. It is simple to move a ruler so as to be parallel to the coasting line and count the squares between the ruler and the line $v = 35$ m.p.h.: the position required is where the area is $1\,605 - 1\,550 = 55$ squares, and is shown in Fig. 313.

The run is made as follows. Acceleration for 45 sec. when a speed of 36 m.p.h. is reached, coasting for 55 sec. during which the speed drops to 29·5 m.p.h., and braking for the last 15 sec. The upper part of Fig. 313 shows the current-time curve, which is derived from the speed-time curve and the current-speed curve of Fig. 311.

The r.m.s. current gives an indication of the beating of the motors. To find the r.m.s. we plot (current)^2 against time, find the area under this curve, divide by the total time of 115 sec., and take the square root. It is found that mean square is 38 400 (amperes)^2 and the r.m.s. is 196 A.

Traction Motors. The types used are the series or compound motor for d.c., series for single-phase, and induction motor for three-phase. The conditions of service are very severe, so that the traction motor is built on very robust lines. As it has to be protected against water and mud, it is totally enclosed, and if necessary ventilating ducts are specially arranged in the design. Fig. 314 shows a box-frame tramway motor made by the B.T.-H.: it has cast steel and malleable iron in place of the more usual cast iron, in order to resist the large centrifugal forces at high speeds. The motor is series, has four poles, and interpoles to give sparkless running. The armature is of the two-circuit variety, there are two or more turns per coil, and the coils are placed in open slots in the punchings.

The series and lightly compounded motors have a torque-speed (or current-speed) characteristic that shows a rapid variation of speed with torque, whilst a shunt (or an induction) motor has small variation of speed with torque. Fig. 315 shows the two types of characteristic. Suppose we have two identical motors driving different wheels that are not connected by a rod: if one wheel is smaller, it has a larger angular velocity and the motor driving it will have a higher speed than the other. We represent the speeds by N_0' and N_0'', on either side of a normal speed N_0. If the motors have a series characteristic, the torques are T_0' and T_0'', which are seen to be nearly equal, so that the motors share the load fairly. If the motors have a shunt characteristic, it is seen that they share the load very unequally; as shown in the figure, the higher speed motor is acting as a generator and is doing less than no useful work, whilst the other is doing more than the total mechanically necessary work. Whilst

FIG. 314. BOX-FRAME TRAMWAY MOTOR
(*B.T.-H.*)

this would hardly occur in practice, it is nevertheless true that the motors would share the work very unequally, and for this reason motors with a shunt-characteristic cannot be used on individual drive. Induction motors on three-phase systems have their driving wheels linked by connecting rods. It is also impossible to use induction motors for a multiple-unit system, in which two or more locomotives draw a heavy train.

Starting and Speed Control of D.C. Motors. The speed of a d.c. series motor can be varied either by varying the field or the voltage applied to it. The field can be varied by tappings on the field or by a shunt across it; the latter is called "field weakening." The voltage applied can be varied in three main ways, by means of series resistance or the series-parallel method, by the metadyne, or by the Ward-Leonard system; the last method is not used in traction.

For the purpose of starting, field control is clearly useless, and recourse is had either to series resistance (notching), the series-parallel method in conjunction with series resistance, or the metadyne.

Notching. The current in a series motor is given by

$$E - kNI = IR,$$

or

$$I = E/(R + kN),$$

where E is the voltage, N the speed, k a constant of the field and armature, and R the d.c. resistance of the armature and external resistance. At zero speed the current is E/R: external resistance is inserted to limit I to some predetermined value, and as the motor speeds up the current drops and the external resistance can be reduced eventually to zero. The current (or torque curve) is like that shown in Fig. 315. The method of calculating the steps of resistance is given in *Electrical Technology*, by H. Cotton (Sixth Edition, pages 143–6). In this method of starting, known as *notching*, the resistance is put in series with the motor so that the current has a certain maximum value, I_1 say, and remains until the speeding up reduces the current to a certain minimum value, I_2 say. The resistance is then reduced so that the current regains the value I_1, and so on until no resistance is left.

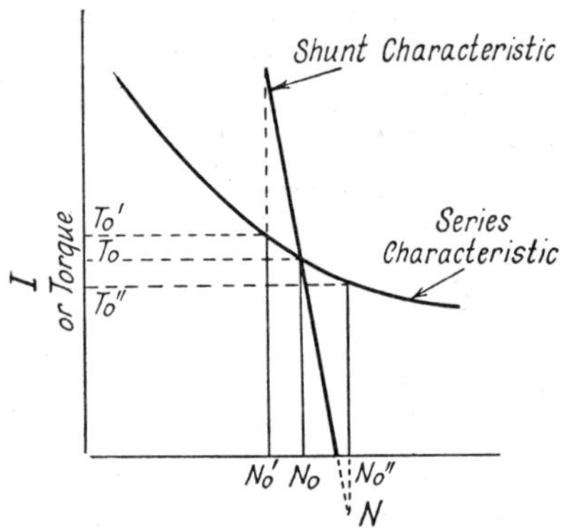

FIG. 315

MOTOR CHARACTERISTICS

Series-parallel Control. The current in the series resistance of the last method of starting a series d.c. motor involves a great waste. Part of this waste can be avoided by the series-parallel method, when there are two or more motors.

If there are two motors, they can be started in series with a limiting resistance, run up to half speed (when the series resistance is zero and their *total* back e.m.f. is equal to the supply voltage), switched over into parallel with limiting resistance again, and then run up to full speed, when the back e.m.f. of each is equal to the supply voltage.

If there are four motors, more combinations are possible, viz. series, series-parallel, and parallel.

There are two main methods of effecting the change from series to parallel, the shunt-transition and the bridge-transition methods, which are shown in Figs. 316A and B. In the former method the motors are run up to the full series position, when the series resistance is cut down to zero. Then some series resistance is reinserted, and one motor is short-circuited. Then this motor has one end opened, and this end is connected across so that the motors are in parallel. The series resistance is then cut out as the motors speed up. There is a jerk in this system as one motor is shorted and ceases to act, and then another jerk when it is reinserted.

In the bridge-transition method, a resistance is put across each motor after the full series position is reached, and then the shorting bar between the motors is removed, leaving the motors (each in series with a resistance) in parallel with each other. If the resistances across the motors have the correct value, the shorting bar has no current, since the arrangement is that of a Wheatstone bridge, and the transition is perfectly smooth.

Fig. 317 shows a tramway controller using the series-parallel

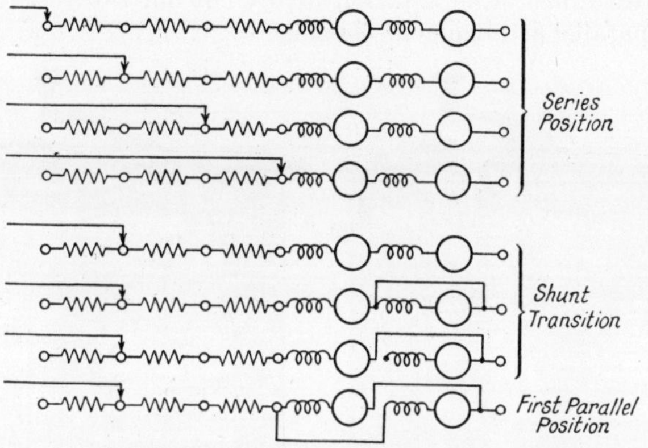

FIG. 316A. SHUNT-TRANSITION METHOD

method. The controller uses magnetic blow-outs; the blow-out coil system consists of flat, iron inserts in the movable part shown on the extreme right of the figure.

Alternate plates are connected to the iron core of one of the two blow-out coils, and the remaining plates to the other coils, so that the flux is upwards in one compartment and downwards in the next. The contact fingers are situated between the iron plates, and the directions of the various currents are so arranged that the arc is always blown outwards on to the metal cover of the controller, the inside of the cover being coated with an asbestos material.

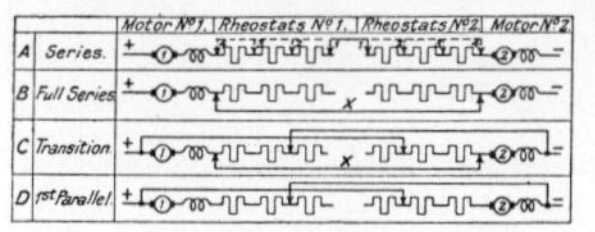

FIG. 316B
BRIDGE-TRANSITION METHOD

Fig. 318 shows a simplified form of the power diagram of a 1 500 volt d.c. train equipment. L_1, L_2, L_3 and L_4 are line switches and are used in series pairs. Motors 1 and 2 are in series, and have three breaks L_3, L_4, and P; motors 3 and 4 are in series and have the breaks L_1, L_2, and M. Bridging contactors S_1 and S_2 have full line voltage of 1 500 volts across them when the motors are in parallel, and for this reason there are two breaks. Resistance W protects the system at switch-on in case there is a fault in motor 1.

On the first notch, L_1, L_2, and S close, and then L_3 and L_4. The motors are then in series with full limiting resistance. Contactors R then operate on the following notches, and the motors are running in series on the full line voltage. Then S_1 and S_2 close and S opens, and the motors are still in series. Contactors R then open, M and P close, S_1 and S_2 open, and the motors are then in parallel with full limiting resistance. Contactors R finally cut out this resistance and the final parallel arrangement is reached.

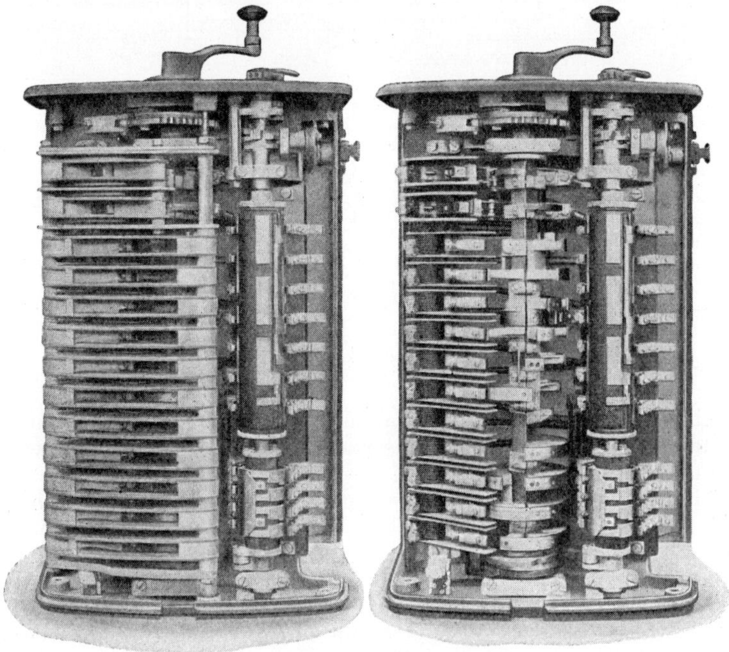

FIG. 317. TRAMWAY CONTROLLER
(*English Electric*)

ENERGY SAVED BY SERIES-PARALLEL CONTROL. Let us consider the cases (i) where the motors are started in parallel with limiting resistance, and (ii) where they are started by the series-parallel method. In both cases we assume that the series limiting resistances are continuously varied so that the current through each motor, whether in series or parallel, is equal to the maximum permissible value. It follows that each motor produces a constant torque, in whatever combination it finds itself, and thus there is the same constant acceleration in both methods. We will assume further that armature and series field resistances can be ignored, as they are small compared with the limiting resistances.

Fig. 319 (*a*) shows the electrical conditions in the first case. The total current drawn from the supply is $2I$, where I is the maximum permissible value per motor. The speed, and with it the back

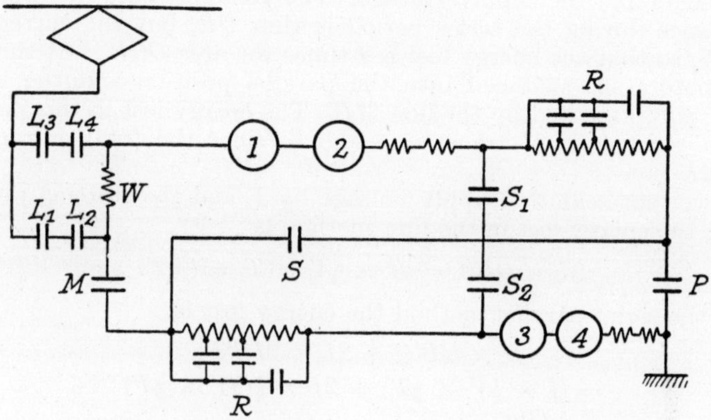

FIG. 318. POWER DIAGRAM OF 1 500 V. D.C. TRAIN EQUIPMENT

e.m.f., increases uniformly with time, until the back e.m.f. is equal to the supply voltage at time t_A. At time t_P the back e.m.f. is PQ and the voltage drop in the limiting resistance is QR.

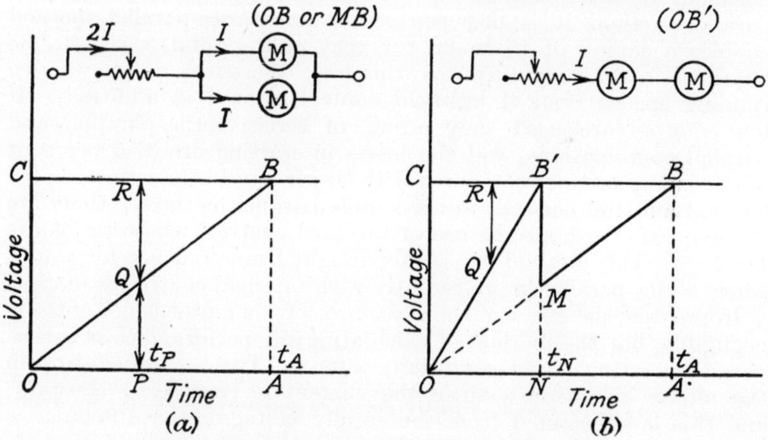

FIG. 319. ENERGY LOSSES DURING NOTCHING

The power dissipated in the resistance is thus $2I \cdot QR$, and the total energy lost in the starting process is therefore $2I$ times the area OBC.

Fig. 319 (*b*) shows the conditions in the second case. The motors speed up at the same rate as before and therefore the back e.m.f.

of each motor is represented by the line OB, as before. The back e.m.f. of the series combination, however, is twice this value and is represented by OB' ,where N is mid-way between O and A and NM represents half the supply voltage. The voltage across the limiting resistance during the series period is thus QR, but the current is only I, so that the energy lost is I times the area OCB'. At time t_N the motors are switched into the parallel position and the back e.m.f. is represented by the line MB. The energy lost in this period of the starting is $2I$ times area $MB'B$, since the total current is now $2I$.

If we represent the supply voltage by V and the starting period by T, the energy lost in the first method is

$$2I \times OBC = 2I \times \tfrac{1}{2}V \times T = IVT.$$

In the series-parallel method the energy lost is

$$\begin{aligned} &I \times OB'C + 2I \times MB'B \\ &= (I \times \tfrac{1}{2}V \times \tfrac{1}{2}T) + 2I \times \tfrac{1}{2}(\tfrac{1}{2}V \times \tfrac{1}{2}T) \\ &= \tfrac{1}{2}IVT. \end{aligned}$$

The energy input to the motors in either method is

$$2I \times \tfrac{1}{2}V \times T = IVT,$$

since each receives a current I at a mean voltage $\tfrac{1}{2}V$. Thus the efficiency of the first method is 50 per cent, whilst in the series-parallel method it is $66\tfrac{2}{3}$ per cent. The series-parallel method enables a saving of 15 to 20 per cent of the total energy to be obtained in ordinary tramway running; moreover, it allows two running speeds (four if tap-field control is used in addition). If four motors are used they admit of series, series-parallel, and parallel combinations, and the losses in starting are 37·5 per cent of the energy used (as compared with 50 per cent in the series-parallel method and 100 per cent in the simple parallel method); there are three speeds (without the use of tap-field control) whose ratios are 1 : 2 : 4. This method is used for freight trains but not for trams, where series-parallel (more recently with tap-field control) is used.

In practice the effect of the resistance of the motor is not entirely negligible, but the method of calculating the performance of series-parallel starting is not essentially altered. The resistance drop in the motor is constant, since the current is considered constant, and this is subtracted from the supply voltage; the difference is available for back e.m.f. and voltage drop in the limiting resistance and the method is then as shown in Fig. 319.

The Metadyne.* In the methods of control described above, resistance is put in series with the motor and slowly cut out. During this process of notching, which is jerky, a great deal of energy is

* The author's thanks are due to the Metropolitan-Vickers Electrical Co. for the information on this development.

wasted in the resistances. The metadyne achieves smooth control without dissipating energy in a resistance. It is, in essence, a rotating transformer for d.c. power with a transformation ratio that can be varied (continuously, if desired). Thus it can draw power from a constant (d.c.) voltage source and deliver it at a

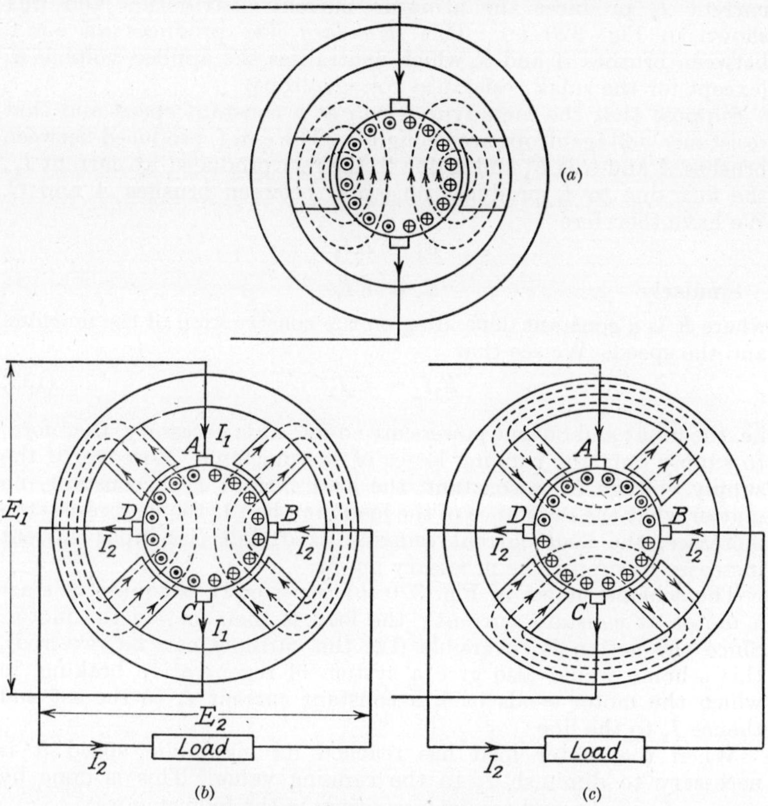

FIG. 320. THEORY OF METADYNE

constant current and varying voltage to an accelerating motor; this is clearly the best way in which the motor can receive the power.

The metadyne has a d.c. armature, but twice as many poles and brushes with the given armature as an ordinary d.c. machine. Fig. 320 (*a*) shows an ordinary d.c. machine with two poles and two brushes: a current flowing in the direction shown causes the armature current distribution shown in the figure with the corresponding cross-flux, which is mainly restricted to the pole faces. Fig. 320 (*b*) shows the metadyne using the same armature. There are four poles and four brushes, as shown, and a current I_1 produces an armature current distribution as in Fig. 320 (*a*). The flux due to the armature current is now provided with a path through the yoke by the four poles in the way shown. This *primary flux* produces an e.m.f. in the armature between the brushes B and D, so that a current I_2 flows through the load in the direction shown. The load current I_2 produces the armature-current distribution and flux shown in Fig. 320 (*c*). This *secondary flux* produces an e.m.f. between brushes A and C, which neutralizes the applied voltage E_1 (except for the small resistance voltage-drops).

Suppose that the metadyne is run at a constant speed and that resistance voltage-drops are negligible. The e.m.f. produced between brushes A and C is E_1 and is due to the flux produced by current I_2; the flux due to I_1 produces no e.m.f. between brushes A and C. We have therefore

$$E_1 = KI_2.$$

Similarly $$E_2 = KI_1,$$

where K is a constant depending on the construction of the machine and the speed. We see that

$$E_1I_1 = E_2I_2, \quad . \quad . \quad . \quad . \quad (133)$$

i.e. the input and output powers are equal. It is necessary, therefore, to supply only the running losses of the machine. Moreover if the supply voltage E_1 is constant, the load current I_2 is constant, no matter what the resistance of the load may be. If the load resistance increases, the load current remains fixed, but the input current increases to supply the necessary power.

The scheme shown in Fig. 320 (*b*) is perfectly adequate to start a motor at constant current; the load is then merely the motor. Since the action is reversible (i.e. the currents can be reversed), this scheme would also give a system of regenerative braking, in which the motor sends back a constant current I_2 to the set and thence I_1 to the line.

When the motor load has reached its maximum speed it is necessary to diminish I_2 to the running value. This is done by means of *variator* and *regulator windings* in the following way.

The variator winding is wound round the poles so that the flux lines are like those due to the secondary current, i.e. as shown in Fig. 320 (*c*). The variator excitation is said to be positive if its flux is in the same direction as those due to I_2, and negative if the flux is in the opposite direction. The use of the variator windings destroys the transformer property of the metadyne, as expressed in equation (133). For if the variator winding were given enough current to produce the flux yielding the back e.m.f. E_1, the current I_2 would fall to zero, and then there would be an input but no output. The metadyne would then speed up. Conversely if the variator flux were equal to the flux produced by I_2 in the absence of the variator winding and opposed it, I_2 would have to increase by 100 per cent to overcome this flux. We should then have an output equal to twice the input, and the metadyne would need mechanical power, equal to its input electrical power, to keep it running.

The metadyne is maintained as a transformer by means of a *regulator winding*, which produces a flux as in Fig. 320 (*b*). This flux affects the output current and power, and if the current in the regulator winding is correctly adjusted, the output power remains equal to the input power.

The effects of the armature, variator, and regulator currents can be summarized in the form of the equations

$$\left.\begin{aligned} E_1 &= KI_2 + k_vI_v \\ \text{and} \quad E_2 &= KI_1 + k_rI_r \end{aligned}\right\} \qquad (134)$$

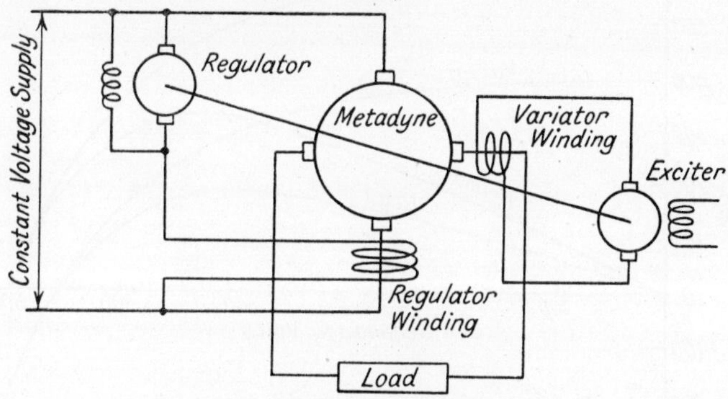

FIG. 321. COMPLETE METADYNE SET

where I_v and I_r are the variator and regulator currents, and k_v and k_r are constants of the machine, the windings, and the speed. The input and output powers are

$$\left.\begin{aligned} P_i &= E_1I_1 = KI_1I_2 + k_vI_1I_v \\ \text{and} \quad P_o &= E_2I_2 = KI_1I_2 + k_rI_2I_r \end{aligned}\right\} \qquad (135)$$

The condition for transformer action is

$$k_vI_1I_v = k_rI_2I_r \qquad (136)$$

Fig. 321 shows a complete metadyne set, which may be used for motor starting and regenerative braking. The method of exciting the field of the exciter is determined by the required current-voltage curve of the load (i.e. secondary). Fig. 322 shows a set of characteristics. The secondary current-secondary voltage curve is determined by the requirements of the load. The primary current curve is calculated from the fact that the input power equals the output power, i.e. $I_1 = I_2E_2/E_1$, E_1 in this case being 600 volts and E_2 is along the abscissa. It is assumed that in the set a secondary current of 200 amperes is required to produce a back e.m.f. of 600 volts in the primary circuit. From this assumption and the curve of secondary current, the curve of variator current is drawn for a variator winding having the same number of turns per pole as the armature winding (i.e. for $k_v = K$). Then from these curves and equation (136) the regulator current is calculated and the curve drawn, again with $k_r = K$.

Modifications are required, of course, to allow for resistance drop, iron saturation, windage losses, etc.

The metadyne has applications wherever control of d.c. motors is required. The control is smooth and requires no switching, so that switchgear and arcing are avoided. In some cases it is cheaper

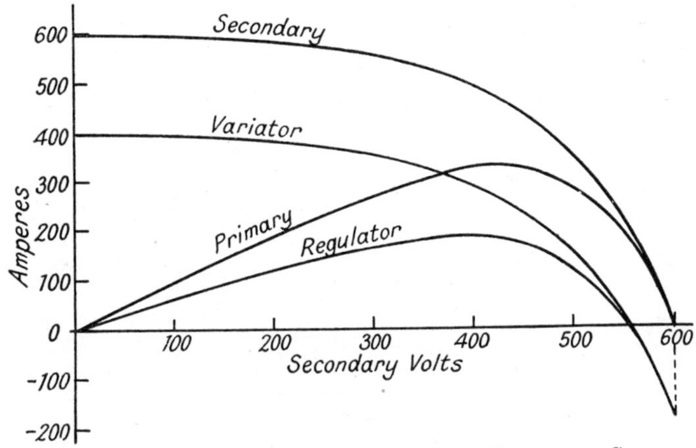

FIG. 322. CURRENT-CHARACTERISTICS OF METADYNE SET

than the Ward-Leonard system in first cost. In traction it provides smooth acceleration, without skill on the part of the driver, and regenerative braking down to very low speeds. It is already being used on the Underground railway.

Field Weakening or Tapped Field Control. When the motors have run up to full speed, an increase of speed is still possible by cutting out some of the field turns by means of tappings or by a shunt. It is usual to have not more than two tappings giving 15 and 30 per cent increase in speed. The results are best explained by an example.

EXAMPLE. The following figures relate to the series-wound motors of an electric locomotive—

Current per motor (amperes)	200	300	400	500
Train speed (m.p.h.)	41·5	33·5	28·5	28·0
Tractive effort per motor (lb.)	1 300	2 460	3 660	4 870

Calculate the values of speed and tractive effort for the same range of armature current when the series field current is reduced 20 per cent by a field diverting resistor.

By means of diagrams illustrate not more than three methods employed in modern electric traction to transmit torque from the motors to the driving wheels, discussing briefly their relative advantages.

(*Lond. Univ.*, 1939.)

A clear picture of the action of a field diverter is obtained by plotting speed and tractive effort against armature current: Fig. 323

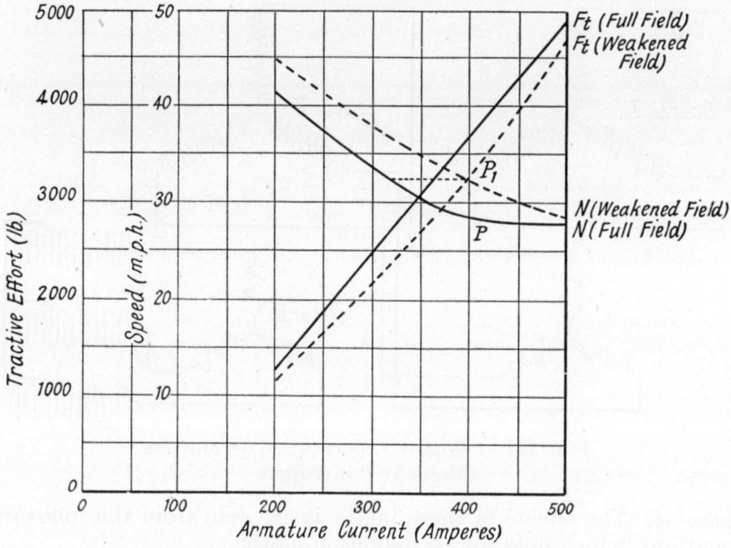

FIG. 323. FIELD-WEAKENING CALCULATION

shows the curves for the values given above in full line. If the field current is reduced 20 per cent by a shunt resistance, the field current is 0·8 of the armature current. Thus when the armature current is 400 A., the flux is that produced by a normal field coil with $0{\cdot}8 \times 400 = 320$ A.; and as we ignore the resistance drop in the armature and assume that the supply voltage is constant, being equal to $N\varphi$, the speed with the reduced field is that which occurred previously at 320 A. Thus the point P on the speed-current curve becomes the point P_1. In this way we get the speed-current curve for the weakened field. The tractive effort is proportional to $I\varphi$, i.e. to I/N at constant voltage supply. The tractive effort at 400 A. with the weakened field is thus

$$3\,660 \times \frac{28{\cdot}5}{32} = 3\,260 \text{ lb.}$$

Fig. 323 shows the new tractive effort-current curve. It is simple to derive the new tractive effort-speed curve from the two new curves.

Starting and Speed Control of A.C. Motors. The methods adopted differ considerably according as to whether the motor is three-phase or single-phase.

Three-phase Motors. Starting is done by means of liquid or metallic rheostats in the rotor circuit.

Speed control is effected in two ways, by *cascading* and *pole*

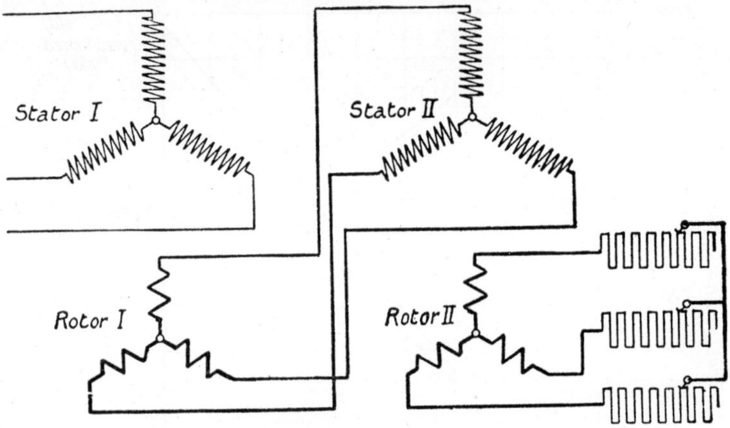

FIG. 324. CASCADE CONNECTION OF MOTORS
(*Electric Traction* (Dover))

changing. The effects of these methods are seen from the following equations, which hold for an induction motor.

$$\text{Speed} = f(1 - s)/p \qquad (137a)$$

and

$$\text{Rotor power dissipated} = ksf\mathrm{T}/p, \qquad (137b)$$

where f is the frequency, s the slip, p the number of pairs of poles, k a constant, and T the torque.

If power were dissipated in the rotor by means of resistance, the slip would increase by equation (137*b*) and hence the speed fall by equation (137*a*). Instead of wasting this power it may be used to drive another induction motor; and thus we achieve the *cascade connection* of Fig. 324, in which power is taken by slip rings from the rotor of the first motor to drive the second motor. The rheostat in the second rotor enables speed regulation up to the cascade synchronous speed, which is $f/(p_1 + p_2)$, where p_1 and p_2 are the pairs of poles in the two machines, which are coupled mechanically. If the motors have equal numbers of poles, the cascade synchronous speed is half that of a single motor. The two motors provide approximately equal mechanical power. A disadvantage is the low power factor of the combination.

Equation (137*a*) shows that if the number of poles is changed the speed is changed. The ways in which the number of poles can be changed are numerous and complicated; but the principle may be illustrated simply as in Fig. 325, where a winding is shown as giving 4 poles and 8 poles by altering the supply connections.

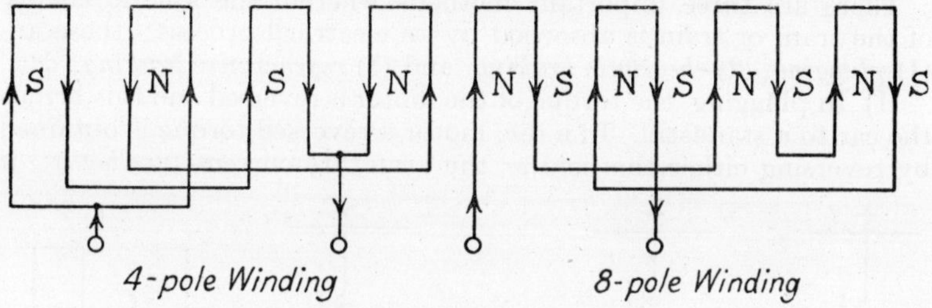

FIG. 325. POLE CHANGING

Single-phase Motors. The voltage can be reduced on starting without the use of resistances, and this gives a large saving of energy; it should be noted, however, that this advantage of the a.c. system has been to a large extent neutralized by the introduction of the metadyne.

In the a.c. system, as the power is supplied from the line by a transformer, all that is necessary is a number of tappings on the

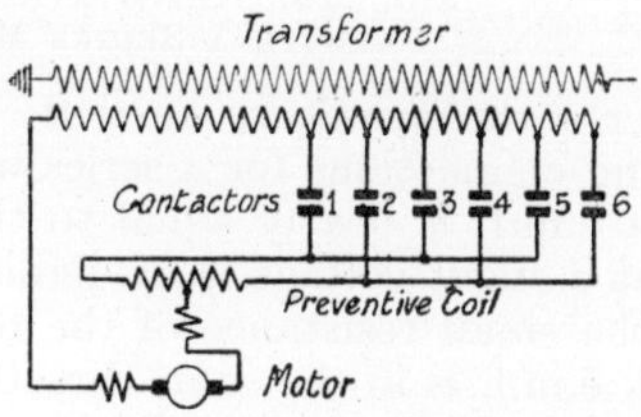

Sequence of Contactors

Notch	Contactors 1	2	3	4	5	6
1	●	●				
2		●	●			
3			●	●		
4				●	●	
5					●	●

FIG. 326. CONNECTIONS FOR CONTACTOR METHOD OF TAP-CHANGING
(*Electric Traction* (Dover))

secondary of this transformer. A preventive coil is used to ensure satisfactory operation, in a manner shown in Fig. 326. In the case shown there are five notches, and at each position two adjacent contactors are closed. The preventive coil ensures that the part of the transformer secondary between the two contactors is not shorted. A very important advantage of this method is that each notch is a running position, so that there are available many speeds of running.

Electric Braking. On trains, trams, and trolley buses there are available mechanical and electrical brakes. The wheel brakes, which are mechanical, are worked by compressed air on trains and by hand on trams or trolley buses. On trams the mechanical track brake consists of one or more pairs of wooden blocks, which are pressed on to the track rails by means of levers; this brake is for use on steep gradients and utilizes the weight of the car. The magnetic track brake consists of electromagnets, which are suspended normally to clear the track; when they are energized they are attracted to the track.

There are three important methods wherein the kinetic energy of the tram or train is absorbed by an electrical process; these are (1) *plugging*, (2) *rheostatic braking*, and (3) *regenerative braking*.

(1) In plugging, the torque of the motor is reversed and this brings the car to a standstill. In a d.c. motor a reversed torque is obtained by reversing either the field *or* the armature current (*not both*); it

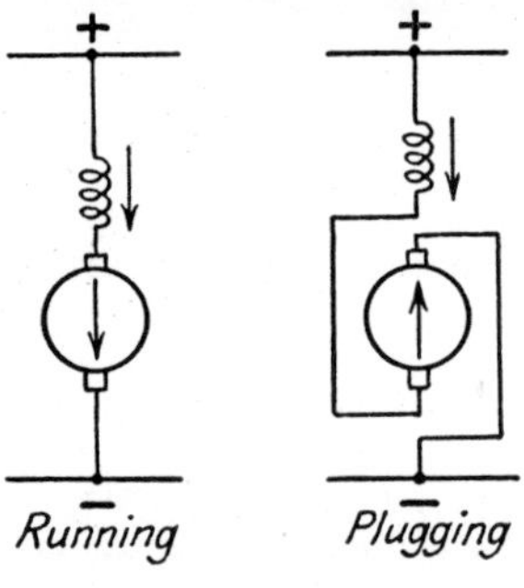

FIG. 327. PLUGGING A SERIES MOTOR

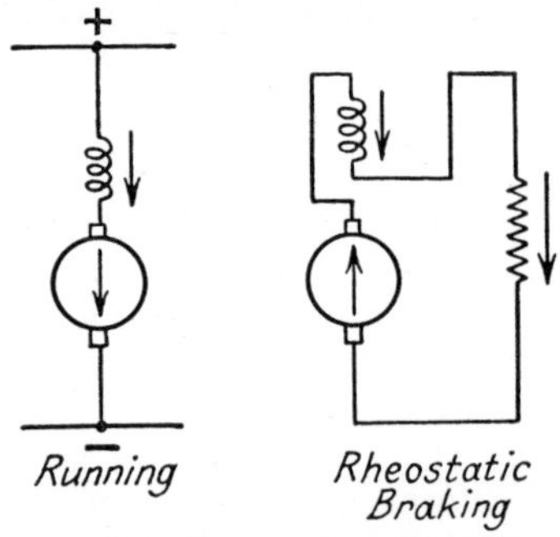

FIG. 328. RHEOSTATIC BRAKING WITH A SERIES MOTOR

is usually convenient to reverse the armature current. Fig. 327 shows the running and plugging connections for a series motor. In the running position the back e.m.f. is nearly equal to the supply voltage and opposes it, so that a small voltage is available to drive the normal current through the small resistance of the motor. In the plugging position the back e.m.f. is in the same direction as the supply voltage; so that at the instant of switching twice supply voltage is available, and an enormous rush of current would take place (about twice the current taken by the stationary motor on full voltage). Limiting resistance has therefore to be inserted in series with the motor. During the braking period the supply has to give energy (at the rate of VI watts), and this energy plus the kinetic energy of the car has to be dissipated in the series limiting resistances. The method is thus wasteful of energy, although it is efficient for braking purposes.

Plugging can be achieved in an induction motor by reversing the direction of rotation of the magnetic field, and this is easily done by reversing the connections to two of the three phases. In this case the current does not increase to an excessive value. By using different values of rotor resistance, any desired speed-torque braking curve can be obtained.

(2) In rheostatic braking the motor is disconnected from the supply and connected to a resistance. The kinetic energy of the car drives the motor which then acts as a generator and dissipates energy in the resistance. This method can be used for d.c. and synchronous motors.

In the case of d.c. shunt and synchronous motors the field is kept across the supply, but the armature is switched from the supply to across a resistance; if the supply fails, the field disappears and there is no braking.

Fig. 328 shows the running and braking connections of a series motor; the direction of the current in the armature is reversed, and so to facilitate the build-up of the series field the connections to the field winding are reversed. This ensures that the residual field is in the correct direction at the beginning of build-up, and the flux will not pass through a zero position. The braking resistance is low enough for the generator to be self-exciting.

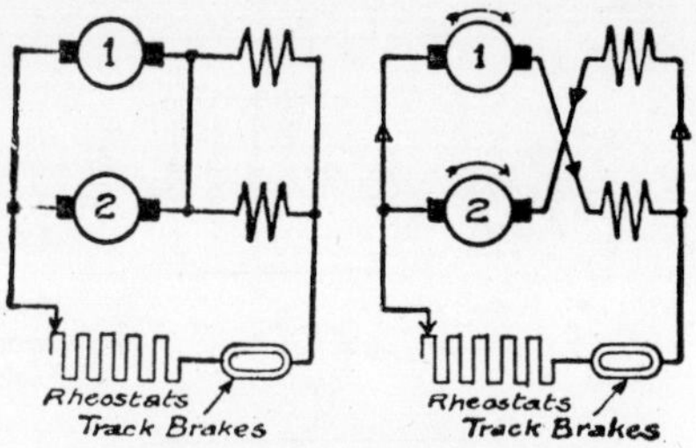

FIG. 329. AVOIDING SHORT-CIRCUIT IN PARALLEL OPERATION DURING RHEOSTATIC BRAKING
(*Electric Traction* (Dover))

In traction work there are always two or four motors. These could be used in series for rheostatic braking, but the resulting voltage would be too high. In practice they are put in parallel for braking; in this case it is essential to ensure that the machines share the load; otherwise the machine that builds up first would send a large current through the other and cause it to build up in the opposite direction, with the result that the machines would short-circuit one another. Fig. 329 shows two methods of avoiding a short-circuited condition: in the first method an equalizing bar is used; in the second a cross-connection of the fields makes the armature current of the more powerfully excited machine strengthen the field of the less powerfully excited machine, with the result that equal excitation is achieved in both machines.

Rheostatic braking cannot be used with induction motors.

(3) Plugging and rheostatic braking involve the wasting of the kinetic energy of the tram or train, whilst the former even draws more wasted energy from the supply during the braking period. A worth-while economy is effected if the kinetic energy of the vehicle can be turned into electrical energy and pushed back into the supply. This method is known as *regenerative* braking.

The induction motor acts automatically as a regenerative brake at speeds above the synchronous speed, and is of special advantage on mountain railways. It is found that the motor returns up to 20 per cent of the total energy on certain railway runs, and saves a great deal of brake shoe wear.

The series d.c. motor cannot be used for regenerative braking without modification. For if the motor is to act as a generator, its armature current reverses and the series field connections must be reversed, otherwise the field flux will be neutralized and the build-up will not occur. But even if the driver were skilful enough to reverse the field connections at the exact moment, the method would still be useless. For at the instant of reversal the e.m.f. generated by the motor is small and is completely overpowered by the supply voltage, which drives current through the field in the

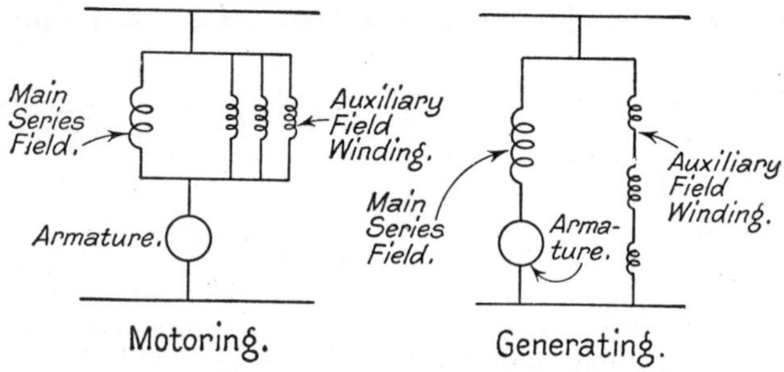

Fig. 330. French Method of Regenerative Braking

wrong direction, reverses the field and causes the e.m.f. of the motor to aid the supply voltage. The result is a short-circuit of the supply. The main trouble associated with regenerative braking by series motors is seen to be due to the lack of control of the field. There are various methods of overcoming this difficulty, either by modification of the windings or by supplying the machine with separate excitation.

Fig. 330 shows a well-known French method. During motoring the machine acts as a series motor, but has a main series field winding and auxiliary windings in parallel with it. During generation the auxiliary windings are switched (in series) across the supply, and the machine acts as a shunt generator slightly and differentially compounded. If there are several motors, there need not be any auxiliary windings. During motoring the field windings are in series with their respective armatures, and the motor circuits are in parallel. But during regeneration the circuit is as shown in the right-hand side of Fig. 330, except that in place of a single armature we have all the armatures in parallel, and what are shown as auxiliary field windings are the ordinary series windings of all the motors but one.

Fig. 331 shows the Metropolitan-Vickers regenerative system, which uses an auxiliary generator; this can be either one of the train motors or a special machine. The magnitude of the regenerated current is controlled by varying the field strength of the auxiliary generator, and thus the regeneration does not depend wholly upon the speed of the tram. The stabilizing resistance is used to prevent current surges when the tram crosses from one section of the supply to another, and to compensate for variable line voltage (towards which regeneration is very critical). Suppose that the line voltage rises, so that the regenerated current tends to decrease. The current in the stabilizing resistance, being the sum of the auxiliary generator and regenerated currents, tends to decrease also. The voltage drop

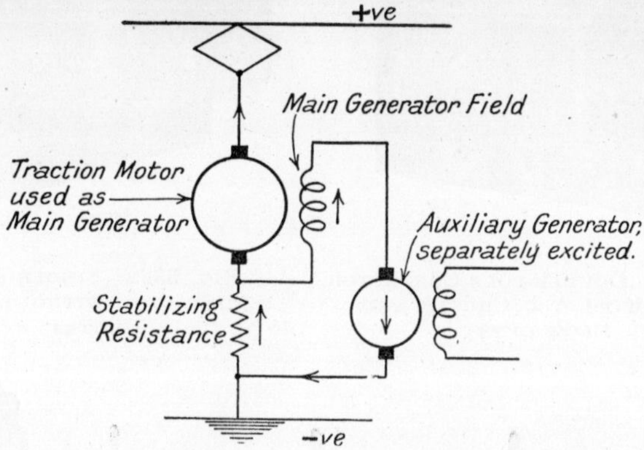

FIG. 331. METROPOLITAN-VICKERS REGENERATIVE SYSTEM

in this resistance decreases and thus the voltage across the main generator field increases, so that the e.m.f. in the main generator increases and thus compensates for the rise of the line voltage.

The modern tendency is to use regenerative braking down to about 10 m.p.h., then rheostatic braking down to 4 m.p.h., and finally mechanical braking to a standstill. This diminishes wear on the brake shoes.

Trolley-bus Control Equipment. Trams and small locomotives use drum controllers, of the type shown in Fig. 317, but where the current is too large for such equipment a master controller is used. The contacts instead of being made or broken by hand, are closed or opened by contactors, which are operated by solenoids. Figs. 332A and 332B show a double-pole and a single-pole contactor used for trolley-bus control. Fig. 333 shows a master controller for a trolley bus.

In a trolley-bus it is usual to use one motor of rating 80/95 h.p. at 500/550 volts, as the adhesion between a rubber-tyred wheel and the road is high. To permit of regenerative braking a compound-wound motor is used. A foot-operated master controller is used, to

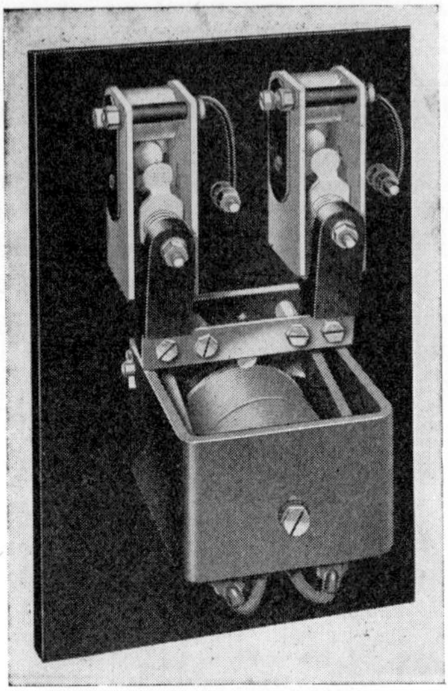

FIG. 332A. DOUBLE-POLE CONTACTOR FITTED WITH ARC CHUTES AND BLOW-OUTS

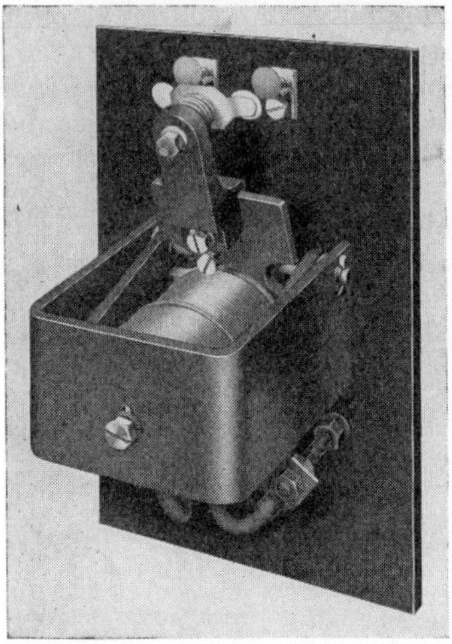

FIG. 332B. SINGLE-POLE CONTACTOR WITHOUT ARC CHUTES

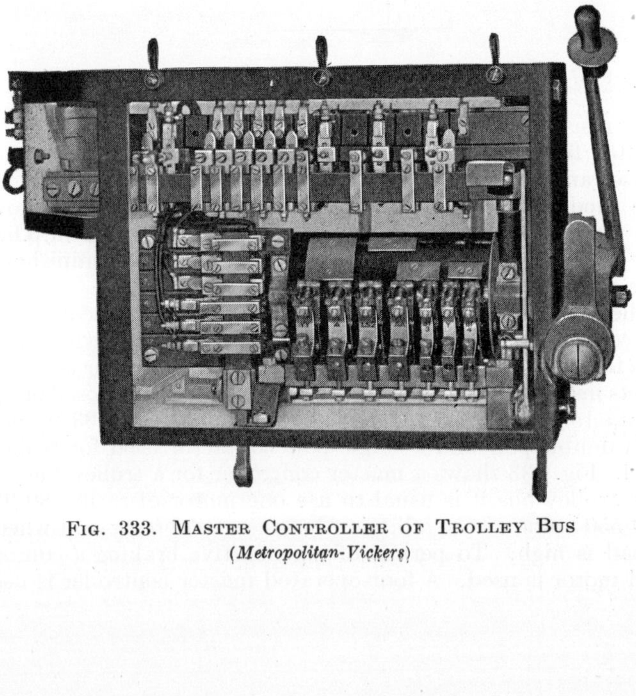

FIG. 333. MASTER CONTROLLER OF TROLLEY BUS
(*Metropolitan-Vickers*)

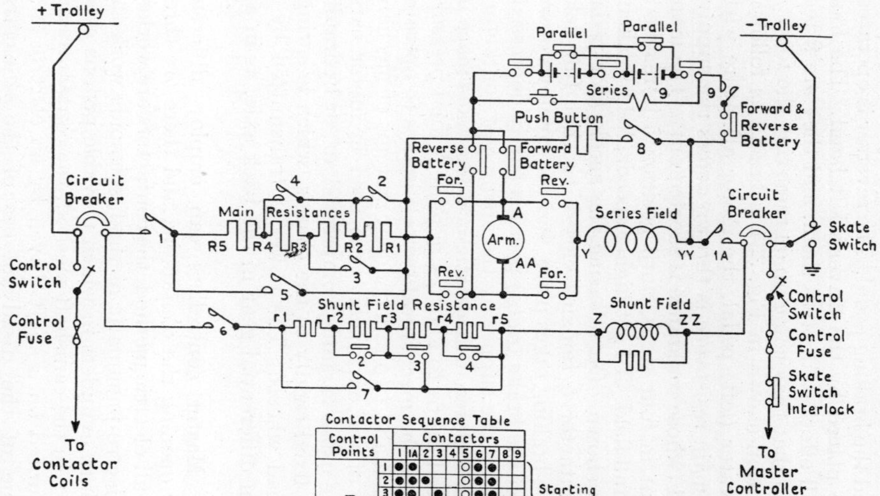

Contactor Sequence Table

Control Points		1	1A	2	3	4	5	6	7	8	9	
Accelerating	1	●	●				○	●	●			Starting Points
	2	●	●	●			○	●	●			
	3	●	●		●		○	●	●			
	4	●	●		●	●	○	●	●			
	5	●	●	●	●	●	○	●	●			
	6	●	●	●	●	●	●	●	●			Full Field
	7	●	●	●	●	●	●	●				Field Weakening Points
	8	●	●		●	●	●	●				
	9	●	●			●	●	●				
	10	●	●				●	●				
	11	●	●				●					
Rheostatic Braking	1				●			●		●		
	2				●			●	●	●		
Battery											●	

● Contactor closed.
○ Contactor closed on notching back only.

FIG. 334. KEY DIAGRAM OF REGENERATIVE-RHEOSTATIC EQUIPMENT WITH BATTERY MANOEUVRING (B.T.-H.)

ensure that the driver's hands are available for steering and the using of the hand (mechanical) brake. Fig. 334 shows the diagram of operation of the trolley-bus controller made by the B.T.-H. Co. with regenerative and rheostatic control. The controller has an accelerating controller and reverser, and a braking controller for rheostatic braking. The two controllers are operated by foot pedals, the accelerator pedal and the braking pedal. The reverser is operated by an insulated handle placed by the driver's left hand. The foot pedals need only a small effort to depress them, and they are fitted with restoring springs so that they come up on removing the foot.

The method of operation shown in Fig. 334 is briefly as follows. On pressing the accelerator (left) pedal, the motor armature with its series field and full main resistance is thrown across the supply, as is also the shunt field (whose resistance is shorted out). The main resistance is shorted out in four steps, until position 5 is reached. This is a position with full field. To get higher speeds, field weakening is used in the way shown. On releasing the accelerator pedal back to position 1, contactor 5 remains closed, so that full field obtains and regenerative braking takes place, and the car slows down to about 12 m.p.h. Then the brake (right) pedal is partly depressed, the motor is removed from the supply by the opening of contactors 1 and 1A, and contactor 8 closes. Rheostatic braking takes place in two steps, until the speed drops to about 4 m.p.h. A final pressure on the brake pedal causes the mechanical (wheel) brakes to operate. Statutory regulations demand two independent systems of mechanical brakes, which are of the expansion type used in motor cars, and are operated by the usual vacuum, compressed air, or hydraulic devices. The motor is fixed rigidly to the chassis, which is spring-supported from the road wheels. The power is transmitted by a shaft to a worm-driven differential gear in the back axle, as in an ordinary motor-bus.

Master Controllers. Master controllers can employ different methods for closing or opening the contactors, and there are three methods in use. In the all-electric method, the contactors are worked by solenoids; in the electro-pneumatic method, solenoids work the valves that admit air into cylinders in which movable pistons close or open the contactors; in the camshaft method, the contactors are operated by cams mounted on a shaft driven by an electric or a pneumatic motor. One of the disadvantages of the all-electric method is that supply variations may cause uncertainty of action of the contactors. In the latter two methods, large mechanical forces can be put on the contactors, so that vibration can cause no harm: moreover, the systems are electrically simple and mechanically very robust, so that they are favoured.

Multiple-unit Control. Electric trains on suburban service or on the Underground are varied in length according to the traffic requirements. The number of motor and trailer coaches thus varies, and it is necessary to arrange for the simultaneous control of motors at different points of the train.

In the multiple-unit method of control, each group of two or four motors in every motor coach is provided with a series-parallel controller, reverser, starting resistances, and current collecting gear. Such a group is called a *unit* of the train equipment. The controllers and reversers are remote controlled by a master controller situated in any motor coach, and multicore cables couple the controlling circuits in parallel. Fig. 335 gives a diagrammatic sketch of three motor coaches with multiple-unit control. The controller in use is

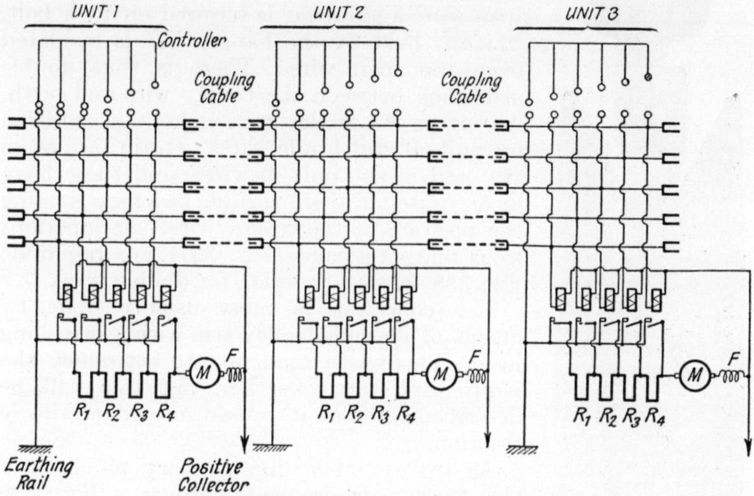

FIG. 335. MULTIPLE-UNIT CONTROL

that on the left, but it may be any one of the three. In the position shown the first three contacts are made; the making of the first contact on the left of the controller earths the controller, making the second energizes the first (left) contactor and connects the motor, field, and all the resistances across the supply, whilst making the third contact energizes the second contactor and shorts out the resistance R_1. This is the position shown.

Overhead Construction for Tramways and Trolley Buses. Statutory regulations demand that the trolley wire be placed at a minimum height of 17 ft. above the street surface (except under bridges), and supported at intervals not greater than 120 ft.

The trolley wire is of hard drawn copper, and the most usual cross-section is grooved circular, as shown in Fig. 336, which illustrates an ear, trolley wire, and trolley wheel. The advantages of the grooved wire over a circular wire are (1) elimination of soldering during erection, resulting in a quicker and cheaper erection, and (2) smoother running, since clipped fittings offer less obstruction than soldered fittings.

The most usual method of supporting the trolley wire is the following. Poles are erected on both sides of the road, and span wires of stranded galvanized steel are run between the poles. The span wires are insulated from the poles by span-wire insulators. The globe insulator has two malleable iron castings, one compressed over the other, which take the strain; they are insulated from each other by mica. A hanger is attached to the span wire, and an ear is screwed on by a bolt, which is held by the hanger but is insulated from the span wire. There is thus double insulation between the trolley wire and earth. At curves the trolley wire is kept in position by pull-off and bridle wires; strain insulators are used in the pull-off wires, and these have incorporated in them turnbuckles for adjusting the position of the trolley wire. At junctions frogs and crossings (Fig. 337) are required; Fig. 338 shows a junction for double track.

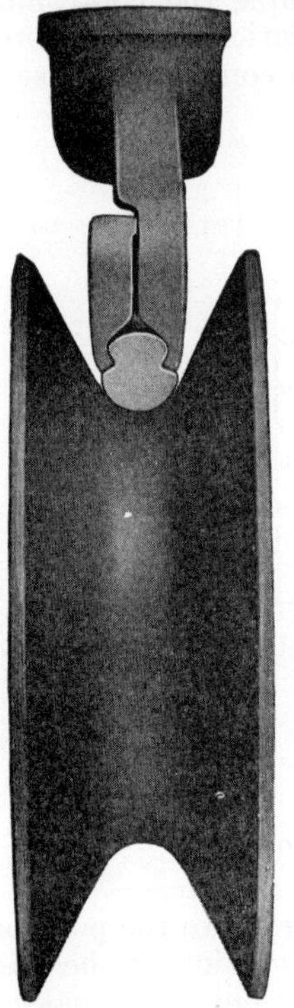

FIG. 336. EAR, TROLLEY WIRE AND TROLLEY WHEEL
(*British Insulated Cables*)

The collection is most usually done by means of a sprung trolley arm with a swivelling head. There is a tendency to introduce the pantograph for collection, but this will be described later as it is used mostly in railway traction.

As trolley buses diverge very much from the track, dewirements occurr. Fig. 339 shows a hydraulic arrangement to overcome dewirement; the hydraulic shock absorber acts in the downward direction only. Note the carbon current collector.

The overhead construction is as for tramways, except for the modifications entailed by the double line required. At junctions and cross-overs the trolley wires of opposite polarity must be insulated from each other at the crossings, and a special form of section insulator has been developed for this purpose. The current is collected by two trolley arms, and these are so constructed that the bus can deviate by as much as 13 ft. from the centre line of the overhead wires.

Feeding and Distributing System for Tramways. The regulations demand that the voltage of the trolley wire shall not exceed 550 volts and the generating voltage 650; the potential difference between any two points of the earthed return must be less than 7 volts, and the potential of any point must not be more than 4·2 volts above earth. These conditions necessitate separate feeding systems for the trolley wire and the track rails.

Fig. 340 shows how the regulations are obeyed on a long system by the use of boosters. The negative bus-bar is earthed by two

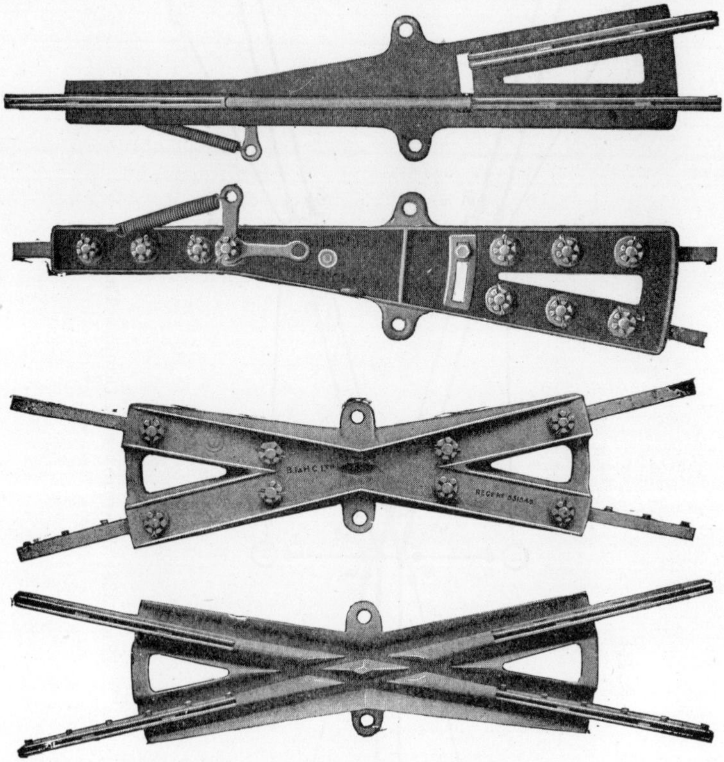

FIG. 337. FROGS AND CROSSINGS
(*British Insulated Cables*)

buried plates. One lead from the negative bus-bar is run to the track near the generating station, and one from the positive bus-bar to the trolley wire. If no other connections were used, the potential along the trolley wire would decrease with the distance away from the station; whereas the potential of the track would rise (because of the voltage drop of the current in it), so that a point would be reached where the potential would be greater than 4·2 volts. The potential along the trolley wire is kept constant within narrow limits by feeding the sections, which are isolated from each other, by

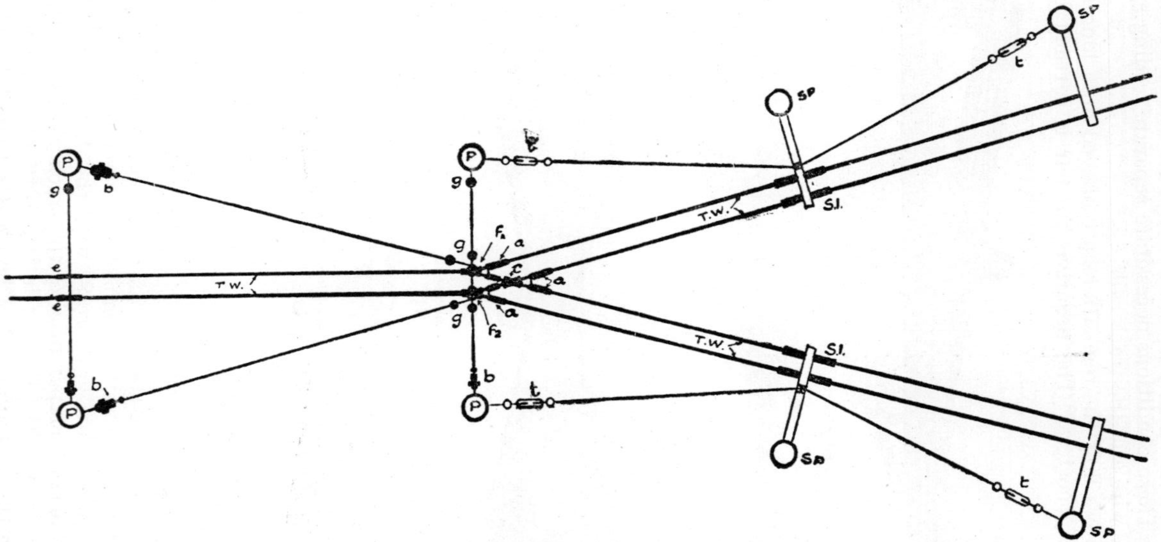

FIG. 338. JUNCTION FOR DOUBLE TRACK
(*Electric Traction* (Dover))

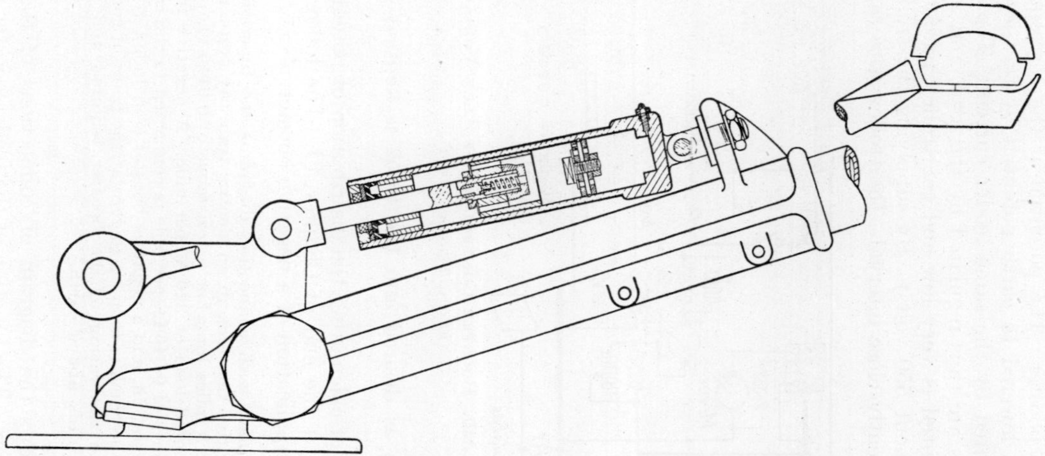

FIG. 339. HYDRAULIC SHOCK ABSORBER FOR TROLLEY BUSES

feeders in series with positive or *feeder boosters*; in Fig. 340 one such booster is shown feeding the right-hand end of the trolley wire. The booster voltage regulates itself by the current in the feeder in series with the booster. The potential at the corresponding point of the track is lowered to earth potential by the *negative booster*, which is regulated by the same feeder current. The positive and negative boosts are thus regulated by the load, and are such that the track potential is very low and the trolley potential is nearly constant (at about 550 volts). By supplying the trolley wire and track at sufficiently close intervals, the trolley wire can be kept as

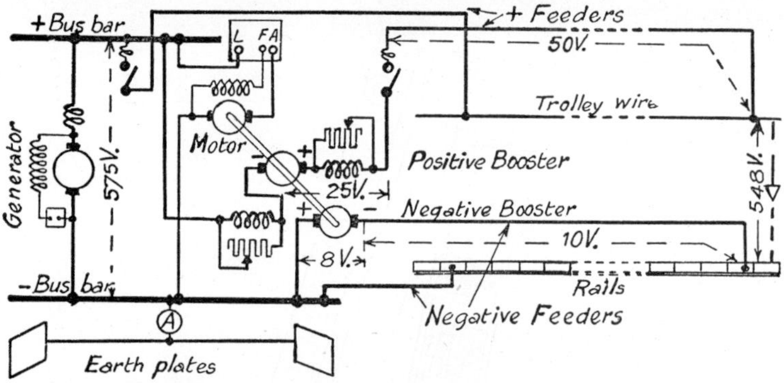

FIG. 340. BOOSTER CONTROL OF TROLLEY VOLTAGE
(*Electric Traction* (Dover))

near 550 volts as desired and the track is kept at nearly earth potential.

The method of calculating the distribution of potential for given loads is given at the end of Chapter VI. The following example illustrates the application of a negative booster.

EXAMPLE. Explain with connection diagrams the function of (*a*) feeder boosters, (*b*) negative boosters, in an electric tramway system. A section of a tramway track 3 miles long has a resistance of 0·0145 Ω. per mile, and a uniformly distributed load of 320 A. per mile. A negative feeder having a conductor resistance of 0·046 Ω. per mile is connected to the track at a point 2 miles from the station, and a negative booster is included in the circuit. If the potential of the track is reduced to zero at the point of connection to the booster, calculate the rating of the booster required and the maximum potential of the rails above earth. (*Lond. Univ.*, 1931.)

Fig. 341 shows the diagram of connections of the feeder and negative boosters. The function of the feeder booster is to keep the trolley wire at constant potential, whilst that of the negative booster is to keep any point of the track within 4 V. of earth potential and any two points within 7 V. potential difference.

We assume that in this case there is no feeder booster, and the scheme is as shown in Fig. 341. Fig. 342 shows the conditions of current and voltage. In the last mile 320 A. enter the trolley wire and 320 A. come in from the track (point P). At a point distance x from the generating station a current of $320(3 - x)$ goes along the

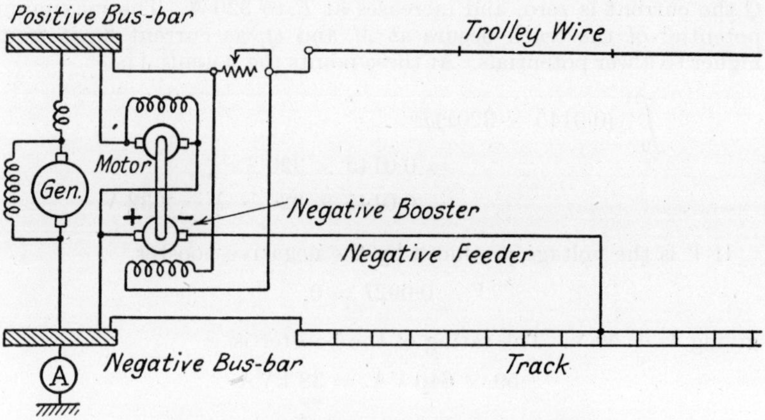

FIG. 341. NEGATIVE BOOSTER FOR TRACK

trolley wire: let i_x be the current returning along the track, and let I be the current in the negative feeder. Then

$$i_x + I = 320(3 - x) \text{ or } i_x = 320(3 - x) - I.$$

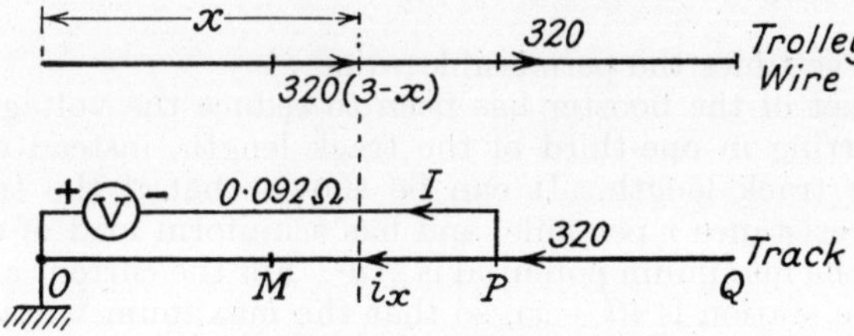

FIG. 342

As the point P has been reduced to zero voltage, the voltage drop along OP is zero, i.e.

$$\int_0^2 0{\cdot}0145 i_x dx = 0.$$

Substituting for i_x we find

$$\int_0^2 [320(3 - x) - I]dx = 0$$

$$= [320(3x - \tfrac{1}{2}x^2) - Ix]_0^2$$

$$= 1\,280 - 2I,$$

i.e. $$I = 640.$$

The distribution of the current along the track is then the following. At M, one mile out, there is no current. Between M and O the current flows towards O, at which point the current is 320 A.; between M and P the current is towards P, where it is 320 A. At Q the current is zero, and increases at P to 320 A. The maximum potential of the track occurs at M and Q, as current flows from higher to lower potentials. At these points the potential is

$$\int_0^1 (0{\cdot}0145 \times 320x)dx$$
$$= 0{\cdot}0145 \times 320[\tfrac{1}{2}x^2]_0^1$$
$$= 0{\cdot}0145 \times 320 \times \tfrac{1}{2} = 2{\cdot}32 \text{ V}.$$

If V is the voltage produced by the negative booster,

$$V - 0{\cdot}092I = 0,$$

giving $V = 59$ V. The rating of the booster is

$$59 \times 640 \text{ VA.} = \underline{\underline{38}} \text{ kVA.}$$

If the negative booster were not used, the maximum potential would occur at Q and be

$$\int_0^3 [0{\cdot}0145 \times 320\,(3 - x)]dx = \underline{\underline{20{\cdot}8}} \text{ V.},$$

which is five times the permissible value.

The effect of the booster has been to reduce the voltage drop to that occurring in one-third of the track length, instead of that in the whole track length. It can be shown that if the track is of length l, resistance r per mile, and has a uniform load of i amperes per mile, the maximum potential is $\frac{1}{2}ril^2$. For the current at distance x from the station is $i(l - x)$, so that the maximum voltage, which occurs at the far end, is

$$\int_0^l ri(l - x)dx = ri[lx - \tfrac{1}{2}x^2]_0^l = \tfrac{1}{2}ril^2.$$

Reducing the effective track length to one-third therefore reduces the maximum voltage to one-ninth. For example, in the case worked out above the maximum voltage is reduced from 20·8 to 2·32.

If the track is 5 miles long and negative feeders are run out to points at 2 and 4 miles from the station, the maximum voltage on the track is reduced to one twenty-fifth.

Overhead Collection for Locomotives. The trains collect their current from an overhead wire or wires and use the bonded rails as return.

Pantographs are used for the collection of the current. Fig. 343 shows a pantograph in the normal operating position. The contact is made by two pressed-steel pans of channel section, each fitted with two renewable copper strips. The pans are attached to light aluminium castings, which are designed to break if the pans foul an obstruction. The pans are allowed small swivelling and vertical motions by the castings, which are spring supported, so that small

Fig. 343. Pantograph
(*Metropolitan-Vickers*)

variations in the level of the overhead line do not cause the pantograph to move. The pantograph framework is of light steel tubing with four sections which are joined at the corners. The lower sections are fitted to shafts which are carried in ball bearings fixed to the main framework. The latter is fixed on the supporting insulators. Each shaft is fitted with two cranks, one is connected to springs which balance the dead weight; and the other by springs to cranks fitted to countershafts, which are operated by single-acting pneumatic cylinders. When air is admitted to the cylinders, the springs are stressed and the pantograph rises; when the air is released the pantograph is lowered. The working pressure is 27 lb., and does not vary by more than 10 lb. Adjustable collars inside the cylinders enable the pressure to be varied from 15 to 35 lb.

EXAMPLES XII

1. Discuss the advantages of series motors for electric traction on suburban railways.

The bogie of a motor-coach is equipped with two motors each having characteristics as given in Example 16, which apply to 42 in. wheels. If the wheels driven by one motor (A) are 41 in. diameter and those driven by the other motor (B) are 40 in. diameter, determine, when the motors are operating in parallel and the train speed is 30 m.p.h., (i) the current input to each motor, (ii) the total tractive effort, (iii) the output at each driving-axle. (*Lond. Univ.*, 1949.)

2. Distinguish between rheostatic and regenerative braking in electric traction. Under what conditions would regenerative braking be employed? Explain, with the aid of a diagram, how the d.c. series motors of a locomotive are connected and operated during regenerative braking and how stability is obtained.

The characteristics of a d.c. series motor at 525 V. are as follows—

Current (A) .	100	150	200	250
Speed r.p.m.) .	2 060	1 585	1 300	1 175
Torque (lb. ft.) .	135	290	450	612

Determine the braking torque when the machine is operating as a self-excited series generator at a speed of 1 400 r.p.m. and loaded with a resistor of 2·5 Ω. Resistance of motor, 0·3 Ω. Ignore core and friction losses. (*Lond. Univ.*, 1953.)

3. An electric train maintains a schedule speed of 28 m.p.h. between stations situated 3 miles apart, with station stops of 30 sec. The acceleration is 1·5 m.p.h.p.s. and the braking retardation is 2 m.p.h.p.s. Assuming a simplified rectilinear speed-time curve, calculate (a) the maximum speed of the train, (b) the energy output of the motors in Wh. per ton, if the tractive resistance is 10 lb. per ton. (*Lond. Univ.*, 1932.)

4. Explain how an actual speed/time curve for an electric-train service can be replaced by a curve having a simple geometric shape. Deduce from first principles the relationship between acceleration, retardation, maximum speed, running time, and distance between stops, assuming a simplified speed/time curve. (*Lond. Univ.*, 1933.)

5. A 250-ton electric train has an average speed of 32 m.p.h. between stations on the level situated 1·25 miles apart. The acceleration at starting is 1·25 m.p.h.p.s. and the braking retardation 2·3 m.p.h.p.s. Assuming a trapezoidal speed-time curve, and a free-running speed 30% higher than the average speed, calculate the specific energy consumption for the run in Wh. per ton mile, and the power in kW. usefully employed at the end of the period of acceleration and during coasting. Assume a train resistance of 12 lb. per ton and allow 10% for the effect of rotational inertia. (*Lond. Univ.*, 1931.)

6. Describe, with the aid of diagrams, the process of arc extinction in (i) a 1 500 V. d.c. contactor, (ii) a 33 kV., a.c. air-blast circuit-breaker, (iii) a 33 kV., a.c. oil-break circuit-breaker. Explain why arc control is necessary in case (iii). Sketch one form of arc-control device and explain the action. (*Lond. Univ.*, 1949.)

7. A 200-ton train has a gross tractive effort of 25 000 lb. which may be assumed constant during the acceleration period. The speed is then maintained constant at its maximum value until braking begins. It is required to travel a distance of 0·8 mile between two stations in 120 sec. on an up-gradient of 1 in 100. The train resistance may be taken as 10 lb. per ton and the rotational inertia is 10 per cent. Braking retardation (including the effect of train resistance and gradient) is 2·2 m.p.h.p.s. Determine the maximum speed which must be attained in order to carry out the schedule, and the total energy supplied to the driving wheels. (*Lond. Univ.*, 1954.)

8. Discuss briefly the reasons why (i) series motors are employed on motor-coach trains for suburban electric railways, (ii) compound motors are usually employed on trolley buses.

The speed-current characteristic of a compound trolley-bus motor at 500 V. and with full shunt field is as follows—

Armature current (A.)	– 200*	– 150*	– 100*	0*	50	100	150
Speed (m.p.h.) . .	25·5	20	17	14	13·1	12·4	11·8

* Operating as a differentially-compounded generator at 500 V.

Determine the characteristic, over the range 75–200 A., when operating with the series-field winding alone. Resistance of the armature and series-field windings, 0·2 Ω.; resistance of shunt-field winding 100 Ω.; number of turns per pole in series-field winding 15; number of turns per pole in shunt-field winding 900. (*Lond. Univ.*, 1947.)

9. Obtain expressions for the voltage drop and I^2R loss in each core of a 2-wire d.c. distributor, when fed from both ends at equal voltages and carrying a uniformly distributed load of I amperes per unit length. The cross-section of each conductor is a and its length l: the specific resistance of the material is ρ.

A 2-wire d.c. distributor 300 yd. long, for which ρ is $\frac{2}{3}$ microhm-inch, is to supply a load of 2·5 A. per yard when fed at 240 V. at each end; not less than 220 V. is to be available at any loadpoint. Determine the minimum cross-section, and the efficiency. (*Lond. Univ.*, 1947.)

10. Explain with complete connection diagrams a method of reducing the rail drop in (*a*) a heavily loaded single-phase railway line, and (*b*) a d.c. tramway system.

11. Explain with a diagram of connections one arrangement of feeding and sectionalizing an overhead tramway system, having regard to the statutory regulations governing such distribution systems.

Find the length of the distribution section corresponding to a drop of 22 V., given: average current per car, 20 A.; schedule speed in section, 10 m.p.h. with an interval of 2 min. between cars; resistance of trolley wire, 0·46 Ω. per mile. (*Lond. Univ.*, 1932.)

12. Explain the principle and show the advantages of the series-parallel control of traction motors. Draw the development and connections of a series-parallel controller for two tramway motors (omitting the reversing drum), and sketch the magnetic blow-out arrangement of a typical controller. What modifications are needed for a motor-coach train? (*Lond. Univ.*, 1933.)

13. Explain briefly the disadvantages of using regenerative braking on a d.c. traction system fed only by mercury-arc rectifier substations. How is the surplus energy absorbed in such a case?

A 400-ton train has its initial speed of 38 m.p.h. reduced to 25 m.p.h. by regenerative braking in travelling 6 500 ft. down a uniform gradient of 1 in 75. Calculate the energy in kWh. returned to the system. Tractive resistance is 10 lb. per ton, overall efficiency 75% and the allowance for rotational inertia may be taken as 10%. (*Lond. Univ.*, 1949.)

14. Give a simple explanation of the action of a metadyne. Sketch the arrangement of the machine and a diagram of connections.

Discuss the relative advantages of rheostatic and metadyne control of the motors of electric trains in conditions which call for frequent stops. (*Lond. Univ.*, 1949.)

15. An electric train (dead weight, 130 tons; accelerating weight 143 tons) is equipped with four 600 V. motors, which are arranged in two pairs for series-parallel control. If, during series-parallel starting the current per motor is maintained at 400 A., calculate (i) the duration of the starting period, (ii) the speed of the train at transition, (iii) the rheostatic losses during (*a*) the series, and (*b*) the parallel, steps of the starting period.

At this current (400 A.) and normal voltage (600 V.) the tractive effort per motor is 4 330 lb. and the train speed is 23·9 m.p.h. Assume the specific train resistance to be 10 lb. per ton, and the resistance of one motor to be 0·1 Ω. (*Lond. Univ.*, 1948.)

16. Why are motors having series characteristics generally used for traction purposes?

A d.c. series traction motor has the following characteristics—

Current (A.) .	100	200	300	400
Speed (m.p.h.) .	41	28	23·5	21·3
Tractive effort (lb.)	650	1 900	3 400	5 000

Determine the approximate speeds and tractive efforts for the above currents if the motor is operated at the same voltage but with 30% of the field turns cut out. (*Lond. Univ.*, 1950.)

17. A 130-ton train is equipped with four motors; the characteristics of each at normal voltage are as follows—

Current (A.) .	100	200	300	400
Train speed (m.p.h.)	65	36·5	29·8	26·5
Tractive effort (lb.).	330	1 450	2 740	4 100

Calculate and draw the speed-time curve from the start to a speed of 36 m.p.h. for a run on level track, and determine the r.m.s. current per motor during this period. Assume constant normal supply voltage, a constant motor current of 400 A. during rheostatic acceleration, a constant tractive resistance of 10 lb. per ton, and the accelerating weight of the train to be 10% greater than the dead weight. (*Lond. Univ.*, 1947.)

18. Give a diagram of connections and explain the action of a negative booster. A section *ABC* of an uninsulated rail return system is 3 miles long. *A* is earthed, and *B* is 2 miles from *A*. A negative feeder, with booster in circuit, is tapped to the rail at *B*. The loading is 400 A. per mile and may be assumed to be uniform.

Determine the maximum p.d. between any two points on the rail system, assuming no leakage, if the potential of *B* is 2·5 V. below earth. Determine also the output of the booster. The resistance of the rail system is 0·035 Ω. per mile. The resistance of the negative feeder is 0·03 Ω. (*Lond. Univ.*, 1936.)

19. Explain why the accelerating weight of an electric train is greater than its dead weight.

An electric train weighing 200 tons starts from rest up a gradient of 2%, and attains a speed of 30 m.p.h. in 25 seconds, the acceleration being constant. If the train resistance is 10 lb. per ton and the accelerating weight is 10% greater than the dead weight, calculate the tractive effort and the total power output from the driving axles at the end of the accelerating period. (*Lond. Univ.*, 1950.)

20. Sketch a typical speed-time curve for an electric train operating on suburban service, assuming level track. Explain the shapes of the various parts of this curve.

Two stations on an electric railway are situated on a uniform gradient of 1 in 100. A 130-ton train (143 tons accelerating weight) makes the run down the gradient to the following schedule: uniform acceleration of 1·3 m.p.h.p.s. until the speed is 32 m.p.h.; power is then cut off and the train coasts for 100 sec., when the brakes are applied to bring the train to rest with uniform retardation.

Draw the speed-time curve and calculate the energy output (kWh.) from the driving axles. Assume the train resistance to be 12 lb. per ton and the braking effort due to the brakes alone to be 24 000 lb. (*Lond. Univ.*, 1953.)

21. Distinguish between rheostatic and regenerative braking in electric traction. Under what conditions would regenerative braking be employed? Explain, with the aid of a diagram, how the d.c. series motors of a locomotive are connected and operated during regenerative braking and how stability is obtained.

The characteristics of a d.c. series motor at 525 V. are as follows—

Current (A.) .	100	150	200	250
Speed (r.p.m.) .	2 060	1 585	1 300	1 175
Torque (lb. ft.) .	135	290	450	612

Determine the braking torque when the machine is operating as a self-excited series generator at a speed of 1 400 r.p.m. and loaded with a resistor of 2·5 Ω. Resistance of motor, 0·3 Ω. Ignore core and friction losses.

(*Lond. Univ.*, 1953.)

CHAPTER XIII

INDUSTRIAL UTILIZATION

Introduction. Electrical drive of industrial and other machinery is in most cases more economical than drive by any other method. It is invariably cleaner, more convenient, and more flexible. Electric drive is so convenient that in some cases it is used although a prime mover is already installed; one example is in diesel-electric traction. The diesel engine runs best at a certain speed, and in the absence of electric drive variable, and very awkward, gears would be necessary. The insertion of a generator followed by a motor enables the prime mover to work at a constant speed, whilst the speed of the electric motor is easily varied in any desired manner. In some cases, as on board ship, the use of electric drive does away with the necessity for a long and expensive shaft from the engine room to the propellers; in place of the shaft there is an electric cable which transmits the power, in the form of electrical power, from the engine room to the motor which drives the propeller.

This fact that the power room need not be in the vicinity of the power equipment is a great advantage in works where some of the heat of combustion is used for heating services. The power room can be placed in a position most convenient for the heating services, and cables will take the electrical energy to wherever it is required.

Group and Individual Driving. The utmost use is made of the flexibility of electrical transmission when every tool and piece of apparatus is driven by its own small motor; this is *individual drive*, and the tendency is in this direction, as each machine can be put where it is most effectively used.

In group drive the power from one large motor is transmitted via shafts and belting to a number of machines or tools. Fig. 344 shows a motor driving a lathe through a shaft and belting. A number of lathes may be driven from the same overhead shaft in any combination. Any lathe is put into operation by a lever which shifts its driving belt from a loose to a fast pulley. Belt drive is cheapest for powers up to 250 h.p. provided the distance between pulleys is not too small and the speed high. For ordinary drive a speed of 3 500 ft. per min. is efficient; for machine tools 1 500 ft. per min. The ratio for a distance between pulley centres of 20 ft. is 6 to 1, for 5 ft., $1\frac{1}{2}$ to 1. The belt slip is of the order of 1 per cent. In heavy work, rope drive is most suitable, and the diameter of the rope drum should be more than forty times the diameter of the rope, otherwise the life of the rope is short. An efficient rope speed is 5 000 ft. per min.

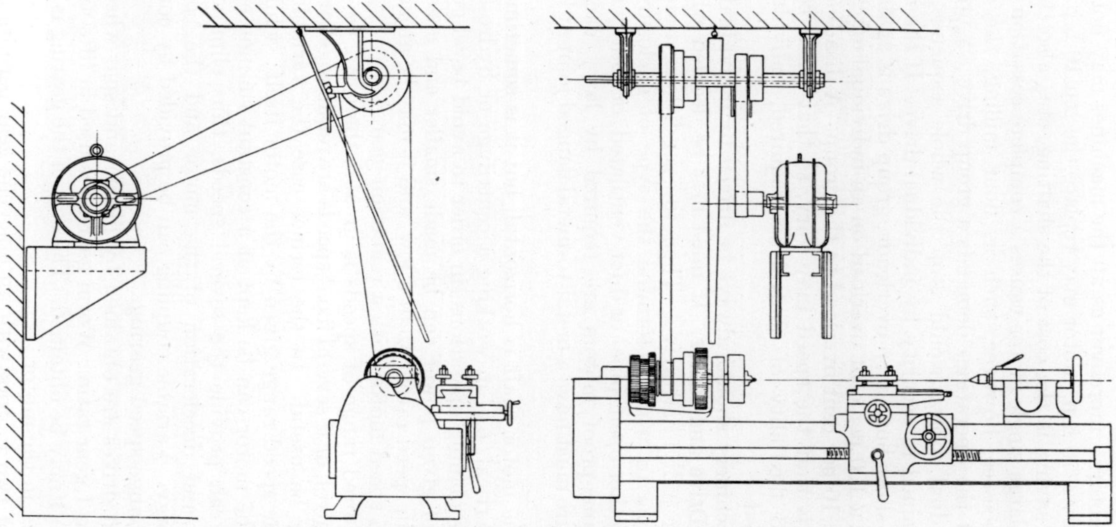

Fig. 344. Motor Driving Lathe through Shaft and Belting

A main advantage of group drive is a saving in first cost of the motors and control gear, as a 100 h.p. motor costs much less than ten 10 h.p. motors; moreover as all the machines may not be used simultaneously a 75 h.p. motor may replace the ten 10 h.p. motors. Against this saving is the cost of the shafting, etc., also the breakdown of the large single motor causes a complete cessation of work. In some processes, however, such as flour milling, the sequence continuity of the operations demands a group drive; a breakdown in an individual motor would stop the whole sequence, so that nothing is gained in security by individual drive. If the machines are liable to short but sharp overloads, group drive is again advantageous; for a 100 per cent overload on an individual machine may mean only a 10 per cent overload in the group. A disadvantage in group drive is that the speed may vary, so that when constancy of speed and flexibility of control are required, individual drive should be used.

Heavy machinery should always be driven by individual motors.

Choice of Drive and Motor. It must first be decided whether to use group or individual drive. Then the conditions under which the motor has to work determine the type of enclosure of the machine, whether open, pipe- or duct-ventilated, or totally enclosed. In mines flame-proof motors are required by law; each flame-proof enclosure must have a metal-to-metal flanged joint at least 1 in. wide.

Usually the motor shaft is horizontal, but it is sometimes better to have it vertical, e.g. in working a centrifuge or hydro-extractor and for some planing operations, in order to avoid bevel gears.

A motor of given power can be made smaller and cheaper if it runs at a high speed rather than a low, and it may be an advantage to use a high speed motor with a reduction gear.

If the required range of speeds is greater than can be given by electrical control, or if several fixed speeds are required, change-speed gearing may be useful. In the former case the variable gearing multiplies the speed range given by the motor itself; whilst in the latter case the motor can be kept at a constant running speed and the gearing can provide the several speeds, thus eliminating the acceleration and deceleration of the motor and the consequent waste of energy. Variable coupling can be provided by conical belt pulleys or change-speed gearing.

Slow speed drives are obtained by pinion and spur-wheel with a 7 to 1 or 5 to 1 gear ratio. Worm gearing is used in lifts and a gear ratio of 30 to 1 may be obtained. The cost of the gearing is balanced by the saving in the motor.

The use of two motors with an epicyclic gear permits of a wide variation of power and speed, and is useful for large lifts. For full speed and power both motors work, and the epicyclic gear serves merely as a mechanical coupling. For low speed and small power one motor is disconnected from the supply, and the other acts through the epicyclic gear which then becomes operative.

Friction clutches enable the motors to pick up the load gradually, and cushion them from excessive shocks during running. Fig. 345 shows a magnetically operated clutch combined with a coupling; the clutch is made in two sections, one of which has a magnetizing coil supplied with current by two slip rings shown on the left. Push-button control is common, and this is a great advantage in systems where operation is frequent. The ease of operating the magnetic clutch makes it useful for the selection of speeds from different

FIG. 345. MAGNETICALLY OPERATED CLUTCH
(*Electrical Engineer*)

gears; thus a motor-driven capstan has two clutches connected to two gears of different ratios, and the motor can be switched from one to the other by pressing a button.

In many industrial processes it is required to have a reversal of driving at fixed intervals. This may be achieved by mechanical means: one way is to use a crank and connecting rod driven by the motor to give a reciprocating motion; another way is to switch over from an open-belt to a crossed-belt, but this is inconvenient as the oncoming belt slips and gets scorched, and the reversing cannot be timed accurately. In general it is more satisfactory to reverse the direction of rotation of the motor than to rely on mechanical reversing gear. Trips near the ends of the runs can be used to cause the necessary switching for reversal. The kinetic energy of the motor can be absorbed by regenerative braking or rheostatic braking. Very quick and accurately timed reversal can be obtained in this way, and the method is used for rolling mills, planing, etc.

A very important thing is to choose a motor whose characteristics resemble as closely as possible the type of the load. This is specially important in individual drive, but in group drive a load equalizer makes this of less consequence. The motor characteristics must suit the load both for starting and running, although a bad starting characteristic may be overcome by using a friction clutch, gear box, or belt. The following table gives the starting torque, starting current, speed-load characteristic, and range of speed control for the motors most commonly used. A constant speed-load characteristic is one in which the speed at full-load is only a few per cent less than that at light load; an inverse characteristic is one in which the speed falls rapidly as the load increases. Full load torque is written T and full load current *I*.

COMPARATIVE TABLE OF DIFFERENT TYPES OF ELECTRIC MOTORS

Type of Motor	Starting Torque	Starting Current	Speed-load Characteristic	Range of Speed-control
D.c. shunt	2T	2*I*	Constant	4:1
D.c. series	3T	2*I*	Inverse	3:1
D.c. compound . . .	2T	2*I*	Constant	As d.c. shunt
Single-phase a.c. series .	3T	2*I*	Inverse	Several fixed speeds
Repulsion	3T	2*I*	Inverse	4:1
Synchronous . . .	¾T	3*I*	Absolutely constant	None
Squirrel-cage induction .	T	6*I*	Constant	None
Slip-ring induction . .	2½T	2*I*	Constant	None
Pole-changing induction .	T	6*I*	Constant	Four fixed speeds
Cascade induction . .	T	6*I*	Constant	Two fixed speeds
Induction cascade with a.c. commutator . . .	2T	2*I*	Constant or inverse	2:1
Synchronous induction .	2T	2*I*	Constant	None
Schrage	2½T	2*I*	Constant	3:1

We shall consider the properties and applications of these types of motors in the next section.

There is now the question of the rating of the motors to be installed. The *continuous rating* of a motor is the output which the motor can give for an indefinitely long period without exceeding a specified, safe temperature rise with the specified voltage and speed. The *short-time rating* is the output which the motor can give for $\frac{1}{2}$ or 1 hour (specified) without exceeding a safe, specified temperature with the quoted voltage and speed.

When the load is continuous, there is no difficulty about the rating, which is clearly the continuous rating. The use of too large a motor results in an inefficient use of the electrical energy. A large machine may be run at 25 per cent overload for a few hours.

When the load fluctuates, it is difficult to decide upon the rating, and experience is the best guide. If the load cycle is periodic although fluctuating, as in the case of traction motors, the r.m.s.

value of the motor current can be calculated (see pp. 352–355), and this gives an indication of the heating of the motor and thus of the rating required. If the load cycle is not periodic, a short time rating may be chosen; but, as stated above, experience is the best guide in this case.

Recently a great deal of research has been directed into the question of noise in motors, and a fair amount of information is now available. Extreme silence is necessary in motors for domestic purposes, for use in theatres, etc. But even in factories a reduction of noise is advantageous, as it reduces the fatigue of the workers and increases their output. It is worth while paying more for a noiseless motor than for a noisy one of the same efficiency; how much more is a question of economics and humanitarianism.

PROPERTIES OF MOTORS AND APPLICATIONS

D.C. Shunt Motor. 0–25 000 h.p. and up to 3 000 volts.

For starting, see *Electrical Technology*, by H. Cotton.* The speed is controlled by a rheostat, in series with the shunt field winding, which varies the field current. The motor can be reversed by reversing the armature current *or* the field current. The speed does not fall much with the application of load.

Applications are the driving of lineshafts, lathes, milling machines, conveyers, fans.

It is not suitable for use with flywheels, or with fluctuating or heavy loads, or for parallel operation.

Taking the cost of a squirrel-cage induction motor of the same power as unity, the cost of the d.c. shunt motor is 1·5 to 3·0, depending on the voltage and the range of speed-control required.

D.C. Series Motor. 0–3 000 h.p. and up to 1 500 volts.

For starting, a resistance is put in series with the motor and cut out as the motor speeds up; when two motors are available the series-parallel method effects a saving of energy. (See pp. 370–3, Chapter X.) The metadyne affords even a better way of starting. The speed is controlled by a diverter which is placed across the series field. The rotation can be reversed by reversing the armature current *or* the field current. The speed rises rapidly as the load falls off, so that a series motor must never be used where a light load is likely to be met.

Applications are traction, haulage, cranes, and moving heavy slides.

Where a very delicate speed-control is required from zero to full speed, as in colliery winders, the Ward-Leonard method is much used (see *Electrical Technology*†), as are also metadyne and amplidyne methods.

* Sixth Edition, p. 142 † Sixth Edition, pp. 156–8.

The series motor is suitable for parallel operation and for load equalization by a flywheel (see the Ward-Leonard-Ilgner Control, *Electrical Technology**).

The cost is 1·5.

Compound-wound D.C. Motor. Up to 3 000 h.p. and 1 500 volts. The characteristics are intermediate between those of the shunt and series motors, and the motor does not race on light load. Starting is as for the series motor. The speed is controlled as in the shunt or series motor, usually by a rheostat in the shunt field circuit. By proper proportioning of the differential-compounding the speed can be kept constant with load variation over a wide range, and this is specially useful in driving tandem rolling mills.

The cumulatively-compounded motor, with flywheel, is suitable for driving planers, shears, guillotines, and other loads subject to large peaks.

The cost is 2·0.

Single-phase Series Motor. 0–3 000 h.p.

Starting is by means of a transformer with tapped secondary (see Fig. 326). There are several fixed speeds, one for each tapping.

Its principal application is traction, and the frequency of supply is $16\frac{2}{3}$ or 25 cycles, as commutation troubles are difficult at higher frequencies. The cost is 2·5.

Repulsion Motor. Up to 100 h.p.

Starting, reversing and speed-control are done by shifting the brushes from the neutral axis; alternatively it can be started with a series limiting resistance. Although the speed can be controlled from zero to 1·3 times synchronous speed by brush shifting, the method is not very satisfactory because of the large change of speed for a small brush movement.

Applications are the drive of centrifugal pumps, printing machines, hoists and spinning machines. The cost is 2·5.

Synchronous Motor. Up to 10 000 h.p. and 15 kV.

The motor used to be started by running it up to a speed just greater than synchronous speed by means of a pony motor and then synchronizing on to the a.c. supply. The present method of starting employs a number of copper bars in the pole face, connected at their ends; these short-circuited loops form a squirrel-cage winding and the motor behaves like a squirrel-cage induction motor on starting. During running these *damping windings* cause little loss as there is no induced e.m.f. in them at synchronous speed; moreover, they damp out any oscillations that may occur due to sudden increases or decreases of load. In order to keep the starting current, due to the squirrel-cage action, low, the stator voltage must be reduced to about 60 per cent of the normal value; this may be done by limiting resistance or reactance, or better by an auto-transformer or star/delta switching.

* Sixth Edition, p. 157

A three-phase synchronous motor can be reversed by reversing the connections of two phases. Pole-changing is sometimes used to give two fixed speeds.

One of the outstanding advantages of the synchronous motor is its ability to work at unity or leading power factor. In fact it is used, running light, to improve the power factor of a transmission system, and it is then called a *synchronous condenser*.

The applications are motor-generator sets, frequency changers, fans, compressors, pumps, lineshaft drive, rubber calenders and rolling mills. Very small motors are used for clocks.

There is no need for a flywheel with a synchronous motor, as peak loads less than the pull-out torque are satisfactorily dealt with by the machine. 100 per cent overloads are easily withstood.

The good starting characteristics of the synchronous motor have caused its introduction into rolling mills where frequent reversals are required. A certain 600 h.p. motor takes 5 sec. from forward to reverse full-speed.

The cost is 2.

Squirrel-cage Induction Motor. Up to 300 h.p. and 11 kV.

This is the simplest and cheapest of all motors and requires the least maintenance as it has no commutator or brushes. The starting torque is high, but the starting current is so great that all but the smallest motors must start with a reduced voltage, and hence have a reduced torque. The voltage is reduced by an auto-transformer or star/delta switching.

The double squirrel-cage induction motor (Boucherot type) has two windings. One winding is made of brass and has a high resistance, and it is placed near the periphery of the rotor so that it has a low leakage inductance; the other winding is of copper, has a low resistance, and is placed at the bottom of the slots so that it has a high leakage inductance. At starting, the main current is in the high resistance winding, since it has a low inductance, and so there is a good starting torque with a reasonably small starting current; the current in the low resistance winding is very small because of its high inductance. When the motor runs up to speed the reactance of the low resistance winding is small (s times the inductance at standstill) so that the main current is in this winding. The efficiency can be high, because the resistance of this winding can be made very low, without the accompaniment of a high starting current. The motor can be reversed by reversing the connections of two phases.

There is no possible speed control in these motors. Because of the absence of sliding parts the motor is useful in explosive atmospheres. It is suitable for most industrial drives of small power where speed control is not required. A clutch can be used to start the motor on light load.

Slip-ring Induction Motor. Up to 11 000 h.p. and 11 kV. without the use of transformers.

Starting is by means of resistances between the slip-rings and neutral. The slip-rings are short-circuited when the machine has reached full speed. There is no speed control of any appreciable range.

The motor is suitable for most industrial drives up to the highest power where speed control is not required. It is the most used of motors, for driving lineshafts, generators, lifts, pumps, mills, winding machines, and haulage.

It may be provided with sufficient resistance between the slip-rings to give 10 per cent or more slip on overloads, in order to avoid a slip-regulator.

The cost is 1·5.

Pole-changing Induction Motor. These are almost always squirrel-cage motors. By changing the number of poles several speeds are possible; this is done by switching.

The motor is of use in certain cases where a fast constant speed is required in one direction and a slower constant speed in the other. Thus in crane service a heavy load may be lifted slowly and a light load quickly. Another application is in lifts, where a high-speed winding is used for acceleration and normal running and a low-speed winding for retardation and landing.

The cost is about 2.

Cascaded Induction Motors. Two motors are coupled mechanically and the wound rotor of one is connected via the slip-rings to the stator of the other. The effect is the same as changing the number of poles; if the motors tend to run in the same direction the effective number of poles is the sum of the poles in the motors; if in the opposite direction, the difference. By switching, several fixed speeds are possible.

The use of this type of motor is restricted to higher-power drive, viz. rolling mills, colliery winders, pumps, etc.

The cost is 2.

Induction Motor Cascaded with A.C. Commutator Motor. The leads from the slip-rings feed an a.c. commutator motor, so that the slip-energy is not wasted. A speed range of 2 : 1 is obtainable, and the applications are in high-power drive.

The cost is 2.

Synchronous-induction Motor. Up to 5 000 h.p. and 11 kV.

The motor is described in *Electrical Technology*.* The motor starts as a slip-ring induction motor with the aid of resistances. When it has run up to speed, d.c. excitation is switched on and it runs as a synchronous motor. If the load is too heavy for the machine to run as a synchronous motor, its speed drops and it runs as an induction motor. The power factor is unity or leading.

This motor is replacing the slip-ring induction motor in many applications. It is used for large fans, compressors, lineshafts,

* Sixth Edition, pp. 391–3.

pumps, and generally for machinery where a constant speed is normally required, but a small decrease in speed is permissible with overloads.

The cost is 2.

Schrage Motor. Up to 1 000 h.p.

This is a most useful a.c. commutator motor; the power factor and speed can both be controlled, the latter having a range of 3 : 1.

Fig. 346 shows a diagram of the motor. The mains feed a primary winding on the rotor via slip-rings, and there is a regulating

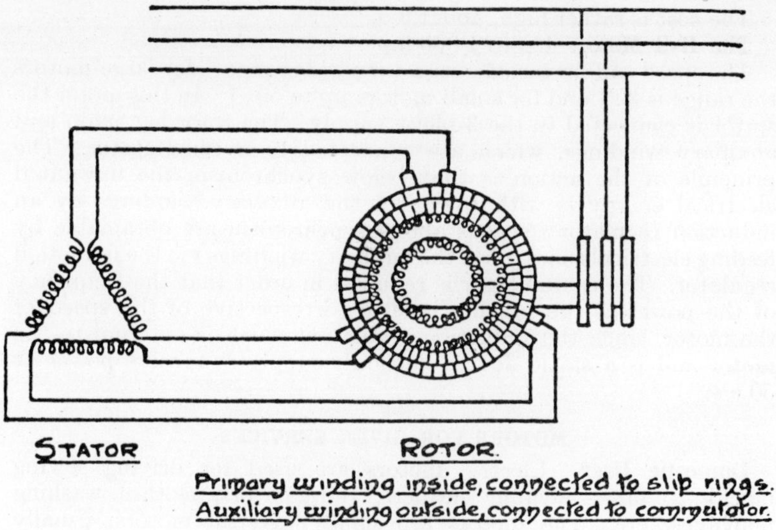

FIG. 346. SCHRAGE MOTOR
(Electrical Times)

or auxiliary winding on the rotor connected to the commutator. The auxiliary winding is in the same magnetic field as the primary winding, and has thus an induced current at line frequency. Three pairs of brushes collect this current and deliver it to the stator (secondary) windings at slip frequency; the secondary windings themselves have an induced current of slip frequency, and the currents are thus of the same frequency and suitable for combination. The magnitude of the e.m.f. collected by a pair of brushes depends upon the number of commutator bars between them, and the phase by their position on the commutator. It is therefore possible to vary the magnitude and the phase of the e.m.f. by moving the brushes. (When the brushes are together in pairs, no e.m.f. is transferred from the commutator to the stator windings, and the motor acts as an induction motor.) By varying the brush positions the power factor and the speed can be varied, the latter from $\frac{1}{2}$ to $1\frac{1}{2}$ synchronous speed. Slight variations of the speed (*inching*) can be obtained by resistances in series with the secondary windings.

At a given brush position the speed-load characteristic is like that of a shunt d.c. motor.

Applications are paper-making machines, printing-presses, and textile works. There is good reason to use them more generally, for lifts, pumps, machine tools, belt-conveyers, etc., as the starting torque is high and the control is convenient.

The cost is rather high, about 3·5.

The N-S Motor. Up to 1 000 h.p.

The speed can be varied over a very wide range: for large motors the range is 3 : 1 and for small motors up to 50 : 1. In this motor the stator is connected to the 3-phase supply. The rotor has main and auxiliary windings, which are connected to a commutator. The principle of the action is that below synchronism the unwanted electrical energy is withdrawn via the auxiliary windings by an induction regulator: speeds above synchronism are obtainable by feeding electrical energy into the auxiliary windings via the induction regulator. The commutator is required in order that the frequency of the power at the brushes is 50 c/s irrespective of the speed of the motor, since the induction regulator, which is external to the motor and is a static device, can only supply or receive power at 50 c/s.

MOTORS FOR GIVEN SERVICES

Domestic Uses. Electric motors are used for driving sewing machines, fans, vacuum cleaners, refrigerators, clothes washing machines, etc. The motors are small universal motors, usually series type.

Machine Tools. A great economy is effected in large works, especially of the mass production type, by the use of electrically driven portable tools, such as drills, spanners, etc. The motors are high speed, and gearing is often used. D.C. motors are series or compound wound; in the former case gear-resistance and windage prevent racing. Universal motors are series type, and squirrel-cage motors are used. Larger portable tools are driven by d.c. compound motors to limit the racing.

LATHES, MILLING AND GRINDING MACHINES. The motors are made up to 50 h.p., and are d.c. shunt motors, or induction motors with slip-rings or pole changing.

PLANERS. These consist essentially of a bed having attached to it a platen which moves forwards and backwards on the bed. The work is clamped to the platen and is planed by a stationary clamped tool; at the end of the stroke the tool moves slightly to make a new cut. Most planers work on one stroke, and a reversing motor is used with a slow cutting stroke and a quick return stroke.

The "Lancashire" planer drive consists of a d.c. motor generator set (with a.c. or d.c. motor) and the planing machine motor, which is supplied from the d.c. generator. When a.c. is used, a small direct-coupled exciter is required for exciting the generator. The generator field is increased, decreased or reversed for controlling the planer motor. Regenerative braking is used to decelerate and reverse. There are 18 cutting speeds, each with a range of 4 : 1, and the return speeds have a 2 : 1 range. Automatic reversal is controlled by *tappets* on the platen, and these operate a totally enclosed master switch. The push-button station has four buttons for "start," "stop," "inch-cut," and "inch-return."

Punches and Shears. This type of machine is provided with a heavy flywheel, which gives up its energy at the actual time of cutting. Fig. 347 shows a double punching machine with side shears driven by a 25 h.p. slip-ring induction motor.

Cranes. D.C. motors are preferred because of the smooth speed-control; and they are of the series and compound types, because of the high starting torque.

Slip-ring induction motors are often used for conveying and hoisting.

Where d.c. is used, the motors for lifting are compound wound, and for travelling and reversing series wound. Drum-type controllers are used, operated by contactors which are controlled by press-buttons.

Lifts. Hydraulic lifts are suitable for dealing with heavy loads over short distances, otherwise electric lifts are more economical and convenient. The car is usually counterbalanced to reduce energy consumption. In the simplest arrangement the winding engine is at the head of the shaft.

The winding-drum system, which is being superseded, has one rope attached from the lift to the drum and another from the counterweight to the drum. One rope winds up whilst the other unwinds. The drum cannot be made standard, as its size depends upon the length of travel of the lift.

In the wedge drive system, used in this country, the ropes wedge themselves into parallel V-grooves in the driving sheave, which is usually driven by the motor via worm-reduction gearing.

Five feet per sec. per sec. is considered the maximum safe acceleration for passengers, although greater accelerations are comfortable if they are sufficiently smooth. Lifts speeds vary from 100 to 350 ft. per min.

The electric circuit for operating the lift must be independent of the lighting circuit, and all cables armoured or put in metal conduits.

A lift motor must start and stop easily, and so its armature must be light and run at a moderate speed. Motors used are the d.c. compound, slip-ring induction, induction-repulsion and the variable

FIG. 347. DOUBLE PUNCHING MACHINE WITH SIDE SHEARS
(Bennie & Sons)

speed a.c. commutator motors; the first two are the most common. Speed-control of the d.c. motor is by shunt field variation or by variable voltage control (Ward-Leonard); the Ward-Leonard method can be used with a.c. to drive a d.c. lift motor, and this is specially convenient as rheostatic losses are avoided. Speed-control of the induction motor is by rotor resistance, and involves I^2R loss.

In single-phase installations a shunt type commutator motor is used. (See *Electrical Technology*.*)

Textile Machinery. The types of drive used are group drive from line shafts driven directly by motors, semi-group drive in which the motor shaft has several pulleys driving a number of machines, and individual drive. All motors must be totally enclosed because of the fluff from the material being manufactured, and insulation must be moisture-proof because of the damp atmosphere.

Three-phase motors are used as their speed is fixed by the frequency of the supply; d.c. motors are not suitable as their speed varies with voltage and temperature. The motors must be of high efficiency and have a good starting torque.

FIG. 348. MATRIX ROLLING PRESS DRIVEN BY SQUIRREL-CAGE MOTOR
(*Metropolitan-Vickers*)

Printing Machinery. Squirrel-cage motors are used for driving guillotines, driving platens, and other small machines at constant speed. Fig. 348 shows a matrix rolling press driven by a 4 h.p. 725 r.p.m. squirrel-cage motor; this type of machine is often required to have a slow forward stroke and a quick return stroke, and two squirrel-cage motors can be used, one for each stroke.

Rotary presses and other machines require a variable speed drive. Platens and flat-bed machines require speed variation down to half normal speed, and slip-ring motors with rotor resistance are suitable. This speed variation may be insufficient for high speed rotary presses, and then a d.c. compound or an a.c. commutator

* Sixth Edition, Chapter XXII.

motor is used with a speed range of 4 : 1. A steady crawling speed for inching the press is obtained by a *pony* or *barring* motor; this motor is d.c. compound or slip-ring, and drives through worm-reduction gearing and an overrunning clutch, such that when the main motor takes over the drive the pony motor can be brought to rest.

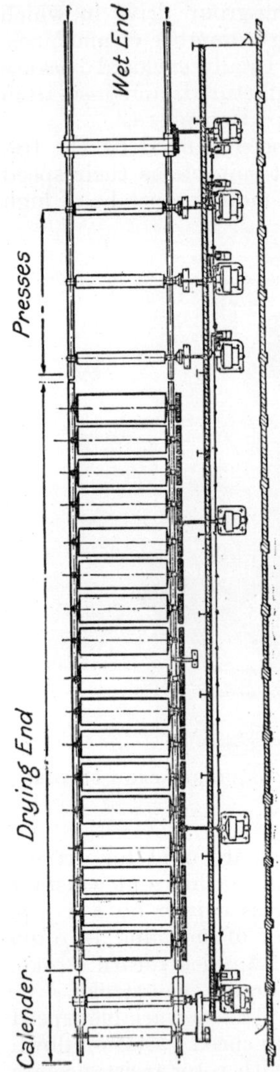

FIG. 349. HARLAND DRIVE FOR PAPER-MAKING MACHINE

Paper-making Machinery. In the past, d.c. was generally used because of the speed variation required, but a.c. is being introduced. Modern practice uses a combination of a.c. and d.c.: a.c. motors for the constant speed drives of the choppers and beaters; for these squirrel-cage motors with clutches, or slip-ring or synchronous-induction motors are used.

The paper machine requires very delicate control and the Ward-Leonard system is often used. Fig. 349 shows a Harland drive for a 234 in. paper machine at 1 000 ft. per min. The process is continuous; the pulp is sent in at one end, is squeezed by presses, dried, and calendered at the other end. The problem of the drive is very difficult, as not only must the speed of the whole machine be capable of control up to a range of 25 : 1 according to the class of paper, but also the speed must be absolutely constant or there will be a variation of quality, or even breakage. Crawling speeds are required for inspection and renewals. Nowadays sectional drive is used, with Ward-Leonard control (preferably driven by an a.c. motor to give the best possible regulation of the generator). The *Harland interlock drive* ensures that the various motors driving the sections have the correct corresponding speeds. In this system a master or control shaft is driven by a separate master motor, and runs the whole length of the machine. At every section the master shaft carries a light conical pulley, from which a belt is taken to a corresponding cone pulley on one of the three shafts

FIG. 350. HARLAND INTERLOCK DRIVE
(The Engineer)

of a differential gear (Fig. 350). A second shaft of the differential is driven by the section motor, and the third shaft is coupled to a regulator which controls the speed of the section motor. When the first and second shafts are moving at the same speed, the third shaft is stationary; if they have different speeds, the third shaft moves, in one direction if the section motor is faster than the master motor

FIG. 351. 1 000 H.P., D.C. MOTOR FOR ROUGHING MILL (*G.E.C.*)

(with the requisite reduction factor), and in the opposite direction if it is slower. The regulator then provides the necessary control for correcting the speed of the section motor.

Iron and Steel Works. The iron and steel industry is the largest single consumer of electrical energy, and it is estimated that 300 kWh. are used per ton of ingots, 200 kWh. being used in rolling.

The mill motors are generally d.c. shunt wound with light stabilizing windings and compensating and interpole winding. The speed range is 3 : 1, and speed control is effected by a rheostat in the shunt field. Fig. 351 shows a 1 000 h.p., d.c. motor, 165/500 r.p.m., for driving a roughing mill. The motors are designed with great mechanical strength to withstand the heavy duty of rolling. The driving bearings are fitted with special thrust collars to take the end thrust which may occur if the intermediate torque shaft fails. The supply is 500 volts direct current.

Merchant mills produce sections of widely varying value, and the motor speed must be varied to suit requirements. D.C. motors are used; also induction motors with Scherbius speed-control.

In tandem mills the load may vary because of the varying temperature of the steel, but the speed must be kept fixed. D.C. motors are preferred; but induction motors, with auxiliary commutator motors for speed control, are also used.

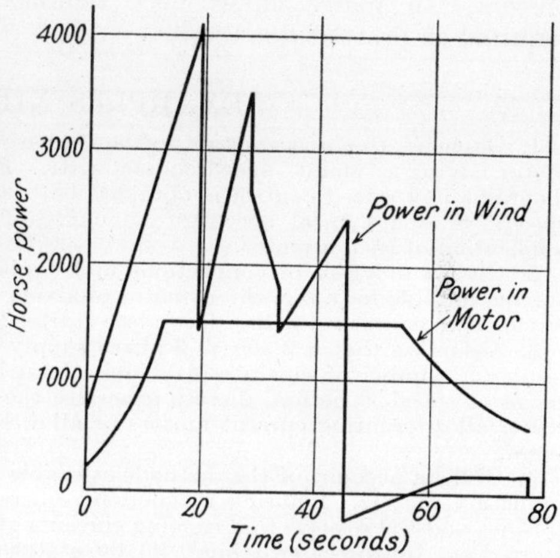

FIG. 352. POWER-TIME DIAGRAM OF A WINDER

Mining. The collieries are supplied with three-phase at high voltage, and step-down transformers are used at the various points of tapping.

The motors must be flame-proof, and all commutators and slip-rings must be in flame-proof chambers with an inch metal-to-metal flange. Squirrel-cage motors are used up to 10 h.p. Larger motors are slip-ring or d.c. motors.

Ventilating fans are of the utmost importance in mines, and it is desirable to install duplicate equipment. Fans in a fully developed section operate at full speed the whole time; in less developed sections they may drive at a reduced speed at the week-end. A shunt or compound d.c. motor may be used in both cases, shunt field control varying the speed in the latter case. Slip-ring induction motors are also used; the cascade system may be used for speed control.

Quick and safe winding is essential for the work of a mine. Flywheel storage is used to relieve the winding motor of the very heavy peaks during the accelerating period. Fig. 352 shows the power-time diagram of a winder for a shaft of 2 610 ft.; during the latter part of the wind the motor delivers energy to the flywheel, which returns the energy during the early part of the wind. In this way the motor works at 1 500 h.p. maximum, but the cage experiences a power of 4 200 h.p. maximum. The motor-generator consists of a slip-ring induction motor of 1 540 h.p. driving two 1 065 kW., 500 volt, d.c. generators in series, and a steel flywheel of 23 tons weight and 12 ft. diameter. The two motors driving the cage are d.c. motors with separately excited shunt field, and are in series across the 1 000 volt d.c. supply from the generators.

The control system is the Ward-Leonard.

Coal-cutting and drilling machines prepare the face for blasting and shot-firing. The band conveyers are driven by constant speed motors. In many mines very reliable pumping installation is required to prevent flooding.

EXAMPLES XIII

1. Describe the construction of a 3-phase variable-speed commutator motor having a "shunt" speed characteristic. Explain with simplified vector diagrams how speed control is obtained both above and below synchronous speed. Sketch typical speed-torque curves for the motor and give one application of such a motor. (*Lond. Univ.*, 1950.)

2. Give a diagram of connections and explain the action of a contactor starter suitable for a d.c. shunt motor of about 100 h.p. rating. What are its advantages compared with a face-plate starter? (*Lond. Univ.*, 1950.)

3. Assuming that a 3 300 V. 3-phase supply is available in a factory containing a number of separate machines, about half of which require a 2 to 1 range of speed variation, discuss generally the relative merits of the use of either all alternating current motors or all direct current motors for driving them. (*Lond. Univ.*, 1932.)

4. Give an account of the methods available for starting slip-ring and cage induction motors. Ignoring magnetizing current, find the relative starting torques and the relative line starting currents of a cage induction motor when started (*a*) by direct switching, (*b*) by a star/delta switch, (*c*) by an auto-transformer with a 50% tap. Give the connection diagram for a drum-type starter for either the star/delta or the auto-transformer method. (*Lond. Univ.*, 1934.)

5. The motor driving a planing machine is required to operate at a low speed on the cutting stroke and at a high speed on the return stroke. Give details of a suitable scheme when the factory supply is direct current. (*Lond. Univ.*, 1936).

6. What are the most suitable types of motors for operating passenger lifts when the supply is (*a*) direct current, (*b*) single-phase, (*c*) three-phase? For each case describe the method of starting and speed control used in practice. (*Lond. Univ.*, 1932.)

7. Describe in detail a method of control for a direct current lift motor. What is a suitable type of single-phase motor for lift duty and how is it controlled? (*Lond. Univ.*, 1936.)

8. Draw a diagram of connections showing the internal circuits of a Schrage 3-phase adjustable-speed motor and describe briefly the construction. Explain the action when running at (i) lowest speed, (ii) synchronous speed, (iii) highest speed.

In a particular 6-pole motor the speed is 900 r.p.m. when the motor is developing a gross output of 3 h.p. and the frequency is 50 c/s, the brushes being in the position to give straight induction-motor action. With the brushes set for a speed of 650 r.p.m., the voltage between adjacent brushes is 5 V. when the motor is developing a gross output of 2 h.p. Calculate the current in each phase of the secondary winding at this speed, assuming the power factor of these circuits to be 0·8 at a speed of 650 r.p.m., and unity at a speed of 900 r.p.m. (*Lond. Univ.*, 1948.)

9. Explain how the current-limit principle may be applied to control the operation of a contactor-type starter for a d.c. shunt motor when the operating coils of the contactors are (i) shunt wound, (ii) series wound. What are the relative advantages and disadvantages of shunt- and series-wound operating coils for such starters?

A current-limit starter for a 10 h.p., 460 V., d.c. shunt motor has three sections of starting resistance. Calculate the grading of these sections and the current settings of the contactors if the initial starting current is to be equal to full-load current and the subsequent peaks are to be twice full-load current. The resistance of the armature circuit of the motor is 1·3 Ω., the field current is 0·6 A. and the full-load efficiency is 88%. (*Lond. Univ.*, 1947.)

10. Give an account of the special conditions imposed upon the electrical equipment of (*a*) rolling mills in a steel works, *or* (*b*) a large planing machine. Describe fully how these conditions are fulfilled in the case selected. Assume an alternating current supply, the driving motors being either direct or alternating current. (*Lond. Univ.*, 1930.)

11. Describe with connection diagrams one method of obtaining a wide speed range on a large steel-rolling mill supplied from a 3-phase, 50-cycle, 6 600 V. system. Make an approximate estimate of the efficiency of the method described. (*Lond. Univ.*, 1932.)

12. Explain (i) the time-limit, and (ii) the current-limit methods of controlling the starting current of a d.c. motor. Describe one method of applying the time-limit method to an automatic starter.

An automatic starter for a 10 h.p., 460 V. shunt motor has four sections of starting resistance. Calculate the resistances of these sections if the initial starting current is equal to the full-load armature current and the subsequent peaks are twice this current. Resistance of armature circuit, 1·5 Ω.; resistance of field winding, 920 Ω.; full-load efficiency of motor, 85%.

(*Lond. Univ.*, 1948.)

13. Explain the principle of the cascade method of controlling the speed of a 3-phase induction motor. What advantages has a 3-phase commutator machine over an induction machine for operating in cascade with an induction motor?

A 500 h.p., 16-pole, 3-phase induction motor, which normally runs at a speed of 370 r.p.m. when the supply frequency is 50 c/s, is cascaded with a 3-phase commutator machine to obtain a speed of 300 r.p.m., the torque being equal to the normal full-load torque. Calculate the kVA. rating of the commutator machine, assuming the power factor of the rotor circuit of the induction motor to be 98% at full load, normal speed, and 92% under cascade operation. The open-circuit voltage between slip-rings at standstill is 650 V. Friction, windage and rotor core losses may be ignored. (*Lond. Univ.*, 1949.)

14. A variable-speed d.c. motor is to operate from a 480 V. supply against a constant load torque, equivalent to 500 h.p. at 450 r.p.m., over the range 300 to 600 r.p.m. The flux is to be maintained constant.

A separately-excited booster, rigidly coupled to the motor shaft and working with equal flux maxima in either sense, is used to effect the required speed change. Explain, with the aid of a diagram of connections, how the control will be effected, and give the armature e.m.f. and current of each machine at speeds of 300, 450 and 500 r.p.m. Ignore all losses.

(*Lond. Univ.*, 1949.)

15. Compare 3-phase induction motors and d.c. shunt motors from the point of view of ease of speed control.

A 400 V., 50 c/s, 3-phase, 6-pole, squirrel-cage induction motor has a stator resistance per phase of 0·5 Ω. The stator is normally delta-connected, and when started by direct switching the motor takes 50 kW. and a line current of 200 A. Determine the values of supply current and torque (in lb.-ft.) when started by (*a*) a 3-phase, 400/280 V. step-down auto-transformer, and (*b*) a star-delta starting switch. Ignore the no-load current of the transformer and the power loss in the stator of the induction motor. (*Lond. Univ.*, 1950.)

CHAPTER XIV

ECONOMIC CONSIDERATIONS

Introduction. In all engineering applications the question of cost is of first importance. In most cases the cost decides whether a certain project will be carried out, although political and other considerations may intervene; thus Hungary, having been deprived of most of its coal, water power and gas by the Treaty of Trianon, has been forced to a policy of railway electrification.

The calculation of the cost of any scheme is often difficult, as the cost varies considerably with time, tariff conditions, and even with convention. Thus schemes involving the use of aluminium were very expensive some fifty years ago and are very cheap now, because of the improved methods of producing aluminium and the consequently very much lower cost. Tariff walls render a scheme economic which under free trade would not be so; the advantages and disadvantages of protection and free trade are explained (and exaggerated) in the daily press, so that no further discussion is necessary, but the situation must always be considered. Convention, as well as the law of supply and demand, fixes the rate of interest of loans, and thus has a very large influence on the cost, real or calculated, of a system. A very interesting example of this occurs in the electrical circles of the U.S.A. It is well known that the cost per kilowatt of installing a hydro-electric plant is much greater than that of a steam power station, but the running cost per kilowatt-hour is much less. It can, then, be easily shown that the steam power station is cheaper if the rate of interest is above a certain value, and the hydro-electric plant is cheaper if the rate is less than this value. The utility companies, which own the steam power stations, claim that the hydro-electric plants are uneconomic and base their calculations on a high rate of interest; the Federal Government, which sponsors the hydro-electric installations and is more concerned with the welfare of the people during the depression than with the rate of profit, claims that the hydro-electric schemes are economic, and bases its calculations on a low rate of interest. Both claims are correct, given their basic assumptions.

In some instances one scheme may be cheaper than any other under all circumstances, and then there is no question of its not being undertaken.

Economic problems occur in the fields of generation, transmission, distribution and utilization of electrical energy. The electrical engineer has to find the cheapest and most convenient electrical method, and then he has to persuade the consumer that it is worth while using this method. If the method is cheaper than the non-electrical method, he has to persuade the consumer that it is worth his while to scrap his present plant and go over to the electrical method; sometimes this is easy, sometimes, as in the case of the railways, it is very difficult. If the electrical method is dearer, as in electrical heating for domestic purposes, he can point out certain indirect savings, such as cleanliness and the non-spoiling of decoration, and persuade the consumer that, taking all the considerations into account, the electrical method is more economical.

In all cases it is a great advantage to be able to present a concrete calculation to the consumer or user, giving definite costs together with the advantages of use.

Interest and Depreciation. Suppose that a capital outlay P is required for a certain installation, and the interest per unit is r p.a. (5 per cent is equivalent to $r = 0{\cdot}05$). Then the installation must provide rP per annum as interest, and this is added to the yearly running cost.

If the installation were to last for ever, this is the only charge that would have to be made. In practice an installation lasts for a finite period, from one to forty or fifty years, and it is necessary to provide a *sinking fund* to produce a sufficient sum at the end of the probable life to replace the installation by a new one. Let the cost of replacement be Q; Q is equal to P if the used installation has zero value and is easily removed, less than P if it has a scrap value, and greater than P if it has small scrap value and requires expensive removal. The problem is to find the yearly charge q to provide a sinking fund such that the amount Q is available at the end of the n years. An amount q earns interest rq in one year, so that it is worth $q + rq = q(1 + r)$ at the end of a year. In the same way its value is multiplied by the ratio $(1 + r)$ every year, so that it is worth $q(1 + r)^2$ at the end of two years, $q(1 + r)^3$ at the end of three years, and so on. Thus the first payment is worth $q(1 + r)^n$ at the end of the n years. The second payment is made at the beginning of the second year and is in $(n - 1)$ years, so that its value at the end of the period is $q(1 + r)^{n-1}$. The total sum at the end of the period is therefore

$$q(1 + r)^n + q(1 + r)^{n-1} + \ . \ . \ . + q(1 + r)^2 + q(1 + r)$$

$$= q\frac{(1 + r)^{n+1} - (1 + r)}{(1 + r) - 1} = q\,\frac{1 + r}{r}\,[(1 + r)^n - 1].$$

This must be equal to the cost of renewal, Q, so that

$$Q = q\,\frac{1 + r}{r}\,[(1 + r)^n - 1],$$

giving

$$q = Q\frac{r}{1+r} \div [(1+r)^n - 1]. \qquad (138)$$

The total charge on the installation is then $rP + q$.

Example. Find the total annual charge on a cable system costing £150 000 to buy and lay, the cables lasting 30 years and having negligible scrap value. Interest is at 6% compounded annually.

The interest charge is 0·06 × 150 000 = £9 000 per annum. The replacement value is £150 000 so that the sinking fund charge is

$$q = 150\,000 \times \frac{0{\cdot}06}{1{\cdot}06} \div [1{\cdot}06^{30} - 1]$$

$$= \frac{150\,000 \times 0{\cdot}06}{1{\cdot}06 \times 4{\cdot}75}$$

$$= £1\,790.$$

The total charge is thus £10 790.

If the interest were at 3 per cent, the charges would be £4 500, £3 050 and £7 550 respectively. If the interest were zero the charges would be 0, £5 000 and £5 000 respectively. It is seen that a rise in the interest rate causes a rise in the interest charge, a fall in the sinking fund charge, and a rise in the total charge.

GENERATION

In Chapter I we have discussed the problem of generation and the cost involved in the various systems. The question of load factor was discussed in detail, and it was shown how the cost of generation of 1 kWh. decreases as the load factor increases; this is due to the fact that the standing charges are distributed over more units of energy if the load factor is higher.

The cost of energy is composed of standing charges which are independent of the output, and running charges which are proportional to the output. The standing charges consist of interest on capital and depreciation, and the main part of salaries, wages, repairs, and maintenance. The running charges consist of the main part of the fuel and stores, a part of the repairs and maintenance, and part of the wages.

Interest and depreciation are calculated as follows. Interest is at 6 per cent, so that for every £100 of plant the charge is £6. Depreciation varies considerably with the type of installation; thus it is likely to be much less in a hydro-electric plant than in a diesel station. A customary allowance is 1 per cent on the first £25 and 3½ per cent on the remaining £75, giving £2·88 per £100 plant. The capital charge is thus £8·88 per £100 of plant. Now suppose that the plant costs £15 per kW. and that 25 per cent spares and standby is allowed, so that the cost per working kW. is £18·75. Then the capital charge per working kW. is

$$(18{\cdot}75/100) \times 8{\cdot}88 = £1{\cdot}67.$$

If a kW. generator works for a year at a load factor of L per cent, the fuel cost is found to be 0·28 + 0·036L and other costs 0·845 + 0·0081L, so that the total cost is

$$\begin{aligned} 1{\cdot}67 + 0{\cdot}28 + 0{\cdot}036L + 0{\cdot}845 + 0{\cdot}0081L \\ = 2{\cdot}79 + 0{\cdot}044L. \end{aligned}$$

The total units produced are

$$24 \times 365 \times (L/100) = 87{\cdot}6L,$$

and the cost per unit (kWh.) is

$$\begin{aligned} \frac{2{\cdot}79 + 0{\cdot}044L}{87{\cdot}6L} &= £[0{\cdot}00050 + 0{\cdot}0318/L] \\ &= (0{\cdot}12 + 7{\cdot}63/L) \text{ pence.} \end{aligned}$$

The effect of the load factor is quite clear from this expression If the load factor is 20, 40, 60, 80, 100 per cent, the cost per kWh. is 0·50, 0·31, 0·25, 0·22, 0·196 pence respectively.

In comparing the cost of energy produced by steam power and hydro-electric stations, it is essential to take into account the effect of the load factor. Since the standing charges for the hydro-electric plant are great, it is essential that a high load factor be employed. The best scheme is to use the hydro-electric plants and the large efficient steam power plants to provide the base load, and thus obtain a load factor of 70 per cent or higher. The peak loads can be supplied by diesel stations or steam accumulator plants, since these can be brought into action very quickly and have very small or no standby losses.

Maximum Demand. The maximum demand made upon a generating station determines the size and cost of the installation. Given a fixed total energy output in a year, a low load factor entails a high maximum demand and a high cost per unit. The matter can be considered more easily if the total cost of the energy is considered. The total cost consists of the standing charges, which are proportional to the maximum demand, and the running charges, which are proportional to the energy used.

For this reason a two-part tariff is often fixed, a charge being made for the maximum demand (integrated over 15 min. or half an hour) and for the energy consumed.

A consumer at the end of a long chain of generation, transmission and distribution should pay more per kW. and per kWh., because of the extra standing charges in the transmission and distribution systems and the power loss on route; but a certain effect to be explained later causes a radical modification.

In many cases alternative tariffs are available, and then the consumer can calculate, if he has the necessary information, which of the two tariffs is cheaper.

Large consumers have maximum demand indicators, and they are warned by a bell when they are approaching or exceeding their maximum demand; they may often arrange to drop some dispensable part of their load to keep within their limit. Small consumers may find it uneconomical to install a maximum demand indicator, and then the supply company estimates their maximum demand by the size of house and the connected appliances, and offers a tariff accordingly. The estimate is sometimes inequitable, but usually rough justice is attained.

The following example illustrates the calculation of alternative tariffs.

EXAMPLE. Explain why two-part tariffs are used in charging for electrical energy. A consumer has a maximum demand of 12 kW. There are two alternative tariffs: (1) a fixed charge of £5 4s. 0d. per annum plus a running charge of ½d. per unit; (2) a charge of 4½d. per unit for the first 150 kWh., and for all units in excess of this 1d. per unit. Calculate the cost per unit under each tariff for load factors of 10, 30, and 100 per cent

(*Nat. Cert.*, 1932.)

At 100 per cent load factor the units used are

$$12 \times 24 \times 365 = 105\,120 \text{ kWh.}$$

At 10 and 30 per cent the units are 10 512 and 31 536. According to the first tariff the charge of £5 4s. is to be borne by 10 512, 31 536 or 105 120 units, so that the charge per unit is ½d. plus 0·142, 0·047 or 0·014, i.e. 0·64, 0·547 or 0·514d. per unit. According to the second tariff the cost is 1d. per unit plus 150 times 3½d., and the cost per unit is 1·045, 1·015, and 1·0045d.

Diversity Factor. The maximum demands of a number of consumers are most unlikely to occur all at the same time, so that the maximum demand on the supply or generating stations is much less than the sum of the maximum demands of the separate consumers. The *diversity factor* is the ratio of the sum of the maximum demands of the separate consumers divided by the maximum demand of the station. It is usually much greater than unity.

It is quite clear that the consumer should pay a standing charge per kW. of maximum demand of amount equal to that paid by the station divided by the diversity factor. Thus the average standing charges per kW. of maximum demand are £3 16s. 0d. for generation and £4 14s. 0d. for distribution. If all loads were switched on together, so that the diversity factor were unity, a consumer would have to pay £3 16s. 0d. plus £4 14s. 0d. viz. £8 10s. 0d. per kW.

demand. As the diversity factor in generation is 25 and in distribution about 7, the consumer need pay only

$$\frac{£3\ 16s.\ 0d.}{25} + \frac{£4\ 14s.\ 0d.}{7} = 3s. + 13s.\ 6d. = 16s.\ 6d.$$

per kW. of maximum demand.

This calculation supposes that the utmost use is made in generation and distribution, and in their combination, of the diversity factor.

It follows that the encouragement of the diversification of load is likely to cause a much greater economy than even the existence of a high load factor. This, in fact, is exemplified in the use of water heaters at specified times, during which very low charges are made.

TRANSMISSION

In transmission systems carrying bulk power over long distances, the question of voltage regulation is unimportant; in some very long lines a regulation of 40 per cent is considered satisfactory. The only consideration then is that of the most economical method of working. There are two things to be chosen, the voltage of transmission and the size of the conductors.

Size of Conductors. Kelvin's Law. Let us suppose that the voltage is fixed and the current is I per conductor. We need consider only unit length of the line, as the argument holds for any length. Let A be the cross-sectional area of the conductors. The annual charge on the capital outlay consists of a constant part and a part which is proportional to the cross-section; thus the insulator costs are constant, the copper cost is clearly proportional to A, and the cost of supports, etc., is partly constant and partly proportional to A. We may therefore write the annual charge as

$$P_1 + P_2A. \qquad (139)$$

The resistance of the conductors per unit length is ρ/A, so that the power loss is $3(\rho/A)I^2$ in a three-phase system. We add this up over a year and obtain a money loss of

$$\frac{3\rho c}{A}\int^{\text{year}} I^2 dt = \frac{P_3}{A} \qquad (140)$$

where

$$P_3 = 3\rho c \int^{\text{year}} I^2 dt, \qquad (141)$$

and c is the cost of a unit.

The most economical working is obtained when the annual charge on the capital outlay and the losses are a minimum, i.e. when

$$P_1 + P_2A + P_3/A$$

is a minimum. This condition occurs when

$$(d/dA)\,(P_1 + P_2A + P_3/A) = 0 = P_2 - P_3/A^2,$$

i.e. when

$$A = \sqrt{(P_3/P_2)}. \qquad (142)$$

When this condition holds $P_2A = P_3/A$, i.e. the variable part of the annual charge is equal to the cost of the losses per year. This is *Kelvin's law.* The difficulty in practice is to find P_3 as given in equation (141). Moreover we have assumed that the annual charge is of the form $P_1 + P_2A$, and this need by no means be true. For example, the cable dielectric and sheath in cables do not vary in cost according to this simple law, nor need the cost of laying. The saving circumstance is that accuracy is not required, as the curve is flat in the region of the minimum cost and an error of 20 per cent either way is not likely to cause an appreciable loss of efficiency.

It can be seen from equation (141) that the load factor is insufficient to determine the value of P_3; for the load factor is given by

$$L = \frac{\int^{\text{year}} VIdt}{VI_{max}} = \frac{\int^{\text{year}} Idt,}{I_{max}}$$

so that $\int Idt$ is given by the load factor but not $\int I^2dt$. For example, a load factor of 50 per cent may be due to a load which draws a current i all the year except during a very short interval, say 1 hour, when the load is $2i$. The value of P_3 is then

$$3\rho c \int^{\text{year}} i^2dt = 3\rho c \times 8\,760 \times i^2.$$

The same load factor of 50 per cent with the same maximum demand may occur with a load that draws current $2i$ for half a year and zero current the other half-year. The value of P_3 is then

$$3\rho c \int^{\text{half-year}} (2i)^2dt = 3\rho c \times 4\,380 \times 4i^2,$$

which is double the value for the former case.

Equation (141) may be written in the form

$$P_3 = 3\rho c \times 8\,760 \times \bar{I}^2 \quad . \quad . \quad . \quad (141a)$$

where $\bar{I}$ is the r.m.s. current. In practice the load is distributed in fairly typical ways and it is possible to express the r.m.s. current as a multiple (k) of the average current for a given load factor.

Load Factor (%) . .	10	20	30	40	50	70	100
R.m.s. current/Average current . . .	2·2	1·7	1·45	1·3	1·2	1·08	1·00

P_3 is then given by

$$\left.\begin{aligned} P_3 &= 3\rho c \times 8\,760 \times k^2 I_{av}^2 \\ &= 3\rho c \times 8\,760 \times k^2 \times (L/100)^2 I_{max}^2 \end{aligned}\right\} \quad (141b)$$

EXAMPLE. State Kelvin's law for transmission.

An 11 kV. 3-core cable is to supply a works with 500 kW. at 0·9 p.f. lagging, for 2 000 hours p.a. Capital cost of the cable per core when laid is £600 + 1 800*s*. per mile where s = cross sectional area of core in in.^{2}

If the energy losses cost 0·6d. per unit and the Interest and Sinking Fund is covered by a charge of 8% p.a., calculate the most economical current density and state the conductor diameter. (*Nat. Cert.*, 1936.)

The annual charge on capital outlay per mile is

$$(8/100)\,(600 + 1\,800s.) = £[48 + 144s.].$$

The current per conductor is

$$\frac{500\,000}{(\sqrt{3}) \times 11\,000 \times 0{\cdot}9} = 29{\cdot}2 \text{ A.},$$

and

$$\int^{\text{year}} I^2 dt = 2\,000 \times 29{\cdot}2^2 \text{ ampere}^2 \times \text{hours}.$$

The resistance per mile of a conductor of 1 in.^{2} cross-section is 0·0415 Ω., so that the cost of the losses is

$$£[3 \times 0{\cdot}0415 \times (0{\cdot}6/240) \times 2\,000 \times (29{\cdot}2^2/1\,000)] \div s$$
$$= £(0{\cdot}53/s.).$$

The most economic cross-section is given by

$$s = \sqrt{(0{\cdot}53/144)} = 0{\cdot}061 \text{ in.}^2,$$

the current density is 480 A./in.^{2}, and the conductor diameter is 0·278 in.

Current Density. We have supposed in the previous section that the voltage was fixed, and the current was then found from the load. If we consider equation (142), which is the condition of most economical working, and if we note that P_3 is proportional to I^2 (let $P_3 = p_3 I^2$) we get

$$A = \sqrt{(p_3 I^2/P_2)}$$

or

$$I/A = \sqrt{(P_2/p_3)} \quad . \quad . \quad . \quad . \quad (143)$$

Now P_2 and p_3 are determined by considerations of the construction and materials, so that they are constant. It is thus seen that Kelvin's law determines the *current density* of the conductors, independently of the voltage of the system.

Transmission Voltage. We have seen in Chapter II that as the voltage is raised n times the copper required is diminished n^2 times; but the cost of the system increases rapidly with rising voltage because of many other factors, and there is an optimum voltage of transmission.

The method of finding the optimum voltage is the following. We are given the voltage of generation, the power to be transmitted, and the length of the transmission system. We assume a certain voltage of transmission and work out the following costs.

(1) *The transformers* at both ends. For a given power this cost increases very slowly with voltage.

(2) *Switchgear.* This cost increases also slowly with voltage, but rather faster than for transformers.

(3) *Lightning arresters.* This cost increases rapidly with rising voltage.

(4) *Insulators and supports.* This cost increases very rapidly with voltage.

We then work out the current density by Kelvin's law. From the assumed voltage and power, the current and hence the size of

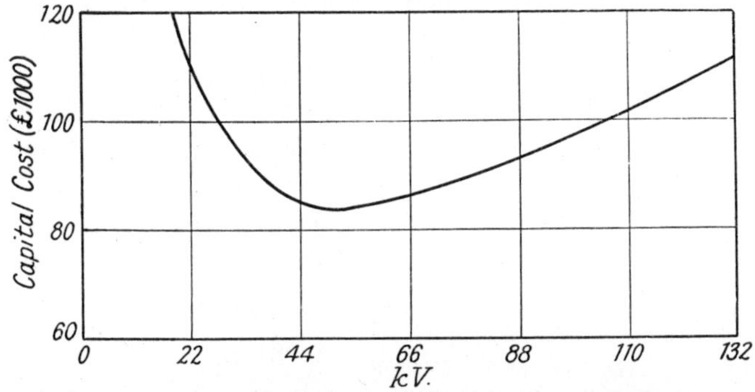

FIG. 353. CAPITAL COST OF A TRANSMISSION SCHEME, TEN MILES LONG, TO CARRY 20 000 kW. AT 0·8 POWER FACTOR

conductor can be calculated. In this way we have the cost of the copper. We add all the items and obtain the total cost of the system. We do this for various voltages and plot the curve of cost against the transmission voltage.

Fig. 353 shows the capital cost of a transmission system, ten miles long, to carry 20 000 kW. at 0·8 power factor lagging, with various voltages. Any voltage between 44 and 88 kV. would be satisfactory, and in fact the higher voltage would be chosen as large voltages are easier to control than large currents, and the load can be increased if required.

We may discuss the problem approximately in general terms in the following manner. Suppose the system is of length l and has to carry a power P. The cost of transformers and switchgear is $(A + BV)P$, where A and B are constants and V the voltage; BV is of the same order of magnitude as A. The cost of the insulators is $(C + DV)l$, where DV is much greater than C.

We have now to consider the cost of the line. The current per conductor is proportional to P/V. By Kelvin's law the current density is determined by factors independent of the voltage and length, so that the cross-section is proportional to the current, viz. proportional to P/V. The cost of the copper is thus proportional to lP/V; ElP/V say. The total cost is thus

$$(A + BV)P + (C + DV)l + ElP/V.$$

For minimum cost

$$(d/dV)\,[(A + BV)P + (C + DV)l + ElP/V] = 0,$$

i.e. $$BP + Dl - ElP/V^2 = 0,$$

giving $$V = \sqrt{[ElP/(BP + Dl)]} \quad . \quad . \quad . \quad (144)$$

According to American practice the economic voltage between lines of a three-phase a.c. system is

$$V = 5{\cdot}5\sqrt{(l + P/100)},^*$$

where V is in kV., l in miles, and P in kW. Thus if $l = 10$ and $P = 20\,000$,

$$V = 5{\cdot}5\sqrt{(10 + 200)} = 79{\cdot}5 \text{ kV}.$$

It is seen that this is well above the optimum value given in Fig. 353.

The following table gives values used in normal practice—

l (miles) .	10–20	20–50	50–75	75–100	100–150	150–250
V (kV.) .	11–22	33–66	66–110	110–132	110–154	132–220

The voltage in kV. is approximately the route length in miles.

Power Factor. If a balanced three-phase system is supplying a load P at voltage V and power-factor $\cos\phi$, the conductor current is $P/(\sqrt{3}\,V\cos\phi)$. A low power-factor therefore means a high current, and this has three important effects. First, the line losses are proportional to the square of the current, so that they are proportional to $1/\cos^2\phi$. Thus the losses at 0·8 power factor are $1/0{\cdot}8^2 = 1{\cdot}57$ times the losses at unity power-factor. Secondly, the ratings of the generators and transformers are proportional to the current and thus to $1/\cos\phi$; larger generators and transformers are thus required. Thirdly, low power factors are usually lagging, and this causes a large voltage drop; extra regulating equipment is therefore required.

It has already been shown in Chapter VII how the power factor can be improved by synchronous condensers and static condensers. The supply authorities use synchronous condensers for voltage control, and large consumers (or small distributing authorities) are encouraged to keep a high power factor by the above methods through a tariff arrangement.

* Cable Research Handbook gives $5{\cdot}5\sqrt{(l + \text{kVA}/150)}$.

The tariff may consist of a standing charge per maximum kVA. demand plus a charge per kWh. used; or a charge per kWh. with a penalty for bad average power factor; or it may consist of a standing charge per kW. of maximum demand, plus a charge per kWh., plus a charge per kVAR.

Fig. 354 shows how the power and reactive volt-amperes are measured in a three-phase circuit. The vector diagrams are self-

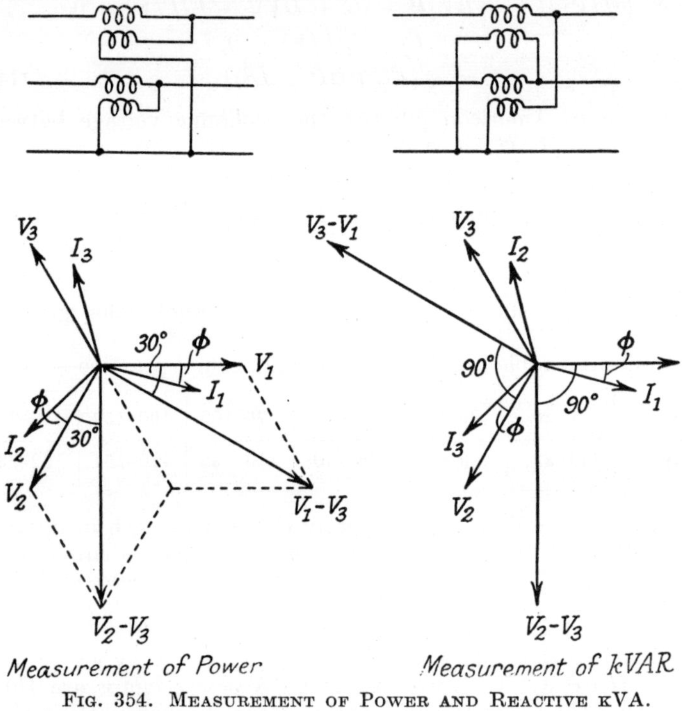

FIG. 354. MEASUREMENT OF POWER AND REACTIVE KVA.

explanatory, and it is assumed that the load is balanced. The wattmeter registers the scalar product of I_1 and $V_1 - V_3$, and of I_2 and $V_2 - V_3$. The magnitudes of $V_1 - V_3$ and $V_2 - V_3$ are both $(\sqrt{3})V$, whilst the angles they make with I_1 and I_2 are $30° - \phi$ and $30° + \phi$ respectively. The wattmeter registers therefore

$$
\begin{aligned}
& I\,(\sqrt{3})V\cos(30° - \phi) + I\,(\sqrt{3})V\cos(30° + \phi) \\
&= (\sqrt{3})IV\,(\cos 30° \cos\phi + \sin 30° \sin\phi \\
&\qquad + \cos 30° \cos\phi - \sin 30° \sin\phi) \\
&= 3IV\cos\phi.
\end{aligned}
$$

The reactive volt-ampere meter reads the scalar products of I_1

and $V_2 - V_3$, and I_2 and $V_3 - V_1$. The magnitudes are $(\sqrt{3})V$ and the angles both $90° - \phi$ so that the meter reads

$$(\sqrt{3})IV\,[\cos(90 - \phi) + \cos(90 - \phi)] = 2(\sqrt{3})IV \sin\phi,$$

which is the reactive volt-amperes apart from a constant factor.

It can be shown that the wattmeter reads correctly even on unbalanced loads; but the reactive meter will be correct only if the common end of the voltage coils is at the neutral point and one coil is reversed, as shown in Fig. 355.

The kW. and kVAR. can be compounded to give the kVA. in the way shown in Fig. 356. An instrument with such an arrangement has been constructed to read kW., kVAR. and kVA. The ratio of kWh./kVAh. over a given period does not give the average of the power factor, because the magnitude of the kVA. usually varies.

Power Factor Improvement. The consumer is encouraged

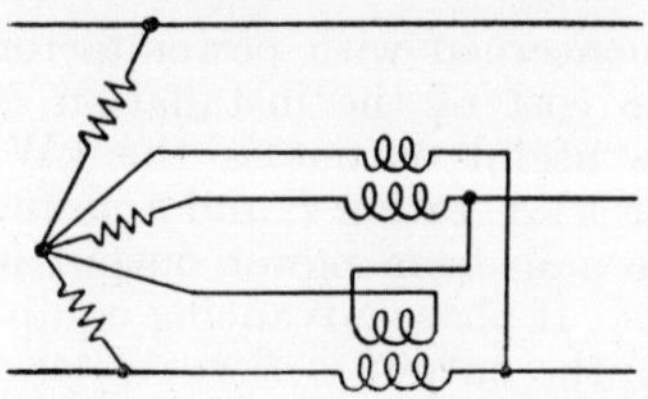

FIG. 355. MEASUREMENT OF kVAR. ON UNBALANCED LOAD

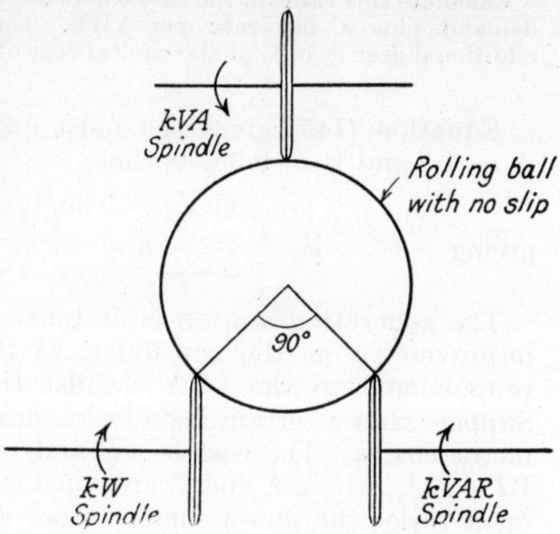

FIG. 356. MEASUREMENT OF kVA.

to improve the power factor of his plant by the tariff, and he will do so up to the economic limit. Suppose that the consumer is charged a fixed sum A per kVA. maximum demand plus a flat rate per kWh. Suppose that he is taking a load P at a power factor $\cos\phi$; his kVA. is $P/\cos\phi_1$. He may install condensers or a synchronous condenser to increase his power factor to $\cos\phi_2$, when his maximum demand is $P/\cos\phi_2$. He will save in this manner

$$A(P/\cos\phi_1 - P/\cos\phi_2)$$

per annum. The kVAR. to be supplied by his power factor installation is $(P\tan\phi_1 - P\tan\phi_2)$, and suppose that the cost is B per kVA. per annum. The cost is

$$B(P\tan\phi_1 - P\tan\phi_2).$$

His net saving is

$$A(P/\cos\phi_1 - P/\cos\phi_2) - B(P\tan\phi_1 - P\tan\phi_2).$$

He will install equipment so that his saving is a maximum, and this occurs when the differential coefficient with respect to ϕ_2 is zero. This occurs when

$$- AP \sec \phi_2 \tan \phi_2 + BP \sec^2 \phi_2 = 0,$$

or

$$\sin \phi_2 = + (B/A) \quad . \quad . \quad . \quad . \quad (145)$$

From this ϕ_2 and $\cos \phi_2$ are easily found. An interesting point is that the most economical angle of lag is independent of the original value of the installation.

Example. Deduce an expression for the most economical angle of lag at which a consumer should take power when the charge is based on a fixed sum per kVA. of maximum demand plus a flat rate per kWh.

Calculate this value in the case where the tariff is £4·5 per kVA. of maximum demand plus a flat rate per kWh. The annual cost of condensers and additional gear is 10% of the capital cost of £3·5 per kVA. of such plant.

(*Lond. Univ.*, 1930.)

Equation (145) gives the most economical angle. In this case A = 4·5 and B = 0·35, so that

$$\sin \phi_2 = 0{\cdot}35/4{\cdot}5 = 0{\cdot}0778,$$

giving

$$\phi_2 = \underline{\underline{4^\circ\ 28'}} \text{ and } \cos \phi_2 = \underline{\underline{0{\cdot}997.}}$$

The generating station is as much concerned with power factor improvement as the consumer, as the cost of the installation is proportional to the kVA., whilst the useful output is the kW. Suppose that a certain installation has a kVA. rating P and a power factor $\cos \phi_1$. The cost is AP and the gain from power output is $BP \cos \phi_1$, where A and B are constants. If phase-advancing equipment raises the power factor to $\cos \phi_2$, the output is $P \cos \phi_2$ and the extra gain is $B(P \cos \phi_2 - P \cos \phi_1)$. The kVAR. to be provided is $(P \sin \phi_1 - P \sin \phi_2)$ and costs $C(P \sin \phi_1 - P \sin \phi_2)$. The next gain is thus

$$B(P \cos \phi_2 - P \cos \phi_1) - C(P \sin \phi_1 - P \sin \phi_2) \quad . \quad (146)$$

It is often asserted, wrongly, that this will be done up to a value of ϕ_2 for which the extra gain is greater than the cost. What should be done is to make the extra gain a maximum, not just positive. The best angle ϕ_2 is given by putting the differential coefficient of expression (146) equal to zero,

i.e.

$$- BP \sin \phi_2 + CP \cos \phi_2 = 0,$$

giving

$$\tan \phi_2 = C/B. \quad . \quad . \quad . \quad . \quad . \quad (147)$$

Again we see that the value of ϕ_2 is independent of ϕ_1.

Distributors. The cross-sectional area of a feeder is chosen for the minimum cost of conductor and losses, and given by Kelvin's law; the consideration of voltage-drop is not important, as regulation can be allowed for.

In distributors, however, the permissible voltage drop determines the cross-section, and the problem is not one of economics entirely. A problem which arises is the cross-sections of a series of distributors across which the total voltage-drop is limited. Suppose that distributors carry currents I_1, I_2, . . . I_n in series, and the voltage-drop across them is limited to V. (See Fig. 357.) It is required to find the cross-sections of the most economical arrangement. Let the resistances be R_1, R_2, . . . R_n. The condition of voltage-drop gives

$$I_1R_1 + I_2R_2 + \ldots + I_nR_n = V. \qquad (148)$$

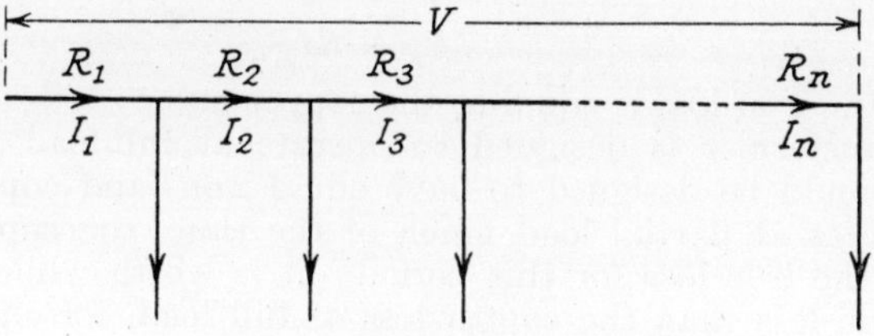

FIG. 357. DISTRIBUTORS IN SERIES

As the length of each part is fixed the cost is

$$C = k(1/R_1 + 1/R_2 + \ldots + 1/R_n) \qquad (149)$$

since the cross-section of any part is inversely proportional to its resistance. Let the resistances be varied by dR_1, dR_2, . . . dR_n. The condition (148) gives

$$I_1dR_1 + I_2dR + \ldots + I_ndR_n = 0. \qquad (150)$$

For minimum cost we put $dC = 0$, so that

$$(1/R_1^2)dR_1 + (1/R_2^2)dR_2 + \ldots + (1/R_n^2)dR_n = 0. \qquad (151)$$

Equations (150) and (151) can hold for all values of dR_1, dR_2, . . . dR_n only if they are of the same form, so that

$$\left.\begin{aligned} &\frac{I_1}{1/R_1^2} = \frac{I_2}{1/R_2^2} = \frac{I_3}{1/R_3^2} = \ldots \frac{I_n}{1/R_n^2}. \\ \text{or} \quad &\frac{I_1}{A_1^2} = \frac{I_2}{A_2^2} = \ldots = \frac{I_n}{A_n^2}, \end{aligned}\right\} \qquad (152)$$

where the A's are the cross-sectional areas. In the case of a uniformly loaded distributor of length l, the current I_x at distance x is $i(l - x)$, so that A_x is proportional to $\sqrt{(l - x)}$.

Transformer Losses. Suppose that a transformer has primary current and voltage I_1 and E_1, and secondary values I_2 and E_2. If the output power factor is ϕ, the output is $E_2I_2 \cos \phi$. The copper losses are $R_1I_1^2 + R_2I_2^2 = RI_2^2$, where R is the resistance transferred to the secondary, and the iron loss is P_i, a constant. The efficiency is thus

$$\eta = \frac{E_2I_2 \cos \phi}{E_2I_2 \cos \phi + RI_2^2 + P_i}$$

$$= \frac{E_2 \cos \phi}{E_2 \cos \phi + RI_2 + P_i/I_2}. \quad . \quad . \quad . \quad (153)$$

The efficiency is a maximum with respect to I_2 when the denominator is a minimum, and this occurs when

$$R - P_i/I_2^2 = 0,$$

or

$$P_i = RI_2^2,$$

i.e. when the iron-loss is equal to the copper-loss.

If a transformer is designed to operate at full-load the whole time, it should be designed to have equal iron- and copper-losses. If it operates at partial load much of the time, the copper-loss is less than the iron-loss for this period; it is worth while having a smaller iron-loss than the copper-loss at full load, for the iron-loss is constant.

The all-day efficiency is the output in kWh. divided by the input in kWh. for the day, so that it is given by

$$\eta_{all\text{-}day} = \frac{\int E_2I_2 \cos \phi dt}{\int E_2I_2 \cos \phi dt + \int RI_2^2 dt + \int P_i dt}.$$

If the load variation is known, P_i and R should be chosen to make this a maximum.

EXAMPLE. The average daily load on a transformer which is permanently connected to the supply mains is: full load at 0·9 p.f. for 5 hr., half load at 0·8 p.f. for 9 hr., and no load for 10 hr. Find the ratio of the full load copper-loss to iron-loss in the economical design of the transformer.

(*Lond. Univ.*, 1933.)

Let the full load current be I. Then

$$\int E_2I_2 \cos \phi dt = \int^5 E_2I \times 0{\cdot}9 \times dt + \int^9 E_2(\tfrac{1}{2}I) \times 0{\cdot}8 \times dt$$

$$= 8{\cdot}5E_2I \text{ W}.$$

$$\int RI_2^2 dt = \int^5 RI^2 dt + \int^4 R(\tfrac{1}{2}I)^2 dt$$

$$= 6RI^2,$$

and

$$\int P_i dt = 24P_i.$$

The all-day efficiency is thus

$$\eta_{all\text{-}day} = \frac{8{\cdot}5E_2I}{8{\cdot}5E_2I + 6RI^2 + 24P_i}$$

$$= \frac{8{\cdot}5E_2}{8{\cdot}5E_2 + 6RI + 24P_i/I}.$$

The maximum efficiency is obtained when

$$24P_i = 6RI^2$$

or

$$P_i = \tfrac{1}{4}RI^2,$$

i.e. the iron-loss is one-quarter of the copper-loss at full load.

Actually the matter is not as simple as stated above, which is the usual simple explanation. There are really two quite different problems to be considered.

If it is required to design a single transformer to deal with a given load, the question is one of economics. It cannot be said that the most economic design is that in which the iron- and copper-losses are equal, which in this case is quite an irrelevant and arbitrary criterion. The procedure, which is lengthy, is to take a certain amount of iron and copper and make the best geometrical design. The annual cost of the iron and copper is estimated, and also the total losses. Suppose we have the total cost as

$$A \times C + B \times I,$$

where C = copper, I = iron, and A and B are constants. Let the losses have an annual cost of

$$D/C + E/I.$$

The latter expression is only approximate and requires careful evaluation. The problem is then to find the minimum value of

$$(AC + D/C) + (BI + E/I)$$

for varying values of C and I. The condition is found to be

$$C = \sqrt{(D/A)} \text{ and } I = \sqrt{(E/B)}.$$

This determines the amount of material and also the losses.

The second problem is to find how many transformers, of many identical transformers which are available, should be switched in to deal with a given load, whether to switch in a few working at high current or a large number working at low current. It is seen that, in this case, a number should be switched in so that each has equal copper- and iron-losses. For here the cost of the transformers is fixed, whether they are used or not, and the only cost to be considered as variable is that of the losses. The losses are least, for a given power output, when the efficiency of the combination is a maximum; and as the transformers are identical the efficiency of the combination is equal to the efficiency of each.

EXAMPLES XIV

1. Explain carefully why the generator costs per unit of electricity are reduced if the load factor on a system is improved.

Describe briefly the various methods which are employed for dealing economically with peak loads on an alternating-current system supplying an industrial area. **(*Lond. Univ.*, 1932.)**

2. Explain briefly why the total cost of generation may be separated into a fixed and a running charge, and give a short representative list of costs which would appear under each heading.

How does the diversity factor and the load factor influence the cost of generation? **(*Nat. Cert.*, 1933.)**

3. Define (*a*) load factor and (*b*) diversity factor and discuss their effects upon the economics of power supply.

In the case of a two-part tariff consisting of a kVA. charge and a kWh. charge, how is the choice of phase-advancing equipment affected by these charges?

Derive an expression relating such tariff charges and the annual costs of the phase-advancing equipment with the most economical degree of power-factor improvement. Assume the total power taken by the load to remain unaltered, and neglect losses in the phase-advancing equipment.

(*Lond. Univ.*, 1954.)

4. Define the terms: "maximum demand"; "demand factor"; "load factor"; "plant load factor"; "diversity factor."

A generating station had a connected load of 43 000 kW. and a maximum load of 20 000 kW., the units generated being 61 500 000 for the year. Calculate the load factor and the demand factor for this case.

(*Nat. Cert.*, 1933.)

5. State Kelvin's law, and explain why in practice the law is usually not strictly observed.

In given circumstances, the economic current density of a single-core cable is 300 A/in.^{2}. Energy costs 0·8d. per kWh., and that portion of the cable cost which is independent of size is £600 per mile, on which the annual charge is 8 per cent. The cable is used at a constant current of 200 A., although the economic optimum is 250 A. Compare the total annual cost of using this cable with the corresponding cost for a similar cable for which the economic current is 200 A. The resistance of the conductor material having a cross-sectional area of 1 in.^{2} is 0·045 Ω. per mile. (*Lond. Univ.*, 1953.)

6. Discuss the general principles of charging for electrical energy.

A load has a maximum demand of 2 000 kVA. at a lagging power factor of 0·8, and the tariff is £5 per kVA. maximum demand plus 0·4d. per kWh. Assuming a load factor of 50%, calculate the saving in the annual charge if the power factor is corrected to 0·95 lagging. Interest and depreciation on the correcting apparatus amount to 5s. per kVA. rating, and losses in the correcting apparatus may be assumed to average 2·5 kW. throughout the year.

(*Lond. Univ.*, 1948.)

7. The load on a certain installation may be considered constant at 1 200 kVA., 0·75 lagging power factor for 2 500 hr. per annum. The tariff is 0·65d. per kWh. plus £5 10s. 0d. per kVA. maximum demand.

(*a*) Determine the annual charge for electrical energy.

(*b*) Power-factor improving apparatus is installed to improve the power factor to 0·95 lagging.

Determine the kVAr. required and the new annual charge if the power-factor improving apparatus costs £5 per kVAr., annual interest and depreciation charges are 10 per cent of the capital cost, and the losses in the apparatus are 4 per cent of the kVAr. rating. (*Lond. Univ.*, 1950.)

8. State and prove the law of economy as applicable to an insulated feeder cable. The cost of each of the copper conductors for a mile of an overhead

transmission line is £100 plus £700 per in.^{2} of cross-sectional area. The load factor of the main load is 25% and the losses 11%. The combined rate of interest and depreciation is 9% per annum, and the cost of the energy wasted is 1d. per kWh. Calculate the economical maximum current density for the line. A mile of copper wire of 1 in.^{2} cross-sectional area has a resistance of 0·046 Ω. If the section of the conductor differed by 10% from the most economical value, by what percentage would the total cost of running be increased? (*Lond. Univ.*, 1933.)

9. Determine the economic voltage for a three-phase power system transmitting an equivalent load of 3 000 kW. for 3 500 hr. per annum at an average power factor of 0·9, the load being equally divided among five substations at which there is the usual transformer, arrester and switch-gear equipment. The real length of line is 450 000 ft. whilst the equivalent length may be taken as 350 000 ft., the towers being spaced 600 ft. apart.

The specific resistance of the copper material of the line is 2/3 μΩ. per inch cube, the cost 2·875d. per in.^{3} and the cost per unit of energy 0·5d.

The annual charge on the components at the stations may be taken as follows—

For single-phase transformers of normal full load

P kW. at V kV. . . . . £$(0{\cdot}012V + 1{\cdot}56)\sqrt{P}$.

For the arresters and switchgear at V kV. . . £$1{\cdot}25V$

For the poles and insulators for a V kV. system . £$(0{\cdot}0144V + 1{\cdot}92)$

The annual depreciation of the line may be taken as 6% of the initial cost. (*Lond. Univ.*, 1931.)

10. Describe the principle of a two-element meter for giving the wattless kVAh. of a three-phase power system, giving vector diagrams showing the phase relationships.

If the kWh. output of the system is also known, why is the power factor determined from the two instruments not necessarily the true average power factor over the period of time concerned? (*Lond. Univ.*, 1931.)

11. What factors determine (a) the first cost, and (b) the running cost, of a low-voltage cable network?

A two-core feeder cable 880 yd. long, costing £(2 500 + 10 000 a) per mile, where a is the cross-section of one core in sq. in., supplies 100 kW. at 400 V. d.c. for 5 000 hours per annum. If the cost of energy at the supply end is 0·4d. per kWh., and interest, depreciation, etc., account for 12% per annum on the capital invested, determine the most economical value of a and the average energy cost per kWh. at the load. (Specific resistance, $\frac{2}{3}$ microhm-inch.) (*Lond. Univ.*, 1947.)

12. Explain the economic basis for tariff charges in which the consumer's power factor and maximum demand are involved. Assuming a 440 V., 3-phase supply, show how these quantities are measured or estimated. A load having a maximum value of 1 000 kW. at power factor 0·8 has an annual load factor of 40%. Find the cost per unit to the consumer if the tariff is £5 a year per kW. demand plus 0·5d. per unit plus 0·25d. per reactive kVAh. (*Lond. Univ.*, 1931.)

13. Show that the economical limit to which the power factor of a lagging load can be raised is independent of the original value of the power factor when the tariff consists of a fixed charge per kVA. plus a flat rate per unit.

A consumer is charged £3 per annum per kVA. maximum demand plus a flat rate per kWh. Phase advancing plant can be purchased for £3 per kVA., the cost for depreciation, housing, and interest on capital being 12%. Find the most economical angle of lag to which the load can be improved under these conditions. (*Lond. Univ.*, 1932.)

14. A factory takes a load of 200 kW. and 0·85 p.f. (lagging) for 2 500 hr. per annum, and buys energy on a tariff of £7 per kVA. plus 0·75d. per kWh. consumed. If the power factor is improved to 0·9 (lagging) by means of condensers costing £25 10s. 0d. per kVA. and having a power loss of 100 W.

per kVA., calculate the annual saving effected by their use. Allow 8% per annum for interest and depreciation on the condensers. (*Nat. Cert.*, 1935.)

15. An industrial load has an average value of 500 kW. at 0·7 p.f., lagging, for 2 000 h. per annum. The maximum demand is 30% in excess of the average—also at 0·7 p.f., lagging. Energy costs £5 per annum per kVA. maximum demand, plus 0·5d. per unit.

Power factor correcting plant is installed, raising the power factor of the average load to 0·9, lagging, and costing £4 per kVA. Find the annual saving in cost. Allow 10% for interest and depreciation on the capital cost of the additional plant. (Neglect the energy losses in the phase advancing apparatus, and assume its kVA. input is constant.) (*Lond. Univ.*, 1936.)

16. State Kelvin's law for the most economical size of conductor and explain why this law may be modified in practice.

Determine the most economical area of conductor if the maximum value of the current is 1 000 A. and the load factor is 50%. The conductor costs 15d. per lb., the cost of energy wasted is 0·7d. per kWh., and the annual interest and depreciation is 10% of the capital cost. The conductor weighs 0·32 lb. per in.^{3} and has a resistance of 0·8 microhm per inch cube. Explain any assumptions made. (*Lond. Univ.*, 1947.)

17. Give reasons for the approximate relationship $C = A \times$ (kVA.) $+ B$ $+$ (kWh.) in which C, the annual operating cost of a generating station, is expressed in terms of the maximum kVA. demand and the total units generated; A and B are constants for a particular station. Account for differences between the values of the constants used when contrasting steam and hydro-electric stations of comparable size.

What regard should be paid to these constants in controlling the power taken from various stations feeding into a common network? In particular, distinguish between "base load" and "peak load" working. (*Lond. Univ.*, 1949.)

18. Explain the purpose of power factor correction by an industrial consumer. Describe briefly three methods by which it may be effected.

A synchronous condenser, which may be assumed to work at zero power factor, costs £5 per kVA. If a consumer is charged £4 per annum per kVA. of maximum demand, and interest, depreciation, etc., amount to 10% on capital invested, determine his most economical operating power factor. The uncorrected power factor is 0·8 lagging, and the load is steady. (*Lond. Univ.*, 1948.)

19. Explain the terms "load factor" and "diversity factor," as applied to an electricity undertaking. Indicate how they are likely to be affected by "staggering the load."

A connected load of 20 000 kVA. is fed from a substation having an installed capacity of 5 000 kVA. The annual output of the substation is as follows—

Hours duration	4 380	2 190	1 460	730
kVA. . .	1 000	2 000	3 000	5 000
Power factor .	0·5	0·8	0·95	0·9

Determine the kW. load factor and the kVA. diversity factor.

If the energy is obtained in bulk at a cost of £3 per annum per kVA. of maximum demand plus 0·3d. per kWh., determine the mean cost per kWh. of *output*, if the substation works with an all-day efficiency of 90% and a peak-load apparent efficiency of 97%. (*Lond. Univ.*, 1948.)

20. Deduce the condition for minimum volume of copper in a 2-wire distributor when the overall voltage drop has a fixed value. Assume a distributor fed from one end and loaded at two points.

A 2-wire distributor, AD, 300 yd. long, is fed at A with 240 V. and loaded as follows: 60 A. at *B,* 100 yd. from *A*; 70 A. at *C,* 200 yd. from *A*; 50 A. at *D.* Determine the cross-sections of the conductors for minimum volume of copper and a voltage of 235 V. at *D.* Assume the resistance of 1 000 yd. of copper of 1 in.^{2} cross-section to be 0·023 Ω. The condition for minimum volume of copper for three load points is similar to that derived for two load points. (*Lond. Univ.,* 1949.)

APPENDIX I

HYPERBOLIC SINE AND COSINE

It is perhaps easiest to grasp the properties of the hyperbolic functions by comparing them with the trigonometric functions, viz. the sine and cosine. It is shown in textbooks on trigonometry that

$$\left.\begin{aligned} \sin z &= z - \frac{z^3}{3!} + \frac{z^5}{5!} - \frac{z^7}{7!} + \dots \\ \text{and} \qquad \cos z &= 1 - \frac{z^2}{2!} + \frac{z^4}{4!} - \frac{z^6}{6!} + \dots \end{aligned}\right\} \quad . \quad . \quad (1)$$

The hyperbolic sine and cosine are defined as the corresponding series in which all the signs are positive. Thus

$$\left.\begin{aligned} \sinh z &= z + \frac{z^3}{3!} + \frac{z^5}{5!} + \frac{z^7}{7!} + \dots \\ \text{and} \qquad \cosh z &= 1 + \frac{z^2}{2!} + \frac{z^4}{4!} + \frac{z^6}{6!} + \dots \end{aligned}\right\} \quad . \quad . \quad (2)$$

If we put jz in place of z in equations (2) we get

$$\begin{aligned} \sinh (jz) &= jz + \frac{(jz)^3}{3!} + \frac{(jz)^5}{5!} + \frac{(jz)^7}{7!} + \dots \\ &= j\left(z - \frac{z^3}{3!} + \frac{z^5}{5!} - \frac{z^7}{7!} + \dots\right) \end{aligned}$$

$$\left.\begin{aligned} \text{so that} \qquad \sinh (jz) &= j \sin z \\ \text{Similarly} \quad \cosh (jz) &= \cos z. \end{aligned}\right\} \quad . \quad . \quad . \quad . \quad . \quad . \quad (3)$$

In the same way we find that

$$\left.\begin{aligned} \sin jz &= j \sinh z \\ \text{and} \qquad \cos jz &= \cosh z. \end{aligned}\right\} \quad . \quad . \quad . \quad . \quad . \quad (3a)$$

It is clear from equations (2) that

$$\left.\begin{aligned} \sinh (-z) &= -\sinh z \\ \text{and} \qquad \cosh (-z) &= +\cosh z. \end{aligned}\right\} \quad . \quad . \quad . \quad . \quad . \quad (4)$$

We deduce immediately the formulae

$$\left.\begin{aligned} \sinh (A + B) &= \sinh A \cosh B + \cosh A \sinh B, \\ \sinh (A - B) &= \sinh A \cosh B - \cosh A \sinh B, \\ \cosh (A + B) &= \cosh A \cosh B + \sinh A \sinh B, \\ \cosh (A - B) &= \cosh A \cosh B - \sinh A \sinh B. \end{aligned}\right\} \quad . \quad (5)$$

The method of proof for the third of these equations will be given as an example.

$$\begin{aligned}\cosh(A+B) &= \cos j(A+B)\\ &= \cos jA\cos jB - \sin jA\sin jB\\ &= \cosh A\cosh B - j\sinh A\,.\,j\sinh B\\ &= \cosh A\cosh B + \sinh A\sinh B.\end{aligned}$$

The hyperbolic sine or cosine of a complex number $a + jb$ can be expanded in the following way.

$$\left.\begin{aligned}\sinh(a+jb) &= \sinh a\cosh jb + \cosh ja\sinh jb\\ &= \sinh a\cos b + j\cosh a\sin b\\ \text{Similarly}\qquad & \\ \cosh(a+jb) &= \cosh a\cos b + j\sinh a\sin b.\end{aligned}\right\} \quad . \quad . \quad (6)$$

From equations (1) it follows that

$$\begin{aligned}\cos z + j\sin z &= 1 - \frac{z^2}{2!} + \frac{z^4}{4!} - \dots\\ &\quad + j\left(z - \frac{z^3}{3!} + \frac{z^5}{5!} - \dots\right)\\ &= 1 + jz + \frac{(jz)^2}{2!} + \frac{(jz)^3}{3!} + \frac{(jz)^4}{4!} + \dots,\end{aligned}$$

so that $\qquad \cos z + j\sin z = \varepsilon^{jz}$

Similarly $\qquad \cos z - j\sin z = \varepsilon^{-jz}$. $\quad . \quad . \quad . \quad . \quad (7)$

By addition and subtraction we get

$$\left.\begin{aligned}\cos z &= \tfrac{1}{2}(\varepsilon^{jz} + \varepsilon^{-jz})\\ \sin z &= \tfrac{1}{2j}(\varepsilon^{jz} - \varepsilon^{-jz})\end{aligned}\right\} \quad . \quad . \quad (8)$$

In exactly the same way we find that

$$\cosh z + \sinh z = \varepsilon^{z}$$

and $\qquad \cosh z - \sinh z = \varepsilon^{-z},$

so that

$$\left.\begin{aligned}\cosh z &= \tfrac{1}{2}(\varepsilon^{z} + \varepsilon^{-z})\\ \text{and}\qquad \sinh z &= \tfrac{1}{2}(\varepsilon^{z} - \varepsilon^{-z}).\end{aligned}\right\} \quad . \quad . \quad . \quad (9)$$

The hyperbolic sine and cosine can thus be expressed in terms of simple exponential functions.

Linear Differential Equation. A linear differential equation of the form

$$d^2y/dz^2 = ky \quad . \quad . \quad . \quad . \quad (10)$$

occurs very often in electrical and mechanical systems. We try as a solution

$$y = A\varepsilon^{mz}.$$

This gives $$dy/dz = mA\varepsilon^{mz}$$

and $$d^2y/dz^2 = m^2A\varepsilon^{mz},$$

since the differential coefficient of ε^{mz} is $m\varepsilon^{mz}$.

Equation (10) is satisfied provided

$$d^2y/dz^2 = m^2A\varepsilon^{mz} = ky = kA\varepsilon^{mz},$$

i.e. if $$m^2 = k, \text{ or } m = \pm\sqrt{k}.$$

There are thus two solutions, $\varepsilon^{+(\sqrt{k})z}$ and $\varepsilon^{-(\sqrt{k})z}$, so that

$$y = A_1\varepsilon^{+(\sqrt{k})z} + A_2\varepsilon^{-(\sqrt{k})z}, \qquad (11)$$

where A_1 and A_2 are constants. This is the complete solution. It may be put in the form

$$y = A_1[\cosh(\sqrt{k})z + \sinh(\sqrt{k})z] + A_2[\cosh(\sqrt{k})z - \sinh(\sqrt{k})z]$$
$$= (A_1 + A_2)\cosh(\sqrt{k})z + (A_1 - A_2)\sinh(\sqrt{k})z.$$

As A_1 and A_2 are constants, so also are their sum and difference, which we may replace by C_1 and C_2. Thus

$$y = C_1\cosh(\sqrt{k})z + C_2\sinh(\sqrt{k})z. \qquad (12)$$

This also is the solution of equation (10).

If k is negative, $-K$ say, then $(\sqrt{k}) = j\sqrt{K}$. The solution is then

$$y = C_1\cosh j(\sqrt{K})z + C_2\sinh j(\sqrt{K})z$$
$$= C_1\cos(\sqrt{K})z + jC_2\sin(\sqrt{K})z$$

or $$C_1\cos(\sqrt{K})z + C_2'\sin(\sqrt{K})z. \qquad (12a)$$

Table of Hyperbolic Functions. A brief table of the hyperbolic sine and cosine is given below.

HYPERBOLIC FUNCTIONS

z	cosh z	sinh z	z	cosh z	sinh z
0·00	1·0000	0·0000	0·55	1·1551	0·5781
0·02	1·0002	0·0200	0·60	1·1855	0·6366
0·04	1·0008	0·0400	0·65	1·2188	0·6967
0·06	1·0018	0·0600	0·70	1·2552	0·7586
0·08	1·0032	0·0801	0·75	1·2947	0·8223
0·10	1·0050	0·1002	0·80	1·3374	0·8881
0·12	1·0072	0·1203	0·85	1·3835	0·9561
0·14	1·0098	0·1405	0·90	1·4331	1·0265
0·16	1·0128	0·1607	0·95	1·4862	1·0995
0·18	1·0162	0·1810	1·00	1·5431	1·1752
0·20	1·0201	0·2013	1·10	1·6685	1·3356
0·22	1·0243	0·2218	1·20	1·8107	1·5095
0·24	1·0289	0·2423	1·30	1·9709	1·6984
0·26	1·0340	0·2629	1·40	2·1509	1·9043
0·28	1·0395	0·2837	1·50	2·3524	2·1293
0·30	1·0453	0·3045	1·60	2·5775	2·3756
0·32	1·0516	0·3255	1·70	2·8283	2·6456
0·34	1·0584	0·3466	1·80	3·1075	2·9422
0·36	1·0655	0·3678	1·90	3·4177	3·2682
0·38	1·0731	0·3892	2·0	3·7622	3·6269
0·40	1·0811	0·4108	2·5	6·1323	6·0502
0·42	1·0895	0·4325	3·0	11·0677	10·0179
0·44	1·0984	0·4543	3·5	16·5728	16·5426
0·46	1·1077	0·4764	4·0	27·3082	27·2899
0·48	1·1174	0·4986	4·5	45·0141	45·0030
0·50	1·1276	0·5212	5·0	74·2099	74·2032

APPENDIX II

ALTERNATING CURRENTS

THE reader who is not familiar with the theory of alternating currents will find a detailed discussion in a companion volume of this series.* A résumé of the more important results will be given here for reference.

A current or voltage, such as shown in Fig. 358 (A), which is repeated at intervals of time of T seconds is said to be *periodic*, the period being T. The fundamental frequency of the wave is

$$f = 1/T \text{ cycle per sec.}$$

It can be shown that such a wave consists of a sum of waves with frequencies which are multiples, even and odd, of the fundamental. There may be a d.c. component also.

In electrical power engineering, current and voltage waves are usually of the form shown in Fig. 358 (B), in which the value at time $(\frac{1}{2}T + t)$ is the negative of that at time t. In this case, the waves consist of the fundamental and odd harmonics only.

Fig. 358 (C) shows a pure sine wave, which is represented by $E \cos (\omega t + \theta)$. At the instant $t = 0$, the wave is equal to $E \cos \theta$. The wave repeats itself in time T where

$$\omega T = 2\pi,$$

so that the period is $T = 2\pi/\omega$, and the frequency is $f = 1/T = \omega/2\pi$.

The quantity ω is 2π times the frequency and is called the *pulsatance* or angular frequency.

Mean Value. The mean values of the waves of Figs. 358 (B) and 358 (C) are clearly zero, since they are as much and as often negative as positive. The mean value of the wave in Fig. 358 (A) may be something different from zero. We may get some such a wave by considering the power expended in a pure resistance R across which the voltage $E \cos (\omega t + \theta)$ is applied. The current is $(E/R) \cos (\omega t + \theta)$, and the power is

$$E \cos (\omega t + \theta) \times (E/R) \cos (\omega t + \theta)$$
$$= (E^2/R) \cos^2 (\omega t + \theta)$$

and is shown in Fig. 358 (D). In this case the power is always positive and has a mean value which is positive. The mean value is found in the following way. The power is

$$(E^2/R) \cos^2 (\omega t + \theta)$$
$$= \tfrac{1}{2}(E^2/R)\,[1 - \cos 2\,(\omega t + \theta)].$$

* *Applied Electricity*, by A. T. Starr (Pitman, 1957).

The mean value of $\cos 2(\omega t + \theta)$ is zero, since this is a pure sine wave of double the frequency, so that the mean value of the power is $(E^2/2R)$. The power wave is in fact a pure sine wave of double frequency, with axis a line at height $(E^2/2R)$.

There is, however, a use for an extended idea of mean value, which is obtained as follows. Imagine the wave (any of those

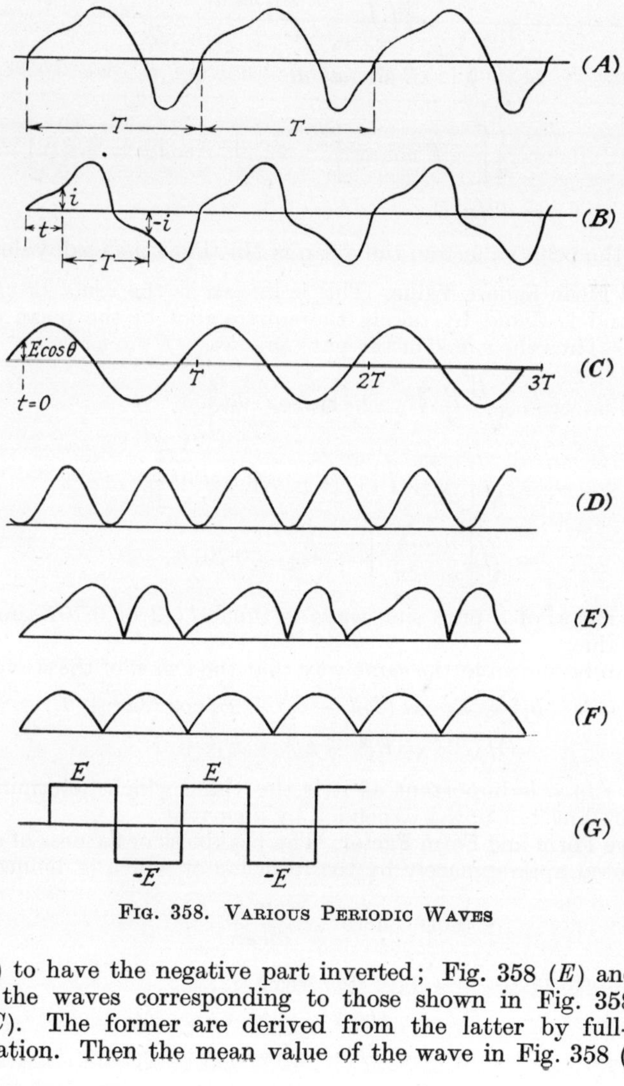

FIG. 358. VARIOUS PERIODIC WAVES

shown) to have the negative part inverted; Fig. 358 (*E*) and (*F*) shows the waves corresponding to those shown in Fig. 358 (*A*) and (*C*). The former are derived from the latter by full-wave rectification. Then the mean value of the wave in Fig. 358 (*A*) is the mean of that shown in Fig. 358 (E); the mean value of the sine wave is the mean of that shown in Fig. 358 (F).

The mean value of a sine wave is found in the following way. The wave is repeated in half-cycles, so that we need find only the mean over a half-cycle. The mean is

$$\frac{1}{\frac{1}{2}T}\int_0^{\frac{1}{2}T} E \sin \omega t \, dt$$

$$= \frac{\omega}{\pi}\int_0^{\pi/\omega} E \sin \omega t \, . \, dt$$

$$= \frac{1}{\pi}\Big[- E \cos \omega t\Big]_0^{\omega/\pi} = \frac{E}{\pi}\Big[- \cos \pi + \cos 0\Big]$$

$$= (2/\pi)E.$$

E is the peak value and the mean is $2/\pi$ times the peak value.

Root Mean Square Value. This is known as the *r.m.s.* or *effective value* and is found by taking the square root of the mean of the square. Thus the r.m.s. of the pure sine wave $E \sin \omega t$ is

$$\sqrt{\left[\frac{\omega}{\pi}\int_0^{\pi/\omega} E^2 \sin^2 \omega t \, . \, dt\right]}$$

$$= \sqrt{\left[\frac{\omega E^2}{2\pi}\int^{\pi/\omega} (1 - \cos 2\omega t) dt\right]}$$

$$= \sqrt{\left[\frac{\pi}{\omega}\,\frac{\omega E^2}{2\pi}\right]} = \frac{E}{\sqrt{2}} = 0{\cdot}707E.$$

The r.m.s. of a pure sine wave is thus $1/\sqrt{2}$ or 0·707 times the peak value.

It can be shown in the same way that the r.m.s. of the wave

$$E_1 \cos (\omega t + \theta_1) + E_2 \cos (2\omega t + \theta_2) + E_3 \cos (3\omega t + \theta_3) + \ldots$$

is $$(1/\sqrt{2}) \sqrt{(E_1^2 + E_2^2 + E_3^2 + \ldots)}.$$

The r.m.s. is important as it is the value which determines the heat produced or power expended by a current.

Wave Form and Form Factor. The peakiness or flatness of a wave is denoted approximately by the form factor which is defined by

$$\text{Form factor} = \frac{\text{r.m.s.}}{\text{mean}}.$$

The form factor of a pure sine wave is

$$\frac{1/\sqrt{2}}{2/\pi} = \frac{\pi}{2\sqrt{2}} = 1{\cdot}11.$$

The form factor of a square wave, shown in Fig. 358 (G) is clearly 1, since the mean, r.m.s., and peak values are all equal.

The form factor of the wave $\sin^2 \omega t$, shown in Fig. 358 (D), can be shown to be $\sqrt{1{\cdot}5} = 1{\cdot}225$. Thus the peakier a wave the greater is its form factor.

Since the mean value of the pure sine wave of Fig. 358 (C) is the mean of that in Fig. 358 (F), and since also their r.m.s. values are the same, it follows that they have the same form factor.

EXAMPLE. Draw the waves ($100 \sin \omega t + 10 \sin 3\omega t$) and ($100 \sin \omega t - 10 \sin 3\omega t$). Find their effective and mean values and hence also their form factors.

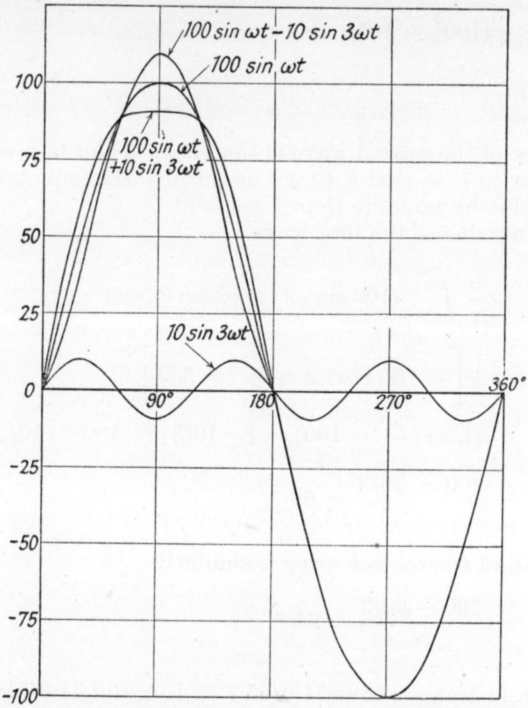

FIG. 359. WAVE WITH THIRD HARMONIC

Fig. 359 shows the various waves. The r.m.s. value of the former is

$$\sqrt{\left[\frac{\omega}{\pi}\int_0^{\pi/\omega} (100 \sin \omega t + 10 \sin 3\omega t)^2 dt\right]}$$

$$= \sqrt{\left[\frac{\omega}{\pi}\int_0^{\pi/\omega} (100^2 \sin^2 \omega t + 10^2 \sin^2 3\omega t + 2\,000 \sin \omega t \,.\, \sin 3\omega t)dt\right]}$$

$$= \sqrt{\left[\frac{\omega}{\pi}\int_0^{\pi/\omega} \tfrac{1}{2}(100^2 - 10\,000 \cos 2\omega t + 10^2 - 100 \cos 6\omega t + 2\,000 \cos 2\omega t - 2\,000 \cos 4\omega t)dt\right]}$$

$$= \sqrt{\left[\frac{\omega}{\pi} \times \frac{\pi}{\omega} \times \tfrac{1}{2}(100^2 + 10^2)\right]}$$

$$= \frac{1}{\sqrt{2}}\sqrt{(100^2 + 10^2)}$$

$$= 71{\cdot}0.$$

The r.m.s. of the second wave is the same. The r.m.s. of the wave 100 sin ωt is 70·7, so that a 10 per cent third harmonic increases the effective value by no more than $\frac{1}{2}$ per cent.

The mean value of the first wave is

$$\frac{\omega}{\pi}\int_0^{\pi/\omega} (100 \sin \omega t + 10 \sin 3\omega t)dt$$

$$= \frac{1}{\pi}\left[- 100 \cos \omega t - \frac{10}{3} \cos 3\omega t\right]_0^{\pi/\omega}$$

$$= (1/\pi)\,[-(-100) - (-10/3) + 100 + 10/3]$$

$$= \frac{200 + 20/3}{\pi} = 65{\cdot}7.$$

The mean of the second wave is similarly

$$\frac{200 - 20/3}{\pi} = 61{\cdot}5.$$

The form factors are thus 71·0/65·7 = 1·08 and 71·0/61·5 = 1·156.

Vector Representation of Alternating Currents. There is a very helpful, graphical method of representing alternating quantities by means of a rotating vector. Thus the alternating voltage $E \cos (\omega t + \theta)$ can be represented by the rotating vector OP of Fig. 360. The vector is considered to have started from the position OP_0 at time $t = 0$, where $\angle XOP_0 = \theta$. At time t, $\angle XOP = \omega t + \theta$, and the horizontal component or step of the vector OP is $E \cos (\omega t + \theta)$, provided the length of OP is made equal to E.

If the waves $E_1 \cos(\omega t + \theta_1)$ and $E_2 \cos(\omega t + \theta_2)$ are added, the resultant wave can be represented by the vector sum of the vectors representing the separate waves. For let the single vectors be OP_1 and OP_2 in Fig. 361 The vector sum of these vectors is obtained (from the definition of vector sum) by completing the parallelogram OP_1PP_2, when the sum is the vector OP. Now $P_1P = OP_2$, so that

$$ON_2 = ON_1 + N_1N_2 = E_1 \cos(\omega t + \theta_1) + E_2 \cos(\omega t + \theta_2).$$

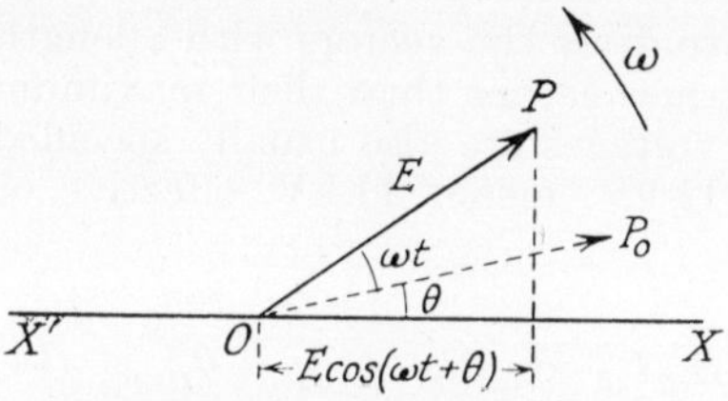

FIG. 360. ROTATING VECTOR REPRESENTING A.C. WAVE

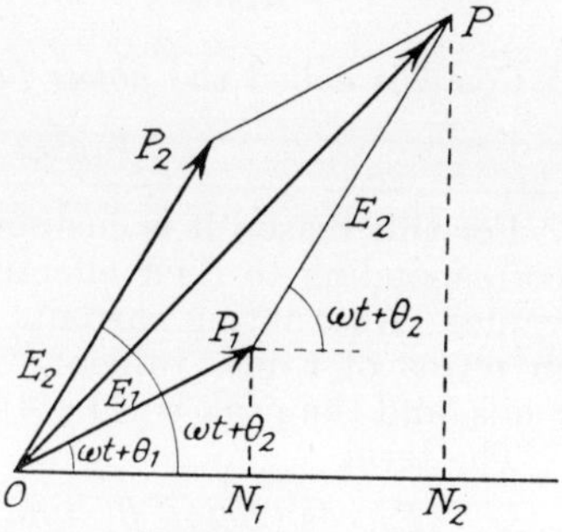

FIG. 361. SUM OF TWO VECTORS

Thus the horizontal step of OP is the resultant wave, and the resultant wave is therefore represented by the vector sum of the separate vectors. All three vectors, OP_1, OP_2, and OP rotate in the positive (anti-clockwise) direction with angular velocity ω. For this reason, when waves of only one frequency are dealt with, it is unnecessary to consider the rotation, as the waves keep the same relative positions.

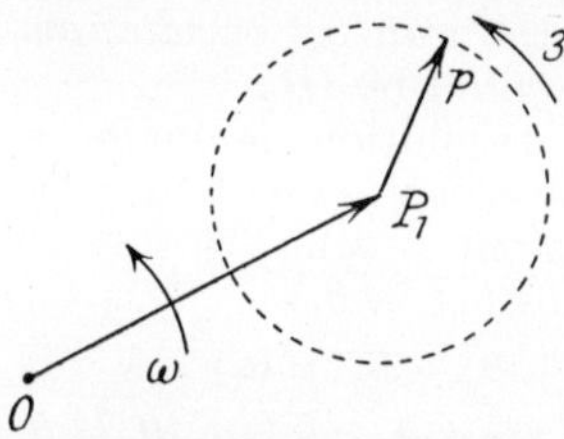

FIG. 362. ROTATING VECTORS FOR WAVE AND THIRD HARMONIC

If, however, waves of different frequencies occur, the vectors change their relative positions and the rotations cannot be ignored. Thus the wave of the previous example can be represented by OP_1 of length 100 rotating with angular velocity ω added to P_1P of length 10 rotating with angular velocity 3ω (Fig. 362).

Power. Suppose that the current is $I \cos(\omega t + \theta_1)$ and the potential difference is $E \cos(\omega t + \theta_2)$; these are represented in Fig. 363. The phase difference between the voltage and current is

$$\phi = (\omega t + \theta_2) - (\omega t + \theta_1) = \theta_2 - \theta_1.$$

The power is

$$P = I\cos(\omega t + \theta_1)E\cos(\omega t + \theta_2)$$
$$= \tfrac{1}{2}EI\,[\cos(\theta_2 - \theta_1) + \cos(2\omega t + \theta_1 + \theta_2)],$$

and has a steady component

$$\left.\begin{aligned}&\tfrac{1}{2}EI\cos(\theta_2 - \theta_1) = \tfrac{1}{2}EI\cos\phi\\ &= E_{eff}I_{eff}\cos\phi.\end{aligned}\right\} \quad . \quad . \quad . \quad (13)$$

Cos ϕ is called the *power factor*, so that the mean power is

$$E_{eff}I_{eff} \times \text{power factor.}$$

For this reason it is customary to draw the vectors with a length corresponding to their effective values rather than their maximum values. Alternating currents and voltages are also usually specified in terms of r.m.s. values; thus 11 kV. means 11 kV. effective or r.m.s. and the peak is $(11\sqrt{2})$ kV.

The term

$$\tfrac{1}{2}EI\cos(2\omega t + \theta_1 + \theta_2) = E_{eff}I_{eff}\cos(2\omega t + \theta_1 + \theta_2) \quad . \quad (14)$$

represents fluctuating power which averages out to zero.

The current can be resolved into two parts; one, in phase with the voltage, has magnitude $I\cos\phi$ and is called the *wattful component* of the current, and is represented by OI_1 in Fig. 363; the other has a phase difference of $\pi/2$ with respect to the voltage, and has magnitude $I\sin\phi$ and is called the *wattless* or *idle component*, and is represented by OI_2.

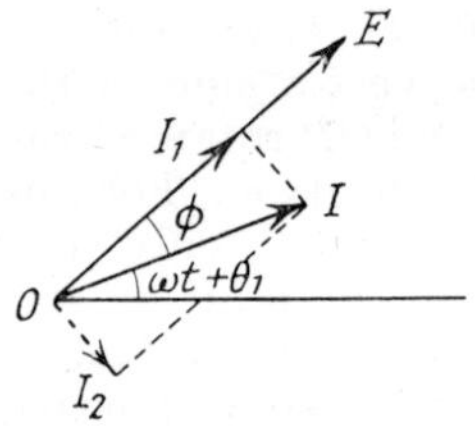

FIG. 363. POWER AND WATTLESS COMPONENTS OF CURRENT

Algebraically the resolution is made as follows.

The e.m.f. is $E\cos(\omega t + \theta_2)$.

The current is $I\cos(\omega t + \theta_1)$

$$= I\cos(\omega t + \theta_2 - \theta_2 + \theta_1)$$
$$= I\cos(\omega t + \theta_2)\cos(\theta_2 - \theta_1) + I\sin(\omega t + \theta)\sin(\theta_2 - \theta_1)$$
$$= I\cos\phi\cos(\omega t + \theta_2) + I\sin\phi\sin(\omega t + \theta).$$

The first term represents the in-phase current, which is represented by OI_1 and has magnitude $I\cos\phi$, whilst the second term is the quadrature or wattless current and has magnitude $I\sin\phi$.

The power is

$$P = E\cos(\omega t + \theta_2)\,[I\cos\phi\cos(\omega t + \theta_2)$$
$$+ I\sin\phi\sin(\omega t + \theta_2)]$$
$$= EI\cos\phi\cos^2(\omega t + \theta_2) + \tfrac{1}{2}EI\sin\phi\sin 2(\omega t + \theta_2).$$

The first term represents a power which varies from zero to a maximum of $EI \cos \phi$, and has a mean value of

$$\tfrac{1}{2}EI \cos \phi = E_{eff} I_{eff} \cos \phi.$$

The second term represents a purely fluctuating power, such as flows through an inductance or capacitance, varies between $+\tfrac{1}{2}EI \sin \phi$ and $-\tfrac{1}{2}EI \sin \phi$, and has a mean value of zero. It is called the *reactive power*, and may be represented by a vector of double frequency and amplitude

$$\tfrac{1}{2}EI \sin \phi = E_{eff} I_{eff} \sin \phi. \qquad (15)$$

Vector Method Applied to Simple Circuits. Suppose that an alternating e.m.f. $E \cos (\omega t + \theta)$ is applied across a resistance R. The current that flows is the e.m.f. divided by R so that

$$I = (E/R) \cos (\omega t + \theta),$$

and is thus represented by a vector of magnitude (E/R) lying on the voltage vector. The power factor is unity, since ϕ is zero, and the power is (E^2/R), E being the effective value.

Suppose the e.m.f. is placed across a coil of inductance L with no losses. The current I is given by

$$L(dI/dt) = E \cos (\omega t + \theta),$$

so that

$$\begin{aligned} I &= \int^t \frac{E}{L} \cos (\omega t + \theta) dt \\ &= (E/\omega L) \sin (\omega t + \theta) \\ &= (E/\omega L) \cos (\omega t + \theta - \pi/2). \end{aligned}$$

The constant of integration must be zero, as an alternating e.m.f. cannot produce a current with a d.c. component. The current is thus represented by a vector of magnitude $(E/\omega L)$ lagging the voltage by a phase of $\pi/2$ or 90°, and is shown in Fig. 364. ϕ is here 90°, so that the power factor is zero. The current is entirely wattless. Fig. 365 shows the current, voltage, and power (the last is shown shaded).

If the e.m.f. is placed across a condenser of capacitance C, let Q and $-Q$ be the charges on the plates. The current is

$$\begin{aligned} I = dQ/dt &= (d/dt) [CE \cos (\omega t + \theta)] \\ &= -\omega CE \sin (\omega t + \theta) \\ &= \omega CE \cos \left(\omega t + \theta + \frac{\pi}{2} \right). \end{aligned}$$

The current vector has magnitude ωCE and leads the e.m.f. vector by $\pi/2$ (see Fig. 364). ϕ is $-90°$ and the power factor is zero.

SIMPLE SERIES CIRCUIT. Suppose the e.m.f. is applied to the series combination of an inductance, resistance, and capacitance,

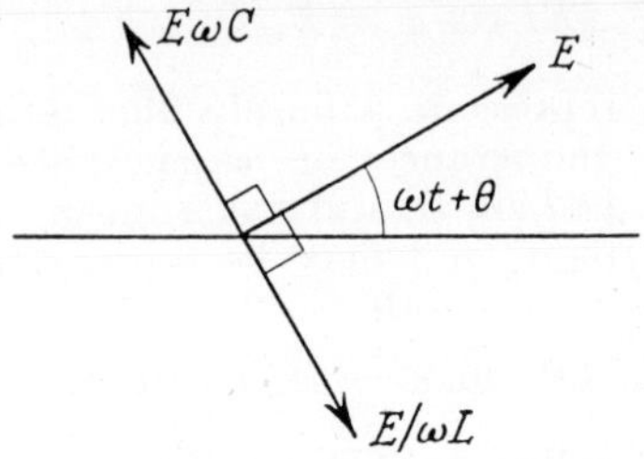

FIG. 364. CURRENTS IN COIL AND CONDENSER

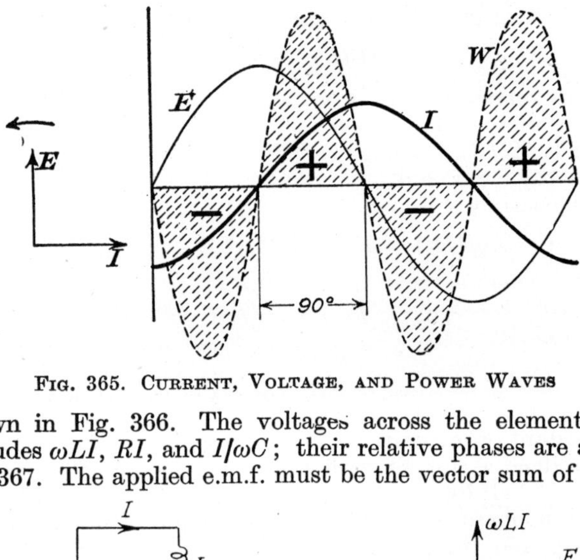

FIG. 365. CURRENT, VOLTAGE, AND POWER WAVES

as shown in Fig. 366. The voltages across the elements are of magnitudes ωLI, RI, and $I/\omega C$; their relative phases are as shown in Fig. 367. The applied e.m.f. must be the vector sum of these, as

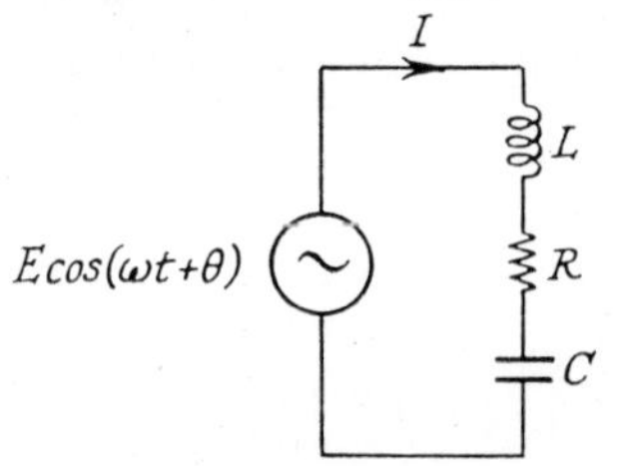

FIG. 366. SERIES CIRCUIT

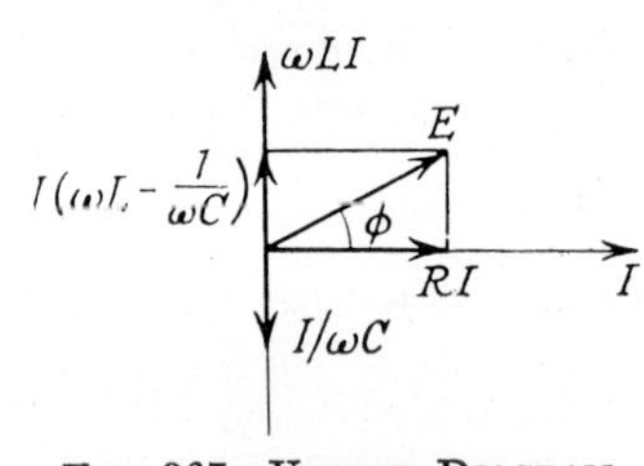

FIG. 367. VECTOR DIAGRAM OF SERIES CIRCUIT

shown in the figure. We have chosen ωLI as greater than $I/\omega C$, but the method is the same when it is less. It is clear that

$$E = \sqrt{\{[I(\omega L - 1/\omega C)]^2 + (RI)^2\}}$$
$$= I\sqrt{[R^2 + (\omega L - 1/\omega C)^2]}.$$

ωL is called the *reactance* of the coil, $-1/\omega C$ the reactance of the condenser, and the expression under the square root sign is the *impedance* of the circuit. If we put the total reactance equal to X and the impedance Z, we have

$$E = IZ, \; I = E/Z,$$

where $$Z = \sqrt{(R^2 + X^2)},$$

and $$X = \omega L - 1/\omega C.$$

It is clear that

$$\tan\phi = \frac{(\omega L - 1/\omega C)}{R}$$

and the power factor

$$\cos\phi = \frac{RI}{E} = \frac{R}{Z} = \frac{R}{\sqrt{[R^2 + (\omega L - 1/\omega C)^2]}}.$$

The current in the circuit is a maximum for a given e.m.f. when the impedance is a minimum. This occurs when

$$\omega L - 1/\omega C = 0, \text{ or } \omega^2 LC = 1.$$

This is the condition of *resonance*; the current is then simply E/R. The voltages across the inductance and capacitance are equal and opposite, and of magnitude $\omega LI = (\omega L/R)E$. The quantity $(\omega L/R)$ is written Q, and is the ratio of the voltage across L or C to that applied.

EXAMPLE. An alternating voltage of 230 V. and frequency 50 is placed in series with a coil of resistance 5 Ω., and inductance 0·4 H., and a condenser. Find the capacitance for resonance, the current at resonance, and the voltage across the coil and condenser.

$$C = 1/\omega^2 L = 1/[4\pi^2 \times 50^2 \times 0{\cdot}4]$$
$$= \underline{\underline{25{\cdot}4}}\ \mu\text{F}.$$

The current is $I = E/R = 230/5 = \underline{\underline{46}}$ A.

The voltage across the condenser is

$$\frac{I}{\omega C} = \frac{46}{2\pi \times 50 \times 25{\cdot}4 \times 10^{-6}} = \underline{\underline{5\,770}}\ \text{V}.$$

The reactance of the coil is $\omega L = 125{\cdot}6$, so that the impedance of the coil is

$$\sqrt{(5^2 + 125{\cdot}6^2)} = 125{\cdot}7.$$

The voltage across the coil is thus

$$125{\cdot}7 \times 46 = \underline{\underline{5\,782}}\ \text{V}.$$

If a circuit is supplied with an e.m.f. which contains harmonics, it may happen that resonance occurs at an harmonic frequency. The result is that the current wave has quite a different form from the voltage wave. An example will show the effect and how it is calculated.

EXAMPLE. An e.m.f. ($100 \sin \omega t - 10 \sin 3\omega t$) is applied to a coil of resistance 5 Ω. and inductance 0·4 H. in series with a condenser of capacitance 3·0 μF. Calculate the current. The fundamental frequency is 50 cyc.

The reactance at 50 cyc. is

$$\omega L - \frac{1}{\omega C} = (2\pi \,.\, 50 \,.\, 0{\cdot}4) - \frac{1}{2\pi \,.\, 50 \,.\, 3 \times 10^{-6}}$$
$$= 125{\cdot}6 - 1\,061 = -\,935.$$

The 50 cyc. current is thus

$$\frac{100}{\sqrt{(5^2 + 935^2)}} \sin (\omega t - \phi_1) = 0{\cdot}107 \sin (\omega t - \phi_1)$$

where $\phi_1 = \tan^{-1} (-\,935/5) = \tan^{-1} (-\,197) = -\,89° \, 43'$.

The reactance at 150 cyc. is

$$3\omega L - 1/3\omega C = 376{\cdot}8 - 353{\cdot}7 = 23{\cdot}1,$$

so that the 150 cyc. current is

$$-\frac{10}{\sqrt{(5^2 + 23{\cdot}1^2)}} \sin (3\omega t - \phi_3) = -\,0{\cdot}424 \sin (3\omega t - \phi_3),$$

where $\phi_3 = \tan^{-1} \dfrac{23{\cdot}1}{5} = \tan^{-1} (4{\cdot}62) = 77° \, 48'$.

The total current is thus

$$\underline{\underline{0{\cdot}107 \sin (\omega t + 89° \, 43') - 0{\cdot}424 \sin (3\omega t - 77° \, 48').}}$$

If the voltage and current waves are drawn, it is seen that they are very dissimilar.

SIMPLE PARALLEL CIRCUIT. Suppose the circuit in Fig. 368 has an e.m.f. $E \cos (\omega t + \theta)$ as shown. The currents in the inductance, resistance, and capacitance are as indicated. The total current has a wattfull component (E/R) and a wattless component

$$E(1/\omega L - \omega C),$$

so that the total current is

$$I = E\sqrt{[1/R^2 + (1/\omega L - \omega C)^2]}.$$

$1/R$ is called the *conductance* of the circuit and $(1/\omega L - \omega C)$ the *susceptance*. The former is usually denoted by G and the latter by B. The expression under the square root sign is called the *admittance* and is denoted by Y. Then

$$Y = \sqrt{(G^2 + B^2)},$$

and

$$I = EY.$$

The current is a minimum when Y is a minimum, i.e. when

$$1/\omega L - \omega C = 0 \text{ or } \omega^2 LC = 1.$$

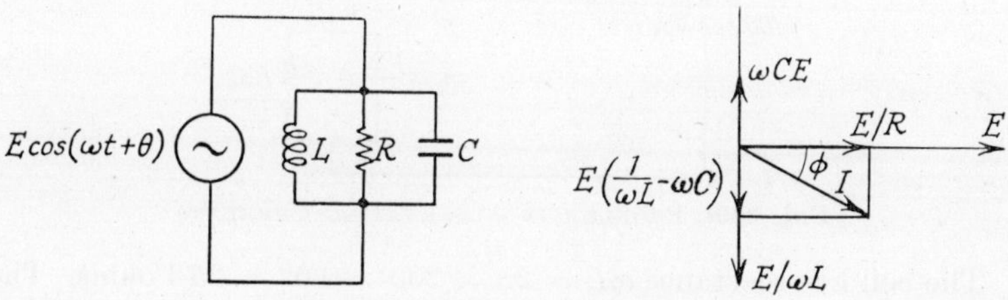

FIG. 368. PARALLEL CIRCUIT

This is said to be the condition for *parallel resonance* or *anti-resonance.* When this condition holds the current is simply E/R or EG. The impedance of the parallel combination is

$$Z = 1/Y = \frac{1}{\sqrt{[1/R^2 + (1/\omega L - \omega C)^2]}}.$$

The phase difference between the current and voltage is ϕ, where

$$\tan \phi = EB/EG = B/G = R(1/\omega L - \omega C).$$

The power factor is

$$\cos \phi = EG/I = G/Y = G/\sqrt{(G^2 + B^2)}$$
$$= [1 + R^2 (1/\omega L - \omega C)^2]^{-\frac{1}{2}}.$$

The susceptance of an inductance L is $(1/\omega L)$, of a capacitance C it is $-\omega C$. Susceptances in parallel add up.

EXAMPLE. An alternating current network consists of a coil of inductance 0·02 H. and resistance 12 Ω. in series with a non-inductive resistance of 8 ohms, across which resistance is shunted a condenser of capacitance 50 μF. A potential difference of 100 V., frequency 200 cyc., is established across the network.

Draw a vector diagram to scale showing the current and voltage in each part of the network. (*Lond. Univ.*, 1933.)

Fig. 369 represents the network. The reactance of the condenser is

$$-1/\omega C = -1/[2\pi \times 200 \times 50 \times 10^{-6}] = -15{\cdot}9\ \Omega.$$

Let i be the current through the 8 Ω. resistance. The voltage across it is $8i$, so that the current through the condenser is

$$8i/15{\cdot}9 = 0{\cdot}503i,$$

and leads this current by 90°. The total current I is the vector sum of these, as shown in Fig. 370, is of magnitude $1{\cdot}12i$, and leads the voltage across the 8 Ω. resistance by an angle of 26° 42′.

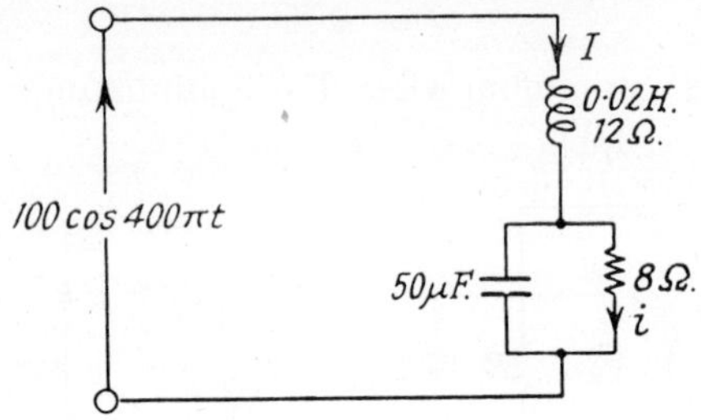

Fig. 369. Problem in Alternating Currents

The coil has reactance $\omega L = 2\pi \times 200 \times 0{\cdot}02 = 25{\cdot}1$ ohms. The voltage across the coil is $12I$ along I and $25{\cdot}1I$ at right angles. This compounded with the $8i$ gives the applied e.m.f. of 100 volts. The voltage across the coil is found to be 98 volts, and that across

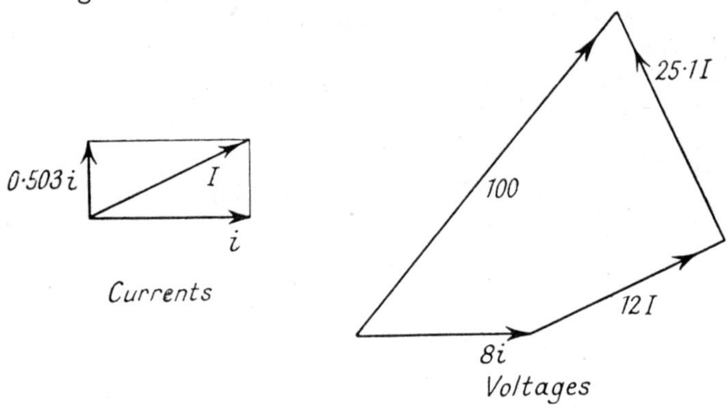

Fig. 370. Vector Diagrams for this Problem

the condenser and resistance is 25 volts. Fig. 370 shows the vector diagrams for the currents and voltages.

Complex Numbers. The graphical method is not very accurate, and becomes very involved for complicated networks. An algebraic method is available for these more advanced problems, and uses *complex numbers*.

A positive or negative number is called a *real* number. A number of the form $y\sqrt{(-1)}$ is called a *pure, imaginary number*; it is customary to replace $\sqrt{(-1)}$ by j. It is clear that

$$j^2 = -1, j^3 = j^2 \times j = -j, j^4 = (j^2)^2 = (-1)^2 = 1,$$

and $$1/j = j/j^2 = j/-1 = -j.$$

A number of the form $x + jy$, where x and y are real, is called a *complex number*. The laws of addition, subtraction, multiplication, and division for complex numbers are exactly the same as those for real numbers, bearing in mind the relations just given above. Thus

$$(a + jb) + (c + jd) = a + c + j(b + d),$$
$$(a + jb) - (c + jd) = a - c + j(b - d),$$
$$(a + jb)(c + jd) = ac + jad + jbc + j^2bd = (ac - bd) + j(ad + bc).$$

$$\frac{a + jb}{c + jd} = \frac{(a + jb)(c - jd)}{(c + jd)(c - jd)} = \frac{ac + bd + j(bc - ad)}{c^2 - j^2d^2} = \frac{ac + bd}{c^2 + d^2} + j\frac{bc - ad}{c^2 + d^2}.$$

This last process is called *rationalization.*

Complex numbers with equal real parts but equal imaginary parts of opposite sign are said to be *conjugate numbers.* Thus $(a + jb)$ and $(a - jb)$ are conjugate numbers; a and b may be positive or negative. The sum and product of conjugate numbers are both real. For

$$(a + jb) + (a - jb) = 2a$$

and

$$(a + jb)(a - jb) = a^2 - j^2b^2 = a^2 + b^2.$$

An important theorem on complex numbers states that if two complex numbers are equal, their real and imaginary parts must be separately equal. For if

$$a + jb = c + jd,$$
$$a - c = j(d - b),$$
$$(a - c)^2 = + j^2(d - b)^2 = - (d - b)^2,$$

so that $(a - c)^2 + (d - b)^2 = 0$.

But as a, b, c and d are real, $(a - c)^2$ and $(d - b)^2$ must be positive. The sum of two positive numbers can be zero only when each is zero, so that

$$a - c = 0 \text{ and } d - b = 0,$$

i.e.

$$a = c \text{ and } b = d.$$

Complex numbers are helpfully represented on the Argand diagram of Fig. 371. The complex number $x + jy$ is represented by the line OP, where $ON = x$ and $NP = y$. Thus a real number is represented by a step along the axis $X'OX$, and an imaginary number by a step along the axis $Y'OY$. x is taken positive to the right and negative to the left, y positive upwards and negative downwards. Thus x and y are both positive as shown, x' is negative and y' positive. The form $x + jy$ is the Cartesian, as x and y are the Cartesian or rectangular co-ordinates of the point P.

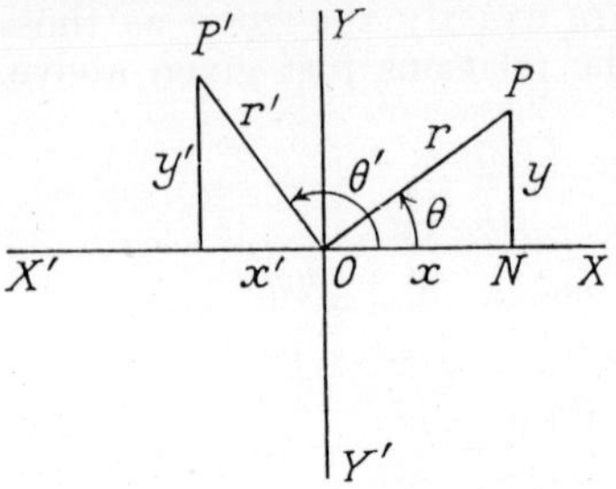

FIG. 371. ARGAND DIAGRAM

The polar co-ordinates of the point P are r and θ, where $r = OP$ and $\theta = \angle XOP$. The r is taken as positive always, and θ is positive in the anti-clockwise direction and negative in the clockwise. The number $(x + jy)$ can be expressed in terms of r and θ, and is then said to be in the *polar form*, by the following method.

$$x/r = \cos\theta,\ y/r = \sin\theta,$$

so that

$$x + jy = r\cos\theta + jr\sin\theta$$
$$= r(\cos\theta + j\sin\theta).$$

It is known that

$$\sin\theta = \theta - \frac{\theta^3}{3!} + \frac{\theta^5}{5!} - \ldots$$

and

$$\cos\theta = 1 - \frac{\theta^2}{2!} + \frac{\theta^4}{4!} - \ldots,$$

so that

$$\cos\theta + j\sin\theta = \left(1 - \frac{\theta^2}{2!} + \frac{\theta^4}{4!} - \ldots\right) + j\left(\theta - \frac{\theta^3}{3!} + \frac{\theta^5}{5!} - \ldots\right)$$
$$= 1 + j\theta - \frac{\theta^2}{2!} - j\frac{\theta^3}{3!} + \frac{\theta^4}{4!} - \ldots$$
$$= 1 + j\theta + \frac{(j\theta)^2}{2!} + \frac{(j\theta)^3}{3!} + \frac{(j\theta)^4}{4} + \ldots$$
$$= \varepsilon^{j\theta}.$$

We may therefore write

$$x + jy = r\varepsilon^{j\theta}.$$

Since $x = r\cos\theta$ and $y = r\sin\theta$,

$$x^2 + y^2 = r^2,\ \text{or}\ r = \sqrt{(x^2 + y^2)}$$

and

$$y/x = \tan\theta,\ \text{or}\ \theta = \tan^{-1} y/x.$$

Therefore

$$x + jy = \sqrt{(x^2 + y^2)}\varepsilon^{j\tan^{-1}(y/x)}.$$

For example,

$$0{\cdot}1 + 0{\cdot}3j = \sqrt{(0{\cdot}1^2 + 0{\cdot}3^2)}\varepsilon^{j\tan^{-1}(0\cdot3/0\cdot1)}$$
$$= 0{\cdot}316\varepsilon^{j1\cdot25},$$

for tan (1·25 radians) = 3.

Similarly,

$$6{\cdot}2 - 1{\cdot}7j = \sqrt{(6{\cdot}2^2 + 1{\cdot}7^2)}\varepsilon^{j\tan^{-1}(-1\cdot7/6\cdot2)}$$
$$= 6{\cdot}43\varepsilon^{-j0\cdot27},$$

and $$-6{\cdot}2 + 1{\cdot}7j = 6{\cdot}43\varepsilon^{j2\cdot87},$$

since the angle here is $\pi - 0{\cdot}27 = 2{\cdot}87$ radians.

There is another and very convenient way of writing the polar, viz.

$$r\varepsilon^{j\theta} \equiv r\,\underline{|\theta},$$

or $$r\varepsilon^{-j\theta} \equiv r\,\overline{|\theta}.$$

The symbols $\varepsilon^{j\theta}$ and $\underline{|\theta}$ are identical in meaning.

The polar form is the more useful when multiplication or division is to be performed, for

$$r_1\varepsilon^{j\theta_1} \times r_2\varepsilon^{j\theta_2} = r_1r_2\varepsilon^{j\theta_1 + j\theta_2}$$
$$= r_1r_2\varepsilon^{j(\theta_1 + \theta_2)},$$

and similarly

$$r_1\varepsilon^{j\theta_1} \div r_2\varepsilon^{j\theta_2} = (r_1/r_2)\varepsilon^{j(\theta_1 - \theta_2)}.$$

In the angle notation the results are

$$r_1\,\underline{|\theta_1} \times r_2\,\underline{|\theta_2} = r_1r_2\,\underline{|\theta_1 + \theta_2}$$

and $$r_1\,\underline{|\theta_1} \div r_2\,\underline{|\theta_2} = \frac{r_1}{r_2}\,\underline{|\theta_1 - \theta_2}.$$

A very common operation is to multiply or divide by j. The polar method runs

$$r\,\underline{|\theta} \times j = r\,\underline{|\theta} \times 1\,\underline{|(\pi/2)} = r\,\underline{|\theta + (\pi/2)}$$

and $$r\,\underline{|\theta} \div j = r\,\underline{|\theta - (\pi/2)}.$$

Thus multiplication by j turns a vector a right angle in the positive direction leaving the magnitude unaltered, and division turns the vector a right angle in the negative direction. These facts are also obvious from the Cartesian form, for

$$(x + jy) \times j = jx + j^2y = -y + jx,$$

and it is clear from Fig. 372 that $(x + jy)$ and $(-y + jx)$ are at right angles in the sense stated.

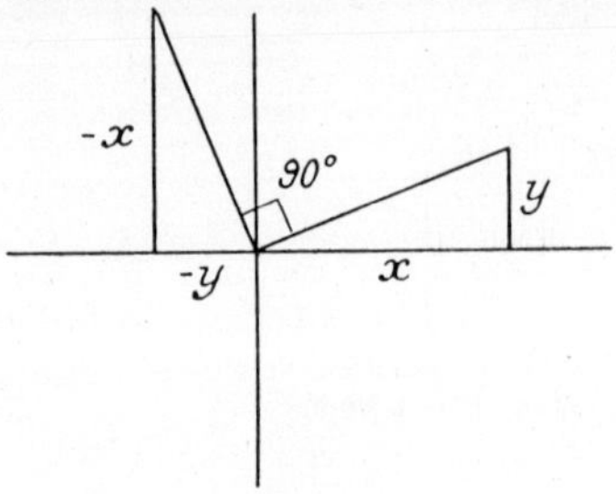

FIG. 372. PERPENDICULAR VECTORS

An important theorem states that if

$$r\,|\underline{\theta} = r'\,|\underline{\theta'},$$

then $\qquad r = r'$ and $\theta = \theta'$.

For if $\quad r\,|\underline{\theta} = r'\,|\underline{\theta'},$

then $$1 = \frac{r\,|\underline{\theta}}{r'\,|\underline{\theta'}} = \frac{r}{r'}\,|\underline{\theta - \theta'},$$

and this can be true only if $r/r' = 1$ and $\theta - \theta' = 0$.

Algebraic Method for Alternating Currents. We have already seen in Fig. 360 that an alternating quantity $E \cos(\omega t + \theta)$ can be represented by the horizontal component of a rotating vector of length E, angular velocity ω, and initial angular position θ. Comparing this figure with Fig. 371, it is seen that the vector OP is represented by the complex number

$$OP = E\varepsilon^{j(\omega t + \theta)} \text{ or } E\,|\underline{\omega t + \theta}.$$

It is obvious that $E \cos(\omega t + \theta)$ is the horizontal component, and hence the real part, of OP; for

$$OP = E\varepsilon^{j(\omega t + \theta)} = E\,[\cos(\omega t + \theta) + j \sin(\omega t + \theta)],$$

so that the real part of OP is $E \cos(\omega t + \theta)$.

The algebraic method of dealing with alternating current circuits is as follows. Suppose that an e.m.f. $E \cos(\omega t + \theta)$ is placed across a coil of inductance L. The current I is given by

$$L\frac{dI}{dt} = E \cos(\omega t + \theta).$$

Let us replace $E \cos(\omega t + \theta)$ by $E\varepsilon^{j(\omega t + \theta)} = E'$ say, which has the former for its real part. In doing so we will have added an imaginary part to the current, which is now I' say. Then

$$L\frac{dI'}{dt} = E\varepsilon^{j(\omega t + \theta)},$$

giving $$I' = \int \frac{E}{L}\varepsilon^{j(\omega t + \theta)}\,dt$$

$$= \frac{E}{j\omega L}\,\varepsilon^{j(\omega t + \theta)} = \frac{E'}{j\omega L},$$

since the integral of $\varepsilon^{j(\omega t + \theta)}$ is $(1/j\omega)\,\varepsilon^{j(\omega t + \theta)}$. By what has been shown on page 455, the complex current I' is represented by a vector of length $(E/\omega L)$ and angle $(\omega t + \theta - \pi/2)$, the $-(\pi/2)$ being due to the division by j. The real current I is the real part of I',

i.e. $$\text{real part of}\left[\frac{E}{j\omega L}\varepsilon^{j(\omega t + \theta)}\right]$$

$$= \quad ,, \quad ,, \quad \left[\frac{E}{\omega L}\varepsilon^{j\left(\omega t + \theta - \frac{\pi}{2}\right)}\right]$$

$$= \quad ,, \quad ,, \quad \frac{E}{\omega L}\left[\cos\left(\omega t + \theta - \frac{\pi}{2}\right) + j\sin\left(\omega t + \theta - \frac{\pi}{2}\right)\right]$$

$$= \frac{E}{\omega L}\cos\left(\omega t + \theta - \frac{\pi}{2}\right) \text{ or } \frac{E}{\omega L}\sin(\omega t + \theta).$$

The method thus agrees with the direct method of page 446. The ratio of the complex e.m.f. E' to the complex current I' is

$$\frac{E'}{I'} = \frac{E\varepsilon^{j(\omega t + \theta)}}{\dfrac{E}{j\omega L}\varepsilon^{j(\omega t + \theta)}} = j\omega L,$$

and is called the *complex* or vector *impedance* of the inductance. It is usual to call it just the impedance, but some slight confusion may arise when, as is often the case, the word impedance is used to denote only the magnitude of the complex impedance. The impedance of an inductance L is therefore $j\omega L$, has magnitude ωL and angle $\pi/2$; when an e.m.f. is divided by $j\omega L$, the resulting current has magnitude $(E/\omega L)$ and lags the e.m.f. by the angle $\pi/2$. It should be noted that the complex e.m.f. and current have the factor $\varepsilon^{j(\omega t + \theta)}$, so that they are represented by vectors which have an angular velocity ω. The vector $j\omega L$, however, is fixed, and is therefore of a different character. In most problems only currents and voltages of a single frequency are dealt with, and it is customary to ignore the rotation of the vectors. It then appears that all the vectors are fixed. This leads to no error as long as voltages and currents are concerned, but it will be shown that error is met when power is considered.

The impedance of a resistance R is clearly R, or $R\varepsilon^{j0}$, and is a vector of length R along the real axis.

The impedance of a condenser C is found in the following way. Let the e.m.f. $E\cos(\omega t + \theta)$ be put across a condenser of capacitance C, the current be I, and the charge on the plates be Q and $-Q$. We have

$$Q = CE\cos(\omega t + \theta)$$

and $$I = \frac{dQ}{dt} = C\frac{d}{dt}[E\cos(\omega t + \theta)].$$

Replacing the e.m.f. by the complex $E' = E\varepsilon^{j(\omega t + \theta)}$ the current becomes I' where

$$I' = C\frac{d}{dt}[E\varepsilon^{j(\omega t + \theta)}]$$
$$= j\omega CE\varepsilon^{j(\omega t + \theta)}$$
$$= j\omega CE'.$$

The impedance of the condenser is thus

$$E'/I' = 1/j\omega C,$$

has magnitude $(1/\omega C)$ and angle $-(\pi/2)$. It is clear that I' has magnitude ωCE and angle $(\omega t + \theta + \pi/2)$, i.e. it leads the voltage vector by $\pi/2$. The real current I is the real part of $j\omega CE\varepsilon^{j(\omega t + \theta)}$

$$= \text{the real part of } j\omega CE\,[\cos(\omega t + \theta) + j\sin(\omega t + \theta)]$$
$$= \quad \text{"} \quad \text{"} \quad \omega CE\,[j\cos(\omega t + \theta) - \sin(\omega t + \theta)]$$
$$= -\omega CE\sin(\omega t + \theta) = \omega CE\cos(\omega t + \theta + \pi/2),$$

agreeing with the value obtained by the direct method of page 446.

Algebraic Method Applied to Simple Series Circuit. The problem of Fig. 366 is solved in the following way. The impedance of the coil, resistance and capacitance in series is

$$Z = j\omega L + R + (1/j\omega C),$$

so that the complex current is given by

$$I' = \frac{E'}{Z} = \frac{E\varepsilon^{j(\omega t + \theta)}}{j\omega L + R + 1/j\omega C}.$$

We have
$$Z = j\omega L + R + (1/j\omega C)$$
$$= R + j(\omega L - 1/\omega C)$$
$$= \sqrt{[R^2 + (\omega L - 1/\omega C)^2]}\,\underline{|\phi}$$

where
$$\tan\phi = \frac{\omega L - 1/\omega C}{R}.$$

I' has therefore magnitude $E/\sqrt{[R^2 + (\omega L - 1/\omega C)^2]}$ and angle $(\omega t + \theta - \phi)$, so that the real current I is

$$I = \frac{E}{\sqrt{[R^2 + (\omega L - 1/\omega C)^2]}}\cos(\omega t + \theta - \phi).$$

Exactly the same result is obtained for I by finding the real part of E'/Z.

Resonance occurs when

$$j\omega L + 1/j\omega C = 0, \text{ i.e. } \omega^2 LC = 1,$$

and then
$$I' = E'/R,$$
so that
$$I = (E/R)\cos(\omega t + \theta).$$

The voltage across the coil at resonance is (in complex form)

$$j\omega LI' = j\omega L(E'/R),$$

so that it has magnitude $E\omega L/R$, and leads the e.m.f. by a right angle. The voltage across the condenser is

$$I' \times (1/j\omega C) = E'/j\omega CR = -j\omega L(E'/R)$$

has magnitude $E\omega L/R$ and lags the e.m.f. by a right angle.

ALGEBRAIC METHOD APPLIED TO SIMPLE PARALLEL CIRCUIT. In such a case we deal with *admittances* in place of impedances, an admittance being the reciprocal of an impedance. Thus the admittance of a coil of inductance L is $1/j\omega L$, of a resistance R it is $1/R$, and of a capacitance C it is $j\omega C$. Admittances in parallel add; admittances Y_1 and Y_2 in series give a single admittance

$$\frac{1}{1/Y_1 + 1/Y_2}.$$

The algebraic method then runs as follows.

$$Y = 1/j\omega L + 1/R + j\omega C,$$

and

$$I' = YE'.$$

$$Y = 1/R + j(\omega C - 1/\omega L)$$
$$= \sqrt{[1/R^2 + (\omega C - 1/\omega L)^2]}\ \underline{|\psi},$$

where

$$\tan\psi = (\omega C - 1/\omega L) \div 1/R$$
$$= R(\omega C - 1/\omega L)$$

Therefore

$$I' = \sqrt{[1/R^2 + (\omega C - 1/\omega L)^2]}E\varepsilon^{j(\omega t + \theta + \psi)},$$

so that the real current has amplitude

$$\sqrt{[1/R^2 + (\omega C - 1/\omega L)^2]}E$$

and *leads* the e.m.f. by the angle ψ. It should be noted that ψ is equal to $-\phi$ of the method on page 451, so that the result is the same.

It is seen that in the work the factor $\varepsilon^{j(\omega t + \theta)}$ has no importance, and for this reason it is frequently omitted. Its existence must not be ignored, however, but kept in mind.

EXAMPLE. Solve the problem of Fig. 369 by the algebraic method.

Here $\omega L = 25{\cdot}1$, and $1/\omega C = 15{\cdot}9$.

The total impedance is

$$j25{\cdot}1 + 12 + \frac{1}{1/8 + j/15{\cdot}9}$$
$$= j25{\cdot}1 + 12 + \frac{1/8 - j/15{\cdot}9}{(1/8)^2 + (1/15{\cdot}9)^2}$$
$$= j25{\cdot}1 + 12 + 6{\cdot}38 - j3{\cdot}21$$
$$= 18{\cdot}38 + j21{\cdot}9.$$

The current is therefore

$$\frac{100}{18{\cdot}38 + j21{\cdot}9} = \frac{100}{28{\cdot}6\,\underline{|49^\circ 57'}} = 3{\cdot}47\,\overline{|49^\circ 57'}.$$

The voltage across the resistance and condenser is

$$\begin{aligned} & 3{\cdot}47\,\overline{|49^\circ 57'} \times (6{\cdot}38 - j3{\cdot}21) \\ &= 3{\cdot}47\,\overline{|49^\circ 57'} \times 7{\cdot}14\,\overline{|26^\circ 42'} \\ &= 24{\cdot}8\,\overline{|76^\circ 39'}. \end{aligned}$$

The graphical method gave 25 V. The remaining currents and voltages can be found in a similar manner. Thus the current through the 8 Ω. resistance is

$$3{\cdot}47\,\overline{|49^\circ 57'} \times \frac{8/j\omega C}{8 + 1/j\omega C} = 3{\cdot}47\,\overline{|49^\circ 57} \times \frac{1}{1 + j(8/15{\cdot}9)}, \text{ etc.}$$

The ease of the method is apparent.

Power in Alternating Current Circuits. We have seen that as far as e.m.f., current and impedance are concerned, we may replace these quantities by complex numbers which are multiples of $\varepsilon^{j\omega t}$ and whose real parts are the actual values. Thus if an e.m.f. $E \cos (\omega t + \theta)$ is placed across a resistance R, we replace the e.m.f. by E', where

$$E' = E\,\varepsilon^{j(\omega t + \theta)},$$

the real part of which is the e.m.f. The complex current is

$$I' = E'/R = (E/R)\,\varepsilon^{j(\omega t + \theta)}.$$

It might be expected that the power will be obtained by forming the product $E'I'$ and taking the real part of this. It is easily seen that this is not true, for

$$\begin{aligned} & \text{the real part of } E'I' \\ &= \quad ,, \quad ,, \quad \left[\frac{E^2}{R}\varepsilon^{2j(\omega t + \theta)}\right] \\ &= \frac{E^2}{R}\cos 2\,(\omega t + \theta), \end{aligned}$$

which is a purely alternating wave and has a mean value of zero. The failure is due to the fact that the real part of $\varepsilon^{j(\omega t + \theta)}$ multiplied by the real part of $\varepsilon^{j(\omega t + \theta)}$ is not equal to the real part of $[\varepsilon^{j(\omega t + \theta)} \times \varepsilon^{j(\omega t + \theta)}]$.

It is not the algebraic method alone that fails with power, but the vectorial (or graphical method) also. For the current vector multiplied by the voltage vector gives a resultant vector which moves with twice the angular velocity of either. The quantity represented by the resultant vector is a pure sine wave of double frequency, and has zero mean value. For the case given above, it is in fact $(E^2/R)\cos 2(\omega t + \theta)$.

The vector and algebraic methods can be extended to allow power to be represented. Instead of representing $E\cos(\omega t + \theta)$ by a single vector with angular velocity ω in the positive direction, we represent it by two vectors, as shown in Fig. 373. These vectors have magnitude $\frac{1}{2}E$, start from angular positions $+\theta$ and $-\theta$, and have angular velocities $+\omega$ and $-\omega$ respectively; the sum of these vectors is actually a stationary vector of magnitude $E\cos(\omega t + \theta)$ along the real axis. The algebraic equivalent is to express $E\cos(\omega t + \theta)$ in the form

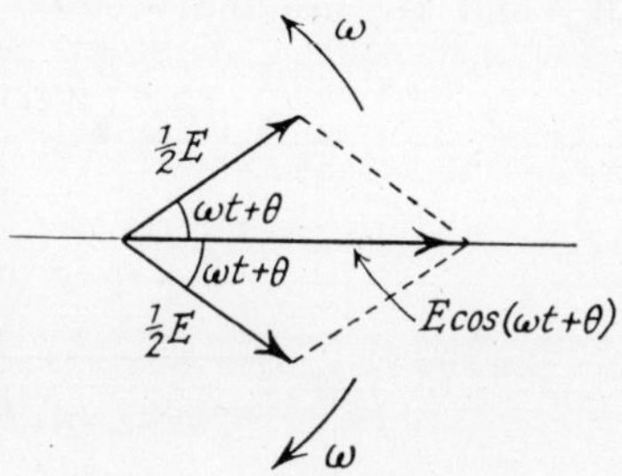

FIG. 373. COMPLETE VECTOR REPRESENTATION OF SINUSOIDAL WAVE

$$E\cos(\omega t + \theta) = \tfrac{1}{2}E\,[\varepsilon^{j(\omega t + \theta)} + \varepsilon^{-j(\omega t + \theta)}].$$

This is an identity.

We deal with the vectors rotating in the positive direction in the way described above. The same treatment is given to the vectors rotating in the negative direction, but all phase angles are given opposite sign. In the algebraic method, the impedance of an inductance, resistance, or capacitance is $j\omega L$, R or $1/j\omega L$ to the e.m.f. $\frac{1}{2}E\varepsilon^{j(\omega t + \theta)}$, but $-j\omega L$, R, or $(1/-j\omega C)$ to the e.m.f. $\frac{1}{2}E\varepsilon^{-j(\omega t + \theta)}$. As an example, consider the application of the e.m.f. $E\cos(\omega t + \theta)$ to a capacitance C. The current is

$$\begin{aligned}
I &= \tfrac{1}{2}E\,[\varepsilon^{j(\omega t + \theta)} + \varepsilon^{-j(\omega t + \theta)}] \div Z \\
&= \tfrac{1}{2}E\left[\frac{\varepsilon^{j(\omega t + \theta)}}{1/j\omega C} + \frac{\varepsilon^{-j(\omega t + \theta)}}{(1/-j\omega C)}\right] \\
&= \tfrac{1}{2}E\omega C\,[j\varepsilon^{j(\omega t + \theta)} - j\varepsilon^{-j(\omega t + \theta)}] \\
&= \tfrac{1}{2}E\omega C\,[\varepsilon^{j(\omega t + \theta + \pi/2)} + \varepsilon^{-j(\omega t + \theta + \pi/2)}] \\
&= E\omega C\cos(\omega t + \theta + \pi/2) = -E\omega C\sin(\omega t + \theta),
\end{aligned}$$

since $-j\varepsilon^{-j(\omega t + \theta)} = \dfrac{1}{j}\varepsilon^{-j(\omega t + \theta)} = \varepsilon^{-j(\omega t + \theta) - j(\pi/2)}$.

Fig. 374 shows how power is adequately represented in this method. The current is $I\cos(\omega t + \theta_1)$ and is represented by two vectors, i_+ and i_-; the voltage is $E\cos(\omega t + \theta_2)$ and is represented by the vectors v_+ and v_-. The power is

$$\begin{aligned}
P = EI &= (v_+ + v_-)(i_+ + i_-) \\
&= [v_+ i_- + v_- i_+] + v_+ i_+ + v_- i_- \\
&= P_0 + P_+ + P_-, \text{ say.}
\end{aligned}$$

v_+i_- is a stationary vector of magnitude $\frac{1}{4}EI$ and angle $\theta_2 - \theta_1 = \phi$, whilst v_-i_+ has the same magnitude and angle $-(\theta_2 - \theta_1) = \phi$. These add to give a stationary vector representing a constant power of $\frac{1}{2}EI \cos(\theta_2 - \theta_1) = \frac{1}{2}EI \cos\phi$, agreeing with equation (13). W_+ and W_- are the vectors, angular velocities $+2\omega$ and -2ω,

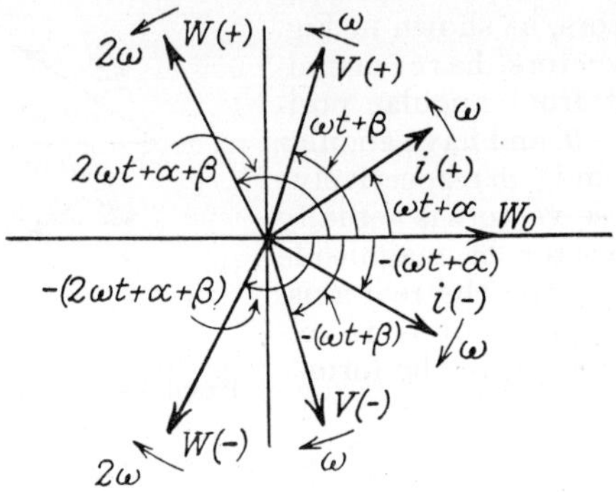

FIG. 374. POWER IN VECTOR DIAGRAM

whose sum is the real fluctuating power $\frac{1}{2}EI \cos(2\omega t + \theta_1 + \theta_2)$, agreeing with equation (14).

VECTOR POWER. If the vector representing the voltage, viz. $E\varepsilon^{j(\omega t + \theta_2)}$, is multiplied by the conjugate of that representing current, viz. $I\varepsilon^{-j(\omega t + \theta_1)}$, we get

$$\begin{aligned} & E\varepsilon^{j(\omega t + \theta_2)} \times I\varepsilon^{-j(\omega t + \theta_1)} \\ &= EI\varepsilon^{j(\theta_2 - \theta_1)} = EI\varepsilon^{j\phi} \\ &= EI\cos\phi + jEI\sin\phi. \end{aligned}$$

The real part of this represents the mean power, whilst the imaginary part is the amplitude of the reactive power.

If the current lags the voltage, θ_1 is less than θ_2, so that ϕ is positive and the reactive power is positive; if the current leads the reactive power is negative.

If the current vector is multiplied by the conjugate of the voltage vector, the only change is that the reactive power has a change of sign.

If a vector is given in the polar form, the conjugate is obtained by changing the sign of the angle. Thus the conjugate of $\varepsilon^{j(\pi/2)}$ is $\varepsilon^{-j(\pi/2)}$, and the conjugate of $6\varepsilon^{-j(\pi/8)}$ is $6\varepsilon^{+j(\pi/8)}$. If the vector is given in the Cartesian form, the conjugate is given by changing the sign of the imaginary part. Thus the conjugate of $3 - j4$ is $3 + j4$, and the conjugate of $-6 + j2$ is $-6 - j2$.

If the e.m.f. is put across an impedance Z, the current is $(E/Z)\varepsilon^{j(\omega t + \theta_2)}$. If the magnitude and angle of Z are $|Z|$ and θ, so that $Z = |Z| \underline{|\theta}$, the current is $(E/|Z|) \underline{|\omega t + \theta_2 - \theta}$. The vector power is then

$$E \underline{|\omega t + \theta_2} \times (E/|Z|) \overline{|\omega t + \theta_2 - \theta}$$
$$= \frac{E^2}{|Z|} \underline{|\theta} = \frac{E^2}{|Z|} \cos \theta + j \frac{E^2}{|Z|} \sin \theta.$$

Example. **Find the real and reactive power fed by a line at 66 kV. into a load impedance of 400–j300.**

Let e.m.f. $= 66\,000 \underline{|\omega t}$.

The current is $$\frac{66\,000 \underline{|\omega t}}{400 - j300} = \frac{66\,000 \underline{|\omega t}}{500 \overline{|36°53'}}$$
$$= 132 \underline{|\omega t + 36° 53'}.$$

The vector power is thus

$$66\,000 \underline{|\omega t} \times 132 \overline{|\omega t + 36° 53'}$$
$$= 8\,712\,000 \overline{|36° 53'} = 6\,970\,000 - j5\,230\,000,$$

so that the real power is 6 970 kW. and the reactive power is 5 230 kW. leading.

Polyphase Currents. For a given sized frame a polyphase generator or motor has a larger output than a single-phase. Polyphase motors have a uniform torque and are self-acting, whilst single-phase motors have a pulsating torque and are not self-starting unless they are of the commutator type. Polyphase transmission requires less copper than single-phase.

In single-phase currents there is one e.m.f. $E \cos (\omega t + \theta)$, although the transmission may be over two or three wires, as described on page 40.

In two-phase working the generator has two windings spaced 90 electrical degrees apart, so that the e.m.f.'s are in quadrature and may be represented by $E \cos (\omega t + \theta)$ and $E \cos (\omega t + \theta - \pi/2)$. In vector diagrams the former is represented by a line of length E (or $E/\sqrt{2}$) making an angle θ with the base line, and the latter by a line of equal length making an angle $(\theta - \pi/2)$ with the base line.

In three-phase working the generator has three windings spaced 120 electrical degrees apart, so that the e.m.f.'s are equal but 120° apart in phase. The e.m.f.'s are $E_1 = E \cos (\omega t + \theta)$, $E_2 = E \cos (\omega t + \theta - 2\pi/3)$ and $E_3 = E \cos (\omega t + \theta - 4\pi/3)$, and can be represented in the vector diagram by three equal lines, making angles of 120° with one another. The e.m.f.'s add up to zero, for

$$E_1 + E_2 + E_3$$
$$= E \cos (\omega t + \theta) + E \cos (\omega t + \theta - 2\pi/3)$$
$$+ E \cos (\omega t + \theta - 4\pi/3)$$

$$= E \cos (\omega t + \theta) [1 + \cos (2\pi/3) + \cos (4\pi/3)]$$
$$+ E \sin (\omega t + \theta) [\sin (2\pi/3) + \sin (4\pi/3)]$$
$$= E \cos (\omega t \times \theta) [1 - \tfrac{1}{2} - \tfrac{1}{2}]$$
$$+ E \sin (\omega t + \theta) [\tfrac{1}{2}\sqrt{3} - \tfrac{1}{2}\sqrt{3}]$$
$$= 0.$$

This fact is obvious from the vector diagram, since the resultant of the three separate vectors is zero.

There are two main methods of transmitting three-phase currents, the star and delta, shown in Fig. 375. The star system can be con-

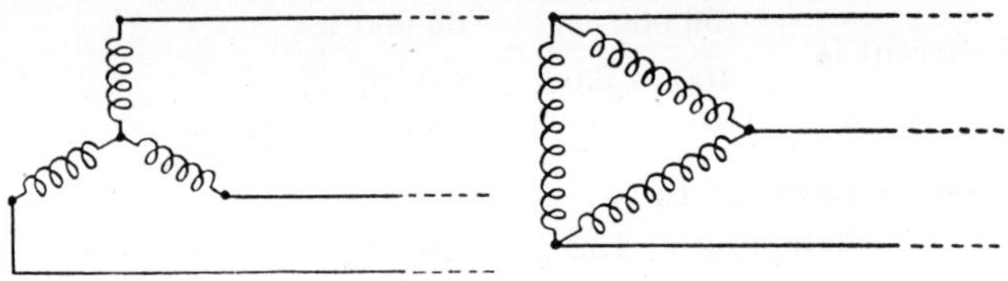

FIG. 375. STAR AND DELTA THREE-PHASE CIRCUITS

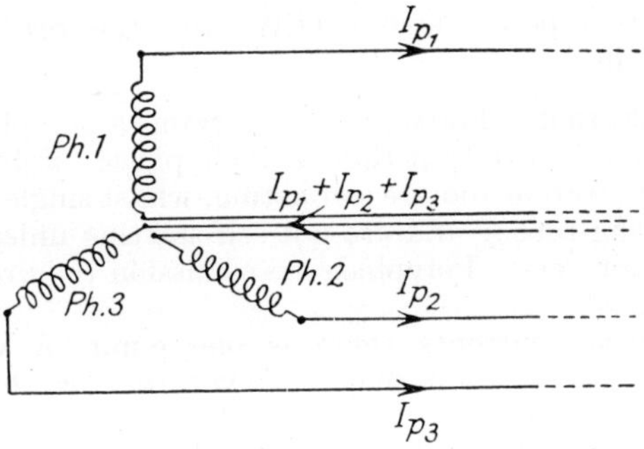

FIG. 376. ARTICULATED FORM OF STAR CIRCUIT

sidered as a condensed form of the system shown in Fig. 376. If the loads were equal so that the terminating impedances had equal magnitudes and phase angle, Z and ϕ say, the currents in Fig. 376 would be

$$I_{p1} = (E/Z) \cos (\omega t + \theta - \phi),$$
$$I_{p2} = (E/Z) \cos (\omega t + \theta - 2\pi/3 - \phi),$$

and
$$I_{p3} = (E/Z) \cos (\omega t + \theta - 4\pi/3 - \phi).$$

These currents add up to zero in exactly the same way as E_1, E_2, and E_3. If the three phases had a common return, the return would carry zero current and could be dispensed with, the resulting system being the star.

The currents in the wires, called the *line currents*, are clearly equal to the phase currents. The voltage between wires is, however, considerably greater than the phase voltage, which is generally the voltage from line to earth. Thus the voltage between wires 2 and 3 is

$$\begin{aligned} E_{23} &= E_2 - E_3 \\ &= E \cos(\omega t + \theta - 2\pi/3) - E \cos(\omega t + \theta - 4\pi/3) \\ &= 2E \sin(-\pi/3) \sin(\omega t + \theta - \pi) \\ &= (\sqrt{3})E \sin(\omega t + \theta) \\ &= (\sqrt{3})E \cos(\omega t + \theta - \pi/2). \end{aligned}$$

Similarly

$$E_{31} = (\sqrt{3})\, E \cos(\omega t + \theta - \pi/2 - 2\pi/3)$$

and

$$E_{12} = (\sqrt{3})\, E \cos(\omega t + \theta - \pi/2 - 4\pi/3).$$

The line voltages are thus $\sqrt{3}$ times the phase voltages and lag behind them by 90°. Fig. 377 shows the phase and line voltages on the vector diagram.

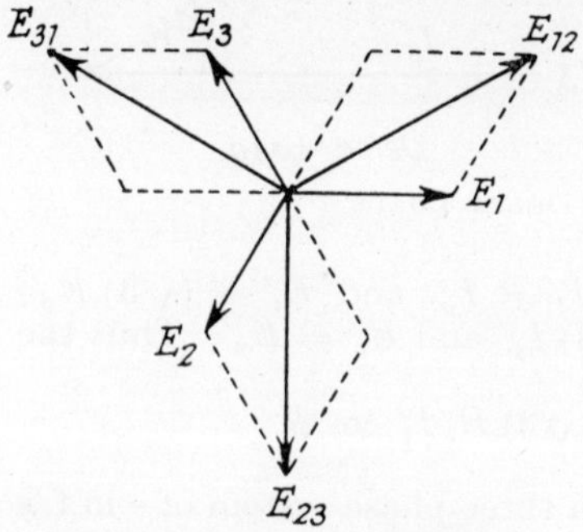

FIG. 377. VOLTAGE VECTORS IN THREE-PHASE CIRCUIT

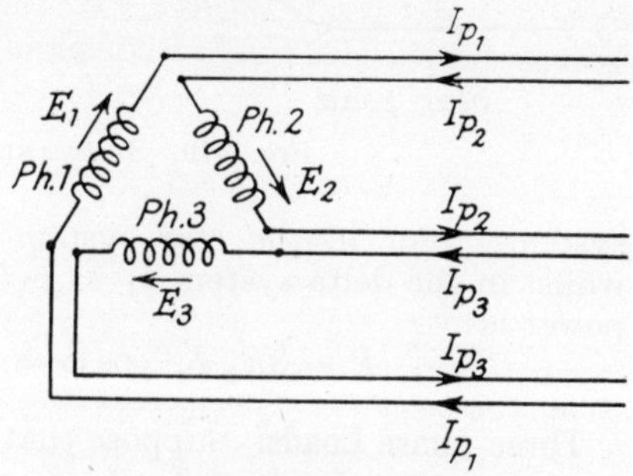

FIG. 378. ARTICULATED FORM OF DELTA CIRCUIT

The delta system can similarly be considered as a condensed form of the system shown in Fig. 378.

In the delta system the line voltages are clearly equal to the phase voltages. The line currents are, however, $\sqrt{3}$ times the phase currents. For when the loads are balanced the phase currents are equal in magnitude and spaced 120° apart in phase, so that

$$I_{p1} = I \cos(\omega t + \alpha),$$

$$I_{p2} = I \cos(\omega t + \alpha - 2\pi/3)$$

and

$$I_{p3} = I \cos(\omega t + \alpha - 4\pi/3).$$

The current in wire 1 is

$$\begin{aligned} I_1 &= I_{p2} - I_{p3} \\ &= I \cos(\omega t + \alpha - 2\pi/3) - I \cos(\omega t + \alpha - 4\pi/3) \\ &= \sqrt{3}\, I \cos(\omega t + \alpha - \pi/2), \end{aligned}$$

as above; thus line current 1 is $\sqrt{3}$ times the phase current and lags behind I_{p1} by 90°. Similarly for the other line currents. The vector diagram for the currents is similar to that for the voltages.

In both the star and delta arrangements the power supplied per phase is $E_p'I_p' \cos \phi$, where E_p' is the r.m.s. of the phase voltage, I_p' the r.m.s. of the phase current, and $\cos \phi$ is the power factor; the total power is $3E_p'I_p' \cos \phi$ assuming balanced loads. The power in terms of line currents and voltages is $(\sqrt{3})\, E_l'I_l' \cos \phi$ for both

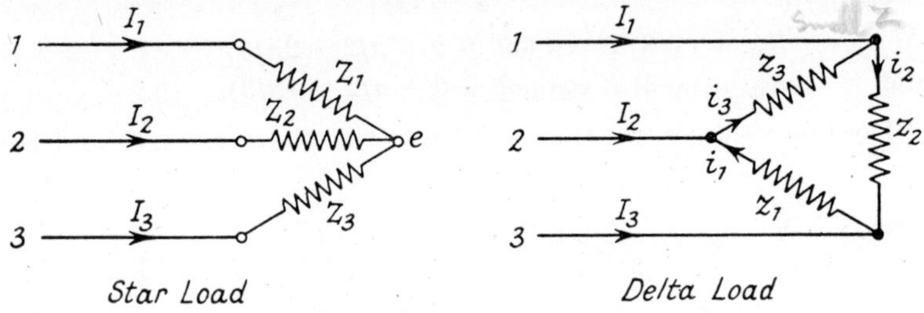

FIG. 379. STAR AND DELTA LOADS

systems. For in the star system $I_l' = I_p'$ and $E_l' = (\sqrt{3})\, E_p'$, whilst in the delta system $I_l' = (\sqrt{3})\, I_p'$ and $E_l' = E_p'$. Thus the power is

$$P = 3E_p'I_p' \cos \phi = (\sqrt{3})\, E_l'I_l' \cos \phi.$$

Three-phase Loads. Suppose that a three-phase system of e.m.f.'s is applied to a star load Z_1, Z_2, Z_3 or a delta load z_1, z_2, z_3 as shown in Fig. 379.

In the star case let the line currents be I_1, I_2, and I_3, whilst the potential of the neutral point is e. We have

$$I_1 = (E_1 - e)/Z_1,$$

$$I_2 = (E_2 - e)/Z_2,$$

and

$$I_3 = (E_3 - e)/Z_3.$$

As there are only three wires $I_1 + I_2 + I_3 = 0$, giving

$$e = \frac{E_1/Z_1 + E_2/Z_2 + E_3/Z_3}{1/Z_1 + 1/Z_2 + 1/Z_3}$$

$$= \frac{E_1Y_1 + E_2Y_2 + E_3Y_3}{Y_1 + Y_2 + Y_3},$$

so that

$$I_1 = Y_1 \frac{Y_2(E_1 - E_2) + Y_3(E_1 - E_3)}{Y_1 + Y_2 + Y_3}.$$

When E_1, E_2, and E_3 are balanced, it follows from Fig. 377 that

$$E_1 - E_2 = (\sqrt{3})E_1 \underline{|30^\circ}$$

and

$$E_1 - E_3 = (\sqrt{3})E_1 \overline{|30^\circ},$$

so that

$$I_1 = (\sqrt{3})E_1 Y_1 \frac{Y_2 \underline{|30^\circ} + Y_3 \overline{|30^\circ}}{Y_1 + Y_2 + Y_3}$$

There are similar expressions for I_2 and I_3. If the loads are balanced, so that $Y_1 = Y_2 = Y_3$,

$$e = \tfrac{1}{3}(E_1 + E_2 + E_3) = 0,$$

and

$$I_1 = (1/\sqrt{3})E_1 Y(\underline{|30^\circ} + \overline{|30^\circ}) = E_1 Y = E_1/Z,$$

since

$$\underline{|30^\circ} + \overline{|30^\circ} = \cos 30^\circ + j \sin 30^\circ + \cos 30^\circ - j \sin 30^\circ$$
$$= 2 \cos 30^\circ = \sqrt{3}.$$

Similarly $I_2 = E_2/Z$ and $I_3 = E_3/Z$. The current I_1 flows along Z_1 where it causes a drop of E_1, so that the power associated with it is $(E'^2/|Z|) \cos \phi$, E' being the r.m.s. voltage, $|Z|$ the magnitude of the impedance and $\cos \phi$ the power factor. The total power absorbed is three times this.

In the delta case let the currents in the load impedances be i_1, i_2, i_3 so that the line currents are $I_1 = i_2 - i_3$, $I_2 = i_3 - i_1$ and $I_3 = i_1 - i_2$.

We have

$$i_1 = (E_3 - E_2)/z_1,$$
$$i_2 = (E_1 - E_3)/z_2,$$

and

$$i_3 = (E_2 - E_1)/z_3,$$

so that

$$I_1 = \frac{E_1 - E_3}{z_2} - \frac{E_2 - E_1}{z_3}$$
$$= E_1\left(\frac{1}{z_2} + \frac{1}{z_3}\right) - \frac{E_2}{z_3} - \frac{E_3}{z_2},$$

with similar equations for I_2 and I_3. These currents are the same as in the star case if

$$\left.\begin{aligned} \frac{1}{z_1} = y_1 &= \frac{Y_2 Y_3}{Y_1 + Y_2 + Y_3}, \\ y_2 &= \frac{Y_3 Y_1}{Y_1 + Y_2 + Y_3}, \\ \text{and} \qquad y_3 &= \frac{Y_1 Y_2}{Y_1 + Y_2 + Y_3}. \end{aligned}\right\} \quad . \quad . \quad . \quad (16)$$

There are the conditions that the star and delta loads be equivalent; this theorem is known as the *star-mesh transformation*. It is possible to express Y_1, Y_2, Y_3 in terms of y_1, y_2, y_3, or rather Z_1, Z_2, Z_3 in terms of z_1, z_2, z_3 in the following way.

$$y_1/y_2 = Y_2/Y_1, \text{ so that } Y_1/z_1 = Y_2/z_2 = Y_3/z_3.$$

Adding numerators and denominators and using (16) we get

$$\frac{Y_1}{z_1} = \frac{Y_2}{z_2} = \frac{Y_3}{z_3} = \frac{Y_1 + Y_2 + Y_3}{z_1 + z_2 + z_3} = \frac{Y_1 Y_2 z_3}{z_1 + z_2 + z_3}.$$

Taking the second and last equalities we get

$$\left.\begin{aligned} \frac{1}{Y_1} = Z_1 &= \frac{z_2 z_3}{z_1 + z_2 + z_3}. \\ \text{Similarly} \qquad Z_2 &= \frac{z_3 z_1}{z_1 + z_2 + z_3} \\ \text{and} \qquad Z_3 &= \frac{z_1 z_2}{z_1 + z_2 + z_3}. \end{aligned}\right\} \quad . \quad . \quad . \quad (17)$$

If the loads are equal $Z = \frac{1}{3}z$.

It is often a great help to transform from star to delta, or conversely.

The following example shows how the problem of a three-phase load is affected by mutual inductance between the loads.

Example. The axes of three identical coils are in the same plane, are 120° apart and meet in a point. The inductance of each coil separately is 0·5 H. and its resistance is 200 Ω., and the mutual inductance between each pair of coils is 0·2 H.

What current will flow in each coil if the three are connected in star to a three-phase 50-cycle supply of 440 V.? (*Lond. Univ.*, 1933.)

Fig. 380 shows the arrangement. If the coils are wound in the same direction, the mutual inductances aid the self-inductances. It is quite clear that the potential of the star point is zero, since the system is symmetrical. Let Z be the self-impedances of the coils and Z_m the mutual impedance. Then

$$Z = 200 + j \,.\, 2\pi \,.\, 50 \,.\, 0{\cdot}5 = 200 + j157$$

and $$Z_m = j \,.\, 2\pi \,.\, 50 \,.\, 0{\cdot}2 = j62{\cdot}8.$$

The r.m.s. voltage between phases is 440, so that $E = 440/\sqrt{3} = 254$. The voltage between A and O is $E_1 = (\sqrt{2})\, E \cos(\omega t + \theta)$ and is given by $E_1 = ZI_1 - Z_m I_2 - Z_m I_3$.

For windings AO and OB, AO and OC aid when the currents flow in the same directions, but I_1 is along AO, I_2 is opposite (i.e. along BO instead of along OB) and so is I_3. In the system $I_1 + I_2 + I_3 = 0$, so that $I_2 + I_3 = - I_1$. Substituting for I_2 and I_3 by this equation we get

$$E_1 = (Z + Z_m)I_1.$$

Similarly $E_2 = (Z + Z_m)I_2$ and $E_3 = (Z + Z_m)I_3$. The currents are each equal in magnitude to the magnitude of

$$\frac{254}{Z + Z_m} = \frac{254}{200 + j219{\cdot}8}$$

$$= \frac{254}{297\,\lfloor 47^\circ\,45'} = 0{\cdot}855\,\overline{|47^\circ\,45'},$$

i.e. the currents have the r.m.s. value of $\underline{0{\cdot}855}$ A. and lag the e.m.f.'s by $\underline{47^\circ\,45'}$.

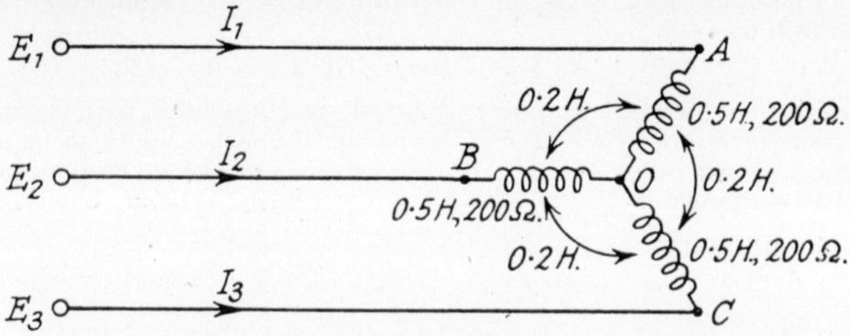

FIG. 380. STAR LOAD WITH MUTUAL INDUCTANCES

EXAMPLES

1. Two currents $I_1 = 100 \sin \omega t$ and $I_2 = 80 \cos (\omega t + \pi/6)$ enter a conductor. Find the total current by the vector and algebraic methods.

2. Voltages $E_1 = 20 \cos \omega t$, $E_2 = 30 \sin \omega t$ and $E_3 = 60 \cos (\omega t - \pi/3)$ act in series. Find the resulting voltage by both methods.

3. Find the currents and voltages in the network of Fig. 381.

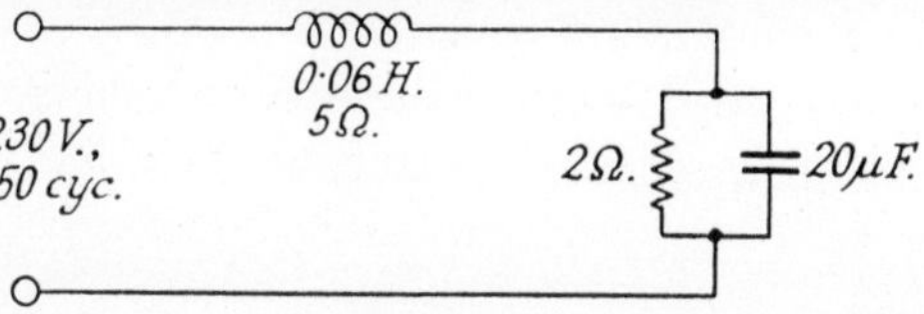

FIG. 381

4. A transmission line 10 miles long consists of two copper wires each 0·30 in. diameter spaced at 3 ft. If the conductors are short-circuited at the far end, what voltage at 50 cyc. must be applied to produce a current of 100 A. in the loop?

5. Repeat the work of the example on page 449, taking 100 Ω. as the resistance of the coil.

6. A 3-phase, 4-wire system with 440 V. between the phase conductors has a motor load of 500 kW. at power factor 0·8. The lamp loads connected between the several lines and neutral are 100, 120 and 250 kW. Find (*a*) the current in each phase conductor; (*b*) the current in the neutral; and (*c*) the power factor of the system. (*Lond. Univ.*, 1931.)

7. A single-phase series railway motor having an impedance $Z = 0{\cdot}1 + 0{\cdot}3j$ is started by a multi-tap transformer. The maximum current on each notch is 400 A., and notching up takes place when the current has fallen to 300 A. The flux at 300 A. is 85% of that at 400 A. Determine the voltage for the first tapping of the transformer, the back e.m.f. generated by the rotation of the armature when the current has fallen to 300 A. on the first tapping, and the voltage for the second tapping of the transformer.

Explain the method employed in your calculations. (*Lond. Univ.*, 1933.)

8. An unbalanced star-connected load is connected to a symmetrical 400-Volt, 3-phase, 3-wire system. The impedances and their connections are—

$1 + j2\ \Omega$. between line R and neutral point of load,
$2 + j3\ \Omega$. between line Y and neutral point of load,
$3 - j3\ \Omega$. between line B and neutral point of load.
Calculate the current in line B. Phase sequence R, Y, B.
(*Lond. Univ.*, 1950.)

9. The impedances, in ohms, of the branches of a 3-phase star-connected load are as follows—

$$\mathbf{Z}_R = 20 + j0,\ \mathbf{Z}_Y = 10 + j10,\ \mathbf{Z}_B = 0 - j25.$$

If this load, with its neutral point isolated, is connected to a symmetrical 3-phase 4-wire system, with 400 V. between line wires, determine the p.d. between the neutral point of the supply system and the neutral point of the load. Phase sequence R-Y-B. (*Lond. Univ.*, 1953.)

APPENDIX III

TRANSFORMERS

A DETAILED description of single and polyphase transformers is given in the companion volume quoted in Appendix II.* This appendix will give very briefly the effects of transformers upon transmission.

Fig. 382 gives the equivalent circuit of a transformer. R_1 represents the resistance of primary and R_2 of the secondary. X_1 and X_2

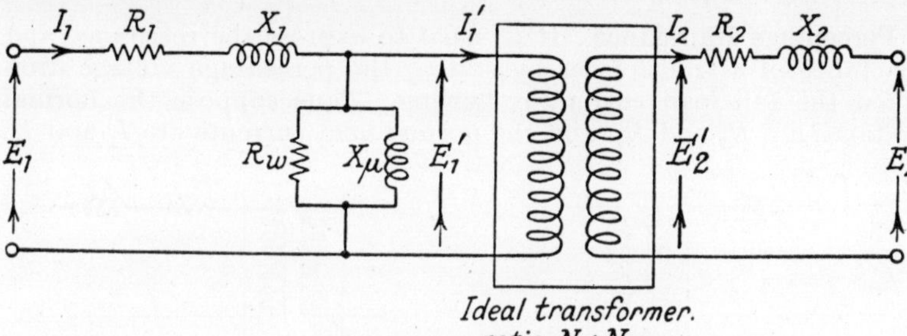

FIG. 382. COMPLETE EQUIVALENT NETWORK OF POWER TRANSFORMER

represent the primary and secondary leakage fluxes. R_w and X_μ represent the iron losses and magnetizing current. The ideal transformer has ratio $N_1 : N_2$, where N_1 and N_2 are the turns; such a transformer has the property that

$$E_1'/E_2' = N_1'/N_2' = -(I_2/I_1').$$

The relation between E_1' and E_2' follows from Faraday's Law that the e.m.f. is equal to $-N(d\phi/dt)$, N being the number of turns and ϕ the flux. The relation between the currents is due to the fact that the magnetizing current has already been allowed for in X_μ, and then $N_1I_1' + N_2I_2 = 0$.

In normal transformers the effect of R_w and X_μ on the regulation is negligible, and the transformer can be represented by R_1, R_2, X_1, X_2 and the ideal transformer. The current I_1' is then equal to I_1; this is very nearly true, for although the magnetizing current is not very small, it is nearly in quadrature with I_1 when the transformer is on load. An impedance in the primary can be transferred to the secondary by multiplying it by $(N_2/N_1)^2$; similarly one can

* See also *The Performance and Design of Alternating Current Machines*, M. G. Say (Pitman).

be transferred from the secondary to the primary by multiplying it by $(N_1/N_2)^2$. For suppose the *total* secondary impedance is Z_2, we have $E_2' = Z_2I_2$. Then the equivalent primary impedance is

$$Z_1 = \frac{E_1'}{-I_1'} = \frac{E_2'(N_1/N_2)}{I_2(N_2/N_1)}$$
$$= Z_2 \times (N_1/N_2)^2.$$

We can therefore represent a transformer on load by either of the circuits in Fig. 383. In the former

$$r_1 = R_1 + R_2(N_1/N_2)^2 \text{ and } x_1 = X_1 + X_2(N_1/N_2)^2,$$

whilst in the latter

$$r_2 = R_1(N_2/N_1)^2 + R_2 \text{ and } x_2 = X_1(N_2/N_1)^2 + X_2.$$

Percentage Impedance. It is usual to express the resistance and reactance of a transformer by stating the percentage voltage drop when the full load current is flowing. Thus suppose the normal voltages are E_1 and E_2, and the normal load currents are I_1 and I_2.

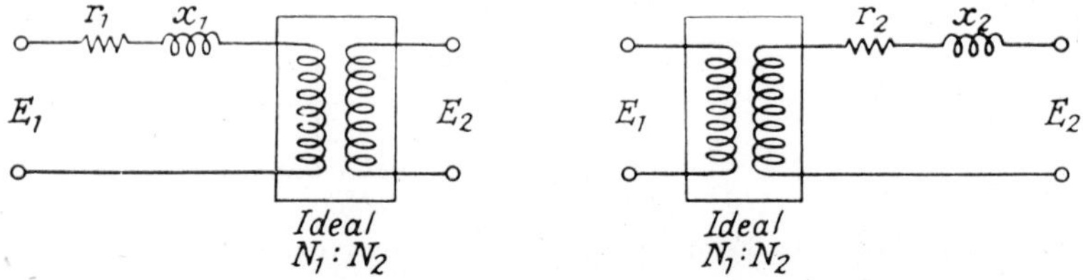

FIG. 383. SIMPLE EQUIVALENTS OF POWER TRANSFORMER

The voltage drop in the primary is $R_1I_1 + jX_1I_1$, so that the primary is said to have

$100R_1I_1/E_1$ per cent resistance and $100X_1I_1/E_1$ per cent reactance.

The secondary has similarly

$100R_2I_2/E_2$ per cent resistance and $100X_2I_2/E_2$ per cent reactance.

If we consider the first equivalent circuit of Fig. 383, we see that the total percentage resistance is

$$\frac{100r_1I_1}{E_1} = \frac{100R_1I_1}{E_1} + \frac{100R_2(N_1/N_2)^2I_1}{E_1}$$
$$= \frac{100R_1I_1}{E_1} + \frac{100R_2I_2}{E_2},$$

which is the sum of the primary and secondary percentage resistances. Similarly the percentage reactance is the sum of the primary and secondary percentage reactances.

It will be found that the same holds for the second circuit of Fig. 383. It is usual to give only the total percentage resistance and inductance. If the working voltages and currents are E_1, E_2, I_1 and I_2, and the percentage resistance and reactance are p_r and p_x, we have

$$r_1 = p_r(E_1/100I_1) \text{ and } x_1 = p_x(E_1/100I_1),$$

or

$$r_2 = p_r(E_2/100I_2) \text{ and } x_2 = p_x(E_2/100I_2).$$

EXAMPLE. A 17 500 kVA. transformer has 1% resistance drop and 3% leakage reactance drop. Find the resistance and reactance transferred to the high voltage winding, which supplies current at 66 kV.

$$E_1 = 66\,000 \text{ V.}, I_1 = \frac{17\,500\,000}{66\,000} = 263 \text{ A.}$$

Therefore

$$r_1 = 1 \times \frac{66\,000}{100 \times 263} = \underline{\underline{2{\cdot}51\ \Omega}}.$$

and

$$x_1 = 3 \times \frac{66\,000}{100 \times 263} = \underline{\underline{7{\cdot}53\ \Omega}}.$$

EXAMPLE. Find the resistance and reactance transferred to the low voltage winding of 11 kV. in the previous example.

$$r_2 = r_1(11/66)^2 = \underline{\underline{0{\cdot}0697\ \Omega}}.$$

$$x_2 = x_1(11/66)^2 = \underline{\underline{0{\cdot}209\ \Omega}}.$$

Transformers in Parallel. Fig. 384 shows two transformers in parallel. It is assumed that the transformers have exactly the same

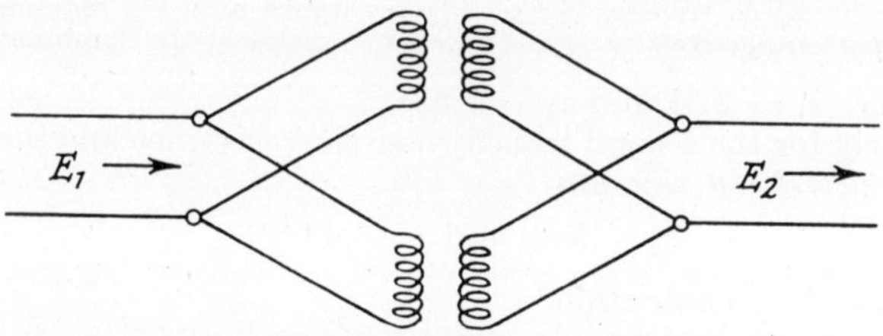

FIG. 384. TRANSFORMERS IN PARALLEL

turns ratio, but different resistances and leakage reactances. We may ignore the voltage ratio, and an equivalent circuit is that shown in Fig. 385. Let Z_1 be the resistance and leakage impedance referred to one side of one transformer, and Z_2 that of the other. If I_1 and I_2 are the currents in the separate transformers,

$$Z_1 I_1 = Z_2 I_2,$$

so that

$$I_1 = \frac{Z_2}{Z_1 + Z_2}(I_1 + I_2)$$

and

$$I_2 = \frac{Z_1}{Z_1 + Z_2}(I_1 + I_2)$$

where $I_1 + I_2$ is the total load current. The combination of the two transformers has an impedance $Z_1Z_2/(Z_1 + Z_2)$, so that the resultant resistance and reactance referred to one side are given by

$$r + jx = \frac{(r_1 + jx_1)(r_2 + jx_2)}{r_1 + jx_1 + r_2 + jx_2},$$

giving
$$r = \frac{r_1(r_2^2 + x_2^2) + r_2(r_1^2 + x_1^2)}{(r_1 + r_2)^2 + (x_1 + x_2)^2}$$

and
$$x = \frac{x_1(r_2^2 + x_2^2) + x_2(r_1^2 + x_1^2)}{(r_1 + r_2)^2 + (x_1 + x_2)^2}.$$

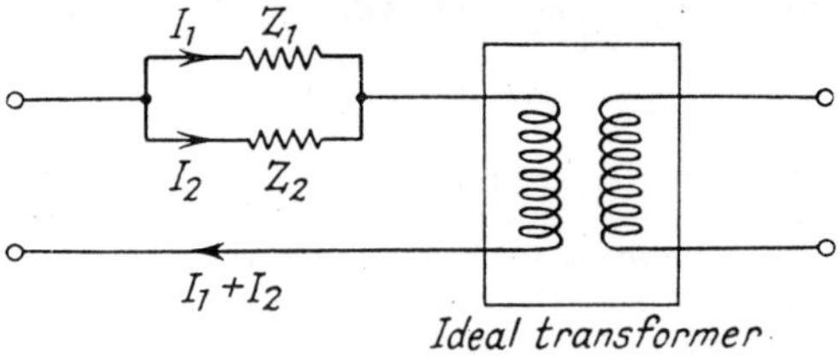

FIG. 385. EQUIVALENT NETWORK OF TWO TRANSFORMERS IN PARALLEL

These equations give the percentage resistance and reactance of the combination in terms of the simple values.

EXAMPLE. The transformer of the two preceding examples acts in parallel with a 15 000 kVA. transformer of 2% resistance and 4% reactance drops. Find the percentage resistance and reactance drops of the combination.

We have $r_1 = 2{\cdot}51$ and $x_1 = 7{\cdot}53$.

Similarly for the second transformer the resistance and inductance in the high voltage side are

$$r_2 = 5{\cdot}80 \text{ and } x_2 = 11{\cdot}60.$$

Thus for the combination

$$r = 1{\cdot}78 \text{ and } x = 4{\cdot}57.$$

The total rating is 17 500 + 15 000 = 32 500 kVA., and the load current 32 500 ÷ 66 = 492 A. The percentage resistance is thus

$$p_r = \frac{100 \times 1{\cdot}78 \times 492}{66\,000} = \underline{\underline{1{\cdot}32}}.$$

and
$$p_x = \frac{100 \times 4{\cdot}57 \times 492}{66\,000} = \underline{\underline{3{\cdot}39}}.$$

Three-phase Transformers. These may be connected star to star, delta to delta, or delta to star as shown in Fig. 386. Three separate transformers can be used, or one transformer with three limbs upon which pairs of windings can be placed.

The star/star connection accentuates the voltage of the third harmonic, since these have the same phase in the three windings and add up. Unless the primary star point is earthed, there is a floating neutral in this system. An extreme case is when only one

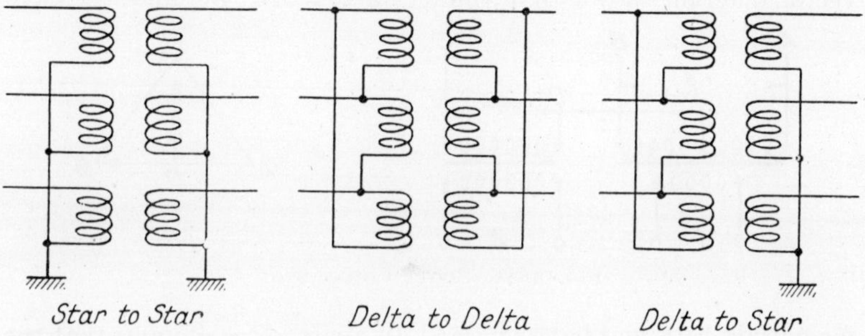

FIG. 386. STAR/STAR, DELTA/DELTA, AND DELTA/STAR TRANSFORMATIONS

phase is loaded and the others are on open circuit; then no load current will flow, for the high primary inductances of the open circuited phases are in series with the loaded phase.

The delta/delta system is much used for moderate voltages, and has the advantage that if one phase is disabled the two good phases can transmit power by the V or open delta system (see Fig. 387).

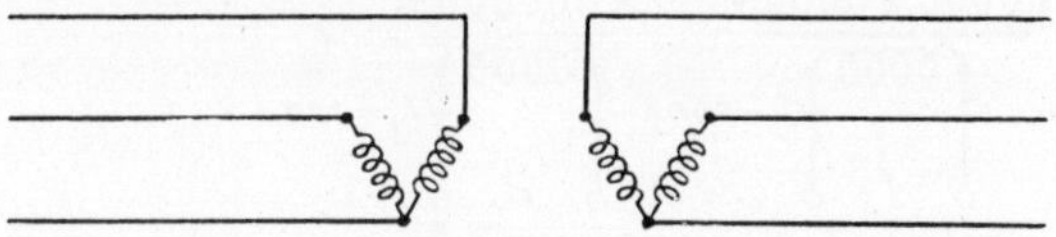

FIG. 387. OPEN DELTA OR V TRANSFORMATION

The voltages are the same as before, but the output rating is only $1/\sqrt{3}$ or 58 per cent of the previous value.

The star/delta connection is very useful for stepping up the voltage, and does not suffer from the wave distortion and floating neutral of the ungrounded star/star system.

For the purposes of calculation we assume that the transformers can be represented by an impedance $r + jx$ in series with the line conductors, the impedance being found from the percentage impedance figures given for the transformers.

EXAMPLE. Give a diagram of connections for Scott-connected transformers linking a three-phase system with a two-phase system.

If the load on the two-phase side is 200 kVA. on one phase and 300 kVA. on the other, both at 400 V., and unity power factor, determine the currents in the transformer windings and the 6·6 kV., three-phase supply lines, neglecting transformer losses. Show by a diagram how the loads are connected; only one alternative need be considered. (*Lond. Univ.*, 1948.)

Fig. 388A shows the diagram of connections. Windings 1 and 2 are supplied with two-phase voltages, i.e. voltages in time quadrature. Let winding 1 induce a voltage $1{\cdot}0E$ across AB and winding 2 a quadrature voltage $(\sqrt{3}/2)E$ across DC. It follows from the vector diagram shown that voltages across AB, BC and CA have

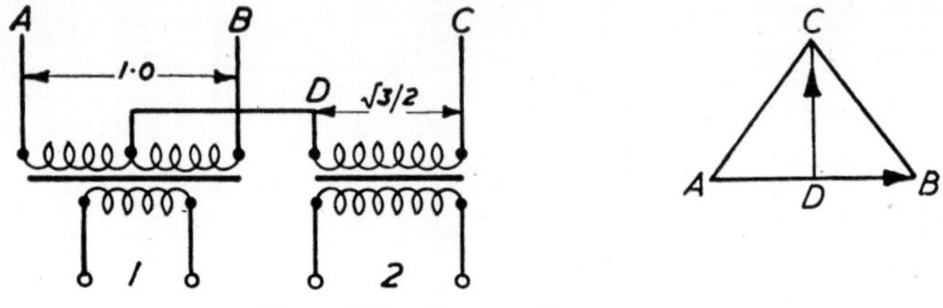

FIG. 388A. SCOTT CONNECTION

magnitude $1{\cdot}0E$ and have phases 120° apart. It is obvious that the connection similarly converts a three-phase supply to a two-phase supply.

Fig. 388B shows the voltages and currents in the system. We indicate voltages with an arrow: this method defines the reference

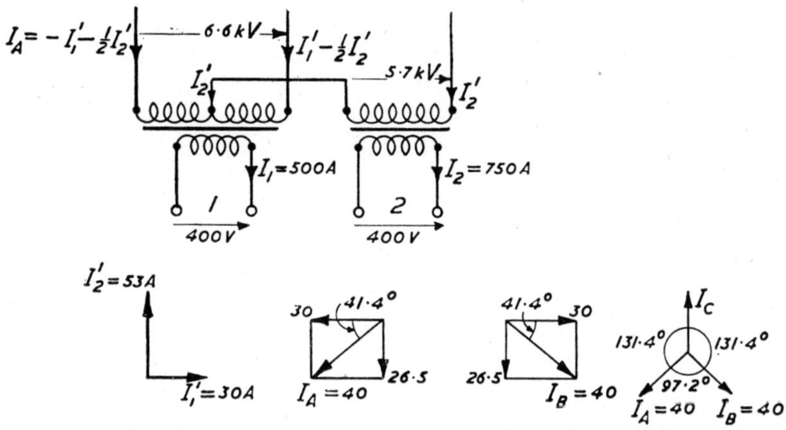

FIG. 388B

phases for voltages and currents. The currents in the two-phase windings are 500 A. and 750 A., each in phase with the voltages of magnitude 400 V.: note that the voltage in winding 2 is in quadrature with that in winding 1, and hence also are the currents.

The voltages AB and DC have magnitudes 6·6 kV. and 5·7 kV. The currents I_1 and I_2 in the two-phase windings induce currents I_1' and I_2' in the upper windings: these currents are in phase with I_1 and I_2, respectively, but flow in the directions shown so as to give zero total ampere-turns if the transformers are considered to be ideal. The amplitudes of I_1' and I_2' are given by

$$I_1' = I_1\,(400/6\,600) = 30 \text{ A.},\ I_2' = I_2\,(400/5\,700) = 53 \text{ A.}$$

Note that these currents are in quadrature.

The three-phase line currents are

$$I_A = -I_1' - \tfrac{1}{2}I_2',\ I_B = I_1' - \tfrac{1}{2}I_2',\ I_C = I_2'.$$

The figure shows the line currents. $I_A = I_B = 40$ A., $I_C = 53$ A., the relative phases being 97·2°, 131·4° and 131·4°.

APPENDIX IV

THÉVÉNIN'S THEOREM

THIS theorem states that *if a network has two terminals A and B between which there is placed an impedance Z, the current through Z is given by*

$$E/(Z + Z_1),$$

where E is the potential difference between A and B when Z is removed, and Z_1 is the impedance of the network between A and B calculated by assuming that all generators are replaced by impedances equal to their internal impedances.

The proof is as follows. Fig. 389 (*a*) shows the condition under load, and Fig. 389 (*b*) when Z is removed. In Fig. 389 (*b*) the potential of A above B is E. Fig. 389 (*c*) shows a system which is

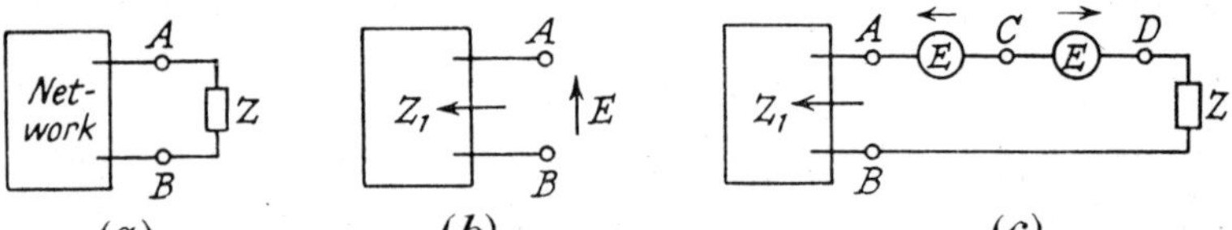

FIG. 389. THÉVÉNIN'S THEOREM

clearly the same as that in Fig. 389 (*a*), since it contains merely an addition of two equal and opposite e.m.f.'s. Now the potential of C is equal to the potential of B if there is no current flowing; the network to the left between B and C is thus equivalent to a passive impedance Z_1. The current is therefore that due to the e.m.f. E acting between C and D through the loads Z and Z_1, so that

$$I = E/(Z + Z_1).$$

The important fact is that E and Z_1 are completely determined by the network and do not depend on Z. Thus if we have a variable feeder in a network, there is no need to work out the current through it *ab initio* for the different impedances of this feeder. We find E and Z_1 for the network, and the current is obtained for any impedance of the variable feeder. The following example illustrates the use of the theorem. Another example is given on page 170.

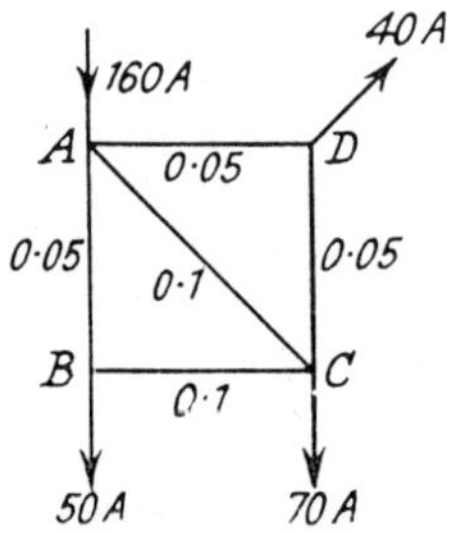

FIG. 390. PROBLEM OF INTERCONNECTION SOLVED BY USE OF THÉVÉNIN'S THEOREM

EXAMPLE. Find the current in the feeder AC for resistances of 0·1 and 0.08 Ω. in the network of Fig. 390.

We imagine that AC is removed, with the resulting network of Fig. 391. Let the current in AB be I_1; the remaining currents can then be easily written down. The voltage drop from A to C by both paths must be equal, so that we have

$$0{\cdot}05I_1 + 0{\cdot}1(I_1 - 50) = 0{\cdot}05(160 - I_1) + 0{\cdot}05(120 - I_1),$$

i.e.

$$I_1 = 76 \text{ A}.$$

The voltage drop from A to C is

$$E = 0{\cdot}05 \times 76 + 0{\cdot}1 \times 26 = 6{\cdot}4 \text{ V}.$$

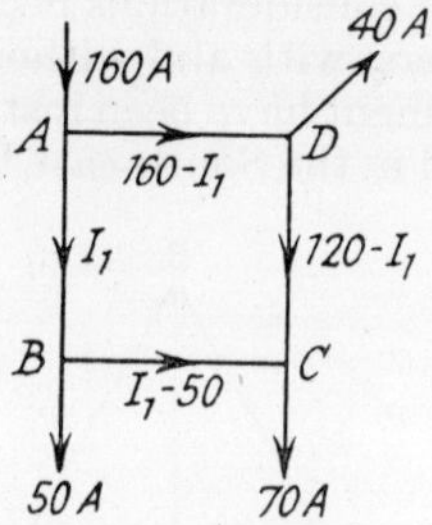

FIG. 391. PROBLEM OF INTERCONNECTION SOLVED BY USE OF THÉVÉNIN'S THEOREM

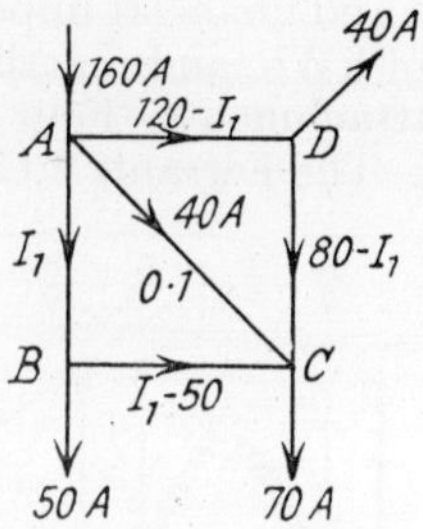

FIG. 392. PROBLEM OF INTERCONNECTION SOLVED BY USE OF THÉVÉNIN'S THEOREM

The impedance of the network between A and C is

$$Z_1 = \frac{0{\cdot}1 \times 0{\cdot}15}{0{\cdot}1 + 0{\cdot}15} = 0{\cdot}06.$$

If $Z = 0{\cdot}1\ \Omega.$, the current in AC is

$$I = \frac{E}{Z + Z_1} = \frac{6{\cdot}4}{0{\cdot}1 + 0{\cdot}06} = \underline{\underline{40 \text{ A}}}.$$

If Z is $0{\cdot}08\ \Omega.$, the current is

$$I = \frac{6{\cdot}4}{0{\cdot}08 + 0{\cdot}06} = \underline{\underline{45{\cdot}7 \text{ A}}}.$$

It is now very easy to solve the network problem in its entirety with a given Z, say $0{\cdot}1\ \Omega$. Fig. 392 shows the network. Let the current in AB be I_1; the remaining currents are then as shown. Equating voltage drops along ABC and ADC we get

$$0{\cdot}05I_1 + 0{\cdot}1(I_1 - 50) = 0{\cdot}05(120 - I_1) + 0{\cdot}05(80 - I_1),$$

so that

$$I_1 = 60 \text{ A}.$$

The currents are thus $AB = 60$, $AC = 40$, $AD = 60$, $BC = 10$, $DC = 20$ amperes.

APPENDIX V

METERING, DISTANCE INDICATION (TELEMETERING) AND REMOTE CONTROL

Measurement of kWh. and Reactive kVAh. In an interconnected system, such as the Grid, the problem of metering is of supreme technical and financial importance. Tariff considerations necessitate the use of kWh. and reactive kVAh. meters, with and without summating attachments. Four types of instrument have been installed on the Grid. The Ferranti F.G.L. type is used in the South-east England

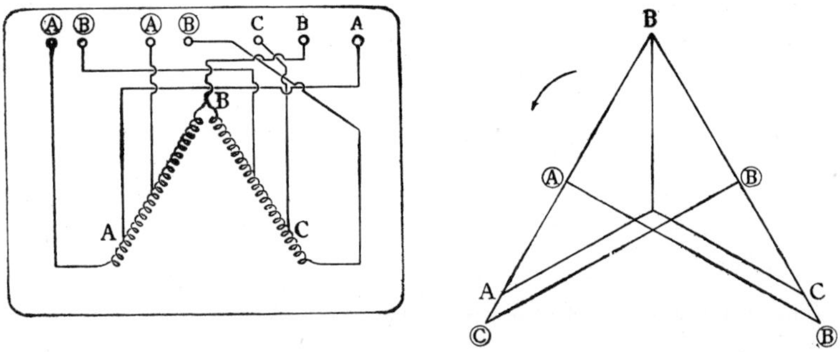

FIG. 393. QUADRATURE TRANSFORMER FOR FERRANTI F.G.L. REACTIVE kVAH. METER

(*I.E.E. Journal*)

electricity scheme. This meter has several novel features. There are two driving elements per disc so placed that side-thrust on the bearings is eliminated, and 80 per cent of the weight of the rotor is borne by permanent-magnet suspension. The reactive kVAh. meters are identical with the energy meters, and a quadrature transformer supplies the potential windings with the potential currents of correct phase. Fig. 393 shows the diagram of the quadrature transformer and Fig. 394 shows the connections of the kWh. and reactive kVAh. meters of this type.

Summation Metering. It is often required to measure the total power in two or more circuits. Fig. 395 shows how this can be done with one wattmeter of ordinary construction, using a number of current transformers with their secondaries in parallel; in a similar method the secondaries of the current transformers form the primary windings of a current transformer, the secondary of which is connected to the current coil of the wattmeter. The accuracy of

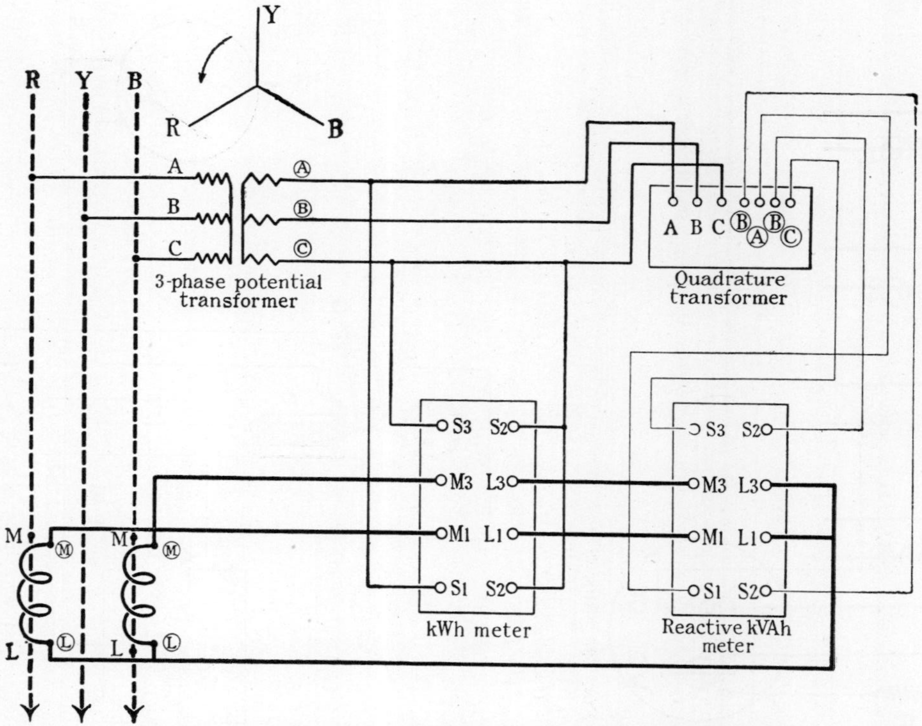

FIG. 394. CONNECTIONS OF kWH. AND REACTIVE kVAH. METERS

(*I.E.E. Journal*)

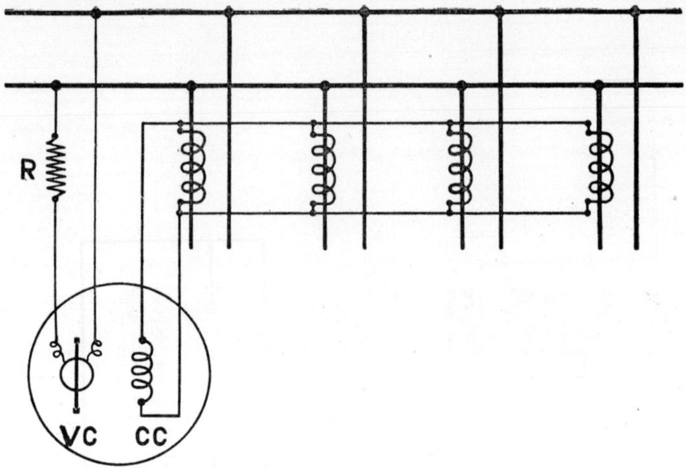

FIG. 395. SUMMATION METERING BY CURRENT TRANSFORMERS WITH SECONDARIES IN PARALLEL

(*Industrial Electrical Measuring Instruments* (Edgcumbe and Ockenden))

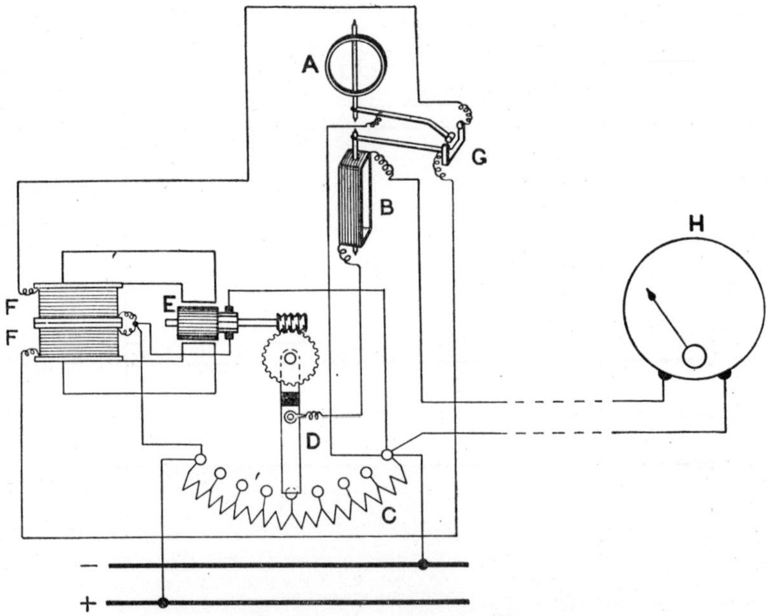

FIG. 396. MIDWORTH DISTANT REPEATER

(*Industrial Electrical Measuring Instruments* (Edgcumbe and Ockenden))

these methods is not good because of the large number of current transformers used. It is also possible that the voltages of the circuits may differ somewhat.

Preference is thus given to mechanical methods of summation, in which the meter index readings are transmitted to some convenient place where they are added mechanically. Each meter of a group is provided with a contact-making device, operated by gravity, a cam or a commutator, and is connected for an interval of time by this device to a transmitter, which sends a signal to the mechanical

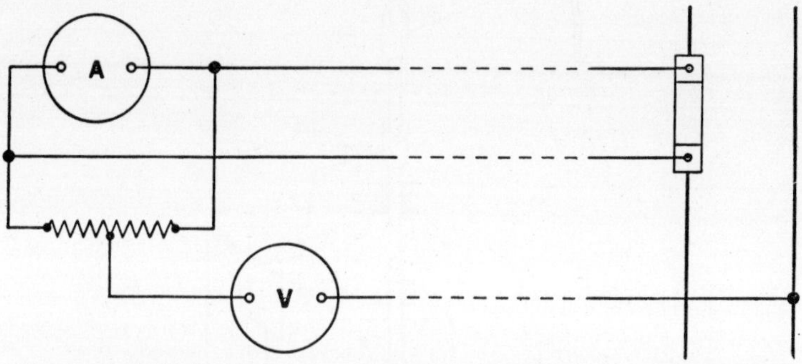

FIG. 397. DIRECT DISTANT INDICATION OF CURRENT AND VOLTAGE
(*Industrial Electrical Measuring Instruments* (Edgcumbe and Ockenden))

summator. This signal registers the power or reactive volt-amperes in the mechanical summator, which adds mechanically the required quantities.

Midworth Distant Repeater. Fig. 396 shows the Midworth *distant indicator*. *A* is the originating movement, the moving coil of a dynamometer wattmeter or the disc of an induction wattmeter, say. This moving part has attached to it a contact arm, which can make contact with the left or the right contact of *G*. These contacts are connected each to an end of the field coils *F F*, the other end of each field coil being connected to the positive of a d.c. supply. When *A* causes the contact arm to touch the left contact of *G*, the lower field coil is put across the d.c. supply; whilst when *A* causes the arm to touch the right contact, the upper field coil is energized. The field coils are wound so that one causes the motor armature *E* to rotate in one direction and the other in the opposite direction. Suppose that *A* moves to the left and makes the contact arm touch the left contact of *G*. The lower field coil is energized and the armature rotates, and in doing so the tapping arm *D* moves. The current through the moving coil *B* is then varied. *B* is the moving coil of a permanent-magnet instrument and experiences a torque in the same direction as *A*. *B*, in moving, takes *G* with it, so that the contact at *G* will be broken when the deflection of *B* is slightly greater than that of A. Thus for any deflection of A, the motor armature will rotate until the deflection of B is just greater than

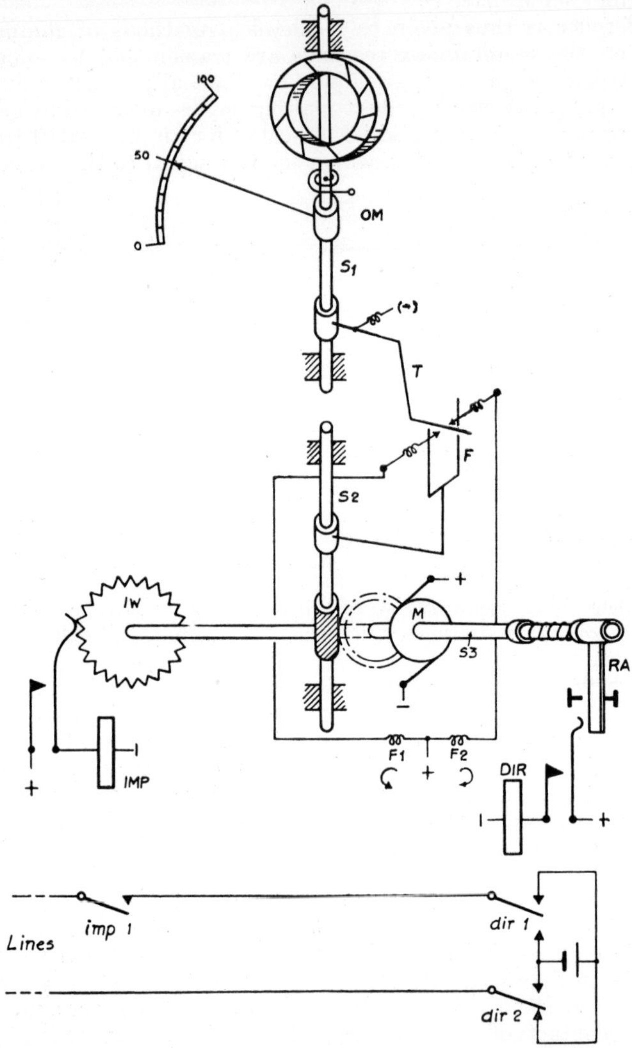

FIG. 398. IMPULSE METHOD APPLIED TO MIDWORTH REPEATER
(*I.E.E. Journal*)

that of A, and will then stop. As B is the coil of a permanent-magnet, moving-coil instrument, its deflection is proportional to the current through it, so that the current through B is a proportional measure of the deflection of A, which represents watts or reactive volt-amperes.

It is now necessary to transmit the current through B to a distant instrument which will register it, and/or add it to similar readings from other meters.

Direct Distant Indication. Fig. 397 shows how the current and voltage can be transmitted to a distant ammeter and voltmeter. The use of a centre-tapped resistance enables the pilot wires for the ammeter to serve as one pilot wire for the voltmeter, without causing the voltmeter current to affect the ammeter. Thus only three pilot wires are required, instead of four. The method of direct indication is limited in d.c. systems, by considerations of leakage and resistance, to a length of 20 miles for voltage measurement and 3 miles for current measurement. In the a.c. systems the limiting lengths are 15 miles and 1 000 yd. respectively.

Indirect Distant Indication. The Midworth system, described above, uses really a direct indication method, for, although any quantity is replaced by a d.c. current or voltage, the latter has to be transmitted direct. There are various methods of transmitting the value of the current or voltage indirectly, examples being the Selsyn, Granat, and impulse methods. Only the application of the latter to the Midworth system will be given. Fig. 398* shows the adaptation. OM corresponds to A of Fig. 396, F to G, M to E, and F_1F_2 to FF. The coil B of Fig. 396 is replaced by a shaft $S2$, which is geared to the motor M. The latter is mounted on a shaft $S3$, which carries a notched impulsing wheel IW and a friction-tight reversing arm RA. As before, if the originating movement moves in either direction, the motor turns $S2$ so as to break the contact at F (formerly G). The motor shaft revolves through an angle proportional to the originating movement, and the wheel IW causes a train of impulses to be sent out, the *number* of the impulses being proportional to the originating movement. The pulses are positive or negative according to the direction of the originating movement, but their magnitudes are fixed. The pulses are sent along a pair of pilot wires. At the receiving end the pulses cause rotary (Strowger) switches to rotate one position per impulse, so that the final positions of the switches represent the originating movement, and hence the kWh. or kVAR.

Data Transmission Systems. For systems using magslips (or selsyns), magsyns and desynns, see section (6.1.1.) of *Electronics* by the author (Pitman).

* This description and diagram are taken from "Remote Control of Power Networks," by G. A. Burns and T. R. Rayner. *Journ. I.E.E.*, Vol. 79, No. 475.

APPENDIX VI

CONVERTING PLANT

It has been found convenient to transmit large blocks of electrical energy by alternating current, because of the ease of transforming up or down in voltage with static equipment, viz. transformers. In some applications it is found desirable to use direct current, e.g. in traction and electro-chemical work. It is then necessary to have an efficient means to convert A.C. to D.C. One common method is to use a rotary converter: this machine is described in detail in Chapter XXI of *Electrical Technology* by H. Cotton (Pitman).

The rotary converter is being displaced by the mercury-arc rectifier for the conversion of A.C. to D.C. (and D.C. to A.C.) for the following reasons. The mercury-arc rectifier is a static device and thus needs less maintenance. Modern types have been designed for the largest powers normally required; they have a very high efficiency and have a small bulk. We shall therefore give a brief description of the use of mercury-vapour diodes and triodes.

Gas-filled Valves. These are used for many purposes and from very low powers up to the highest powers. There are many types and the mode of operation takes many forms.

First there are the types which have a heated cathode that gives a supply of electrons by thermal emission: the main effect of the gas is to neutralize the space charge effects present in a hard tube, and as a main result a large current can flow with a small voltage drop. In this class are gas-filled rectifier diodes and gas-filled triodes or thyratrons.

Then there are the rectifiers with a cathode in the form of a mercury pool. In this type an arc must be started by a special device, and then hot spots are formed on the cathode, from which very large currents can be drawn by field emission. It is estimated that current densities about 4 000 A./cm.^{2} are drawn from the hot spots, each spot carrying about 30 A. The *ignitron* is a member of this class.

Gas-filled Thermionic Diode. Mercury vapour or one of the inert gases is put in a thermionic diode in order to reduce the voltage drop at high currents. The effect of a small gas pressure is to change the current-voltage characteristic in the way shown in Fig. 399. The current increases more rapidly in the gas diode when the ionization potential of the gas is reached: the positive ions neutralize the effect of the space charge and the current increases more rapidly than the three-halves power law. The positive ions do not contribute appreciably to the current directly, as their velocity is small compared with that of the electrons.

If the gas pressure is raised the characteristic changes to that shown in Fig. 400 (*a*), and the potential distribution is as shown in Fig. 400 (*b*). As the voltage is raised from zero a very small current

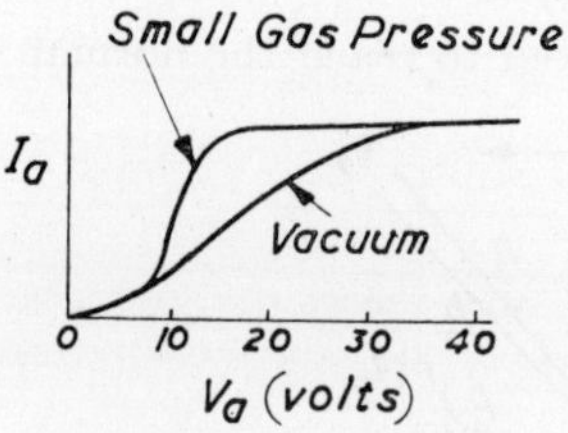

FIG. 399. EFFECT OF GAS IN A DIODE

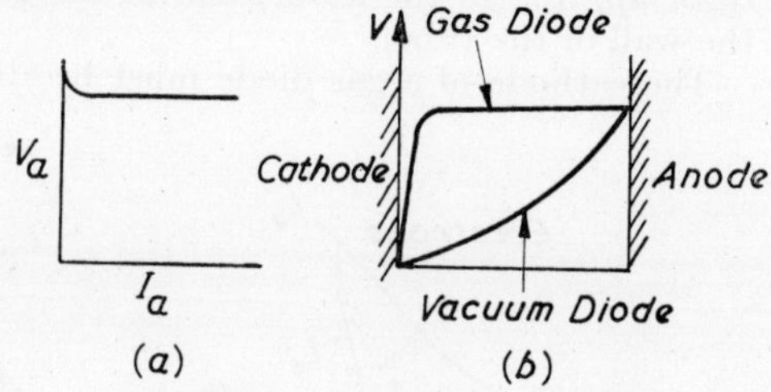

FIG. 400. VOLTAGE IN A GAS DIODE

flows initially, but when the ionization voltage is passed the current rises suddenly to a value limited by the external circuit and the voltage drops (because of the external load). The voltage drop lies between 10 and 20 V. however large the current may be. There is a *positive-ion sheath* near the cathode, which accelerates the electrons from the cathode and injects them with high speed into the *plasma* which stretches from the positive-ion sheath to the anode: in the plasma there are approximately equal and opposite charges, so that there is a small potential drop across it. The main potential drop occurs between the cathode and the positive-ion sheath.

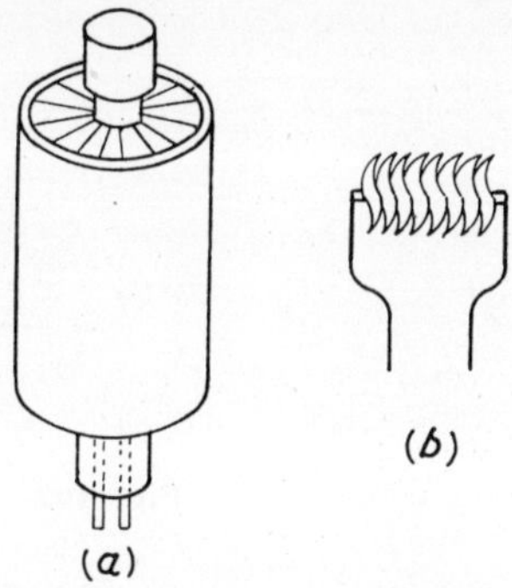

FIG. 401. HEAT-CONSERVING CATHODES FOR THYRATRONS

The cathode of a gas diode can be made more efficient than in a high vacuum tube for the following reason. It is not necessary in a gas tube that the emitting surface be near to the anode since the positive-ion sheath forms automatically near the cathode. It is therefore possible to put the cathode surface inside a metal container, and the radiation of the heat is greatly decreased: alternatively the cathode surface can be made in a corrugated form, which results in a large measure of radiative screening. Fig. 401 (*a*) shows a cathode in which the coated fins are placed inside screening cylinders, and Fig. 401 (*b*) shows a corrugated form of cathode.

In the Tungar rectifier used for battery chargers the gas is 5 cm. of argon. In mercury-vapour diodes the pressure is about 0·1 mm.

In a mercury-vapour tube the pressure varies rapidly with the temperature, so that the voltage drop is larger at lower temperature; this is dangerous, for if the voltage is high, the cathode is liable to be damaged by the bombardment by the positive ions. The rare gases do not suffer from this disadvantage, but they suffer from *clean up*, due to the absorption of the gases on the electrodes and the wall of the tube.

The cathode of a gas diode must be allowed to reach the normal

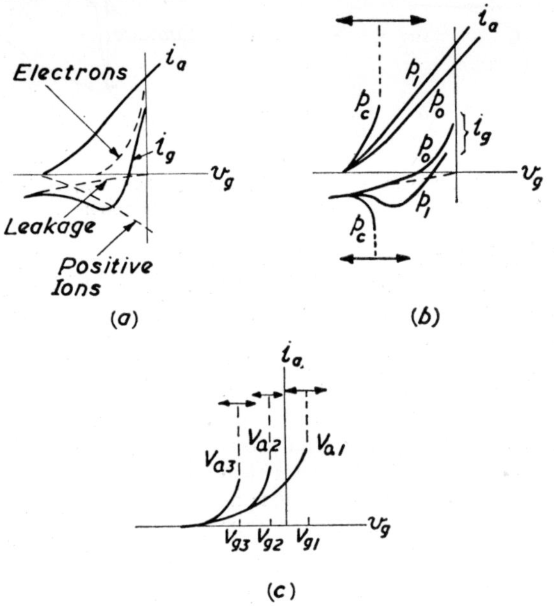

FIG. 402. ACTION IN A GAS TRIODE

operating temperature before switching on the H.T., otherwise the voltage drop may exceed the disintegration value.

Gas-filled Thermionic Triode. This is commonly known as a *thyratron*, provided the gas pressure is sufficiently high to form a plasma so that the grid has a trigger action in the way described in the following.

Fig. 402 (*a*) shows the anode and grid currents in a triode with a small amount of gas as a function of grid voltage. The grid current has three components, one due to leakage, another to electrons (with thermal velocities large enough to reach a grid at a negative potential), and a third due to positive ions. The last contribution is proportional to the number of positive ions present, and hence is proportional to the gas pressure and the anode current. As the gas pressure is raised from 0 through p_1, to a critical value p_c, there is a sudden increase of anode and grid currents in the way shown in Fig. 402 (*b*). The current is limited by the external circuit resistance and is independent of the grid voltage: for this reason these curves have a horizontal line with arrows to indicate that they are independent of the grid voltage. What has happened is that sufficient positive ions are formed to neutralize the electron cloud in the tube to form a plasma: and moreover, positive-ion sheaths are formed round the grid wires and near the cathode surface, in the way indicated in Fig. 403.

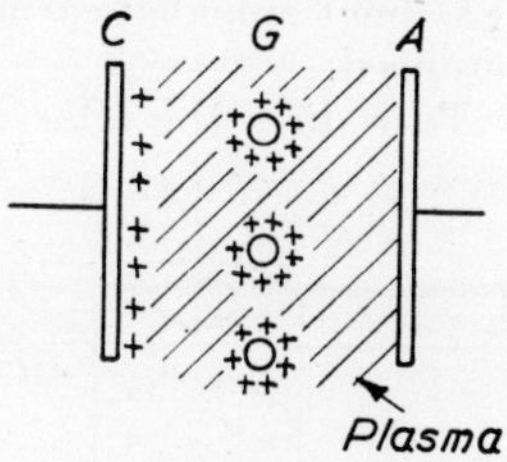

FIG. 403. POSITIVE-ION SHEATHS AND PLASMA

The positive-ion sheath round the grid causes the grid to lose control except at very large negative potentials in a way discussed below, and the plasma neutralizes the space charge effects so that the voltage drop between the anode and the cathode is merely that in the cathode sheath (10 to 20 V.): the current is thus limited by the external circuit. These are the features of a thyratron.

It should be noted that the grid current needs also to be limited by the external circuit.

STARTING ACTION AND CONTROL RATIO. Fig. 402 (*c*) shows the

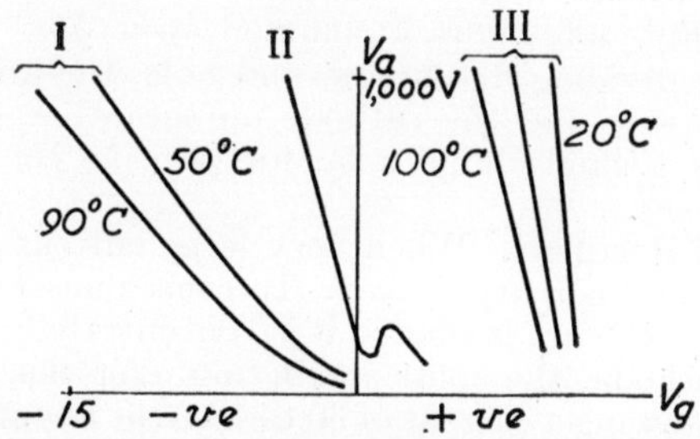

FIG. 404. TYPICAL CONTROL CURVES OF THYRATRONS

anode current as a function of grid potential V_g for various values of anode voltage V_a. The starting grid voltage is nearly a linear function of the anode voltage, and some typical curves are shown in Fig. 404 for three types of tube. In type I the grid voltage is negative, and the thyratron is said to have *negative control*; in type II the grid voltage may vary from negative to positive, and this is intermediate control; in type III the grid potential needs to be positive to strike the tube, and there is *positive control*.

The grid-control ratio is the slope of the lines: thus in type I it is about 1 000/15 = 60.

We may note that because of the plasma, the cathodes may be shielded to reduce heat radiation. The tube cathode should be heated before switching on the voltages in order to prevent cathode disintegration, and in addition the grid and anode currents must be limited for the same reason.

ELECTRODE STRUCTURE OF THYRATRONS. We have already stated that heat-shielded cathodes may be used to reduce the heater power required.

To reduce the effects of wall charges, which would cause erratic

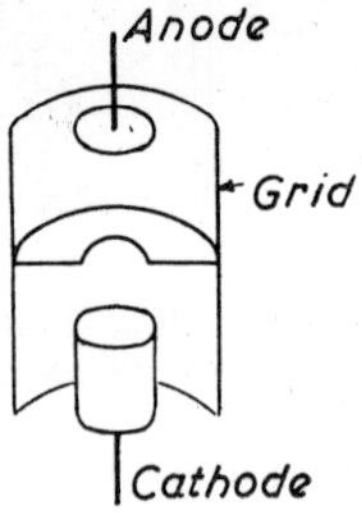

FIG. 405. THYRATRON ELECTRODES

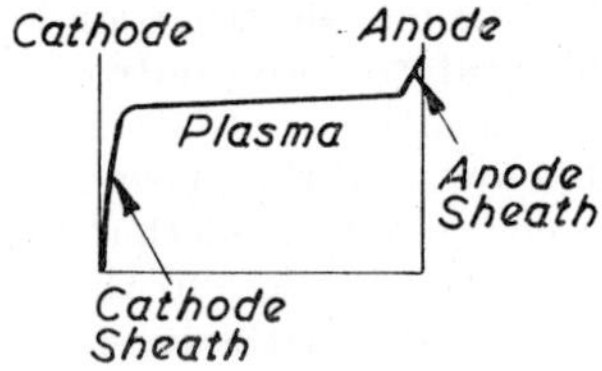

FIG. 406. POTENTIAL DISTRIBUTION IN MERCURY ARC

or variable starting, it is desirable to let the grid structure enclose the anode and cathode, thus screening the inter-electrode space from wall charges. Fig. 405 shows a common structure: the grid is a cylinder with a dividing diaphragm and hole, the anode is on one side of the aperture and the cathode on the other. For negative control there is a single large hole, for positive control there are several small holes.

Mercury-pool Rectifiers. When very large currents are required, a cathode of liquid mercury is used. To cause emission an arc must be started. Once the arc is started it is concentrated at small bright areas of the cathode, the spots wandering over the surface of the pool and each drawing a current of 30 to 40 A. at about 4 000 A./cm.^{2}

The hot spots also produce the mercury vapour for the arc, and the pool is continuously replenished by condensation of the vapour so that there is no problem of disintegration of the cathode.

The potential distribution in a mercury arc is as shown in Fig. 406. The cathode sheath voltage drop is about 9·9 V. over a wide range of currents and pressure: as the work function of mercury is 4·5 V., only 5·4 V. is available for ionizing the gas. Since this is less than the minimum ionizing potential, ionization takes place usually by multiple collisions. The potential gradient in the plasma varies from 0·05 to 0·2 V./cm. in normal operating conditions. The anode sheath contains electrons, and the voltage drop across it increases with current: this is the main reason for an increased voltage drop from anode to cathode at very high currents, which in a typical case may be about 25 V.

A mercury-pool rectifier can stand temporarily very large overload currents, the only limiting feature being overheating of the anode and other parts. To maintain the requisite low pressure of mercury the vapour must be condensed as rapidly as it is produced, and this is done by water or air-blast cooling of the envelope.

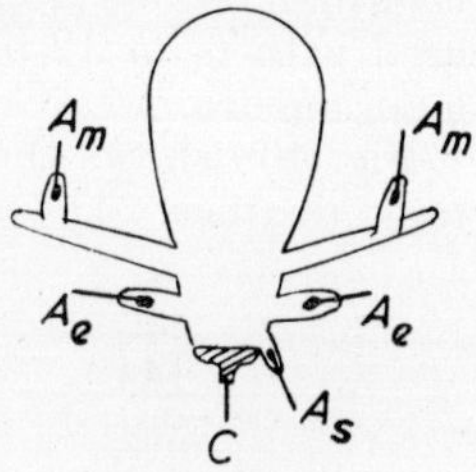

FIG. 407. GLASS-BULB MERCURY-POOL RECTIFIER

GLASS-BULB MERCURY-POOL RECTIFIER. Fig. 407 shows an early type of mercury-pool rectifier in which the envelope is a glass bulb. To start the arc the bulb is tilted to the right and then back again, when an arc is struck between the cathode C and the starting anode A_s (which has a potential with respect to C). The main anodes A_m are fed with a.c. voltages in antiphase, so that current flows to each for a half-cycle. The excitation anodes A_e are connected to a transformer and a small fixed load to prevent the arc from dying out should the load cease temporarily. The large glass dome acts as a condensing surface for the mercury vapour.

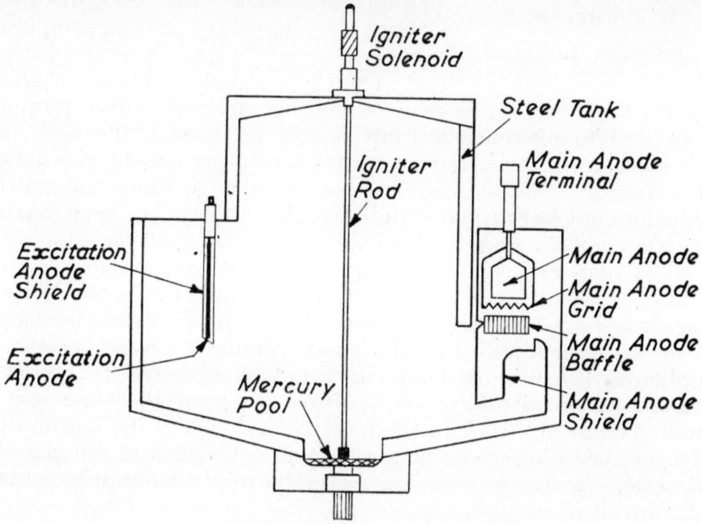

FIG. 408. STEEL-TANK IGNITRON

The main anodes are located on side-arms, as shown, in order to prevent an arc from striking directly between them when one is at a positive maximum and the other at a negative maximum: this phenomenon is known as *arc-back* or *back-fire*.

The disadvantages of this type are the difficulty of cooling and weak mechanical strength in large sizes. Their rating is limited to about 500 A. at 500 V. continuous.

STEEL-TANK RECTIFIERS. These allow adequate cooling facilities (by water jackets) and are mechanically robust. They are made for very high powers (10 MW. or so) and up to 20 kV.: they are generally used with 6- or 12-phase supply, and have the requisite number of main anodes.

Fig. 408 shows the outlines (excluding water jackets) of a steel-tank rectifier. The ignition anode is a rod extending to within a fraction of an inch of the mercury pool, and it has a small direct voltage. To start the arc a solenoid depresses this anode rod on to the mercury and a spring pushes it back. There are excitation and main anodes as in the glass-bulb type, though in small rectifiers the ignition anode may act as excitation anode also. To prevent arc-back, each anode is surrounded by a shield, and the main anode has also a grid and a baffle: the problem of arc-back has not, however, been completely solved.

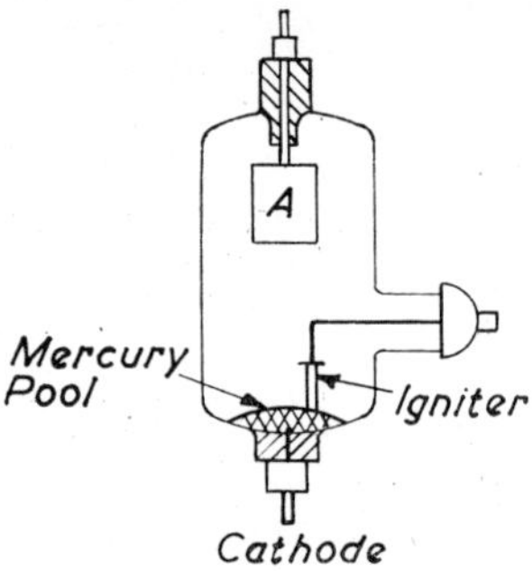

FIG. 409. IGNITRON

The electrodes are of course insulated from the tank, and the seals are somewhat difficult. Often pumping is required to prevent gas from leaking in through the seals, but some modern rectifiers have seals that permit permanent evacuation.

IGNITRONS. This is a mercury-pool rectifier in which the ignition anode does not have to be withdrawn after the arc has been started. Fig. 409 shows a sketch of an ignitron. The ignition anode is a pencil of high resistance material such as silicon carbide (carborundum) or thyrite, one end of which dips into the pool. When a sufficiently large voltage is applied to the ignition anode a cathode hot spot is formed and the main anode can draw a large current.

Ignitrons are usually made in single-phase units to eliminate arc-back. By controlling the ignition potential the tube can be switched on at any time in the positive half-cycle: the ignition has to be supplied each cycle, or the arc will extinguish in the negative half-cycle. In this way the same results may be obtained as in a grid-controlled rectifier.

Ignition anodes may require 100–200 V. and 5–50 A., but the current need flow for a very short time only.

Polyphase Rectifiers. Fig. 410 (*a*) shows a three-phase rectifier system using a Δ-Y transformer and a three-anode, mercury-pool rectifier. The voltages V_1, V_2, V_3 between the windings and the neutral are shown in Fig. 410 (*b*). If it is assumed that the voltage drop in the arc is constant (say 30 V.), it is seen that only one anode can conduct at any instant except when two line voltages are within 30 V. of one another. The voltage across the load is shown by the dotted line in Fig. 410 (*b*), and the current has the same form.

If we ignore the voltage drop in the arc, anode 3 extinguishes and anode 1 takes over at 30°, when their voltage is half the peak.

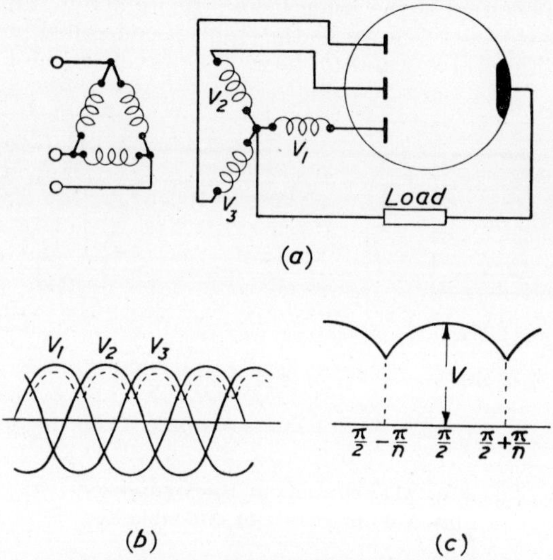

FIG. 410. POLYPHASE RECTIFICATION

The waveform thus varies between half- and full-peak value three times per cycle. If V is the peak value in an n-phase system, the changeover takes place at $(\pi/2 - \pi/n)$ when the voltage is $V \cos (\pi/n)$; thus in a 6-phase system the voltage varies between 0·866 and 1·0 times the peak value.

The mean value of the voltage is seen by Fig. 410 (*c*) to be

$$\bar{V} = [1/(2\pi/n)] \int_{\pi/2 - \pi/n}^{\pi/2 + \pi/n} V \sin \omega t \, d(\omega t)$$

$$= (n/2\pi) \int_{-\pi/n}^{\pi/n} V \cos \omega t \, d(\omega t)$$

$$= V \sin (\pi/n)/(\pi/n).$$

For 3-phase this is $(3\sqrt{3}/2\pi)\, V = 0{\cdot}83\, V.$, whilst for 6-phase it is $(3/\pi)\, V = 0{\cdot}95\, V.$

Fig. 411 (*a*) shows an obvious method of achieving a 6-phase rectifier system. The Δ-double Y system of Fig. 411 (*b*) is the preferred method on large 6-phase installations: in this system the two Y's are 60° apart in phase, and they are connected by a centre-tap transformer *CT*. The action of the centre-tap transformer is to allow the two Y's to act independently by taking up automatically the voltage difference between their maximum voltages, so that one anode of each Y conducts at any time. The result so far as mean current is concerned is the same for a 3-phase sytem, i.e. the mean is 0·83 times the peak. So far as the ripple is concerned both 6-phase systems of Fig. 411 (*a*) and (*b*) have the waveform of Fig. 411 (*c*)

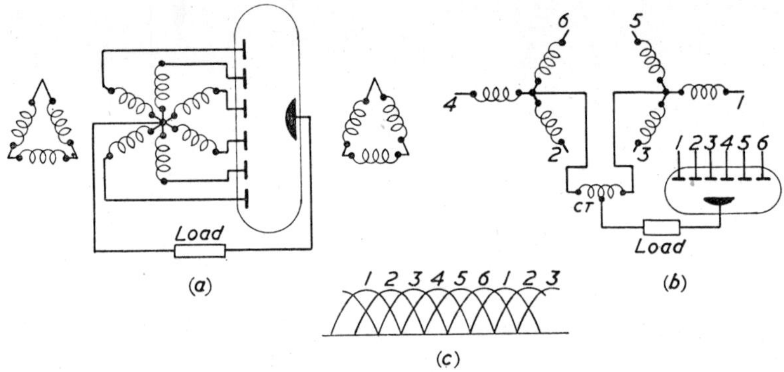

FIG. 411. SIX-PHASE RECTIFICATION
(*a*) Δ-6 phase. (*b*) Δ-double Y.

which varies between 0·866 and 1·0 times the peak. The frequency of the ripple is six times that of the fundamental frequency, and this, together with the low ripple value, enables cheap filters to be used.

The Δ-double Y system possesses the following advantages. Since each anode conducts for one-third of the cycle, the windings of the Y's are used for twice the time as in the system of Fig. 411 (*a*), and this results in a more efficient transformer. Moreover, as each anode takes only half the load current, this results in a smaller arc voltage drop and a smaller loss in the rectifier.

EFFECT OF A LOAD REACTOR. In order to smooth the current in the load resistance it is common practice to include an inductance in series with the load. Because the ripple is at three or six or twelve times the supply frequency the value of inductance is practicable.

Fig. 412 (*a*) shows the case of a Δ-Y system. If the reactor has a very large inductance the current carried by an anode is I_{dc} and occurs for one-third of the time, as shown in Fig. 412 (*b*). It follows that in an *n*-phase system the mean anode current is

$$\bar{I}_a = I_{dc}/n$$

It is also obvious by simple integration that the r.m.s. value of each anode current is $I_{rms} = I_{dc}/\sqrt{n}$.

The power rating per phase of the transformer is $VI_{rms}/\sqrt{2}$, where V is the peak phase voltage, so that the total power rating of the transformer is

$$P_T = nVI_{rms}/\sqrt{2} = \sqrt{n} \,.\, VI_{dc}/\sqrt{2}.$$

The direct-current power (including rectifier losses) is

$$P = VI_{dc}, \text{ where } V = \text{mean value of the voltage.}$$

As $V = V \sin(\pi/n)/(\pi/n)$ it follows that

$$P = VI_{dc} \sin(\pi/n)/(\pi/n).$$

The transformer *secondary utilization factor* is defined as

$$\eta_s = P/P_T = \sqrt{(2/n)} \,.\, \sin(\pi/n)/(\pi/n).$$

For 3-phase $\eta_s = 0{\cdot}67$, for 6-phase it is $0{\cdot}55$ and for 12-phase it is 0·40. Thus the 6-phase star of Fig. 411 (*a*) uses the transformer less efficiently than a 3-phase star. The Δ-double Y is equivalent to a 3-phase system from this point of view, and thus is economical in transformer efficiency.

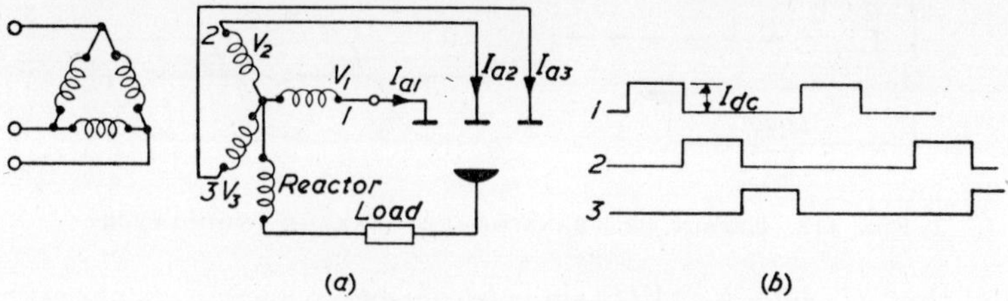

FIG. 412. EFFECT OF A LOAD REACTOR

EFFECT OF TRANSFORMER LEAKAGE. Consider Fig. 413 (*a*) in which V_1 is the first phase voltage, etc. When V_2 begins to exceed V_1 the current I_{a1} should drop at once to zero and I_{a2} should jump to the value I_{dc}: because of the leakage inductance in the windings this does not happen and the currents take the waveforms of Fig. 413 (*b*). The changeover is calculated in the following way. When two anodes pass current at the same time, both are at potential V_a (ignoring the arc drop). There is thus an effective short circuit between the ends of the two phases, and there is an inductance per phase under these conditions which we write as L, the *commutating inductance*. The equations for the changeover of current are thus—

$$LdI_{a1}/dt = V_1 - V_a, \quad LdI_{a2}/dt = V_2 - V_a.$$

Also because of the load reactance

$$I_{a1} + I_{a2} = I_{dc}.$$

Adding the first two equations we get

$$V_a = \tfrac{1}{2}(V_1 + V_2)$$

during the changeover. The anode voltage is V_1 at the point a, drops according to the last formula to the point b during the changeover, and jumps up to the point c which is V_2 at the end of the changeover. The equation for I_{a1} is

$$dI_{a1}/dt = (V_1 - V_a)/L = (V_1 - V_2)/2L,$$

by the last equation. If time is measured from the beginning of changeover, i.e. when $V_1 = V_2$, we have

$$V_1 = V \cos (\omega t + \pi/n),\ V_2 = V \cos (\omega t - \pi/n),$$

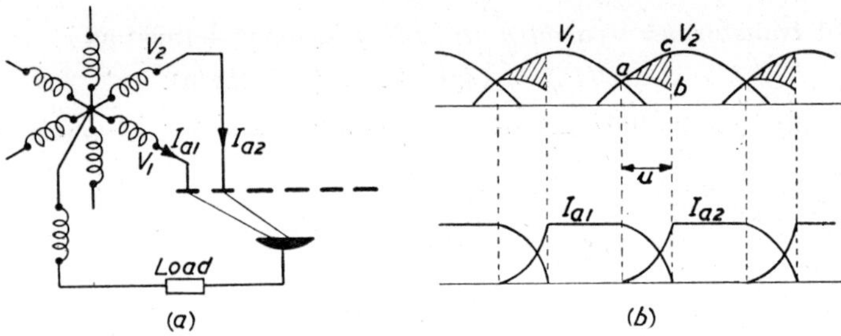

FIG. 413. EFFECT OF TRANSFORMER LEAKAGE INDUCTANCE

and thus $dI_{a1}/dt = -(V/L) \sin (\pi/n) \sin \omega t,$

$$I_{a1} = (V/\omega L) \sin (\pi/n) \cos \omega t + \text{constant}.$$

When $t = 0$, $I_{a1} = I_{dc}$, and hence

$$I_{a1} = I_{dc} + (V/\omega L)(\cos \omega t - 1) \sin (\pi/n),$$

$$I_{a2} = I_{dc} - I_{a1} = (V/\omega L)(1 - \cos \omega t) \sin (\pi/n).$$

These are the waveforms shown in Fig. 413 (*b*). The *commutation angle* u is given by putting $I_{a1} = 0$, so that

$$\cos u = 1 - I_{dc}\, \omega L/V \sin (\pi/n).$$

The average of the rectified voltage is diminished by the shaded areas of Fig. 413 (*b*) and straightforward calculation shows that the mean value is

$$\overline{V} = V(n/\pi) \sin (\pi/n) - (n/2\pi)\, I_{dc}\, \omega L.$$

The first term is the average derived above for the case of no leakage inductance, and thus there is a linear drop with load current in average voltage due to the leakage inductance.

EFFECT OF D.C. MAGNETIZATION. The Δ-Y connection of Fig. 412 (*a*) suffers from the following serious disadvantage. If three single-phase transformers are used the secondary currents, which are the anode currents, have a mean value of $\frac{1}{3}I_{dc}$, and there is a d.c. saturation of the core.

The Δ-zigzag connection of Fig. 414 (*a*) avoids this effect. In this arrangement the Δ arm between b and c is coupled to the winding A of the secondary 2 and the winding B of the secondary 1. It follows that the current in the secondary of winding bc is $I_{a1} - I_{a2}$, and it is seen from Fig. 411 (b) that the average value of this is zero. It can be seen from Fig. 411 (b) that the Δ-double Y is similarly free from d.c. magnetization.

Grid Control of Thyratrons. We have shown that, because of leakage reactance (and also because of resistance losses), the voltage

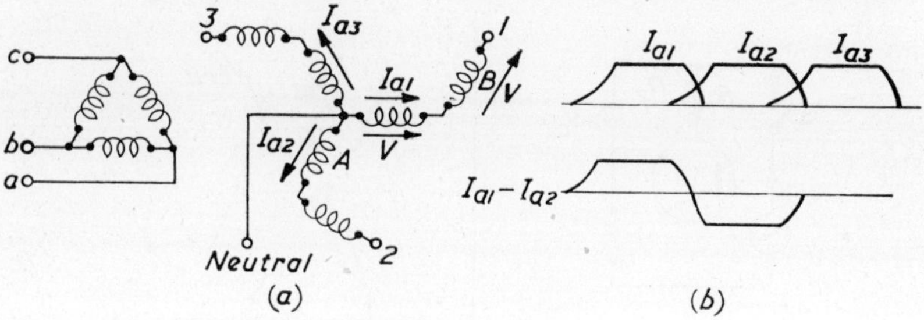

FIG. 414. Δ-ZIGZAG

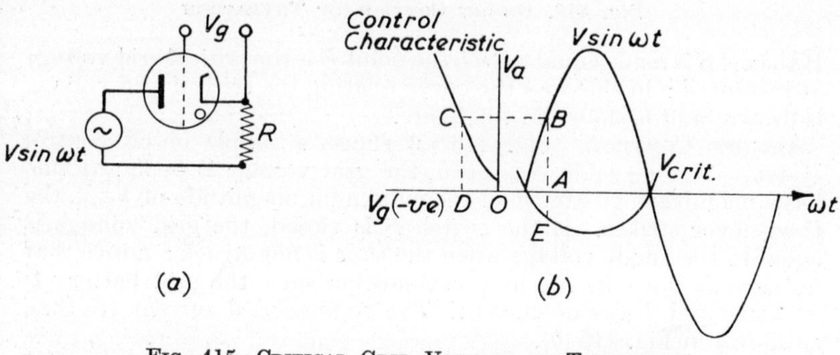

FIG. 415. CRITICAL GRID VOLTAGE OF THYRATRON

across the load falls with increasing current in the load. Where it is desired that the voltage remain constant with the load, as for example in traction systems, it is usual to maintain the output voltage constant by controlling the instant at which the anodes start to conduct. In the case of the ignitron this is done by controlling the instant of the current through the igniter, but in the case of the thyratron the method used is to control the starting instant by a voltage on the grid. We shall describe the case of the grid control of a thyratron, for the simple case of a purely resistive load.

Consider Fig. 415 (a) which shows a thyratron circuit with a resistance load R. When the thyratron is not conducting the anode-cathode voltage is the applied alternating voltage $V \sin \omega t$.

We can now deduce the value of the *critical grid voltage* $V_{crit.}$, which is such that a grid voltage below this does not fire the tube and a voltage above this starts the conduction. The left-hand curve of Fig. 415 (*b*) is the control characteristic (see Fig. 404), and it gives the grid voltage necessary to fire the tube for a given anode voltage. At an instant represented by A, the anode voltage (if the tube is not conducting) is given by AB: it is clear that if BC is horizontal and CD is vertical, OD is the critical grid voltage at the instant A.

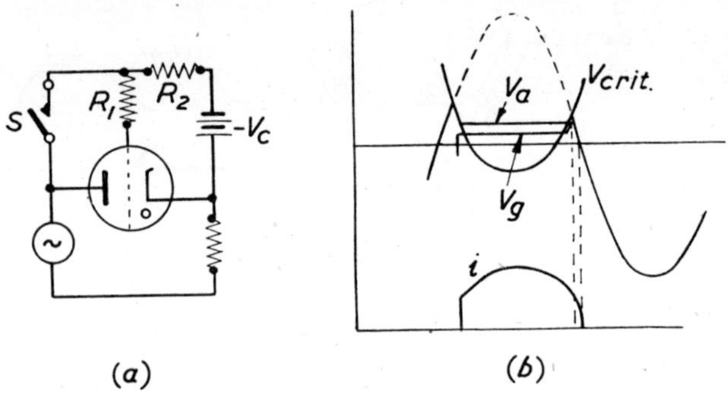

FIG. 416. ON-OFF CONTROL OF THYRATRON

If then AE is made equal to OD, the point E is the critical grid voltage at instant A: in this way the curve relating the critical grid voltage is drawn, and is shown in the figure.

ON-OFF CONTROL. Fig. 416 (*a*) shows a simple on-off control system. If the switch S is open, the grid voltage is $-V_c$; if this has a magnitude greater than the maximum magnitude of $V_{crit.}$, the tube never strikes. If the switch S is closed, the grid voltage is equal to the anode voltage when the tube is not struck: notice that no current flows in R_1, in this condition since the grid battery is negative and draws no current. The voltages and current are then as shown in Fig. 416 (*b*).

V_a and V_g follow the supply voltage (sinusoid) until they cross the curve of $V_{crit.}$. The tube then fires, the anode voltage drops to V_b (the arc drop), and V_g drops still further because of the voltage drop in R_1. Current continues to flow until the supply voltage reaches the value of V_b, when the arc extinguishes. V_a then follows the supply voltage, and V_g rises quickly to V_b and then follows the anode voltage.

CONTINUOUS CONTROL. Continuous control can be achieved by the method of Fig. 416 (*a*) provided the switch S is open and V_c is above the maximum negative value of $V_{crit.}$. Fig. 417 (*a*) shows the case for a small value of V_c and (*b*) for a value nearly equal to the maximum magnitude of $V_{crit.}$. The curve of $(I_{dc}/I_{dc\ max.})$ as a function of V_c is shown in (*c*).

When V_c is a small negative or a positive voltage, the anode conducts for about a full half-cycle, and the value of the mean current I_{dc} is the maximum value $I_{dc\ max.}$. When V_c becomes greater and negative the anode starts to conduct later in the cycle and I_{dc} decreases. The latest start is when V_c is just less than the magnitude

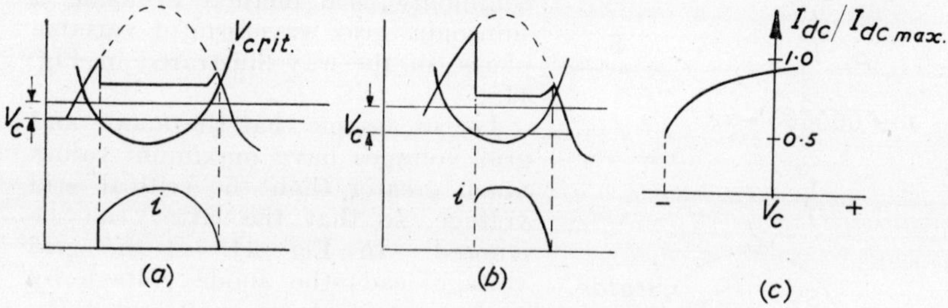

Fig. 417. Continuous Control of Thyratron

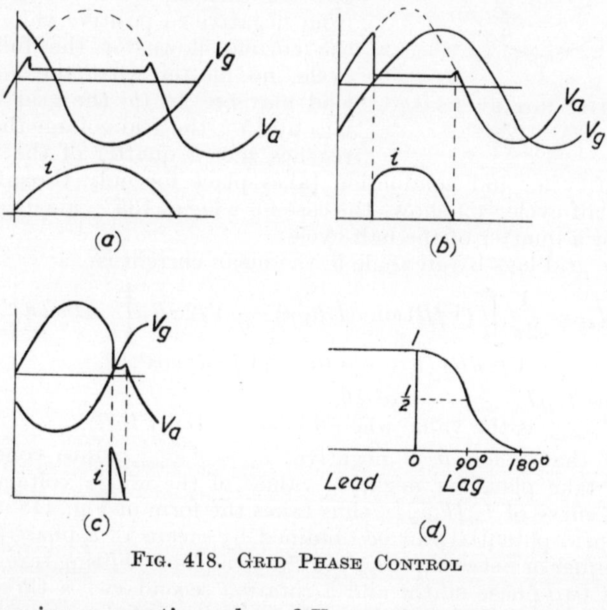

Fig. 418. Grid Phase Control

of the maximum negative value of $V_{crit.}$: the conduction starts 90° later and it is clear that I_{dc} is about $\frac{1}{2}$ $I_{dc\ max.}$. If V_c is below $V_{crit.}$, there is no conduction and hence the curve of I_{dc} drops abruptly to zero, in the way shown in Fig. 417 (*c*).

This method of control is poor, because the horizontal line of grid voltage V_c cuts the curve of critical grid voltage $V_{crit.}$ at a small angle. The result is that small changes of V_c and the characteristics of the thyratron will cause considerable variations of the output voltage. It is better to use a controlling grid waveform which cuts the curve of $V_{crit.}$ sharply: this is a general rule to observe in all coincidence or timing circuits. This object is achieved in the commonly used method of using a sinusoidal grid waveform of variable phase, in the way illustrated in Fig. 418.

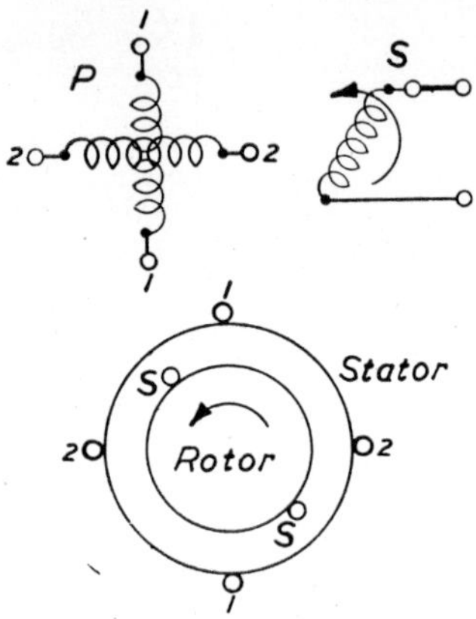

FIG. 419. ROTARY PHASE TRANSFORMER

Let us assume that the anode and grid voltages have maximum values much greater than the critical grid voltage, so that the latter can be ignored. In Fig. 418 (*a*) the grid voltage leads the anode voltage by a phase angle between 0 and 180°: this means that the grid voltage is positive when the anode voltage passes from negative to positive values, and the current flows for the full half-cycle, no matter what the angle of lead may be. In (*b*) the grid voltage lags by 45°: the grid voltage therefore reaches zero a quarter of the way in the half-cycle, and conduction takes place for only three-fourths of the half-cycle. (*c*) shows the case for a lag of 135°: anode current flows for a quarter of the half-cycle.

If the grid lags by an angle θ, the mean current is

$$I_{dc} = \frac{1}{2\pi}\int_\theta^\pi (V/R)\sin\omega t\, d(\omega t) = (V/2\pi R)\Big[-\cos\omega t\Big]_\theta^\pi$$
$$= (V/2\pi R)(1 + \cos\theta) = (V/\pi R)\cos^2\tfrac{1}{2}\theta,$$

and thus $I_{dc}/I_{dc\ max.} = \cos^2\frac{1}{2}\theta$,
since $I_{dc\ max.}$ is the value when θ is zero, viz. $(V/\pi R)$.

Note that when θ is negative, $I_{dc} = I_{dc\ max.}$ since conduction cannot take place for negative values of the anode voltage. The control curve of $I_{dc}/I_{dc\ max.}$ thus takes the form of Fig. 418 (*d*).

The grid potential can be obtained by means of a phase-shifting transformer or network. Fig. 419 shows a phase-shifting transformer using a two-phase stator and a rotating secondary: a three-phase stator could be used equally well. The phase of the voltage in the secondary is tuned simply by rotating the winding in the rotating field caused by the primary. Thus if the flux due to the winding 1 is $\Phi\cos\omega t$ and that due to winding 2 is $\Phi\sin\omega t$ in the planes perpendicular to their windings, the flux in a plane at angle θ is given by $\Phi\cos\omega t\,.\cos\theta + \Phi\sin\omega t\,.\sin\theta = \Phi\cos(\omega t - \theta)$

so that the voltage induced in the secondary has a phase which varies with its positional angle θ.

Fig. 420 (a) shows a phase-shifting circuit commonly used in thyratron control systems. The voltage V_a between C and B is in phase with V', that between B and A: we take these as at zero

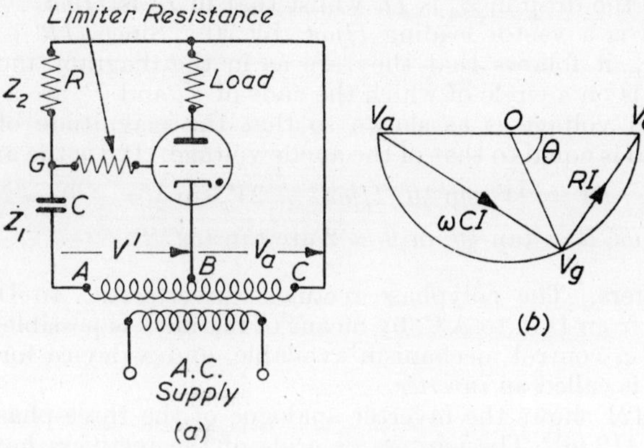

FIG. 420. GRID PHASE-SHIFTER

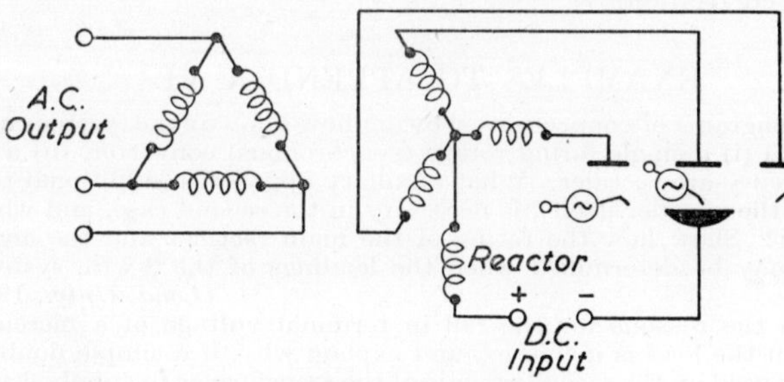

FIG. 421. INVERTER

phase. The potential at the grid with respect to the cathode in the non-conducting state, is given by

$$V_g = (Z_1 V_a - Z_2 V')/(Z_1 + Z_2)$$

since C is at V_a and A at $- V'$. If for simplicity we take $V' = V_a$, we have $V_g = V_a\,(Z_1 - Z_2)/(Z_1 + Z_2)$, and the relative phase is that of the impedance ratio $(Z_1 - Z_2)/(Z_1 + Z_2)$. A very useful case is when $Z_1 = 1/j\omega C$ and $Z_2 = R$,

when
$$\begin{aligned} V_g/V_a &= (1/j\omega C - R)/(1/j\omega C + R) \\ &= (1 - j\omega CR)/(1 + j\omega CR) \\ &= 1 \angle 2 \text{ arc tan } (\omega CR). \end{aligned}$$

Thus the grid voltage has the magnitude of the anode voltage but lags by a phase angle 2 arc tan (ωCR). Fig. 420 (*b*) shows the vector diagram for this case, in which control is possible. Voltages are with respect to the cathode. The potentials V_c and V_a are shown horizontal. If the current I flows downwards through Z_2 and Z_1, the drop in Z_2 is IR whilst that in Z_1 is $I/j\omega C$: it is clear that IR is a vector leading $I/j\omega C$ by 90°. Since $(IR + I/j\omega C)$ is $V_c - V_a$, it follows that they are as in the diagram and the grid voltage is on a circle of which the ends of V_c and V_a are a diameter. The grid voltage is as shown so that the magnitude of the grid potential is equal to that of the anode voltage. If θ is the angle of lag

$$RI = 2V_c \sin \tfrac{1}{2}\theta,\ I/j\omega C = 2V_c \sin \tfrac{1}{2}(\pi - \theta) = 2V_c \cos \tfrac{1}{2}\theta,$$

so that $\omega CR = \tan \frac{1}{2}\theta$, or $\theta = 2 \text{ arc tan } (\omega CR)$.

Inverters. The polyphase rectifier converts A.C. to D.C.: conversion from D.C. to A.C. by means of rectifiers is possible, provided there is a control mechanism available, and a device for this conversion is called an *inverter*.

Fig. 421 shows the inverter analogue of the three-phase rectifier of Fig. 412 (*a*). The igniters or grids of the rectifiers have control waveforms: thus grids will have a three-phase alternating supply of the desired frequency.

EXAMPLES TO APPENDIX VI

1. Draw diagrams of connections showing how a 3-wire d.c. system may be supplied from (i) a single 6-ring rotary (synchronous) converter, (ii) a single 6-anode mercury-arc rectifier. What auxiliary apparatus (additional to that required for the rectifier itself), is necessary in the second case, and what are its functions? Show how the rating of the main rectifier and the auxiliary equipment may be determined when the loadings of the 3-wire system are known. (*Lond. Univ.*, 1947.)

2. Discuss the reasons for the fall in terminal voltage of a mercury-arc rectifier when the load is increased, and explain why, if a simple double-star connection is used on the secondary side of the transformer to supply 6 anodes, the 3-phase primary should be delta-connected.

A rectifier of this type gives a mean output voltage of 480 V. when the effective resistance of the load is 2 Ω. Neglecting losses and overlap, determine the maximum value of the e.m.f. and current in each winding of the transformer. The supply voltage is 6·6 kV. Justify any formula used for calculating the alternating voltage applied to the rectifier. (*Lond. Univ.*, 1949.)

3. Discuss the chief reasons why the mercury-arc rectifier has displaced rotating plant in substations for the conversion of alternating current into direct current.

Draw a diagram of connections for a 6-anode mercury-arc rectifier showing (i) the main connections of the rectifier and transformer, (ii) the arrangements for excitation and ignition.

A 6-anode rectifier is rated at 1 000 kW., 650 V. Calculate the terminal voltage of each phase of the transformer secondary winding at full load, ignoring overlap and assuming the voltage drop in the arc to be 20 V. (*Lond. Univ.*, 1948.)

APPENDIX VII

HEATING, WELDING, AND ELECTROCHEMICAL PROCESSES

High-frequency Heating. There are two forms of heating by an alternating field. In one form an alternating magnetic field is produced, and this causes hysteresis and eddy-current losses in a magnetic material or eddy-current losses only in a non-magnetic conductor: this method is known as *induction heating*. If the material is magnetic, the frequency of the applied magnetic field is relatively low, say 50 c/s to 10 kc/s: in the case of a non-magnetic conductor, the frequency of the field is from 250 kc/s to 1 Mc/s.

In the other form of heating, an alternating electric field is applied to the material, which is a lossy dielectric. The frequency is high, a value of 150 Mc/s being used in practice. An advantage of this form of heating is that the heat is generated in the body of the material: external heat is more difficult to apply as such materials are poor thermal conductors, and the outside gets much hotter than the inside.

Induction Heating. The melting of ferrous and non-ferrous metals by induction heating is a common process. Low frequencies are used, and the supply is by an alternator.

The hardening and tempering of ferrous metals is an important process in the making of tools, gear wheels, shafts, bearings, etc. It is often desired to harden a thin skin of the metal, and in this case a high-frequency field is used with a small *skin depth.*

There are many processes such as the soldering of cans, brazing, sintering of metals, heating for bending, annealing, etc. There is also the very important method of heating the electrodes of a valve during the evacuation process, and the final firing of the getter.

The frequency of supply, the time of application and the power supplied depend upon the job: we can give only the basic principles involved.

SKIN DEPTH. In this case the skin depth depends upon the permeability of the metal, if it is magnetic, and this in turn depends upon the temperature. In the case of iron, if the temperature is below about 780° C., the metal is ferro-magnetic and the skin depth is given by

$$\delta = 0{\cdot}42/\sqrt{f} \text{ inches,}$$

where f is the frequency in c/s.

Above 780° C. the depth is given by

$$\delta = 22/\sqrt{f} \text{ inches,}$$

since the permeability is now unity.

In the non-magnetic case

$$\delta = 1\,950\sqrt{(\rho/\mu f)} \text{ inches,}$$

where ρ is the resistivity in $\Omega. \times$ cm. and μ = permeability = 1.

Thus for iron above 780° C., $\rho = 120 \,.\, 10^{-6}\,\Omega. \times$ cm., and the third formula reduces to the second. For copper $\rho = 2 \,.\, 10^{-6}\,\Omega. \times$ cm., and $\delta = 2{\cdot}3/\sqrt{f}$ inches.

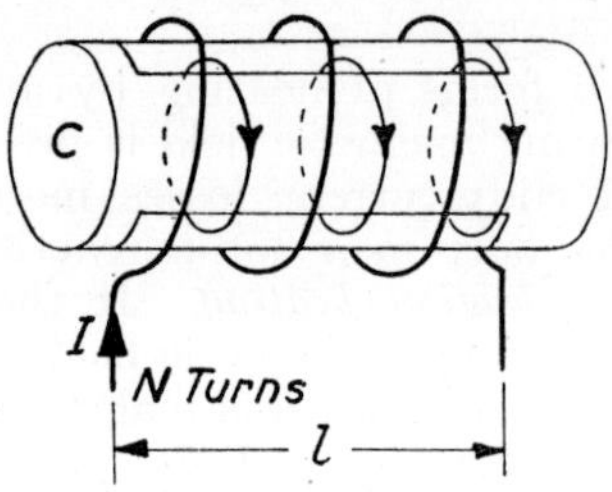

FIG. 422. EDDY-CURRENT HEATING

POWER GENERATED IN THE MATERIAL. A rough idea of the power transferred to the material can be obtained in the following way. Fig. 422 shows a cylindrical billet of material of resistivity ρ and permeability μ. A coil of N turns carrying a current I induces a billet current of approximately $I' = NI$ in the direction shown. This current flows in a path l in. wide, circumference c in. long and skin depth δ in. thick, so that the resistance is $\rho c/l\delta$, and the power is thus

$$W = (\rho c/l\delta)\, I'^2 = (\rho c/l\delta)\,(NI)^2,$$

and the power per unit area is

$$W/\text{in.}^2 = W/cl = (NI)^2\,(\rho/l^2\delta)$$
$$= 5 \,.\, 10^{-4}(NI/l)^2\sqrt{(\mu\rho f)} \text{ watts/in.}^2$$

In this equation ρ is in $\Omega. \times$ cm., f in c/s, l in in. For iron below 780° C.,

$$W/\text{in.}^2 = 4 \,.\, 10^{-6}(NI/l)^2\sqrt{f} \text{ watts/in.}^2.$$

Thus for 1 kW./in.² at f = 50 c/s, the ampere-turns per inch required are about 6 000. At 1 000 c/s the value is 1 300.

In the case of surface hardening, the frequency is of the order of 500 kc/s, the input power is of the order of 10 kW. per square inch (corresponding to about 1 000 ampere-turns per inch), and the heating time is about one second to form a hard skin of a few thousandths of an inch. *Quenching* may be produced by switching on a water spray when the heat is switched off: or in the case of rapid surface heating, the cooling after switching off may be quick enough to cause the required hardening.

In the application of the heating of valve electrodes, the frequency is again about 500 kc/s. The coil is slipped over the valve and the electrodes glow a dull red heat. For large valves, the coil is often a copper tube with water flowing through it.

There is an interesting process, called *tin brightening*, in which a 200 kc/s field is applied to the surface of a metal with an electrolytic tin plating. The power applied is about 250 W./in.² of surface. In an example, the strip runs at 1 000 ft./minute and a power of 1 200 kW. is required.

Fig. 423 shows a Hartley oscillator with a transformer output to the work coil. By choice of transformer ratio, a large current can be produced in the work coil.

Dielectric Heating. A lossy dielectric is shown in Fig. 424, and

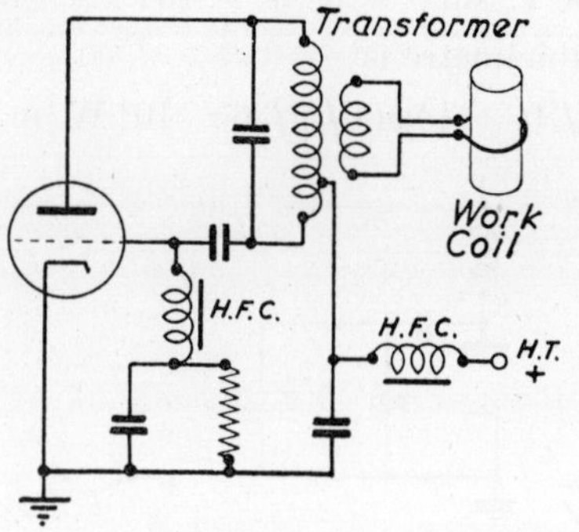

FIG. 423

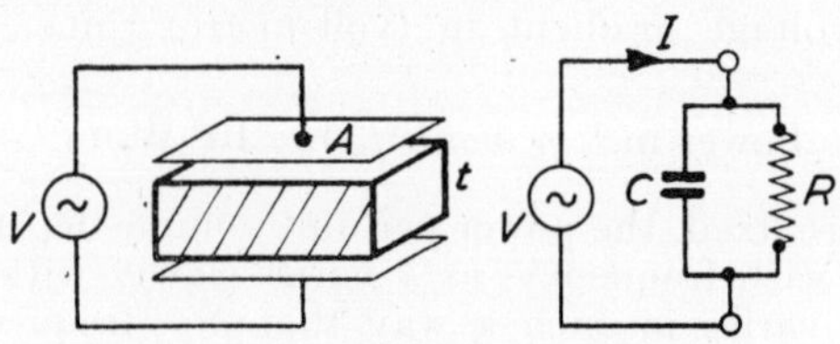

FIG. 424. DIELECTRIC HEATING

it is represented by a parallel combination of a capacitance C and resistance R at a given frequency. If the dielectric is a slab of area A metre² and thickness t metre, the capacitance is

$$C = \varepsilon_0 kA/t = (kA/36\pi \,.\, 10^9 \,.\, t) \text{ farad},$$

where k is the dielectric constant. If a voltage of peak value V is applied, the current is

$$I = j\omega CV + V/R = j\omega CV(1 + 1/j\omega CR),$$

whilst the loss in the resistance is $\frac{1}{2}V^2/R$, the half being present because of the mean value of a sinusoid. The loss angle is defined as δ where

$$\tan \delta = 1/\omega CR,$$

and is illustrated in Fig. 425. It follows that

$$I_R = V/R = \omega CV \tan \delta,\ I = \omega CV \sec \delta = I_c \sec \delta,$$

so that the power developed in the slab is

$$W = \tfrac{1}{2}V^2/R = \tfrac{1}{2}\omega CV^2 \tan\delta = \tfrac{1}{2}VI_c \tan\delta$$
$$= \tfrac{1}{2}VI \sin\delta.$$

When the loss angle δ is small,

$$I \doteqdot j\omega CV,\ \sin\delta \doteqdot \delta,\ W = \tfrac{1}{2}\delta VI = \tfrac{1}{2}\delta\omega CV^2$$

The power per cubic metre is

$$W/At = \tfrac{1}{2}\delta\omega(V/t)^2/36\pi \,.\, 10^9 \text{ W./m.}^3$$

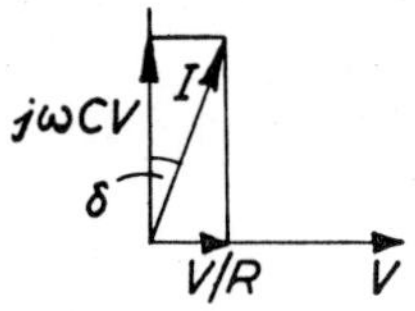

FIG. 425

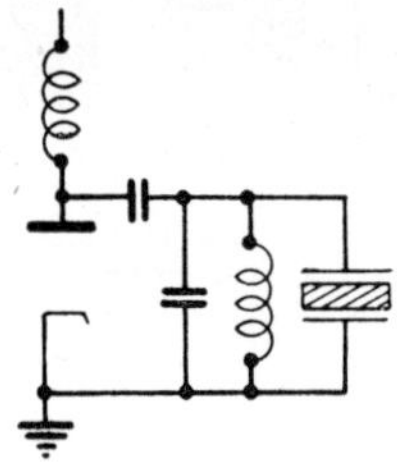
FIG. 426

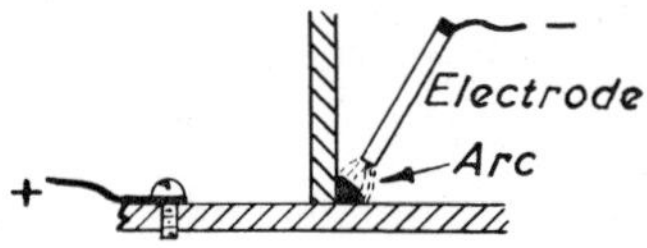

FIG. 427

If E is the voltage gradient in (volt/metre) r.m.s., the formula becomes

$$\text{Power/m.}^3 = \delta\omega kE^2/36\pi \,.\, 10^9 \text{ W./m.}^3$$

If V and t are fixed, the power per unit volume depends upon $\delta\omega$: this increases with frequency, as k varies slowly with frequency if at all, and δ varies in such a way that $\delta \times$ frequency increases fairly rapidly with frequency. In practical units

$$\text{Power/cm.}^3 = \delta kfE^2/18 \,.\, 10^{11} \text{ W./cm.}^3$$

if E is in (volt/cm.) r.m.s.

The range of frequencies used is 2 to 15 Mc/s.

One of the important things is to keep the voltage gradient across the slab at a reasonable value to prevent flashover: a value of 1 to 2 kV./cm. is reasonable. An air gap is a disadvantage, since the voltage gradient in the gap is k times that in the material.

Dielectric heating is applied to annealing of rayon and nylon, bonding of plywood and plastics, curing of rubber and plastics, etc. Fig. 426 shows an output circuit for dielectric heating.

Welding. In welding, metal parts are heated to melting point and adhere on solidification. There are two main ways of supplying the heat by electrical power, and they are known as *resistance* and *arc* welding.

Fig. 427 shows diagrammatically the arrangement in arc welding.

An arc is struck between an electrode and the work and the heat of the arc melts the filling material to form a fillet between the parts to be welded together. The electrode may be a carbon rod up to 1 inch diameter and carries a current up to 800 A.: as the arc voltage is 20 to 30 V., the power is up to 20 kW. The electrode may be of metal and can itself provide the fillet. In welding iron and steel, the electrode is a low-carbon iron; and it often has a covering layer of a compound containing salts of silicon, calcium or magnesium, which forms a slag to prevent oxidation.

In resistance welding a heavy current is passed between the metals to be welded: the contact point between the metals has a relatively high resistance and sufficient heat is produced to cause a weld.

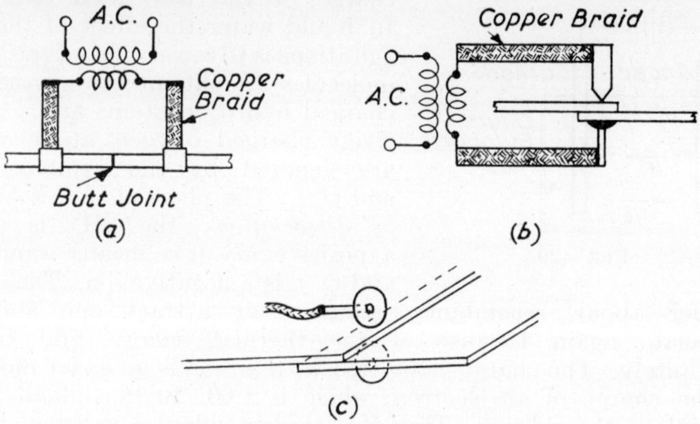

FIG. 428. TYPES OF WELDING

(*a*) Butt welding. (*b*) Spot welding. (*c*) Seam welding.

Fig. 428 shows three common methods of resistance welding. In (*a*) the work is butt welded, in (*b*) there is a spot weld, and in (*c*) a seam weld is formed by two electrodes which take the form of rotating wheels. It is usual for the supply to be A.C., the large currents being produced by step-down transformers.

In many applications of spot welding, e.g. in valve manufacture, it is important to control the heat supplied to the weld within close limits. Circuits have been devised which give one half-wave of current of a controlled valve on the pressing of a switch or pedal.

Electrochemical Processes. The use of electricity to produce chemical changes has many applications, e.g. extraction of pure metals from ores, electrodeposition and refining of metals, electroplating, electrotyping, production of oxygen, chlorine, caustic soda, etc.

The principle of the method was discovered by Humphry Davey, who was thus the first to be able to produce metallic sodium and potassium. The laws of electrolysis were discovered by Faraday, and they are related closely to the modern electrical theory of matter. So far as electrochemical processes are concerned we may consider matter as composed of atoms, which themselves are made up of a positively charged nucleus and a surrounding cloud of electrons. Atoms combine to form molecules, in which the atoms are held together by the attractive forces between the nuclei and the electrons. Thus a water molecule is written as H_2O, and is electrically neutral. This means that the two hydrogen nuclei and the one oxygen nucleus have together a positive charge equal to the negative charge of the associated electrons. In liquid water the effect of thermal agitations is to cause some of the water molecules to split into two positively charged hydrogen atoms and a negatively charged oxygen atom, which are denoted by the symbols $2H^+$ and O^{--}. The phenomenon is known as *dissociation*; the $2H^+$ is called a positive *ion* (ion means wanderer) and O^{--} is a negative ion. These ions wander about, recombine because they attract one another, dissociate again because of the thermal energy, and so on indefinitely. The charge associated with an ion is an exact multiple of the charge of an electron, which is $1{\cdot}60 \, . \, 10^{-19}$ coulomb: the multiple is the valency. Thus O^{--} corresponds to a valency of two, H_2^{++} or $2H^+$ corresponds to a valency of unity. As an example in aluminium chloride, $AlCl_3$, the ions are Al^{+++} and $3Cl^-$, the aluminium being trivalent and the chlorine monovalent.

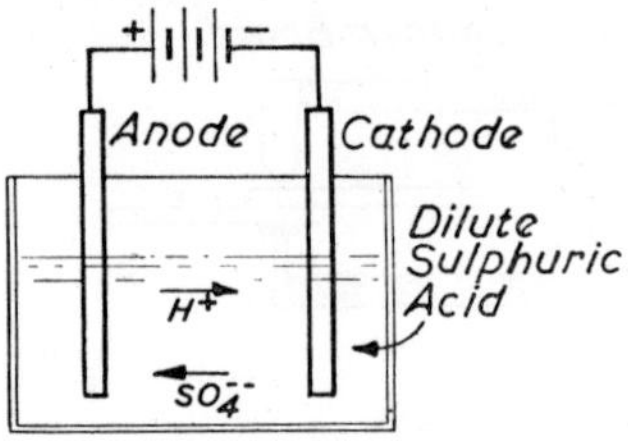

FIG. 429

Because the ions are charged, it is possible to cause a steady drift of the positive ions in one direction and of the negative ions in the other by the application of an electrical potential. Fig. 429 shows diagrammatically an *electrolytic cell* in which this process is produced. Suppose two electrodes dip into acidulated water, which has a copious supply of positive hydrogen ions and negative sulphions (SO_4^{--}). A direct potential is applied between the electrodes; the positive electrode is called the *anode* and the negative is the cathode. The hydrogen ions are repelled from the anode towards the cathode, where they contact the cathode, give up their positive charge and are thus converted to hydrogen atoms and then to hydrogen molecules. The sulphions reach the anode, react with a water molecule to produce a sulphuric acid molecule and an atom of oxygen, which gives up its negative charge. The result is that hydrogen is given up at the cathode and oxygen at the anode, assuming that the electrodes do not take part in the chemical action. The charge necessary to produce an atom of hydrogen is $1{\cdot}60 \,.\, 10^{-19}$ coulomb, and as an atom of hydrogen has a mass of $1{\cdot}66 \,.\, 10^{-24}$ g., it follows that 1 coulomb produces $(1{\cdot}66 \,.\, 10^{-24}/1{\cdot}60 \,.\, 10^{-19}) = 1{\cdot}04 \,.\, 10^{-5}$ g. of hydrogen. This is called the *electrochemical equivalent* of hydrogen.

Since an atom carries as many electron charges as its valency, it follows that the electrochemical equivalent of an atom is obtained by multiplying $1{\cdot}04 \,.\, 10^{-5}$ g. by the atomic weight and dividing by the valency. Thus the electrochemical equivalent of oxygen is $(1{\cdot}04 \,.\, 10^{-5} \times 16/2) = 8{\cdot}3 \,.\, 10^{-5}$ g. The following table gives the electrochemical equivalent of some elements, and also the weight carried in 1 ampere-hour = 3 600 coulombs.

Element	Electrochemical Equivalent (mg./coulomb)	Grammes/Ah.
H^{+}	0·0104	0·0374
K^{+}	0·405	1·459
Na^{+}	0·239	0·859
Al^{+++}	0·094	0·340
Mg^{++}	0·124	0·447
Cu^{++}	0·327	1·177
Zn^{++}	0·337	1·213
O^{++}	0·083	0·305
Cl^{+}	0·367	1·322

EXAMPLE. Find the charge to deposit a one-thou. plating of copper on a sheet of iron of 1 square foot.

The volume of copper is $2 \times 144 \times 0{\cdot}001 = 0{\cdot}288$ in.^{3}, and the weight is $0{\cdot}288 \times 0{\cdot}32 = 0{\cdot}092$ lb. = 41·7 g. The charge needed is thus $(41{\cdot}7/1{\cdot}177) = 35{\cdot}3$ Ah.

The current density in a copper-plating bath is about 30 A./ft.^{2} on each side of the sheet, so that the total current is 60 A. and the time taken is 0·59 hour = 35 min. In practice the time required is 10 or 20 per cent more because of local secondary circuits.

The potential required to pass the current depends on the contact potentials of the ions (or dissociation energy) and the resistance of the electrolytic cell. The dissociation potential for water is 1·47 V.

EXTRACTION OF METALS. The ore is rendered an electrolyte in a way depending upon the ore. Thus bauxite, which contains hydrated alumina, is dissolved in fused cryolite (sodium-aluminium fluoride) and is then electrolysed. Magnesium, sodium and potassium are similarly electrolysed in a fused mixture of salts. Copper is obtained by electrolysis of a copper sulphate solution. The following table summarizes the extraction of some metals.

Metal	Solution	kWh./ton
Aluminium	Fused cryolite	20 000
Copper .	Copper sulphate	2 000
Magnesium	Fused Mg Cl_2	20 000
Sodium .	Fused sodium hydrate, nitrate or chloride	15 000
Zinc . .	Zinc chloride or zinc sulphate	4 000

ELECTRODEPOSITION. The deposition of metal by electrolysis is used in many applications. It includes electroplating, whereby a layer of metal is deposited upon another for the purposes of protection, ornamentation, etc.: the layer may be copper, silver, gold, cadmium, rhodium (for good rubbing contacts), etc. The subject is vast and there are many trade secrets, which give a rapid method of depositing a well-adhering layer, sometimes in acute angles.

In some applications the metal is deposited on a non-metal, e.g. in the making of gramophone records.

An interesting example of electrolysis is the anodic oxidation of aluminium. The aluminium is made by the anode of an electrolytic cell containing sulphuric acid or chromic acid. The oxygen liberated at the anode forms a very hard and adherent film of aluminium oxide, which acts as a very good protecting layer.

PRODUCTION OF CHEMICALS. One of the most important processes is the electrolysis of brine. The action may be represented by the formalae

$$NaCl \rightarrow Na^+ + Cl^-, \; Na^+ + H_2O \rightarrow NaOH + H^+.$$

Chlorine is given off at the anode, hydrogen at the cathode, and caustic soda (NaOH) is left in the cell.

The following list gives some important processes.

Solution	Product	kWh./ton
Brine . .	Caustic soda	3 500
Brine . .	Sodium chlorate, perchlorate, etc.	3 000 to 7 000
Potassium manganate .	Potassium permanganate	75
Water . .	Oxygen and hydrogen	150 kWh. for 1 000 ft.^{3} hydrogen and 500 ft.^{3} oxygen

APPENDIX VIII

PRINCIPLES OF NUCLEAR ENERGY

Structure of Atoms. Matter is composed of atoms which are normally unchangeable. Atoms are the smallest particles characteristic of a given substance, say carbon or iron, and were once thought to be indivisible. There is abundant evidence that atoms are composed of smaller, *elementary particles* which make up all atoms.

The three most important elementary particles are the *electron*, *proton*, and *neutron*; there are also the positron, antiproton, a host of mesons, and the neutrino. The electron in the free state behaves as a particle with a charge $-e$ and a mass m where

$$e = 1{\cdot}60 \,.\, 10^{-19} \text{ coulomb}, \; m = 9{\cdot}1 \,.\, 10^{-31} \text{ kg.} \quad . \quad \text{(viii.1)}$$

The proton has a charge $+e$ and mass $m_n = 1\,849m$, whilst the neutron has no charge and a mass $m_{\text{u}} = 1{\cdot}0012m_{\text{p}}$. If the atom ${}_{8}{}^{16}\text{O}$ is taken as atomic weight 16·000, as is customary,

$$\left.\begin{array}{ll}\text{then} & m_{\text{p}} = 1{\cdot}007593, \; m_n = 1{\cdot}008982 \\ \text{and} & 1 \text{ atomic mass unit} = 1{\cdot}6598{\cdot}10^{-27} \text{ kg.}\end{array}\right\} \quad . \quad \text{(viii.2)}$$

An atom is composed of a *nucleus* of diameter of the order of 10^{-12} cm.: this nucleus is small compared with the atom, which has a diameter of the order of 10^{-8} cm., and contains nearly the whole of the mass of the atom. The nucleus contains protons and neutrons: around it swarm orbital electrons. In the simplest case, viz. the hydrogen atom, the nucleus is simply a proton and there is one orbital electron to give zero total charge. If the *atomic weight* of the atom is A, the number of protons and neutrons together is A; if there are Z protons in the nucleus, Z is the *atomic number* of the atom, and there are $(A\text{–}Z)$ neutrons. The nucleus of an atom of atomic weight A and atomic number Z thus has the weight of about A protons (i.e. about A hydrogen atoms) and a charge $+Ze$. This nucleus is normally surrounded by Z orbital electrons, each having a charge $-e$, so that the neutral atom has no charge and atomic weight A. The electrons associated with a nucleus do not behave as particles but exhibit a wave structure, and wave mechanics was developed to explain their behaviour.

Apart from weight, the chemical and physical properties of an atom are due to the cloud of electrons surrounding the nucleus. An atom X of atomic number Z and atomic weight A is written as ${}_{Z}{}^{A}X$ e.g. ${}_{8}{}^{16}O$ is oxygen with atomic number 8 and atomic weight 16.

As the nucleus is composed of protons and neutrons of about the same mass, it is expected that all atomic weights should be multiples of that of the hydrogen atom. There are three reasons why atomic weights are not integers. First the masses of the proton and neutron are not equal. Secondly it has been found that an element is usually a mixture of atoms with the same chemical properties but of different atomic weights: such atoms are known as *isotopes:* thus chlorine, atomic weight 35·46, is a mixture of ${}_{17}^{35}Cl$ and ${}_{17}^{37}Cl$. Thirdly, the theory of relativity has shown that mass and energy are interchangeable, according to the formula

$$W = mc^2 \quad . \quad . \quad . \quad . \quad \text{(viii.3)}$$

where W is the energy, m the mass, and $c = 3 \cdot 10^8$ m./sec. = velocity of light. If m is in kg., W is in joules. In a nucleus the mass depends upon the binding energy between the protons and the neutrons.

Periodic Table of the Elements. Starting with a nucleus of atomic number 1, i.e. a proton, a neutral (i.e. an uncharged) atom has only 1 electron. This is the hydrogen atom ${}_1^1H$. We say that the electron is in the 1*s* state: electrons are arranged in states or shells, which are labelled progressively *s, p, d, f, g,* . . .: the first three letters stand for *sharp*, *principal*, and *diffuse*, and are names taken over from spectroscopy. The electron is thus written as $1s^1$.

There are isotopes of hydrogen, viz. *deuterium*, which has atomic weight 2 and is thus ${}_1^2H$, and *tritium* of atomic weight 3 and is thus ${}_1^3H$.

If $Z = 2$ the nucleus has two protons and two neutrons, and there are two electrons in the 1*s* state. We write the atom, which is helium, as ${}_2^4He = 1s^2$. If $Z = 3$, there are 3 protons and 4 neutrons in the nucleus: the atom is lithium and is written ${}_3^7Li = 1s^2 2s^1$, two electrons being in the 1*s* state and 1 electron in the 2*s* state. The 2*s* electron is much farther from the nucleus than the 1*s* electrons, and requires only one-quarter of the energy to remove it from the atom. Lithium is therefore more readily ionizable and is a conductor in the solid state. It also has unit valency.

There can be a maximum of 2 electrons in the 1*s* state, 2 electrons in the 2*s* state, 6 in the 2*p* state, 2 in the 3*s* state, 6 in the 3*p* state, and so on. It is sometimes convenient to consider *shells* of electrons, i.e. electrons at about the same energy level. The energy is determined mainly by the number attached to the state: thus we call the first shell K, and it contains the electrons of a 1 state, i.e. the 1*s*, and thus there can be a maximum of 2 electrons in the K shell. Hydrogen has 1 electron in the K shell, helium has 2, and so have all other atoms. The next shell is L and contains all electrons of a 2 state, i.e. 2*s* and 2*p*, so that there is a maximum of 8 electrons in the L shell. Similarly there is a maximum of 8 electrons in the M shell, 18 in the N shell, 18 in the O shell, and so on. Note that the maxima are as follows: L shell 2.1^2, M shell 2.2^2, N shell 2.2^2, O shell 2.3^2, and so on.

The chemical properties of the atom are determined mainly by the electrons in the outermost shell: if there is one electron in this shell the atom is monovalent, if there are two it is divalent, and so on up to 7, when it is septavalent. If the outermost shell is complete, i.e. it has 2, 8, 18 . . . electrons, the atom has no tendency to combine with other atoms and the element is one of the inert gases, He, Ne, A, etc. By arranging the atoms in order of atomic number we achieve the periodic table as shown on p. 514.

Let us take as an example the sodium atom: it has an atomic number 11 and weight 23, and thus has 11 protons and 12 neutrons in the nucleus and is written ${}_{11}{}^{23}\mathrm{Na}$. Its electron structure is ${}_{11}{}^{23}\mathrm{Na} = 1s^2 2s^2 2p^6 3s^1$ or $K_2L_8M_1$.

Radioactivity. It was found at about 1898 that certain atoms gave off radiatons of particles and electromagnetic energy, and it has been established that this is due to changes of the nuclei of these atoms: the phenomenon is called *radioactivity*, and is not affected by temperature, pressure, or similar physical conditions. In these spontaneous changes the main emissions are of three types, α, β, and γ: α emissions are atoms of helium ${}_{2}{}^{4}\mathrm{He}$, β emissions are electrons, and γ emissions are electromagnetic waves. One of the most radioactive atoms is radium, ${}_{88}{}^{266}\mathrm{Ra}$, which disintegrates through a series of atoms to finish up as an isotope of lead. In fact radium itself is produced by a series of disintegrations from uranium ${}_{92}{}^{238}\mathrm{U}$.

Before describing the method of indicating nuclear disintegrations we shall say a word about the rapidity of the phenomenon. The disintegration of the atoms occurs in a random way, but in a given large number of atoms the rate of disintegration is given by a time called the *half life*: this is the time in which half of the atoms disintegrate. In twice this time only a quarter will remain unchanged, in thrice this time only one-eighth will remain, and so on. The half life is a very important property, as it determines the relative abundances of the various products in the mass of material. The half life varies from infinity for a stable atom, to 10^{-14} sec. for a very unstable nucleus. Thus for ${}_{92}{}^{238}\mathrm{U}$ the half life is $4{\cdot}5 \times 10^{10}$ yr., for ${}_{88}{}^{226}\mathrm{Ra}$ it is $1{\cdot}62 \times 10^3$ yr., and for radon (gas) ${}_{86}{}^{222}\mathrm{Rn}$ it is only 3·82 days.

We now describe the change of an atom when it emits an α or a β radiation. Let the atom be ${}_{Z}{}^{A}X$. If A emits an α, which is ${}_{2}{}^{4}\mathrm{He}$, it must lose two protons so that Z decreases to $Z - 2$ and four atomic masses so that A decreases to $A - 4$. The reaction is then written

$${}_{Z}{}^{A}X \rightarrow {}_{Z-2}{}^{A-4}Y + {}_{2}{}^{4}\mathrm{He}\ (\alpha \text{ emission}) \quad . \quad . \quad \text{(viii.4)}$$

If it emits a β, which is an electron of charge $-e$, one of the neutrons in the nucleus must acquire a charge $+e$, so that it becomes a proton: the atomic number therefore increases by one, but the atomic weight is unchanged. The reaction is thus

$${}_{Z}{}^{A}X \rightarrow {}_{Z+1}{}^{A}Y + \beta\ (\beta \text{ emission}) \quad . \quad . \quad \text{(viii.5)}$$

PERIODIC TABLE

Group	I a	I b	II a	II b	III b	III a	IV b	IV a	V b	V a	VI b	VI a	VII	VII	VIII	VIII	VIII	0
Period 1	H1													H1				He2
Period 2	Li3		Be4			B5		C6		N7		O8		F9				Ne10
Period 3	Na11		Mg12			Al13		Si14		P15		S16		Cl17				A18
Period 4	K19	Cu29	Ca20	Zn30	Sc21	Ga31	Ti22	Ge32	V23	As33	Cr24	Se34	Mn25	Br35	Fe26	Co27	Ni28	Kr36
Period 5	Rb37	Ag47	Sr38	Cd48	Y39	In49	Zr40	Sn50	Cb41	Sb51	Mo42	Te52	Tc43	I53	Ru44	Rh45	Pd46	Xe54
Period 6	Cs55	Au79	Ba56	Hg80	La–Lu 57–71	Tl81	Hf72	Pb82	Ta73	Bi83	W74	Po84	Re75	At85	Os76	Ir77	Pt78	Rn86
Period 7	Fr87		Ra88															

Actinide Series											
Ac89	Th90	Pa91	U92	Np 93	Pu 94	Am 95	Cm 96	Bk 97	Cf 98	...103

Rare Earths	La57	Ce58	Pr 59	Nd 60	Pm 61	Sm 62	Eu 63	Gd 64
Lanthanide Series		Tb59	Dy 66	Ho 67	Er 68	Tm 68	Yb 70	Lu 71

Should the nucleus absorb a neutron, the atomic number is unchanged but the atomic weight increases by one, and thus

$$ {}_Z{}^{A}X + {}_0{}^{1}n \rightarrow {}_Z{}^{A+1}X \text{ (neutron absorption)} \quad . \quad \text{(viii.6)} $$

Note that we have written the neutron as ${}_0{}^1n$: also the final atom is of the same kind X since the atomic number is unchanged.

The following is part of the uranium series of radioactivity.

$$
\begin{array}{ll}
{}^{238}U \rightarrow {}_{90}{}^{234}Th + {}_2{}^4\alpha & (4\cdot5\times10^{10} \text{ yr.}) \\
\quad \rightarrow {}_{91}{}^{234}Pa + \beta & (24 \text{ days}) \\
\quad\quad \rightarrow {}_{92}{}^{234}U + \alpha & (1\cdot14 \text{ min.}) \\
\quad\quad\quad \rightarrow {}_{90}{}^{230}Th + \alpha & (2\cdot35\times10^{5} \text{ yr.}) \\
\quad\quad\quad\quad \rightarrow {}_{88}{}^{226}Ra + \alpha & (8\cdot0\times10^{4} \text{ yr.}) \\
\quad\quad\quad\quad\quad \rightarrow {}_{86}{}^{222}Rn + \alpha & (1\cdot62\times10^{3} \text{ yr.}) \\
\quad\quad\quad\quad\quad\quad \rightarrow {}_{84}{}^{218}Po + \alpha & (3\cdot82 \text{ days.})
\end{array}
$$

Transmutation of the Elements. Until 1919 nuclear disintegration was natural and unaffected by physical conditions, but in this year Rutherford and Marsden showed that on bombarding nitrogen by α particles (from radium C^1) a nuclear rearrangement of the following type took place.

$$ {}_7{}^{14}N + {}_2{}^4He \rightarrow {}_9{}^{18}F \leftarrow {}_1{}^1H + {}_8{}^{17}O $$

This reaction is often written as

$$ {}^{14}N\ (\alpha,p)\ {}^{17}O $$

where the left-hand symbol is the target atom, an α particle is made to strike it, a proton $p = {}_1{}^1H$ is ejected and an atom ${}^{17}O$ is left. It is called an (α, p) type of reaction, an α being the incident particle and a proton being ejected.

It was soon realized that transmutations would result if accelerated charged particles were to strike stable nuclei, and in 1932 Cockcroft and Walton were the first to produce nuclear disintegrations by artificially accelerated particles. They accelerated charged hydrogen atoms (protons) by a high voltage rectifier set (up to 250 000 V.) and observed the following reaction.

$$ {}_3{}^7Li + {}_1{}^1H \rightarrow 2\ {}_2{}^4He \quad . \quad . \quad . \quad \text{(viii.6)} $$

For this action to take place the incident proton must have a speed faster than a certain value, or its energy must exceed some value. It is usual to specify energies in nuclear actions in terms of electron-volts: this is very convenient as it gives the voltage that has to be applied to a particle with the electron charge e, which is the same as that of a proton. Thus in the above reaction the proton energy must exceed about 100 000 eV. Note that since

$$ e = 1\cdot60 \,.\, 10^{-19} \text{ coulomb}, $$

$$ 1 \text{ eV.} = 1\cdot60 \,.\, 10^{-19} \text{ joule} = 1\cdot60 \,.\, 10^{-12} \text{ erg} \quad . \quad \text{(viii.7)} $$

It is found that the energy liberated by the nuclear reaction (viii.6) is vastly greater than this: its value can be found by the use of Einstein's formula (viii.3). The atomic mass units are as follows:

${}_3{}^7Li$	7·0182	${}_2{}^4He$	4·0039
${}_1{}^1H$	1·0081		2
	8·0263		8·0078

There is a resultant mass loss of 0·0185 a.m.u., and this corresponds to an energy of

$$(0{\cdot}0185)\ (1{\cdot}6598\ .\ 10^{-27})\ (3\ .\ 10^{8})^{2}\ \text{joule}$$
$$= 2{\cdot}76\ .\ 10^{-12}\ \text{joule} = 1{\cdot}72\ .\ 10^{7}\ \text{eV.} = 17{\cdot}2\ \text{MeV.}$$

and experiment gave precisely this value for the two helium atoms, which have equal and opposite large velocities. This was one of the earliest verifications of Einstein's law of the equivalence of mass and energy.

The reaction just described may be written as ${}_3{}^7Li\ (p, \gamma)\ 2{}_2{}^4He$, since the incident particle is a proton and γ rays (energy) are emitted. This is called a *proton-capture* reaction, and there are many such cases, e.g. ${}_{13}{}^{27}Al\ (p, \gamma)\ {}_{14}{}^{28}Si$. There are also reactions of types (p, n), in which a proton is captured and a neutron liberated, and sometimes (p, α) and (p, d) where d is a deuteron, i.e. ${}_1{}^2H$. An interesting reaction is

$${}_1{}^2H + {}_1{}^2H \rightarrow {}_1{}^3H + {}_1{}^1H.$$

The particle ${}_1{}^3H$ is tritium, the isotope of hydrogen with atomic weight 3.

The following reactions produce neutrons, (p, n), (d, n), and (α, n): the last type was that which led to the discovery of the neutron, and is that in which beryllium is bombarded by α particles, thus

$${}_4{}^9Be + {}_2{}^4He \rightarrow {}_6{}^{12}C + {}_0{}^1n \quad . \quad . \quad . \quad \text{(viii.8)}$$

One of the most important methods of producing new isotopes, usually unstable, is by neutron capture, e.g.

$${}_{29}{}^{63}Cu + {}_0{}^1n \rightarrow {}_{29}{}^{64}Cu + \gamma$$

This reaction is (n, γ), but (n, p) is also common, e.g. ${}_7{}^{14}N\ (n, p)\ {}_6{}^{14}C$ and ${}_{17}{}^{35}Cl\ (n, p)\ {}_6{}^{35}S$: note that in the latter type of reaction the resultant atom has one less atomic number. This method is very important because the neutron, having no electric charge, has great penetrating powers and is more easily able to approach the nucleus of atoms surrounded by a cloud of electrons.

Before leaving this subject, we shall discuss briefly the elements with atomic number greater than 92, viz. neptunium 93, plutonium

94, americium 95, curium 96, berkelium 97, californium 98, einsteinium 99, fermium 100, and mendelevian 101. These are known as trans-uranic elements, and do not normally occur, but are produced by the irradiation of heavy elements by neutrons or α particles, as follows.

$$ {}_{92}^{238}U + {}_{0}^{1}n \rightarrow {}_{92}^{239}U \rightarrow {}_{93}^{239}Np + {}_{-1}^{0}e $$
$$ \hookrightarrow {}_{94}^{239}Pu + {}_{-1}^{0}e \quad . \quad \text{(viii.9)} $$

$$ {}_{92}^{238}U + {}_{2}^{4}He \rightarrow {}_{94}^{241}Pu + {}_{0}^{1}n $$
$$ \hookrightarrow {}_{95}^{241}Am + {}_{-1}^{0}e. $$

A reaction that will be very important in nuclear power production is that in which uranium 233 is produced from thorium.

$$ {}_{90}^{232}Th + {}_{0}^{1}n \rightarrow {}_{90}^{233}Th \rightarrow {}_{91}^{233}Proact + {}_{-1}^{0}e $$
$$ \hookrightarrow {}_{92}^{233}U + {}_{-1}^{0}e. $$

Plutonium and uranium 233 are important because they disintegrate when a slow neutron is incident on them.

Nuclear Fusion. When two light nuclei combine, the action may result in the emission of considerable energy: this is called *nuclear fusion* or a *thermonuclear reaction*. We have considered such a case in formula (viii.6) in which a lithium atom absorbs a proton with the release of 17·2 MeV. of energy.

It is considered that in the centre of the sun 4 protons combine to form an atom of helium with the emission of 2 positrons, thus

$$ 4\,{}_{1}^{1}H \rightarrow {}_{2}^{4}He + 2\,{}_{+1}^{0}e + 27 \text{ MeV}. $$

The chance that 4 protons will collide is very small, and it needs the intense packing and temperature at the centre of the sun to increase the chance for the action to proceed. More likely actions are

$$ {}_{1}^{2}H + {}_{1}^{2}H \rightarrow {}_{2}^{4}He $$

and

$$ {}_{1}^{1}H + {}_{1}^{3}H \rightarrow {}_{2}^{4}He $$

since only two nuclei have to collide. As the mass numbers are ${}_{1}^{1}H = 1{\cdot}0081$, ${}_{1}^{2}H = 2{\cdot}0147$, ${}_{1}^{3}H = 3{\cdot}0170$ and ${}_{2}^{4}He = 4{\cdot}0039$, the mass losses are 0·0255 and 0·0212 a.m.u. respectively, corresponding to 23·8 MeV. and 19·7 MeV. respectively.

So far thermonuclear reactions have not been obtained in a controlled way, although such energy has been released in the hydrogen bomb. Work is actively proceeding in several countries on the subject, and it is hoped that controlled reactions will be achieved. If a method were developed to convert matter entirely into energy, one pound of matter would yield as much energy as the burning of 1·5 megatons of coal.

Nuclear Fission. We have stated that a neutron can be absorbed (or captured) by a nucleus of large atomic number to produce a new nucleus, which may be unstable. In some cases the unstable nucleus splits into two approximately equal nuclei with the emission of several neutrons and considerable energy. The reason why several neutrons are emitted in the action, which is called *nuclear fission* or splitting, is explained by Fig. 430, which shows the curve relating the number of neutrons in a stable nucleus as a function of the number of protons (which is the atomic number Z). It is seen that the number of neutrons increases more rapidly than the

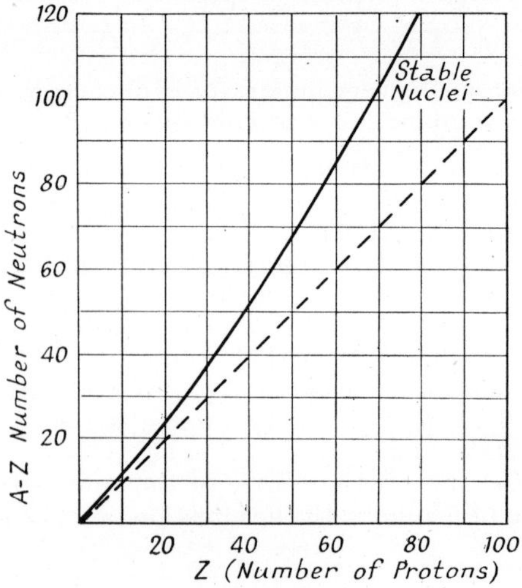

FIG. 430. NEUTRONS AND PROTONS IN STABLE NUCLEI

number of protons. If, therefore, a heavy nucleus splits into two approximately equal parts, each part will need rather less than half the total number of neutrons. Consider as an example ${}_{92}{}^{235}U$, which has 92 protons and 143 neutrons (235 − 92): suppose that it splits into equal halves, which will be ${}_{46}{}^{107}Pd$, and so there are 61 neutrons per half or 122 total, and there would be twenty emitted neutrons. In fact this is very unlikely, and the fragments will lie on either side of atomic number 46: the fragments require more than 122 neutrons, as some elements lie above the curve of Fig. 430 just as some lie below, but several neutrons are emitted. Fig. 431 shows the fission yield for ${}^{235}U$ when capturing slow neutrons. A typical case is

$${}_{92}{}^{235}U + {}_{0}{}^{1}n \rightarrow {}_{42}{}^{95}Mo + {}_{57}{}^{139}La + 2\,{}_{0}{}^{1}n + 7\,{}_{-1}{}^{0}e$$

The mass numbers are ${}^{235}U = 235{\cdot}124$, ${}_{0}{}^{1}n = 1{\cdot}00897$, ${}^{95}Mo = 94{\cdot}945$ ${}^{139}La = 138{\cdot}955$, and we obtain a lost mass of 0·215 a.m.u. (neglecting the electrons): this gives an energy of 198 MeV. This is the order of energy liberated per atom which splits.

It was realized at once that, since more than one neutron is emitted for each neutron absorbed, it should be possible to produce a sustained action or *chain-reaction.*

Reactor with Natural Uranium. Natural uranium contains 99·3 per cent of ^{238}U and 0·7 per cent of ^{235}U. It is found that ^{238}U can be split by fast neutrons with energy greater than 1·1 MeV., but below this energy the ^{238}U captures the neutron without fission. ^{235}U will undergo fission by slow (or *thermal*) neutrons of energy

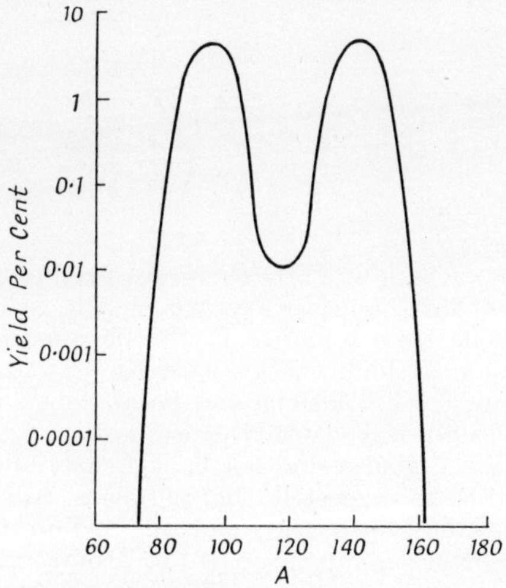

FIG. 431. FISSION-YIELD FOR ^{235}U AND SLOW NEUTRONS

with great readiness: its ability to attract thermal neutrons is expressed by saying that it has a very large capture cross-section for such neutrons, and in fact a nucleus of ^{235}U will attract such a neutron away from a swarm of neighbouring nuclei of ^{238}U.

The behaviour of a nucleus to a neutron (or any other incident particle) is expressed in terms of an effective cross-section, whether it be for fission, capture or scattering: in fission the neutron splits the nucleus, in capture it stays in the nucleus, and in scattering it is reflected away. The effective cross-section is that area over which the incident particle can pass for the effect to take place, and is given in a unit called a *barn* $= 10^{-24}$ cm.^{2}: note that 1 barn is approximately the end-on area presented by a nucleus, which has a diameter of the order of 10^{-12} cm. A cross-sectional area is denoted by σ with a suffix to denote the effect considered: thus σ_f is the cross-section for fission, σ_c for capture, and σ_s for scatter.

A fast neutron has an energy of hundreds of electron-volts upward. A slow or thermal neutron has an energy of the order of

kT, where T is the absolute temperature and k is Boltzmann's constant, and

$$k \simeq 1{\cdot}38 \,.\, 10^{-23} \text{ joule/degree.}$$

At a temperature of 400° A. = 130° C., the energy is about $5{\cdot}5 \,.\, 10^{-21}$ joule, and thus corresponds to $(5{\cdot}5 \,.\, 10^{-21}/1{\cdot}6 \,.\, 10^{-19}) \simeq (1/30)$ eV. The following list gives the properties of ^{235}U and ^{238}U with regard to fission, capture, and scattering for neutrons at 1 MeV. (fast) and (1/40) eV. (slow or thermal).

	^{235}U			^{238}U		
	σ_f	σ_c	σ_s	σ_f	σ_c	σ_s
1 MeV. neutron .	1·4	0·1	—	0·03	0·2	—
(1/40) eV. . .	580	107	8·2	0	2·8	8·2

In the process of fission of ^{235}U there are produced fragments of energy about 200 MeV. and an average of 2·46 fast neutrons. If these fast neutrons were to move in the natural uranium, a few would cause fission at high energy with the ^{238}U, most would be slowed down rapidly by collisions and be absorbed without fission by the ^{238}U, and only a few would cause fission at thermal energy with the ^{235}U. The action would not be self-sustaining.

Fermi and Sziland suggested that a chain reaction could be obtained by slowing down the neutrons before they could be absorbed by the ^{238}U. A substance of low atomic number, such as deuterium, carbon, or beryllium will cause the neutrons to lose energy rapidly by collision, but will not absorb the neutrons to any appreciable extent. Such substances, known as *moderators*, will slow down the neutrons to thermal velocities before they are absorbed in too great numbers by the ^{238}U. When the neutrons have been slowed down to (1/40) eV., say, σ_f for ^{235}U is 580 barns: the capture cross-sections are 107 for ^{235}U and 2·8 for ^{238}U. The latter should be multiplied by $140^{2/3}$, to allow for the greater abundance 140 : 1, and thus the total σ_c is $107 + 2{\cdot}8 \times 140^{213} = 180$. The chance of fission is thus $580/(580 + 180) = 0{\cdot}75$. As there is an average of 2·46 fast neutrons per fission, there can be an average of $0{\cdot}75 \times 2{\cdot}46 = 1{\cdot}84$ further fissions per original. The process can therefore be self-sustaining. The moderators in use are pure graphite, in the form of blocks surrounding the rods of natural uranium, or heavy water. Ordinary water is unsuitable because ordinary hydrogen has too great a capture cross-section to neutrons.

The Calder Hall reactors, shown in Chapter I, use a lattice of pure graphite as moderator. The total structure is designed to produce more neutrons than are necessary to sustain the action, and then the activity is decreased at will by inserting rods of cadmium or boron carbide into the reactor. These materials absorb neutrons very readily without fission, and thus decrease the activity.

One of the by-products of the reactor is ^{239}Pu, by the absorption of neutrons by ^{238}U as shown in equation (viii.9). It so happens that ^{239}Pu is also fissionable with thermal neutrons, and is thus a valuable fuel. It can be separated chemically from the uranium and can be used as a new fissionable fuel. Calder Hall is intended largely to be a source of plutonium produced in this way.

There remains the time constant of the control system. Most of the neutrons are emitted by the fission in 10^{-14} sec. and it is impossible to do anything in this time. But there are many neutrons which are emitted some time later, 1·0 per cent being delayed by 0·01 sec. and 0·07 per cent (i.e. 0·0007) by as much as a minute. If the control rods are placed so that the early neutrons give a reproduction factor of 1, i.e. sustain the action, there is an excess of 0·0007 produced 1 minute or more later. The power of the reactor will thus increase from 1 to 1·0007 in 1 minute or so, which corresponds to a time constant of about 1 500 minutes. If the reproduction factor is set at 1·01, the time constant is still manageable because of the delayed neutrons.

An estimate of the heat equivalent of the fission is obtained as follows. The fraction of atoms of ^{235}U split is about $580/(580 + 107) = 0{\cdot}84$: each atom has a mass $235 \times 1{\cdot}66 \,.\, 10^{-27} = 3{\cdot}9 \,.\, 10^{-25}$ kg. and delivers $0{\cdot}84 \times 200 = 170$ MeV. Therefore 1 kg. of ^{235}U produces $170/3{\cdot}9 \,.\, 10^{-25} = 4{\cdot}3 \,.\, 10^{26}$ MeV. $= 6{\cdot}9 \,.\, 10^{13}$ joule: this corresponds to about $2 \,.\, 10^{7}$ kWh $= 0{\cdot}02$ milliard $= 10\,000$ tons of coal. If a pile contains 30 tons of natural uranium ($\simeq 30\,000$ kg.), the amount of ^{235}U is about 200 kg.: if this is allowed to diminish by 10 per cent before reprocessing, 20 kg. will be used which corresponds to 200 000 tons of coal.

If all the ^{235}U of 1 ton of natural uranium is used, the coal equivalent is 70 000 tons. If all the ^{238}U is finally converted to ^{239}Pu and then used, the coal equivalent is then about 140 times greater, viz. 10 000 000 tons. Efficiencies of 10 per cent of these figures will be more than economical.

Enriched Reactors. If the fraction of ^{235}U in a reactor is increased by the addition of more ^{235}U (produced by a diffusion plant) then the reactor can work with fast as well as thermal neutrons. Such a reactor is said to be a *fast reactor*, and its main advantage is that it can be much smaller than a thermal reactor. "Zephyr" at Harwell is such a type, and the proposed power reactor at Dounreay in north Scotland is to be of the same type. Instead of ^{235}U one can use ^{239}Pu to produce a fast reactor. It is hoped to breed plutonium in the Calder Hall reactors and to extract it chemically. This is a convenient scheme because chemical separation is much easier than the diffusion method of separating ^{235}U from the ^{238}U in natural uranium.

ANSWERS TO EXAMPLES

EXAMPLES I

1. 14·4 per cent.
2. 39·8 per cent: (*a*) 38·0 per cent: (*b*) 38·7 per cent.
3. If both stations are to be built, B takes the base load of 38 600 kW. as its cost per unit is less; A takes the peak load, has a capacity of 11 400 kW., and runs for about 2 000 hours per year The cost under these conditions is £0·00158. If B takes the whole load, the cost is £0·001555 per unit, and this is the cheapest solution.
4. (*a*) 0·375d. per unit: (*b*) 0·370d. per unit.
5. 1·13 lb.: 99·7 tons.
6. 33·4 per cent: 39·3 per cent: 85 per cent.
7. 260 kW. A single, vertical shaft, Francis turbine of maximum rating 600 kW. or 850 h.p.
9. Buy bulk, at a cost of 0·689d. per unit.
10. 0·19d.

EXAMPLES III

1. $\rho = 1{\cdot}694\left[1 + 0{\cdot}00222\,(\theta - 60)\left(1 - \frac{T}{1\,000}\right)\right] \div \left(1 - \frac{T}{1\,000}\right)$ $\mu\Omega$. per cm. cube: 0·424 Ω.
2. 21 300 lb.: $12{\cdot}0 \times 10^6$ lb. per in.^{2}: 0·765 lb. per ft.
3. 3 950 lb.: 22·8 ft.
4. 11′ 4″: 39′ 4″, say 40 ft.
5. 158 ft.
6. 180 ft.
8. 25·4 cycles per second: 114 loops.
10. $13{\cdot}3 \times 10^6$ lb./in.^{2}: $8{\cdot}6 \times 10^{-6}$ per °F.
11. Identical with example 10.

EXAMPLES IV

2. 133 kV.
3. 3·46 A. per line.
5. 0·0126 μF. and 2·36 mH. per mile.

EXAMPLES V

1. 666 MΩ.
2. 8 000 Ω. and 4 000 Ω.
3. 0·206μF. per mile.
6. 47·6 kV./cm. (r.m.s.) or 67·2 kV./cm. (peak).
7. 126 000 Ω. and 76 000 Ω.
10. 4 000 ft.
11. $6{\cdot}9 \times 10^{10}$ cm.

EXAMPLES VI

1. 2 165 V.; 0·768 lagging; 96·4 per cent.
4. 25 per cent regulation, 80 per cent efficiency, 41·3 kV., 0·77 p.f.
5. 30·0 kV., 28·8 kV., 32·0 kV., 33·3 kV.

6. 1 200 kW.
7. 6 845 V.
8. 15·1 A., 233 V.
10. 164 MW.
11. (*a*) 230·1 V.: (*b*) 230·5 V.
12. 6·4 Ω. and 26·3 mH.
13. 123 kV., 203 A., 95·5 per cent.
16. 0·718.
17. 168 kV., 87 per cent.
18. 37·3 kV., 0·92 p.f. lagging.
19. Unity, 0·78 leading.
20. 5 250 kW., 6 750 kW.
23. 62·8 kV., 67·8 kV.
25. *AB*, 30 A.: *AD*, 77·5 A.; *CA*, 7·5 A.; *CD*, 42·5 A.; *CB*, 30 A.; 0·60 V.

EXAMPLES VII

3. 22·9 A.
4. 184 μF.
6. (*a*) 29 MVA.: (*b*) 7 MW. at 0·39.
7. 375 MW.

EXAMPLES VIII

3. 140 kV.: 176 A.: 13°.
4. 200 MVA.
5. − 34, − 24.
6. 975 A.
7. 2 900—3 260 A. (includes doubling and is peak): 790 A. r.m.s. due to decrement.
8. $E_0 = 452 \underline{|77°24'}$, $E_1 = 1\,077\overline{|6°18'}$, $E_2 = 1\,880\ \underline{|5°54'}$.

EXAMPLES IX

11. Connections as in Fig. 211 of text; $4\sqrt{3} : 5 = 1{\cdot}385 : 1$.
17. 1·76 miles from A.

EXAMPLES X

1. 36 kV., 164 kV., 728 A., 328 A.
6. 4 140 V.

EXAMPLES XI

3. $\dfrac{c}{y^2}\left[\dfrac{3}{\left(1+\dfrac{x^2}{3y^2}\right)^{\frac{3}{2}}}\right]$; $\dfrac{c}{y^2}\left[1+\dfrac{2}{\left(1+\dfrac{x^2}{y^2}\right)^{\frac{3}{2}}}\right]$
5. $I = 80 \sec.^3 \theta$; 2 520 lumens.
6. 7·0 f.c.: 0·7 f.c.
7. 4 100 c.p.: 39/1.
8. (*a*) Depends on polar distribution of individual lamps, exceeds 1 500 W. (*b*) 740 W.
9. 1·5 c.p.: 15 100 lumens.
12. 18.
13. 53.
16. 76 ft.-c.
18. 5 ft.-c., 4·3 ft.-c.

EXAMPLES XII

2. 340 ft.-lb.
3. (*a*) 32·0 m.p.h.: (*b*) 60 Wh./ton to overcome resistance and 32 Wh./ton for starting.
5. 61 Wh./ton mile: 2 900 kW.: 248 kW.
7. 27 m.p.h.
11. $2\frac{1}{3}$ miles.
18. 8·3 V.: booster delivers 23·1 kW. at 27·6 V
20. 3·8 kWh.

EXAMPLES XIII

4. Currents $1 : \frac{1}{\sqrt{3}} : \frac{1}{2}$. Torques $1 : \frac{1}{3} : \frac{1}{4}$.

EXAMPLES XIV

4. Load factor is 35 per cent.
7. (*a*) £12 700: (*b*) 500 kVAr, £11 600.
8 370 A./in.^{2}: $\frac{1}{2}$ per cent.
9. 76 kV.
12. 1·03d.
13. 6°54′.
14. Less than £1
15. £928.

EXAMPLES, PAGE 469

1. $91{\cdot}6 \cos (\omega t - 40° \ 54')$.
2. $96{\cdot}0 \cos (\omega t - 58° \ 42')$.
3. Coil 11·45 A., 233 V.: resistance 11·45 A., 22·9 V.: condenser 0·144 A., 22·9 V.
4. 1 160 V.
5. $0{\cdot}106 \sin (\omega t + 83° \ 54') - 0{\cdot}097 \sin (3\omega t - 13°)$. Harmonic much reduced, fundamental not.
6. (*a*) 1 160 A., 1 230 A., 1 712 A.: (*b*) 554 A.
7. 126, 59, 163 V.

INDEX

"A" STROKE, 293
A.C. bridge method of locating a break, 271
Adhesion, coefficient of, 351
α emission, 513
Air-blast circuit-breaker, 240
Alternating-current systems, 40
—— ——, transmission by, 38
Alternators, short-circuit currents of, 214–8
Aluminium, properties of, 50
Arc-back, 491
Arc lamps, 338–9
Arc welding, 506
Arcing earth, 295–6
Atomic mass unit, 511
—— number, 511
—— weight, 511
Auto-transformer tap-changing, 182–3
Auto-valve arrester, 302

"B" STROKE, 293
Back-fire, 491
Balanced vectors, 225
Balancers for three-wire system, 34
Ballistic method of finding a break, 271
Barn, 519
Barrel, 231
Basic rating, 210 *et seq.*
Battersea Power Station, 13
Battery booster, 37–8
Beard-Hunter sheathed-pilot system, 260
Belted-type cable, 101
—— ——, capacitance of three-core, 124
β emission, 513
Biased beam relay, 252
—— protection, 257
Bleeding, 12
Board of Trade Regulation, 37
Bonded sheaths, currents in, 128–32
Booster control, 384
—— transformers, 184
Boosters, 36
Boulder Dam, 22–4
Braking, electric, 375
——, regenerative, 377–9
——, rheostatic, 377
Breakdown, mechanism of, 132–3
Breakdown, voltage of, 132–3
Brightness, 318
Brown-Boveri regulator, 177–9
—— —— linkwork drive, 350
Burke arrester, 300–1
Bus-bar arrangements, 265–6

CABLE joints, 111–14
Cadmium-copper, properties of, 49
Calder Hall reactor, 25
Calorie, 2
Candle-power, standard of, 316
Capacitance, effect of earth on, 88
—— grading, 120
—— of overhead lines, 86–91
Cascade connexions of motors, 374
Catchment area, 17
Cathode-ray oscillograph, high voltage, 306–8
Central Electricity Authority, 27
Chain reaction, 519–20
Circuit-breakers, rating of, 243 *et seq.*
Circulating current protection, 252
—— currents method, 169
Clean-up, 488
Coiled-coil filament, 335
Coiled-filament, 335
Commutating inductance, 485
Compensating sections, tuning by, 161
Conduits, 110
Cone-type joint, 74
Constants of general network, 151
Contactors, 380
Continuous rating, 400
Control ratio, 489
Converting plant, 486 *et seq.*
Co-ordination of insulation, 298
Copper efficiencies, 42
——, properties of, 48–9
Copperweld, properties of, 51
Corona discharge, 91 *et seq.*
—— power loss, 93
Cosine law, 318
Cranes, 407
Critical grid voltage of thyratron, 497
Cross-jet explosion pot, 241–2
Current-limiting reactors, 218 *et seq.*
Cusec, 16

Δ-DOUBLE Y connexion, 494–5
Δ-zigzag connexion, 496–7
Damper, Stockbridge, 78

Dampers, vibration, 78
D.C. high voltage transmission, 43
—— low voltage transmission, 31
Decrement factor, 216–17
Deion circuit-breaker, 243
Depreciation, 417
Deuterium, 512, 516
Dielectric heating, 505
Discharge lamps, 337–40
Disruptive critical voltage, 92
Dissociation, 508
Distance indication, 480 *et seq.*
—— protection, 264–5
Distant repeater, Midworth, 483
Distributors, 31
——, Kelvin's Law, 427–8
——, voltage drop in, 162
Diversity factor, 420
Doubling effect, 215, 277
Drainage coil, 234

EARTHING, 76–8
—— of a line, 287
Eddy-current heating, 504
Effective mass, 352
Electrical transmission of energy, 5
Electricity in Great Britain, 1
Electrochemical equivalent, 509
—— processes, 507 *et seq.*
Electrodeposition, 510
Electrolytic arrester, 301
Electron shells, 512
Energy resources, 4
Energy, sources of, 1, 3
Erection chart, 57–61
Extraction of metals, 508

FALL of potential method, 270
Feeder booster, 36
—— mains, 31
—— reactors system, 219
Feeding points, 31
Ferranti effect, 158
—— surge absorber, 305–6
Field weakening, 373
Fission, nuclear, 23, 517
Fluorescent lamps, 340
Foot-candle, 317
Form factor, 442
Frogs, 385
Fusion, nuclear, 517

γ EMISSION, 513
Gas cable, 104
Gas-filled thermionic diode, 486–7
—— —— triode, 488 *et seq.*
—— valves, 486
General network, constants of, 151
Generator reactors system, 219
Graded time-lag protection, 258
Grading rings, 95
Grid, 27
Grid control of thyratrons, 497*et seq.*
Group drive, 396

H.F. HEATING, 503
"H"-type cable, 101
Half-life, 513
Hams Hall, Birmingham, 13
Harland interlock drive, 411
Heat balance, 7
—— value of fuels, 3
High voltage, advantage of, 38
Hilly country, spans in, 61–4
Holophane light fitting, 325
Horn gap, 299
"H.S.L."-type cable, 101–2
Hunting, 206
Hydro-electric installations, 22–6
—— stations, 16
—— ——, choice of site of, 18
—— ——, operating cost of, 25–7

IGNITER, 492
Ignitron, 486, 492
——, steel-tank, 491
Illumination for various purposes, 319
Impulse protective gap, 301
—— ratio, 298
Incandescent filament lamp, 332–5
Inductance of overhead lines, 81–6
Induction heating, 503
—— regulator, 187–91
Insulation failure, 287
—— resistance, 114–16
Insulators, 67 *et seq.*
——, Hewlett disc type, 72
——, pin type, 70
——, strain type, 72
——, suspension type, 71
Interconnected network, 32
—— systems, calculation of, 165–73
Interest, 417
Interlock system, 259
Internal combustion engine stations, 26 *et seq.*
Intersheaths, 117–20
Inverter, 501–2
Ion sheath, positive, 487, 489
Iron and steel works, 412
Isotopes, 512

JOINT, cone-type, 74
——, McIntyre, 73
——, Pairard clamp type, 74
Joints for overhead lines, 73–6

KELVIN'S law, 421

Kirchhoff's laws, direct application of, 165–7
Klydonograph, 310
kVA., measurement of, 427
kVAh. reactive, measurement of, 480
kVAr., measurement of, 426–7
kWh., measurement of, 480

Laying of cables, 107–11
Lead press, 107
Leakage reactance, 214
Lifts, 407
Light spectrum, 312–3
Lightning, 293–5
Line (or machine) charts, 142–5
—— very long, 158
Load factor, 6
—— reactor in rectifier circuit, 495
Load-end capacitance for medium lines, 147
Locomotive drives, 347
Long transmission line, equivalent networks of, 157–8
—— —— ——, rigorous solution of, 153–7
—— —— ——, very, 158
Loss in head due to pipes, 19
Losses in open-circuited line, 159
Lumen, 315
Luminous flux, 316
—— intensity, 317
Lux, 316

Machine tool drive, 406
McIntyre joint, 73
Magnetically-operated clutch, 399
Maximum demand, 419
—— power with a given regulation, 145
Mean horizontal candle-power, 319
—— spherical candle-power, 320
Mechanical equivalent of heat, 2
—— —— of light, 317
Mechanics of train movement, 351
Medium lines, 147
Mercury-pool rectifiers, 490 *et seq.*
Merz-Hunter, 261
Merz-Price, 253–5, 260
Metadyne, 368
Midworth distant repeater, 483
Milliard, 1, 3
Mining machinery, 413–4
Mixed conditions, 147
Moderator, 24
Mollier diagram, 10
Motors, cascaded induction, 404
——, comparative table of electric, 400
——, compound-wound d.c., 402
Motors, d.c. series, 401
——, d.c. shunt, 401
——, pole-changing induction, 404
——, repulsion, 402
——, Schrage, 405
——, single-phase series, 402
——, slip-ring induction, 403–4
——, squirrel-cage induction, 403
——, synchronous, 402
——, synchronous-induction, 404
Multi-gap arrester, 301
Multiple-unit control, 382
Murray loop test, 268

Negative booster, 37, 388
—— control, 489
—— phase sequence, 225–6
Neon lamp, 338
Network simplification, 168
Networks in tandem, 152
Neutron, 23, 511
Nominal Π-method for medium lines, 149
—— T-method for medium lines, 150
N–S motor, 406
Nuclear energy, 511 *et seq.*
—— fission, 517 *et seq.*
—— fusion, 517
—— power stations, 23 *et seq.*
Nucleus, 511

Oil-filled cable, 103–4
On-load tap-changing, 181 *et seq.*
Opposed voltage protection, 252
Osira lamp, 338–9
Overall efficiency, 5
Overhead line regulations, 46–7
Oxide-film arrester, 302–3

Pairard clamp, 74
Pantograph, 391
Paper-lapping head, 106
Paper-making machinery, 410–12
Parallel operation of generators, 193
Pelton wheel, 18
Percentage reactance, 472–3
—— —— and short-circuit currents, 209 *et seq.*
Periodic table, 512–14
Petersen coil, 296–7
Phase advancers, 198
—— angle control, 186
—— swinging, 206
Philora lamp, 338–9
Phosphor bronze, 52
Planers, 407
Plasma, 487, 489
Plugging, 376

Polar curves, 319 *et seq.*
Pole changing, 374
Polyphase rectification, 493
—— rectifiers, 492 *et seq.*
Positive control, 489
—— ion sheath, 487, 489
—— phase sequence, 225
Power angle diagrams, 200 *et seq.*
—— factor control of voltage, 194
————, improvement 427–8
——, measurement of, 426
Pressure cable, 104
Prime movers, 11
Printing machinery, 409–10
Propeller type turbine, 20
Protection against lightning and overvoltage, 297 *et seq.*
—— of alternators and transformers, 253 *et seq.*
Proton-capture, 516
Pulling-up clamp, 77
Pulsatance, 440
Punches, 407

Q, 3
Quenching, 504
Quill drive, 349

RADIAL system, 31
Radioactivity, 513
Rankine cycle, 10
Rannoch power station, 21
Reactive kVAh., measurement of, 478–9
Reactor enriched, 521
—— fast, 521
Reactors, 24, 218 *et seq.*, 519
Rectifiers, effect of d.c. magnetization in, 496
——, effect of transformer leakage inductance in, 496
——, glass-bulb, 491
——, mercury-pool, 490 *et seq.*
——, polyphase, 492 *et seq.*
——, six-phase, 494
——, three-phase, 493
Regulation, 138
Regulations, overhead line, 46–7
Regulator winding, 371
Relative luminosity factor, 313
Relay, biased beam, 252
——, Buchholz, 257
——, distance, 250–1
——, induction type overcurrent, 247–8
——, instantaneous, 247
——, reverse power, 249
——, solenoid and plunger type, 247
Remote control, 480 *et seq.*
Replaceable starter, 339–40
Resistance heating, 507
Resonance, 275–7
Ring system, 31
—— —— of reactors, 220
Root mean square, 442
Rotary phase-transformer, 500
Rousseau diagram, 320–2
Russell angles, 320–2

SAG at erection, 57
—— of overhead line, 53 *et seq.*
—— —— ——, effects of ice and wind on, 56
Schedule speed, 355
Scott transformer connexion, 476
Self-balance protection, 253
Series-parallel control, 364–8
Service mains, 31
Shears, 407
Sheath eddies, 126–7
—— effects, 126–32
Sheath-circuit eddies, 127
Short time rating, 400
Short-circuit current of alternators, 214–17
Simplification of networks, 168
Single-break switch, 244
Single-core cable, power factor of, 121–4
—— ——, stress and capacitance of, 118
Single-phase systems, 40
Site of steam power station, choice of, 11
Skin depth, 503
"S.L."-type cable, 101
Solid angle, 316
Sources of energy, 1
—— light, 340
Spans in hilly country, 61–4
Specific speed of turbines, 20–2
Spectra of light sources, 315
Spectrum of light, 313
Speed-time curves, 353 *et seq.*
Sphere gap, 299
Split conductor protection, 261
Stability, 204 *et seq.*
Star-delta transformation, 168, 468
Static stability, 205
Steam power stations, 5 *et seq.*
—— —— ——, choice of site of, 11
—— —— ——, heat balance of, 7
—— —— ——, operating costs of, 15
—— —— ——, prime movers, 11
—— —— ——, thermal efficiency of, 5
—— turbines, 11

Steel-cored aluminium, properties of, 50–1
—— copper, properties of, 51
—— tank ignitron, 491
Stockbridge damper, 78
String efficiency of insulator chain, 95
Stringing chart, 57–61
Styrene joint, 114–15
Sudden interruption of a circuit, 286–7
Summation metering, 480–1
Superposition of currents, 167
Surge absorber, 304–5
—— crest ammeter, 310
Switching surges, 281 *et seq.*
Symmetrical components, 224 *et seq.*
—— ——, application of, 228 *et seq.*
—— short-circuit currents, 213–14
Synchronization at the load, 222
Synchronous capacitance of an interconnector, 194
—— condenser, 198

TAP-CHANGING transformers, 180–3
Telemetering, 480 *et seq.*
Telephone interference factor, 232
Textile machinery, 409
Thermal characteristics of cables, 133–5
Thermal efficiency, 5
—— —— of steam power stations, 6
—— resistance, 134
Thévenin's theorem, 170, 478
Third rail, 345
Three-phase systems, 40
Three-wire system, 32–4
Thury system, 44
Thyratrons, 486, 488
——, continuous control of, 498
——, control ratio of, 488
——, critical grid voltage of, 497
——, electrodes of, 490
——, grid phase control of, 499
——, heat conserving cathodes for, 487
——, on-off control of, 498
Thyrite arrester, 303–4
Tie-bar reactor system, 220
Tin brightening, 504
Towers and supports, 64 *et seq.*
Tractive effort, 351, 355
Tramway controller, 366
—— motors and control, 365

Transformer, all-day efficiency of, 430
——, efficiency of, 430
——, equivalent network of, 471–2
—— losses, 429–31
——, percentage impedance of, 472–3
—— ratios, calculation of, 183
Transformers in parallel, 473
——, three-phase, 474–5
Transient reactance, 214
—— stability, 205
Translay system, 261
Transmutation, 515
Transpositions, 231–2
Trans-uranic elements, 517
Travelling waves, 281
—— ——, at junction of lines, 289–90
—— ——, effect of a capacitance on, 291–3
—— ——, energy considerations of, 285
—— ——, reflection and transmission of, 287
Tritium, 512, 516
Trolley bus motors and control, 379–81
—— buses, 380
—— skate, 387
—— wires, 64
Tummel power station, 24
Tuned power lines, 159–61
Tungar rectifier, 487
Two-phase systems, 40

URANIUM, heat value of, 3

VARIATOR winding, 370
Vector power, 462
Visual critical voltage, 92
—— sensitivity of normal eye, 313
Voltage control, 177
—— —— by power factor, 194
—— drop in distribution, 162

WATER circuit-breaker, 240
Wave form, 441
Welding, 506 *et seq.*
Werren overlap method, 269

Z-CONNEXION, 257
Zero phase sequence, 226